GLOBAL AIR POWER

RELATED POTOMAC TITLES

John Warden and the Renaissance of American Air Power

A History of Air Warfare

GLOBAL AIR POWER

EDITED BY JOHN ANDREAS OLSEN

Potomac Books, Inc.
Washington, D.C.

Library of Congress Cataloging-in-Publication Data
Global air power / edited by John Andreas Olsen.
    p. cm.
  Includes bibliographical references and index.
  ISBN 978-1-59797-555-1 (hardcover : alk. paper) — ISBN 978-1-59797-680-0 (pbk. : alk. paper)
  1. Air forces—History. 2. Air power—History. I. Olsen, John Andreas, 1968–
  UG630.G54 2011
  358.403—dc22

                                                    2010040074

Potomac Books, Inc.
22841 Quicksilver Drive
Dulles, Virginia 20166

First Edition

10 9 8 7 6 5 4 3 2 1

CONTENTS

Acknowledgments   vii

List of Abbreviations   ix

Introduction   xv

PART I   1

1   British Air Power, *Tony Mason*   7

2   U.S. Air Power, *Richard P. Hallion*   63

3   Israeli Air Power, *Itai Brun*   137

PART II   173

4   Soviet-Russian Air Power, *Sanu Kainikara*   179

5   Indian Air Power, *Jasjit Singh*   219

6   Chinese Air Power, *Xiaoming Zhang*   259

PART III   295

7   The Asia Pacific Region, *Alan Stephens*   299

8   Latin America, *James S. Corum*   335

9   Continental Europe, *Christian F. Anrig*   373

Afterword: The Future of Air Power, *David A. Deptula*   409

Notes   417

Selected Bibliography   495

Index   513

Biographical Notes   533

Acknowledgments

Global Air Power is a companion to *A History of Air Warfare*, published by Potomac Books in 2010, which examines the use of air power, campaign by campaign, from the First World War to the second Lebanon war. While the two books take different approaches, they are comparable in style and scope. Both books present the professional insight of some of the world's leading experts into their respective topics, and both books should be valuable to a broad range of historians, air power theorists and practitioners, and general readers interested in national defense and international relations. Rather than focus on hardware and tactics, they explore the fundamentals of air power and sociopolitical aspects of its application.

I am most grateful to all the contributing authors who made the editing task enjoyable by submitting excellent drafts, following editorial guidance, and meeting tight deadlines. I owe special thanks to Lt. Gen. David A. Deptula, who retired from the United States Air Force October 1, 2010, after an extraordinary military career. He has provided me with immense insight and inspiration ever since we met in 1997, and I am grateful that he agreed to write the afterword for this book. I also wish to acknowledge my indebtedness, first and foremost, to Margaret S. MacDonald, but also to Simon Moores, Martin van Creveld, and H. P. Willmott for having provided valuable input to the evolving manuscript. In addition, I would like to express my thanks to the team at Potomac Books, especially Elizabeth Demers, Sam Dorrance, Kathryn Neubauer, Kathryn Owens, and Maryam Rostamian.

Finally, I would like to thank the Swedish National Defence College and the Norwegian Defence University College, which provided equal financial support for the book without seeking to question the individual researchers' conclusions or to impose their own viewpoints.

ABBREVIATIONS

2TAF	Second Tactical Air Force
AAA	antiaircraft artillery
AASM	*armement air sol modulaire*
ADF	Australian Defense Force
AEF	air expeditionary force
AEW	airborne early warning
AEW&C	airborne early warning and control
AFDD-1	Air Force Doctrine Document 1
ALFA	AirLand Forces Application
ALO	army air liaison officer
AMRAAM	advanced medium-range air-to-air missile
ASC	air support control
ATO	air tasking order
AWACS	Airborne Warning and Control System
AWPD	Air War Plans Division
BEF	British Expeditionary Force
CAF	Chinese Air Force
CAP	combat air patrol
CAS	close air support
CASF	Composite Air Strike Force
CCP	Chinese Communist Party
CD	confidential document
CENTCOM	Central Command

CFSP	Common Foreign and Security Policy
CIA	Central Intelligence Agency
CONUS	Continental United States
COSC	Chiefs of Staff Committee
DCC	Defence Committee of the Cabinet
DFC	Distinguished Flying Cross
DoD	Department of Defense
DWP 2009	Defense White Paper 2009
EAC	European Airlift Center
EAF	Expeditionary Aerospace Force
EATC	European Air Transport Command
EPAF EAW	European Participating Air Forces' Expeditionary Air Wing
EPW	enemy prisoner of war
ESAF	El Salvadoran Armed Forces
ESDP	European Security and Defense Policy
EU	European Union
EW	electronic warfare
FAA	*Fuerza Aerea Argentina*
FAA	Federal Aviation Administration
FAG	*Fuerza Aerea Guatemala*
FAP	Peruvian Air Force
FAR	Revolutionary Air Force
FARC	Revolutionary Armed Forces of Colombia
FAS	*Fuerza Aerea Salvadorena*
FAS	*Fuerza Aerea Sur*
FEAF	Far East Air Forces
FMLN	Farabundo Martí National Liberation Front
FOFA	follow-on forces attack
FRY	Federal Republic of Yugoslavia
FSCL	Fire Support Coordination Line
GCI	ground control intercept
GDP	gross domestic product
GHQ	general headquarters
GPS	Global Positioning System
IAAFA	Inter-American Air Force Academy
IAI	Israel Aviation Industries

ICBM	intercontinental ballistic missile
IDC	Imperial Defence College
IDF	Israel Defense Forces
IED	improvised explosive device
IFF	identification friend or foe
IFR	in-flight refueling
IJN	Imperial Japanese Navy
IRBM	intermediate-range ballistic missile
ISAF	International Security Assistance Force
ISR	intelligence, surveillance, and reconnaissance
ISTAR	intelligence, surveillance, target acquisition, and reconnaissance
JAAF	Japanese Army Air Force
JASDF	Japan Air Self-Defense Force
JCS	Joint Chiefs of Staff
JDAM	joint direct attack munition
JFACC	Joint Force Air Component Commander
JIC	Joint Intelligence Committee
JNAF	Japanese Naval Air Force
JPC	Joint Planning Committee
JSDF	Japan Self-Defense Force
JSF	joint strike fighter
JTAC	joint tactical air controller
LANTIRN	Low-Altitude Navigation and Targeting Infrared for Night
LDP	Liberal Democratic Party
MAD	mutually assured destruction
MOD	Ministry of Defence
MTW	major theater war
NAC	North Atlantic Council
NASA	National Aeronautics and Space Administration
NATO	North Atlantic Treaty Organization
NCO	noncommissioned officer
NRF	NATO Response Force
NSC	National Security Council
ORS	operational research section
PAF	Pakistan Air Force

PAVN	People's Army of Vietnam
PFF	path finder force
PfP	Partnership for Peace
PGM	precision-guided munition
PLA	People's Liberation Army
PLAAF	People's Liberation Army Air Force
PLO	Palestine Liberation Organization
PRC	People's Republic of China
QDR	Quadrennial Defense Review
QRA	quick reaction alert
RAAF	Royal Australian Air Force
RAF	Royal Air Force
RAN	Royal Australian Navy
RAPTOR	Reconnaissance Airborne Pod for Tornado
RDF	Radio Detection Finding
RFC	Royal Flying Corps
RN	Royal Navy
RNAS	Royal Naval Air Service
ROE	rules of engagement
ROK	Republic of Korea
RSAF	Republic of Singapore Air Force
SAAS	School of Advanced Airpower Studies
SAASS	School of Advanced Air and Space Studies
SAC	Strategic Air Command
SACEUR	Supreme Allied Commander Europe
SADC	Singapore Air Defense Command
SALIS	Strategic Airlift Interim Solution
SAM	surface-to-air missile
SAR	Synthetic Aperture Radar
SCCOA	*système de commandement et de conduite des opérations aériennes* (Air Operations Command and Control System)
SEAD	suppression of enemy air defenses
SHAR	Sea Harrier
SOF	special operations forces
SRAM	short-range attack missile
SSM	surface-to-surface missile

TAC	Tactical Air Command
TFX	Tactical Fighter Experimental
TNI–AD	*Tentara Nasional Indonesia–Angkatan Darat*
	(The Indonesian Army)
TNI–AU	*Tentara Nasional Indonesia–Angkatan Udara*
	(The Indonesian Air Force)
TRADOC	Training and Doctrine Command
UAV	unmanned aerial vehicle
UK	United Kingdom
UN	United Nations
UNPROFOR	UN Protection Force
UNSCR	United Nations Security Council Resolution
URNG	Guatemalan National Revolutionary Unity Party
USAAF	U.S. Army Air Forces
USAF	U.S. Air Force
USAFE	U.S. Air Forces in Europe
USN	U.S. Navy
USSBS	U.S. Strategic Bombing Survey
USSR	Union of Soviet Socialist Republics
VPAF	Vietnamese People's Air Force
VTOL	vertical take-off and landing

A DIFFERENT BREED OF CAT

In the century since manned flight began, the authors of hundreds of books and thousands of essays and articles have attempted to come to grips with air power, with varying results. Many of these works have reflected the strong viewpoints of their authors as well as the passionate debates that have swirled around aviation and air power since the invention of the airplane and its application to military affairs. This is, in the main, hardly surprising. No less a personage than Sir Winston Churchill, a man never at a loss for words, remarked, "Air-power is the most difficult of all forms of military force to measure, or even to express in precise terms."[1] To make air power comprehensible, many have tended to focus on the scientific aspects rather than to view it in its wider context.

For various reasons, much of the existing literature focuses on aircraft—speed, flight range, acceleration, maneuverability, weapon systems, and leading-edge electronics. In this regard, U.S. fifth-generation aircraft currently in development certainly present much that is highly impressive. According to Lockheed Martin, the F-22 Raptor provides "a first-look, first-shot, first-kill capability through the use of stealth, advanced sensors and a lethal mix of advanced air-to-air and air-to-ground weapons." The new aircraft requires "shorter takeoff and landing distances as compared to current frontline fighters," bringing its own "precision ground attack capability to the battlefield," and its pilots "will be able to engage the enemy over its own territory and support long-range air-to-ground assets."[2] The F-35 Lightning II (also known as the Joint Strike

Fighter [JSF]), for its part, is envisaged as the next-generation strike fighter, bringing cutting-edge technologies to the battlespace of the future: "the JSF's advanced airframe, autonomic logistics, avionics, propulsion systems, stealth and firepower will ensure that the F-35 is the most affordable, lethal, supportable and survivable aircraft ever to be used by so many warfighters across the globe."[3]

Such advocacy apart, the fascination with technology and equipment has established bonds among airmen across borders and continents that one rarely finds in other services. To be sure, air forces differ in command arrangements, techniques, and tactics, but the construction and design of air bases, the way ground crews organize themselves and work, the maintenance of aircraft, and the training and manuals for maneuver through the skies are very similar around the world. The principles of ground-based air defense are universal, and the organization of air forces into wings and squadrons is recognizable to all. All pilots speak the same "language": regardless of their mother tongue, they seem to share the same sense of humor and apparently have the same confidence in their own abilities. Even jokes about pilots are the same; whether Christian, Orthodox, Hindu, Muslim, or Jew, every pilot recognizes (and to some extent identifies with) the character of Lord Flashheart in the British comedy *Blackadder*. Mottos are also comparable: the Egyptian Air Force has "Higher and higher for the sake of glory"; the Indian Air Force has "Touch the Sky with Glory"; the British Royal Air Force has *Per Ardua ad Astra* ("Through Adversity to the Stars"); and the South African Air Force has *Per Aspera ad Astra* ("Through Hardship to the Stars"). Airmen are not known for their modesty, and Gen. Carl A. Spaatz spoke for many when several decades ago he opined that "we were a different breed of cat right from the start. We flew through the air while the others walked on the ground."[4]

This bond linking pilots across nationalities also reflects that they have often flown the same aircraft. For example, more than 4,400 F-16s have been sold to 25 countries, from Norway to Morocco, and from Chile to Indonesia. Nations and regions exhibit striking differences in geography, weather conditions, and political regimes, but, as pilots say, at 15,000 feet the world is pretty much the same.

There are other commonalities. Air forces all over the world are purchasing fewer fighter-bombers, and year by year, the order of battle is being reduced. All air forces are becoming more and more conversant with satellites and space operations, and many are incorporating these capabilities. Increasingly, ballistic missiles are taking the place of bombers, and remotely piloted vehicles and drones are making a significant, and growing, impact on battles and campaigns.

AN ANATOMICAL CHART

In addition to undergoing these contemporary changes, Western forces at least share a history of warfare that has primarily involved mastering war's "first grammar"—the principles and procedures related to defeating an opponent by armed force—while struggling to come to grips with war's "second grammar"—dealing with insurgencies, guerrilla warfare, and irregular warfare.[5] In order to handle both grammars, airmen need to get ever better at thinking conceptually and strategically. Strategy, in its most generic sense, is the art of winning by purposely matching ends, ways, and means, but the concept of winning—the concept of victory—is, and has always been, situational. Thus, any campaign must be based on political and military objectives, and on a close analysis of the political system and society in question. As Edward N. Luttwak has reminded us:

> Highly developed by the end of the Second World War, "vulnerability analysis" became a lost art in the aftermath, because the advent of all-destroying nuclear weapons seemingly made it unnecessary. . . . To some extent, vulnerability analysis is a matter of engineering, but to a greater extent it must remain an art. Often it is the processes contained inside structures, rather than the structures themselves, that offer the most profitable vulnerabilities—and those might be managerial and bureaucratic rather than technical processes. Before all the technicalities, there is the strategy of a campaign. . . . Science is everywhere applied in air warfare, but no scientific theory can guide the choice and prioritization of targets, on which the success of bombardment must entirely depend. There is no other recourse but to study as broadly as possible the country and its political culture, its leadership, all that is known of current goals in the conflict at hand, the peculiar strengths and weaknesses of the armed forces and their presumed methods at every level. All that and more are needed to construct "an anatomical chart" that identifies the key elements of the enemy as an operating system, or rather as a combination of operating systems.[6]

Although we still need warrior-warriors, increasingly we also need warrior-scholars and warrior-diplomats who can master all the nontechnical aspects of an air campaign and with them the changing character of sociopolitical and historical insight. Placing increasing emphasis on the "second grammar," as we should, the application of air power must change with it in order to stay relevant. The fundamentals of air power remain the same, but the uses of air power, and the concepts governing its utility, may be very different in the second century of manned flight than they were in its first. To think clearly about the future, we need to know where air power came from and how it developed into what it is today. That is the rationale for this book.

ABOUT THIS BOOK

Like its companion volume, *A History of Air Warfare*, this book is intended as an introductory text for students of air warfare, whether military professionals, scholars, and specialists in the field of military studies, or general readers interested in air power history, theory, doctrine, or strategy. The book is divided into three sections. The first covers the world's most experienced air forces: those of the United Kingdom, the United States, and Israel. The second section encompasses chapters on the air forces of Russia, India, and China, each one a sizeable force following its own path toward modernization and adaptation to national goals. The third section focuses on representative air forces in Pacific Asia, Latin America, and Continental Europe that play significant roles in their respective regions. While the book is intended as a cohesive whole, each chapter is self-contained.

Global Air Power offers a new perspective by examining air power doctrine, roles, and missions in a broad range of regions and states. Rather than focus on how advancing technology has affected aircraft capabilities during the hundred-year history of manned flight, the narratives emphasize the sociopolitical contexts that have shaped air power as an instrument of war. The attention the authors devote to the influence that ideology, as well as social and cultural factors, exerted on air power adds insight into the driving factors that came "from within": why, how, and when did the different air forces do what they did?

Thought (in the proper sense of the word) about how air power evolved from an interesting, but unimportant, technical achievement into an extremely powerful and flexible political instrument requires a careful examination of air power's origins, the key factors that influenced its development, and the records of its use in war. There is, of course, no "punctum Archimedis," no single point of objectivity, as answers depend on the region, nation-state, and period of time in question. All written by highly regarded scholars, the nine main chapters in this book explore how a specific country or region integrated air power into its armed forces and how it has applied air power in both regular and irregular warfare, as well as in peacetime operations. The authors cover professional, organizational, and doctrinal issues that air forces confronted in the past, the lessons they learned from victory and defeat, and emerging challenges and opportunities.

In the final analysis, air forces are measured by air power effectiveness, which, in turn, is the product of various factors that are of different importance at different times. They include *morale*, which reflects motivation and individual degrees of daring and élan, as well as squadron-level elements of *unit cohesion*— in essence, the willingness to stick together and act as a team in the heat of combat. Another factor is *leadership*, which includes the vision and performance of generals and senior leaders at the strategic level; officers' ability to design, plan, and conduct campaigns at the operational level; and, equally important, tactical leadership, which requires the understanding of how to pair the right aircraft with the right weapon systems in an integrated fashion, whether for large regular forces or small special operations. The importance of *logistics* and *maintenance* is often underestimated, yet in all forms of warfare logistics is the linchpin of operations, and special expertise is required to deliver both personnel and materiel on time and on target. Further, *training* is of paramount importance, entailing realistic training scenarios, large-scale exercises, and individual flying hours. Finally, air forces must link training with *manuals*, *doctrine*, and *education* so that action is firmly grounded in theory. Encompassing all this remains the reality that warfare is a competition, and thus its effectiveness will always be measured against a given opponent, who is also changing and adapting.

Global Air Power augments the traditional military perspective with examinations of the factors that give air forces their distinctive character. Each analysis shows how the interplay among internal factors, together with external challenges, determines the structure, roles, missions, joint-force relations, acquisition plans, and, ultimately, the effectiveness of air power. Together, these chapters illuminate universal trends as well as similarities and differences among the world's air forces.

PART I

This section examines the driving factors behind the evolution of air power thought and action in the three most combat-proven air forces in the world: those of Britain, the United States, and Israel. Not a single decade has passed during their respective years of existence in which one or more of these countries did not put air power to the test.

BRITISH AIR POWER

Formed on April 1, 1918, six years after the creation of the Royal Flying Corps (RFC), the Royal Air Force (RAF) is the world's oldest independent air force. Today, more than nine decades after its inception, the RAF is the largest air force in the European Union (EU), the second largest in the North Atlantic Treaty Organization (NATO), and the fifth largest in the world.

British politicians and military officers believed in the efficacy of strategic air power almost from the inception of the airplane. Thus, the primacy of offensive operations, command in the air, and the vulnerability of civilian populations were concepts debated in the House of Commons well before the First World War. In chapter 1, Air Vice-Marshal (Ret.) Tony Mason argues that the RAF was created because of duplication, waste, and omissions associated with the existing army and navy air arms in the RFC, their failure to provide effective air defense of the United Kingdom, and the perceived potential for bombers to attack deep into the enemy's heartland.

Mason characterizes the second Smuts report (1917) as "the most important document in air power history," as it set the agenda for decades to come.

He further identifies Hugh Trenchard, "the father of the RAF," as the most important individual contributor: Trenchard pursued and reinforced the RAF's declared policy of defense and deterrence throughout the interwar years, even though it was the RAF's politically cost-effective role of "imperial policing" that assured the force's independence. But despite inspirational and visionary leaders, doctrine degenerated into unchallenged and unsubstantiated dogma. This degeneration, exacerbated by resource constraints and political uncertainty, resulted in Bomber Command's tragic inability to translate the policy into action at the outbreak of the Second World War. Nevertheless, the United Kingdom had well-prepared defenses and the RAF made its mark: it changed the direction of the war in the Battle of Britain, established permanent principles and practices for air-land cooperation during the battles in North Africa, and eventually expelled the U-boats from the North Atlantic. Finally, the Combined Bomber Offensive crippled the German economy and prompted a massive diversion of German resources from the battlefronts.

Having shown how the experience of the Second World War shaped the RAF, Mason next focuses on its nuclear transformation. After 1945 the British started a considerable downsizing of forces and air power primacy was passed to its principal ally, the United States. The Cold War and NATO membership nominally determined RAF policy for the next fifty years, but the RAF operated as a national force in the remnants of the British Empire, in Egypt, and in the Falklands. The acquisition of nuclear weapons finally gave the RAF the strategic potential sought by Trenchard, but increasingly effective Warsaw Pact air defenses and incessant budgetary pressure meant that the phase was short lived. With the end of the Cold War, British forces again found themselves engaged in combat in Iraq, over the Balkans, and in Afghanistan, in all cases acting in close cooperation with the United States, politically as well as militarily. Mason concludes that the loss of the nuclear deterrent role has stimulated doctrinal emphasis on the inherent flexibility of air power, eschewing dogma and leading to the creation of an expeditionary air force prepared for the unpredictable twenty-first century.

U.S. AIR POWER

It is often said that most countries have some air power, but the United States *is* an air power. The U.S. Air Force (USAF) is also the world's most active flying force, having taken part in large or small operations throughout its existence.

In chapter 2, Dr. Richard P. Hallion examines the American experience, arguing that the USAF has been at the forefront of air and space power development since its creation. However, the USAF's history also illustrates some of the challenges and nuances of America's larger experience with flight, including, at times, its dependence on foreign sources of doctrine, technology, and operational acumen. Like their counterparts in other countries, U.S. airmen have had to struggle periodically with surface warfare traditionalists to ensure that lessons proving the value of independent air forces are neither lost nor distorted.

Hallion's story starts with the first heavier-than-air flight of the Wright Brothers in 1903 and the twenty-three military aircraft that the United States possessed in 1914. He then explains how during the Second World War the United States became the world's supreme air power, traces its convoluted history after 1947 and the creation of the independent United States Air Force, and ends his essay with some thoughts on the current structure of an expeditionary air force, including an examination of ongoing investments in, and controversy surrounding, fifth-generation aircraft, such as the F-22 and F-35. The story highlights the development of aeronautical science and technology by focusing on the military-industrial structure, on the relationships between airmen and civilian government officials, and on inter-service tensions, and examines the development of military doctrine, operational concepts, and tactics. Hallion also emphasizes the importance of leadership, providing examples of disastrous consequences when the leadership is not up to its responsibilities.

Hallion finds that the U.S. development of an independent and sustainable air force was constantly hampered by those who insisted that support of their surface warfare brethren was the only valid mission for air forces. The author next reviews the USAF experience in the Korean and Vietnam conflicts, asserting that while air power was not decisive, U.S. casualties would have been considerably higher without it. He further explains how these experiences influenced the USAF's orientation to warfare by focusing on doctrinal issues, committees, white papers, and the public debate on matters military. Hallion also provides insight into how the United States used air power in Operation Desert Storm, and how air power became an integral and even a decisive instrument for U.S. policy in the Balkans in the 1990s. This new role was expressed in terms of *Global Reach–Global Power* and various versions of that idea, leading to the concept of "expeditionary warfare" that now has become the USAF's institutionalized way of life. Hallion concludes that while it is difficult to forecast the future of any particular service, let alone the characteristics of future wars, the need for the kind of capabilities manifested by robust, full-service air forces will continue as warfare in the twenty-first century becomes more complex.

ISRAELI AIR POWER

The Israeli Air Force was formed during the War of Independence, May 16, 1948, and is currently the largest air force in the Middle East. Complementing the Anglo-American perspective, chapter 3 focuses on how Israel's government and citizens came to view air power as an integral and essential part of survival and national defense. The founding father of the Jewish state and the main designer of its national security strategy, David Ben-Gurion, acknowledged the importance of air power, but it was the 1950s and 1960s that became the formative years for the Israeli Air Force, with the Six Day War marking the successful conclusion of the process that had shaped the air force since its inception. While neither the United States nor the United Kingdom had to fight for its own survival after the Second World War, Israel considered its very existence to be at stake every time the state took up arms, and at no point was this clearer than during the opening hours of the Yom Kippur War.

Brig. Gen. Itai Brun focuses on doctrine and public debate, and reality vs. rhetoric, as he examines how the Israeli Air Force performed in its most significant wars: the War of Independence (1948–1949), the Sinai War (1956), the Six Day War (1967), the War of Attrition (1969–1970), the Yom Kippur War (1973), the first Lebanon war (1982), the First Intifada (1987–1993), the Second Intifada (2000–2004), and the second Lebanon war (2006). While nothing compares in terms of impact to wartime experiences, Brun emphasizes the importance of the lessons-learned process between the wars, and the significance of rethinking operational concepts, updating training and education, and reorganizing the force structure to adapt to new circumstances.

Brun's central thesis is that almost six decades after the establishment of the Israeli Air Force, the second Lebanon war marked a significant change in the balance of power between air and ground forces, with a clear deviation from the Israel Defense Forces' original operational doctrine. The Israeli Air Force has always been a central pillar of Israeli military power, and its men and machines have always been an integral part of the nation's overall military strategy. But over time the operational doctrine changed. The defense of the nation's airspace (air superiority) and the support of the ground and naval forces in their missions (interdiction and close air support) retained priority, and initial surprise attacks on the enemy's airfields and offensives into the opponent's heartland remained accepted missions, but air power was basically organized to facilitate ground operations. As a country without strategic depth, Israel did not consider air power as a war-winning instrument on its own, but over the last three decades, as low-intensity conflicts and protracted warfare have become dominant, Israeli policymakers have discovered a new military tool in the Israeli Air Force and its precision-guided munitions. The challenge, according to Brun, is to incorporate all these elements into a comprehensive conceptual framework where operational and strategic excellence matches tactical proficiency. Although the IDF has been criticized for its handling of irregular warfare, whether in Lebanon or in the Gaza Strip, Israel has learned the hard way and many of its lessons have relevance for other countries involved in various forms of small wars.

1

BRITISH AIR POWER

Tony Mason

The Royal Air Force (RAF) is the oldest independent air force, but ideas and circumstances that shaped and permeated its future existed even before its creation in 1918. Britain's declining economic strength in the twentieth century meant that visions of air power dominance, exploitation of technology, political opportunism, rivalry with the army and navy, and even inspirational leadership were all affected by persistent government funding restrictions. Fortunately, the RAF was given many opportunities to demonstrate the value of a service created to apply air power, rather than remain a subordinate adjunct to sea or land.

ORIGINS

Two ideas dominated British air power from the outset: command of the air and the vulnerability of the enemy's heartland. In 1893 a British army major envisaged future wars starting with a great air battle, in which "command of the air would be an essential prerequisite for all land and air warfare."[1] In 1908 H. G. Wells, too, envisaged the potential of air power: "In the air are no streets, no channels, no point where one can say of an antagonist 'If he wants to reach my capital he must come by here'. In the air, all directions lead everywhere."[2] He foresaw a devastating impact from bombing, "everywhere below were economic catastrophe, starving working people, rioting and social disorder."[3] "Sightings" of mysterious dirigibles caused widespread fear and hysteria in Britain.[4] But when Louis Blériot crossed the English Channel in 1909, the dominant public reaction was the congratulation and celebration of a historic event, despite newspaper headlines warning that England was no longer an island.[5]

7

In 1909 the first House of Commons debate on military aviation addressed the impact of bombing, command of the air, and the need for a separate air service.[6] Secretary of State for War Richard Haldane acknowledged that Britain lagged behind other countries in military aeronautics but could easily catch up. Problems of weather and vulnerability suggested that "in the present state of construction, the use of these instruments of war is not very great."[7] The navy had been allocated £35,000 and the army £36,000 for military aviation in 1909–10.[8] Total navy funding for the same year was £12,677,195.[9] The members of Parliament were unimpressed. They argued that the effectiveness of airships and airplanes could be quickly improved, that their contribution to reconnaissance was already apparent, and that that alone would effect "an enormous change in warfare."[10]

It was suggested that the "moral" effect of air attack could be more serious than the material damage: "by appearing over the capital, centers of mobilization, bases of operations at the beginning or even before war is declared, dropping bombs and explosives at random must have a very demoralizing effect on military operations."[11] Only one member of Parliament (MP) argued that indiscriminate bombing of civilians would be "a reversal of the rules of war . . . ineffective . . . not be carried out by civilized nations. . . . No nation would make peace in those circumstances. . . . [W]e need not contemplate such a brutal and futile proceeding."[12] In fact, Britain had not signed the 1907 Hague Conference agreements on the Laws of War, and its subsequent bombing policies would not be constrained by international law.[13]

The importance of securing "command of the air," especially to a country that had hitherto depended on naval supremacy, was emphasized twice in the debate.[14] Subsequently, the case for a separate service was made: "[It] will become a branch of the service [sic], standing by itself and in the long run will be more powerful than either of the two branches . . . an organization which will increasingly provide for the national defence of the country."[15] Haldane was not convinced, but did reflect with innocent foresight, "What an effect it would have on the Mad Mullah if we had an airship with dynamite."[16]

Slow progress and inadequate funding incurred unfavorable media comparisons with France and Germany. In 1911, the newly formed Air Battalion of the Royal Engineers had only six pilots, whose training was restricted by shortage of funds. The naval aviation unit at Eastchurch had just four.[17]

In April 1912 the first attempt was made to create a unified air service when the Royal Flying Corps (RFC) was established with an army wing, a navy wing, and a Central Air School that would train pilots from both services. The expectation was that "though each service requires an establishment suitable to its own special needs, the aerial branch of one service should be regarded as a reserve to the aerial branch of the other."[18] The expectation was not met. Fearful of army dominance, the navy continued to train its own pilots. There were no preparations for mutual support anywhere, the aviation committee lacked authority, and there was no common strategy or procurement.[19] On July 1, 1914, the Admiralty recognized the de facto separation with the formal conversion of the naval wing to the Royal Naval Air Service (RNAS), as a branch of the navy. The division would handicap the evolution of British air power, but at the time it was seen as a major step in its advancement.[20]

WORLD WAR I

Although air power had only a marginal impact on the war's outcome, by 1918 all future air power roles had been attempted and the roots of British doctrine planted. British effectiveness was determined by rapidly advancing technology, reaction to events, and the influence of a handful of strong-minded individuals in senior positions.

The RFC's primary roles in France were providing the army with artillery spotting, reconnaissance, and short-range interdiction. The denial of similar activities to the enemy swiftly led to air-to-air combat and the continuous battle for air supremacy, which was largely determined by the tactical exploitation of aircraft and weapons.[21] Conversely, the RNAS operated more independently, attacking German ports, airship bases, submarine bases, and industrial targets from its base at Dunkirk, in addition to providing reconnaissance for the fleet and accepting responsibility for the defense of Britain in August 1914.

In 1916 Maj. Gen. Hugh Trenchard, commanding the RFC in France, explained how air power should be applied:

> [T]he aeroplane is an offensive and not a defensive weapon. . . . [I]t is impossible for aeroplanes . . . to prevent hostile aircraft from crossing the line if they have the initiative and determination to do so. . . . British aviation has been guided by a policy of relentless and incessant offensive. . . . It would seem probable that this has had the effect so far on the enemy of compelling him to keep back or to detail portions of his force in the air for defensive purposes. . . . The sound policy, then, which should guide all warfare in the air would seem to be this: to exploit the moral effect of the

aeroplane on the enemy but not let him exploit it on ourselves. Now this can only be done by attacking him and by continuing to attack.[22]

Trenchard would take his belief in the primacy of offensive action with him when he became chief of staff of the infant RAF and imbued his successors with it.

THE SMUTS REPORTS

In Britain, sporadic attacks by German Zeppelin airships from 1914 onward had inflicted little damage but the psychological impact on the civilian population after nine hundred years of insularity was considerable. Gotha bombers attacked central London in daylight on June 13 and July 7, 1917, causing extensive damage, two hundred civilian deaths, many more casualties, and widespread panic.

The dominant reaction from the British public was not capitulation but anger and demands for retaliation. The government, already concerned about the possibility of public unrest, was alarmed. Sir William Robertson, chief of the Imperial General Staff, noted that the agitation displayed in the cabinet on July 7 was such that "one would have thought that the whole world was coming to an end."[23] Four days later, on July 11, the War Cabinet decided that

[t]he Prime Minister and General Smuts, in consultation with representatives of the Admiralty, General Staff and Field-Marshal Commanding-in-Chief, with such other experts as they may desire, should examine—(1) The defence arrangements for Home Defence against air raids; (2) The air organization generally and the direction of aerial operations.[24]

The first report addressed in eight days the issue of home defense. Proposals to create a unified command for British air defense, including antiaircraft batteries and designated fighter squadrons, were swiftly implemented. Although historically overshadowed by Smuts's second report, it was to provide a role for the RAF and laid foundations for Britain's air defense in 1940.

The second Smuts report remains the most important document in air power history, not only determining the future of the RAF but also providing the rationale for independent air forces worldwide. General Smuts was advised by Lt. Gen. Sir David Henderson, commander in chief of the RFC, and air power visionaries on his staff, including Sykes, Grove, Tiverton, and Fullerton.[25]

The report recalled the period when the "air service" appeared merely to be ancillary naval and military operations. The existing Air Board had ably coordinated supplies to both the RNAS and RFC but was powerless to address questions of policy. "Essentially the Air Service is as subordinated to military and naval direction and conceptions of policy as the artillery is, and as long as this state of affairs lasts, it is useless for the Air Board to embark on a policy of its own, which it could neither originate nor execute under present conditions."[26] The committee acknowledged that the most obvious and expensive weaknesses of the earlier higher organizations—the failure to remove duplication and competition between the RFC and RNAS—had largely been resolved. Its revolutionary recommendations were not driven by failure to provide for the two existing air services, but by the vision a small number of influential men held for the future of air power. The vision was boldly and confidently stated:

> The time is however rapidly approaching when that subordination of the Air Board and the Air Service can no longer be justified. Essentially, the position of an air service is quite different from that of the artillery arm, to pursue our comparison: artillery could never be used for war except as a weapon in military or naval operations. It is a weapon, an instrument ancillary to a service, but could not be used as an independent service itself. Air Service, on the contrary, can be used as an independent means of war operations. Nobody that witnessed the attack on London on July 11th [*sic*] could have any doubt on that point. Unlike artillery, an air fleet can conduct extensive operations far from and independently of both Army and Navy. As far as at present can be foreseen there is absolutely no limit to the scale of its future independent war use.[27]

To that point, the report was visionary, debatable, but not provocative. However, the following sentence would ensure that whatever its impact on the outcome of the war, navy and army reactions would be hostile:

> And the day may not be far off when aerial operations with their devastation of industrial and populous centres on a vast scale may become the principal operations of war, to which the older forms of military operations become secondary and subordinate.[28]

From the outset, therefore, the case for an independent RAF came to be perceived as based on unproven theories of "strategic" bombing. Almost as an afterthought, it stated the fundamental, immutable concept of air power:

> It is important for the winning of the war that we should not only secure air predominance, but secure it on a very large scale; and having secured it in this war we should make every effort and sacrifice to maintain it for the future. Air supremacy may in the long run become as important a factor in the defence of the Empire as sea supremacy.[29]

This is the only reference in the Smuts reports to command of the air, which was to become the prerequisite for all operations in the air, on land, and at sea.

The second report elevated the visionary ideas of a handful of enthusiasts to the status of policy. By establishing a "third service," it provided a structure where the potential of air power could be exploited without its subordination to army and navy demands. It ensured that there would be no return to the duplication, waste, and omissions that had accompanied the RFC and RNAS. It created an environment in which airmen and -women could develop an air-mindedness: the expertise and breadth of strategic awareness commensurate with the unlimited horizons of the air. It established the precedent of an independent air force that would become a model for the rest of the world. Unsurprisingly, one distinguished commentator called it the Magna Carta of air power.[30]

At the same time, it hung a millstone around air power's neck. The RAF's independence became associated with operations autonomous of the army or navy—specifically, the role of "strategic" bombing. Yet the real rationale for independence was visible: the ability of air power not just to reach beyond the range of the army or navy, but to provide for home air defense, to maintain air supremacy for all operations, and to secure for the air service the best brains and experience available. The first was emphasized, the second was confined to the first report, the third was an afterthought, and the fourth was mentioned only in part of one recommendation. By asserting the potential subordination of the army and navy, it not only stimulated hostility from both, it also implied that air operations with both were themselves subordinate to the primary task of strategic bombing. That, in turn, would foster continuous attempts by both services to recover their own air arms. It encouraged concentration on a strategy that, for many years, would prove beyond the capacity of air power to deliver, leaving the RAF vulnerable to criticism of failure to fulfill the promise rather than the recognition of actual achievements.

Winston Churchill, now minister of munitions, strongly endorsed the need to achieve mastery of the air after which "all sorts of enterprises which are not now possible would become easy"; he questioned, however, the vulnerability of civilian morale. He argued that air attack should be directed against an enemy's fighting power, when civilian casualties would be inevitable, but that it was un-reasonable to assume that an air offensive would win the war by itself, that it was unlikely that terrorizing the civilian population would compel a strong government to surrender, and that in Britain the civilian reaction had been combative, not submissive.[31]

In September 1917 a Wing was established at Ochey in France, comprising two RFC and one RNAS squadron, to attack industrial targets in Germany. After six months, the material effect was not great, but the morale effect was asserted to be considerable. The small force, despite the constraints of small numbers, training limitations, bad weather, and inadequate navigation aids, had created fear and discontent. "How much greater," asked its commanding officer, "would be the impact of twenty squadrons or more?"[32]

On April 1, 1918, the RAF was formed from the RFC and RNAS. In May the Ochey Wing was expanded by the Air Council and established as an independent force "for the purpose of carrying out bombing raids on Germany on a large scale. This will be organized as a separate command of the Royal Air Force under Major General Sir H M Trenchard, who will work directly under the Air Ministry."[33] Trenchard reluctantly acknowledged that the task could be undertaken in addition to existing commitments: "In my opinion the British Aviation is now strong enough to beat the German Aviation in France and to attack the industrial centres in Germany."[34]

Only two weeks later, however, Trenchard reported that the independent force was unable to carry out its task. Operations were impeded by bad weather, difficulties of navigation and target location, engine failures, shortage of aircraft, and enemy opposition. Consequently, the bombers frequently attacked secondary targets, such as railway junctions closer to the front line, or responded to requests from French general headquarters.[35]

Nonetheless, the Allies were interested in the British initiative. Questions at a meeting of the Supreme War Council's Inter-Allied Aviation Committee in Versailles in July 1918 were copied to Trenchard beforehand. The Italian representative queried the utility of the creation of an Allied "independent air force": How should it be employed? Should a commander in chief be appointed, and if so, to whom would he be responsible and how would he work with other formations? The French representative asked what tonnage of projectiles each of the Allied air services could drop in twenty-four hours during the period August to December 1918; how far they could be carried by day and night; and more controversially, whether it was desirable to establish an inter-Allied plan for the massed use of bombing squadrons.[36]

Trenchard replied that an independent Allied force should be created by attaching units to the British force, with a commander in chief reporting to the Supreme War Council on equal terms with the navy and army, and tasked to attack German industry. His answers to the French questions were starkly realistic. Forecast tonnage figures, estimates of bombs on any particular objective, and distances reached would be meaningless and misleading because the amount would depend on the weather, the state of aerodromes, pilots available, serviceability of machines, etc. To reinforce his assertions, he attached a table to his memorandum titled, "THEORETICAL AVERAGE WEIGHT OF BOMBS IN TONS WHICH MAY BE DROPPED DAILY BETWEEN THE MONTHS OF JULY AND DECEMBER 1918 BY BRITISH MACHINES IN FRANCE." His footnote to the table was unambiguous: "These figures are purely theoretical and can in no way expect to be borne out by fact."[37]

TRENCHARD'S FINAL DISPATCH

Trenchard's final dispatch in December 1918 disclosed those limitations that had not been overcome. Moreover, he had been compelled to drop almost 50 percent of bombs on aerodromes to prevent enemy air attacks on his own and because "it was impractical to deal with them on equal terms in the air." Consequently, he had been unable to completely destroy any single industrial center and instead had attacked as many as were within reach. Thereby, he confidently asserted:

> The moral effect was first of all very much greater, as no town felt safe, and
> it necessitated continued and thorough defensive measures on the part of
> the enemy to protect the many different localities over which my force was
> operating. . . . At present, the moral effect of bombing stands undoubtedly

to the material effect in a proportion of 20 to 1, and therefore it was necessary to create the greatest moral effect possible.[38]

Although Trenchard's assertion had no scientific basis, his staff assiduously collected evidence of the bombing's impact from prisoners of war. The limited damage achievable by bombing prompted a magnification of the comparative impact on morale. Nevertheless, the qualifying "At present" subsequently faded, and the assertion crystallized into a driving force of RAF policy for the next twenty-five years.

FUTURE PLANNING

By the end of the war, the RAF had acquired expertise in home defense, army cooperation, maritime operations, embryonic airlift, and, to a limited extent, "strategic" bombing. Basic concepts about air power, especially command of the air and the importance of the offensive, were identified. Twentieth-century technology was understood and applied. The RAF had the potential to lead the world.

In December 1918 Maj. Gen. Frederick Sykes, then chief of the Air Staff, prepared a memorandum on the Empire's air power requirements.[39] He asserted the tremendous striking potential of distinct and separate aviation in future wars in which entire populations, together with their industrial resources, would be "thrown into the scale." The air force would be the first line of imperial defense. Overseas bases would enable imperial policing. Inter alia, he comprehensively detailed his plans for squadrons for home defense, a striking force, cooperation with the army and navy, deployments in the colonies and territories, and cooperation with Canada, Australia, South Africa, and New Zealand. While it was a complete blueprint for the future of British air power, the country was exhausted by war and was seeking ways to reduce defense expenditure, not expand it. The government rejected the proposals on the grounds of expense.[40] Trenchard replaced Sykes two months later.

The RAF did not acquire a heavy bomber force in 1919 to strike deep into Germany; instead, it shed aircraft and manpower, from 291,170 officers and other ranks in November 1918[41] to 28,280 in March 1920.[42] In August 1919 the cabinet affirmed government policies of disarmament and economy in a principle that became known as the Ten-Year Rule:

It should be assumed, for forming revised estimates, that the British Empire will not be engaged in any great war during the next ten years and that no Expeditionary Force is required for this purpose.[43]

Extended several times, the assumption heavily influenced and restricted all aspects of defense provision. It was formally renewed by the Committee for Imperial Defence in 1928 and not abrogated until 1932.

There was no immediate need for strategic bombing, no funding available for its future provision, and no obvious reason to maintain a third service. Trenchard's day-to-day priority was the survival of the RAF, not the implementation of ideas. A need to justify its continued independence in the face of political doubts and army and navy hostility was to drive his actions for the remainder of his tenure as chief of the Air Staff, but he took every opportunity to publicize his belief in the vulnerability of enemy morale to air attack.

In 1919 Trenchard presented his proposals to Churchill for a permanent organization of the RAF that, unlike those of Sykes, acknowledged the political and economic realities of the time while retaining the underlying rationale of an independent service. The proposals were published as a White Paper approved by the cabinet, which, despite the theoretical constraints of the Ten-Year Rule, authorized expenditure of "approximately 15 million pounds per annum during the next few years." Trenchard explained that RAF personnel, material, and accommodation were wartime products. His guiding principle was that the main portion of the RAF would comprise an independent service, including personnel carrying out aeronautical research. This force might become "more and more the predominating factor in all types of warfare."[44] In addition, small parts would be trained for work with the army and navy.

In accordance with the Ten-Year Rule, he assumed that general mobilization would not be required for some years. Therefore, he concentrated on the needs of the moment and laid the foundation of a highly trained force that could be rapidly expanded. He allocated sixteen squadrons to India, Egypt, and Mesopotamia, with flights elsewhere in the Empire. Based in the United Kingdom were four strike force squadrons, two for army cooperation and five for cooperation with the navy. Eighteen squadrons were designated for training and reserves. No specific provision was made for UK air defense.

He emphasized the importance of the RAF's providing its own training to foster the air force spirit engendered in the war: apprentice technicians, cadets, technical experts, flying instructors, and staff officers. It was insufficient "to make the Air Force officer a chauffeur and nothing more." Nonetheless, the need for a large capital outlay on accommodation and infrastructure drove expenditure on personnel to an absolute minimum, while existing war stocks would reduce equipment costs. The building program was even given priority over research, considered by Trenchard as essential to the air force's future efficiency.

The memorandum had no immediate impact on RAF operations, and financial constraints postponed implementation of several of the proposals. However, it laid a solid foundation that, with cabinet approval, should have ended uncertainties about the new service's permanence. It did not because of lingering political doubts and vigorous, sustained opposition from the older services.

In December 1918 Prime Minister David Lloyd George invited Churchill to become minister for war and air, after the former had decided not to maintain a separate air department.[45] Churchill and others explained that the arrangement was only temporary and that the independence of the RAF was not prejudiced,[46] but suspicions remained that the combination was intended to be permanent, driven by reasons of economy.

Meanwhile, that need for economy created an unexpected opportunity for Trenchard. In his memorandum, Trenchard had asserted the value of aircraft in swift and cost-effective action, "in the class of warfare approximating to police work."[47] He argued successfully, in the face of army derision, that the RAF could suppress a tribal chief in British Somaliland. Referred to as the "Mad Mullah" by Haldane in the 1909 Commons debate, this chief had defied the government for several years, despite numerous expensive army expeditions against him. In January 1920, twelve DH9 bombers attacked the mullah's forts, supported troops from the King's African Rifles, and in three weeks drove the mullah into Ethiopia, where he died. The expedition cost £77,000, and the territory enjoyed peace until the Italian invasion of 1940.

Churchill informed the Cabinet Finance Committee of the great value of the air force's contribution to the "control of vast areas like Mesopotamia"[48] and invited Trenchard to submit a plan by which the internal security of Mesopotamia could be maintained. The plan could include secure air bases and landing strips that would "enable these air forces to operate in every part of the protectorate and to enforce control, now here, now there, without the need of maintaining long lines of communications eating up troops and money." Measures of control could include bombing, machine gunning, and the swift movement and resupply of small numbers of troops, while the use of nonlethal gas bombs should also be considered.[49]

Trenchard replied that the RAF could accept the tasks, subject to the introduction of transport aircraft. The cabinet approved, and when Churchill moved from the Air/War Office to be colonial secretary in 1921, he swiftly convened a conference in Cairo to establish a political structure for administering the ex-Turkish territories placed under Britain's trusteeship by the Treaty of Versailles. It was agreed that the RAF should, in the near future, take over the responsibility for Mesopotamia from the army. In August 1921 Churchill expressed concern to the cabinet that estimates of army garrison costs for that year, given at the conference as £4.5 million, would in fact be £10 million, a variation that was impossible to reconcile with his earlier forecasts to Parliament: "[I]f no other way can be found than this of holding the country, we had much better give up this mandate at once."[50] Churchill received cabinet assent for a number of measures, including the reduction of army strength from the twelve battalions required by the War Office (already being reduced from twenty-two) to four, supported by three armored car companies. Eighteen hundred "high class, armed, white" RAF personnel would be based at Baghdad, capable of protecting themselves and of obtaining food and supplies indefinitely by air. "Radiating from this center, the aeroplane will give support to the political officers and local levies in the various districts and will act against rebellious movements when necessary." Churchill concluded by conceding that his cuts would be considered "drastic," and they undoubtedly involved risk, "but there is no other way in which the promises made to Parliament for reduction in cost can be met." The RAF performed the imperial policing role in the Middle East and the Northwest Frontier of India for the next two decades.[51]

Meanwhile, opposition from the other two services continued. The RAF's fear of dismemberment was well founded. In 1922 Secretary of State for Air Samuel Hoare was appointed but was warned by Prime Minister Bonar Law that his appointment might only be temporary because the independent air force cost too much, that there was every reason in peacetime to revert to army and navy control, that withdrawal from Iraq was inevitable, and that there was no money for a third fighting service.[52]

In his memorandum, Trenchard had not ruled out the return of aviation to the army and navy: "[T]hese two small portions (trained to work with the Army and Navy) probably becoming, in the future, an arm of the older services."[53]

For them, however, the "future" was now. A bitter struggle—ostensibly over command, control, and conditions of service of naval aircrew—dominated Air Ministry–Admiralty relations for a decade, despite several government committees affirming the rationale for RAF independence and attempting to effect a compromise.

In December 1921 the Geddes Committee concluded that division would return to duplication and waste and that the independence of the air service was essential to exploit future possible revolutionary developments that might effect "very large economies in the cost of the Fighting Services as a whole by substituting Air for Land or Sea Force."[54] In March 1923 Prime Minister Bonar Law appointed the Salisbury Committee to examine the relations of the navy and air force in fleet air operations, corresponding relations between the air force and army, and the criteria to be applied to determine the strength of the air force. The French Air Force (*Armée de l'air*), still much larger than the RAF, was perceived by the cabinet to present a latent threat.[55] Salisbury's interim report directed the air staff to draw up detailed proposals for a home air defense force capable of protection against the strongest air force within striking distance of the United Kingdom. As a first stage, all capacity for expansion should provide "600 first-line machines" (subsequently expressed as "52 Squadrons") to equal the numbers of the potential enemy, assumed to be France, funded in the first year by £500,000 plus any air ministry savings from elsewhere.[56]

The directive consolidated the independence of the service and encouraged Trenchard to reassert his belief in the superiority of the offensive and the vulnerability of enemy morale. He chose to interpret home defense as not only "keeping attacking aircraft from flying over this country," but also "the winning of an air war" that would need "a relentless offensive by bombing the enemy's country, destroying his sources of supply of aircraft and engines and breaking the morale of his people." He considered that "in a bombing duel, the French would probably squeal before we did."[57]

The final Salisbury Committee report rejected the War Office's attempt to absorb the Air Ministry and confirmed the administration of the RAF as a separate Department of State. Economy and the need to avoid duplication of training and infrastructure reinforced the need for a single air service. It was not possible to sever naval aviation from an air force that was responsible for home defense and cooperation with the army.[58] The decisions left the RAF in control of naval aviation but naval opposition continued. In 1926 Prime Minister Stanley

Baldwin intimated the cost to British air power of the long-running controversy: "It is impossible to achieve progress if decisions of the Government are put in question at every opportunity."[59]

By 1936 it was considered that the independence of the RAF was no longer threatened by the loss of naval aviation. Rearmament was proceeding apace and the fleet air arm was transferred in all respects to the navy, while shore-based aircraft and infrastructure remained with the RAF. By then, British naval aviation slipped behind its international competitors, but the real loser in the long-running interservice argument was coordinated British air power.

SUSTAINING THE BOMBING POLICY

After 1919 "the rhetoric of the moral effect of bombing"[60] was sustained, but the existence of the RAF was based on home air defense and cost-effective imperial policing. "It was [therefore] imperative that the RAF, at all levels, should be able to clearly and cogently articulate the rationale behind the existence of an independent air force."[61]

In 1922 the RAF compiled a draft operations manual, Confidential Document (CD) 22, which included chapters on army and navy cooperation.[62] Chapter VII, "Aerial Operations and Aerial Fighting," emphasized operations specifically requiring "independent" air power, with a theme of a relentless offensive, directly descended from Trenchard's 1916 memorandum, which would dominate RAF policy until the later stages of the rearmament program. Bombing to influence enemy civilian morale was permissible when "legitimate objectives" were present in the area. Destruction of the enemy air force, such as home air defense, was to be achieved primarily by a "relentlessly offensive policy." Subsequent chapters on army and navy cooperation included emphasis on the need to establish air supremacy—regional command of the air—above one's own forces, thereby denying it to the enemy. Finally, chapter XI examined operations "Against an Uncivilised Enemy."[63] The doctrine was not published until 1927. Meanwhile, a larger number of contemporary official publications and professional articles identified imperial policing as the immediate concern of the RAF rather than the strategic bombing themes in CD 22.[64]

In October 1928 Trenchard described the "war aim" of the RAF to the staff and students of the Imperial Defence College (IDC). His address was subsequently circulated to officers to explain air staff policy.[65] He noted that the three services held different views about the air force's war aim and how best to pursue it, because air attacks in future "would not be confined to the armed forces, but will also be directed, as they were by all sides in the last war, against the centres of production of war material and transportation and communications, and of the war control and organization of the country." His war aim was "in concert with the Navy and Army to break down the enemy's resistance. The Air Force will contribute to this aim by attacks on objectives calculated to achieve this end in addition to direct cooperation with the Navy and Army and in furtherance of the policy of His Majesty's Government at the time."[66]

The aim was not contentious, but the "objectives" were. Trenchard argued that air superiority would be gained by air attacks against the enemy's vital centers powerful enough to force the enemy's population, and even their high command, to defend themselves against air attacks instead of counterattacking, until their air force was thrown onto the defensive altogether. Moreover, he argued, demands for protection against enemy air attacks must be resisted to avoid being thrown onto the defensive ourselves.[67]

The attack objectives would vary, including the enemy fleet in harbors, mobilization, transportation, munitions: whatever "is best at the time." Ideas that the RAF would be directed to fight without relation to the operations of the other two services were erroneous. It was also a misapprehension to believe that the air staff wished to bomb indiscriminately "the civil population as such." Air attacks would be directed against military objectives, using that term in its broadest sense.

Trenchard's audience at IDC included future high commanders from all three services, foreign officers, and senior civil servants. He unambiguously affirmed his belief in the primacy of offensive air power and his relegation of provision for the air defense of the United Kingdom to secondary status. His comments were aimed not only at the army and navy, but also at officers on his own staff who believed that air defense was essential and that enemy air forces should be the primary target.[68] His comments, once again directly descended from his 1916 position, overlooked the experience of the independent force in 1918 and his subsequent postwar enthusiastic proposals for RAF expansion to reinforce the United Kingdom's air defense. Trenchard's insistence that air supremacy would be the product of successful offensive air warfare, and not its precondition, would have disastrous consequences for Bomber Command in World War II.

Unusually, Trenchard did not explicitly mention morale in his address, but his references to pressure on the population, his forswearing of indiscriminate bombing of the civil population "as such," and his intention of attacking military objectives "in the broad sense" left plenty of room for a bombing policy in which little distinction would be made between military objectives and any surrounding civilian population in attacks on "vital centres."

The beginnings of German rearmament, in violation of the Treaty of Versailles, became apparent in 1931. The initial British reaction was to pursue collective security and disarmament through the World Conference on Disarmament at Geneva in 1932, even seeking to curtail bombing. In November 1932 Stanley Baldwin, with all the authority of a government minister, declared in the House of Commons that "the bomber will always get through."[69] As new generations of bombers, with speeds matching those of existing fighters, entered production, Trenchard's belief in the offensive appeared to be justified.

AIR DEFENSE

Fortunately, despite the Air Ministry's emphasis on bombing, the air defense of the United Kingdom had not been neglected. London, only sixty miles from the coast, was especially vulnerable. Its protection remained a political concern regardless of the RAF's emphasis on the offense. By 1918 the London Air Defence Area, comprising guns, communications to fighters on standby, and rudimentary controlled interception, had inflicted unacceptable losses on the German bombers, but early warning of attack to give aircraft the maximum opportunity to climb, locate, and intercept incoming bombers presented a challenge to organization and technology.

Although proposals, in 1923, for a fifty-two-squadron air defense force were subsumed in Trenchard's emphasis on the offensive and delayed by the Ten-Year Rule, the organization of the UK air defense was steadily improved.[70] By 1934 the air defense structure used in 1940 had taken shape, but higher bomber speeds and the emerging threat from a resurgent Luftwaffe were increasing the inherent difficulties of early warning and interception of attacks on London. In night exercises, conducted in 1934, only two out of five bombers were intercepted. The vulnerability of the cities was more obvious than the implication for the bombers of a 40 percent attrition rate.

An Air Fighting Committee was formed in 1934 by the Air Ministry to encourage tactical and technical coordination in bombing and air defense.[71] The results of the summer exercises prompted Air Ministry scientists to propose a small specialist committee that would "consider how far recent advances in scientific and technical knowledge can be used to strengthen their present methods of defence against hostile aircraft."[72] The introduction of scientific method would have a much-needed impact on RAF policy.

The new committee under Professor Henry Tizard met for the first time in January 1935. Robert Watson-Watt, a radio research specialist, submitted proposals on February 12 for detecting enemy aircraft by reflecting radio beams from them back to a network of receivers. He followed up with an experiment in which an RAF bomber flew across a radio beam from a BBC transmitter, creating a reflected signal clearly visible on the oscilloscope receiver. The research was strongly supported by Air Marshal Hugh Dowding, then Air Member for Supply and Research, who authorized funding of £12,300 in April. By June 1935 Watson-Watt had established his research base at an isolated location at Orfordness on the east coast of Britain. The detection process was named Radio Detection Finding (RDF). In March 1936 a Chain Home radar station became operational at Bawdsey: the first of many that would, ultimately, provide early warning beyond the British coastline of Luftwaffe attacks.

Meanwhile, fighter performance and armament continued to improve. In response to an Air Ministry requirement for an eight-gun fighter, production of the Vickers Supermarine Spitfire and Hawker Hurricane began in 1935. Both monoplanes had top speeds well in excess of three hundred miles an hour. By August 1939 both would be deployed in Fighter Command.

In 1936 the Air Defense of Great Britain was divided into four operational commands: Bomber, Fighter, Coastal, and Training. Air Marshal Dowding became commander in chief (CinC) of Fighter Command and continued his support for coordinating scientific research and operations. A series of experiments began at Biggin Hill designed to facilitate the incorporation of RDF into the command and control structure of Fighter Command. A Cambridge engineering graduate pilot, Squadron Leader Arthur MacDonald, was appointed to head the joint RAF–Air Ministry scientist team. He was personally left in no doubt by Air Minister Viscount Swinton about the importance of his task: "I hope you realize that the whole future of this country depends on the results which you obtain at Biggin Hill."[73]

Simulated intercepts, directed from the ground using filtered plots from Bawdsey, increased early warning time beyond the coastline. RDF-assisted calculation enabled daylight interception. Procedures were updated. The Biggin Hill experiments marked the first systematic application of operational research to RAF policy. Although much remained to be refined, the organization and technology, which as "RADAR" would make its winning contribution to the Battle of Britain, were established.

Arthur MacDonald was awarded the Air Force Cross. Many years later, as retired Air Marshal Sir Arthur MacDonald, he was still concerned about the implications for Bomber Command. "Was the report ever sent to Bomber HQ and if so, what was the reaction of the Command to it? It cannot have been popular there, or in Air Ministry. It is quite extraordinary how complacent the Air Staff were at that time about the capacity of Bomber Command to destroy military targets."[74] The air marshal's concern was well founded. The Biggin Hill experiments did not appear to influence the Air Ministry's assumptions about bomber vulnerability. The creation of separate commands had weakened communication between Fighter and Bomber. Air Ministry bureaucracy was alleged to have precluded researchers from talking directly to Bomber Command.[75]

REARMAMENT

Government economic weakness after the Great Depression of 1931 inhibited rearmament expenditure on programs until 1934, when hopes of collective security had been abandoned. The Air Ministry assumed that the Luftwaffe would also adopt a strategy of strategic bombardment, seeking a swift "knock-out blow." "Parity" had to be sought to deter such an attack. In November 1934, Prime Minister Baldwin announced that the British government was determined "in no conditions to accept any position of inferiority with regard to what Air Force may be raised in Germany in the future."[76]

In 1935, as the Luftwaffe grew stronger, RAF expansion was considered necessary. A series of "Schemes" was proposed and progressively overtaken from 1936 onward, increasing squadrons, allocating fighters to a proposed field army, providing reserves, constructing more effective bombers, and creating shadow factories that could be rapidly converted to aircraft production. In October 1936 the Joint Planning Committee of the Chiefs of Staff Committee still concluded that a German attack "might be mitigated by air defence measures but . . . the only real answer lies in a counter-offensive of at least similar effectiveness."[77]

Parity, however, remained an elusive concept. RAF expansion was slow. The Air Ministry was torn between further expansion to catch up with the Luftwaffe and the awareness that it was very difficult to recruit and train sufficient numbers of aircrew to man even the existing aircraft.[78] The RAF's inability to catch up with the Luftwaffe led to the air staff's being directed by Air Minister Viscount Swinton to say what was required to counter the increased threat, not necessarily by exact parity in numbers but at least to be "an effective deterrent."[79] Nevertheless, in October 1937 Scheme "J" proposed to increase the bomber force to 1,442, equipped with the four-engined Stirlings, Halifaxes, and Manchesters/ Lancasters to achieve parity with Luftwaffe bombers, which would probably be deployed against England. Parity was believed, without any scientific evidence, to sustain deterrence.

Sir Thomas Inskip, the Minister for Coordination of Defence, questioned the RAF's argument about parity, observing that inferior numbers might not prove critical and that "we may be forced to consider a smaller striking force."[80] The Air Ministry responded by arguing that counterattack remained the chief deterrent and defense, and that publicly accepting inferiority would adversely affect British attempts to "come to an arrangement with Germany." The air staff stated unequivocally that "the bomber force is fundamentally the basis of all air strategy."[81]

Inskip countered by pointing out that, while the Luftwaffe might be expected to deliver an early knock-out blow, the role of the RAF was not to do the same, but prevent it. Thereafter, a combination of economic pressure and striking force would become the weapons of victory later in the war. Consequently, the fighter force should be made as strong as possible and given full reserves, while the bomber force should be reduced. Parity was more important in fighters than in bombers. In addition, fighters were cheaper and easier to produce than bombers. In December 1937 Prime Minister Chamberlain emphasized the essential importance of economic stability in defensive strength. A new Scheme, "K," retained fighter numbers with full reserves but reduced the bomber force by thirteen squadrons with only nine weeks of reserves. Unsurprisingly, the air staff recorded their dissatisfaction with cuts that had been "made against their will and solely for political and financial reasons."[82]

The assertion was valid. Although the government had, from 1934, consistently increased funds allocated to the RAF and its share of total military expenditure on equipment,[83] financial restrictions consistently precluded the RAF from achieving the parity with the Luftwaffe sought by Prime Minister Baldwin.

Meanwhile, the Air Ministry examined detailed Joint Intelligence Committee studies of the Spanish Civil War. Doubts about the relevance of operations there were sufficient to undermine confidence in bombing policy.[84] Raids were perceived as being sporadic, inaccurate, without a coherent strategy, and largely directed against military targets, which accounted for the failure of bombing to have any significant impact on Spanish civilian morale in Madrid or elsewhere. The analysis of successful attacks on ground forces was discounted.[85] The use of escort fighters was dismissed in favor of bombers carrying more ammunition and guns but fewer bombs.[86]

Conversely, the Luftwaffe used its Spanish experience to enhance night navigation, dive-bombing, escort fighters, and close support of ground forces.[87] The widely publicized Luftwaffe bombing of Guernica was not mentioned, but it influenced British public opinion, which between 1935 and 1939 moved from pacifism to active support for a war with Germany. Images of Guernica endorsed the Air Ministry's deterrent, counteroffensive position, encouraging the government to place rearmament above financial considerations in the last prewar years.

Meanwhile, in November 1937, the new CinC of Bomber Command, Air Marshal Sir Edgar Ludlow Hewitt, reported that his force was entirely unprepared for war, unable to operate except in fair weather, and extremely vulnerable both in the air and on the ground.[88] Because of weaknesses in navigation and navigators, target location, target identification, bomb aiming, reconnaissance, bomb damage assessment, and bomber protection, he would lose his medium bombers within three weeks and his heavy squadrons in less than two months. The commander in chief continued to send similar reports to the Air Ministry until his replacement by Air Marshal Charles Portal in May 1940.

The creation of a bombing development establishment had been proposed in 1934 but to no effect. There was no air navigation office until 1938 and no career incentive for aircrew to become navigators. There was no scientific analysis or any attempt to read across to Bomber Command the results of Fighter Command experiments. Indeed, Tizard commented that "no-one seems very anxious to get our advice on these subjects."[89] New bombs were introduced without airdrop testing, with poor explosive content, little understanding of aerodynamics, and inadequate detonators. Bomb design was strongly influenced by production costs. Consequently, "so far as its bombs were concerned, the RAF was unprepared to fight a European war in 1939."[90]

The one RAF institution created (in 1922) specifically to provide thinking about air power lamentably failed. In his message at the opening of the RAF Staff College at Andover, Trenchard envisaged that institution as "the cradle, as

I may call it, of our brain . . . from which I hope will emanate new and brilliant ideas for the development of the air and its power."[91] An Air Ministry memorandum on the RAF Staff College issued later that year warned against applying the principles of war in 1914–18 because inventions could produce a very different future.[92] The first two courses contributed to Air Ministry policymaking, including an input to the preparation of the Operations Manual and recommendations on the provision of home defense.

In March 1938 the commandant observed how heartening it was to hear of progress at Bomber Command in penetration and evasion, which meant that air defense was weak and air offensive strong and effective.[93] Students, subsequently, testified to the air staff's adherence to a bomber strategy being expounded at the RAF Staff College up to the outbreak of war.[94]

One year later, students at the Staff College were told bluntly just how little had actually been done to convert policy into operational capacity. Significantly, the speaker had recently joined the Directing Staff (DS) from No. 25 Armament Group, not from the Air Ministry.[95] He described how, for many years after the war, research into bombing and gunnery had been sadly neglected. Only in 1939 had an Air Ministry Directorate of Armament Development been established to initiate and coordinate tactical and technical developments for both bombers and fighters.

There was a shortage of observers and air gunners, bombing ranges were inadequate, bombing training had been impeded by squadron conversions to new aircraft, and rapid expansion had severely diluted the experience of pilots. Bomber training stations were insufficient, and bombing accuracy was still inferior to that achieved in the Hart biplane. Long-distance formation bombing lacked practice, and the increased speed of the new bombers appeared to have made identifying and hitting a target even more difficult. Most bomber aircraft were inadequately defended. The speaker concluded, "There seems to be a tendency to hitch our wagon to a very distant star, and for many of the more important problems nearer at hand to be neglected."[96]

Later in the year, in an exercise on docklands targeting, a student who advocated using prominent aiming points in the general target area instead of vital points incurred the displeasure of the directing staff who insisted that "bombing was an accurate business." In 1941 the student, now on the newly formed "policy" staff at the Air Ministry, had the rare satisfaction of receiving a paper from Bomber Command operations staff, signed by an erstwhile member of his

Andover DS, that highlighted the problems of identifying and hitting precise targets in industrial areas and advocated using "identifiable aiming points within the target area."[97]

By 1939 carefully selected, highly intelligent staff college students had become receptacles for current wisdom rather than contributors to it.[98] Somehow, lessons from 1914–18 were selected to fit theory, rather than vice versa. Inventions remained largely unconsidered. In an intellectual vacuum, bombing doctrine had become unrealistic dogma.

The Ten-Year Rule, limited resources, and international disarmament negotiations constrained strategic planning and procurement parameters until the rearmament programs of the 1930s, but should not have impaired thinking. In later years, Air Marshal Sir John Slessor wrote a mea culpa for the air staff's many failures to provide the capability to support a bombing policy between the wars. He admitted that lessons learned from the Great War and air policing had been misleading and that the air staff had no experience of or expertise in air warfare "between first class powers." Insufficient attention had been paid to the "technical difficulties of modern air bombardment."[99]

Policing in the Middle East and Northwest India, not strategic bombing, became the RAF's primary interwar role as a result of the government's financial exigency, Trenchard's opportunism, and Churchill's vigorous support. It secured the RAF's independence, but at a heavy cost to bombing policy. There was no opposition in the air and very little from the ground. Targets were generally small, static, easily identified, and, compared to Western industrial centers, flimsy. The target populations had no defense against air attack and their reaction seemed to support the view that civilian morale was indeed vulnerable to bombing and could influence enemy responses. Visibility for most of the year was good. Targets were usually well within the range of aircraft of World War I vintage. There was no incentive to give priority to longer-range, more expensive aircraft, with all-weather navigation by night or day, better bombs, heavier bomb loads, and improved bomb sights, or to large formation attacks, or to scientific analysis of targets, or to the protection of bombers. By the time the threat from Germany was acknowledged, it was impossible to recover before the outbreak of war.

ARMY COOPERATION

The first requirement of the RAF in World War II was to cooperate with the British Army in France, not to initiate strategic bombing of targets in Germany.

In World War I, the roles, and importance, of army cooperation were established. The Trenchard Memorandum (Scheme C) allocated two squadrons to "Cooperation with the Army" in the event cooperation with the army withered. To a certain extent, the neglect was understandable. A British Expeditionary Force (BEF) was only created in 1938 and rapidly expanded in 1939; army cooperation until then was confined to imperial policing.

The air staff's consistent emphasis on "strategic" bombing and opposition to the allocation of offensive aircraft to army support were motivated by doctrine and ghosts of the RFC's subordination to the army. Close support duties were "uneconomical" and should only be contemplated "to turn defeat into a rout, or to prevent our defeat of our forces becoming a rout . . . no units are really trained in such work."[100] Although the then Wing Commander John Slessor identified the importance, potential, and difficulties of providing army cooperation in 1936,[101] as Director of Plans, he consistently opposed army demands for greater air support. As late as October 1939, he observed that when the army desired bomber support, they would get it, but no bombers would be allocated specially and permanently to support the field force.[102]

Procurement for army cooperation was increased, but below Bomber, Fighter, and Coastal Command in priority. In 1939 the RAF deployed eight fighter squadrons, of which the majority were Hurricanes, five Lysander army cooperation, six Blenheim, and eight Fairy Battle light bomber squadrons, to France. By June 1940 they had been engulfed in the disaster that culminated for Britain in the evacuation of the remnants of the BEF from Dunkirk.

The German offensive began on May 10. On June 18 the last of the defeated BEF were evacuated from France. The major features of the confrontation are well known.[103] French ground and air force deployments were based on the static defenses of the Maginot Line; it was outflanked when the Netherlands and Belgium were quickly overrun. The German combined arms offensive achieved surprise, shock, and Allied paralysis; the main German thrust through the Ardennes was not anticipated or identified; Allied communications collapsed; airfields were overrun; the BEF suffered incessant air attack, relieved only sporadically by the handful of Hurricanes, reinforced from UK squadrons until May 19, when Churchill ordered that no more fighters should leave the country, however great the need in France. British Blenheim and Fairy Battle light bombers were decimated by flak and fighters; Allied command of the air was lost and only partially recovered for the evacuation from Dunkirk. Untold acts of heroism by RAF air and ground crew were in vain.

The Allied armed forces were unprepared for the kind of war that Germany imposed. The RAF was unable to cooperate effectively with an expeditionary force because such a force had only been created in 1938; no priority had been given to the procurement of army cooperation aircraft, and there was no joint doctrine, no joint force command, no integrated alliance planning, no realistic joint force training, and no understanding of the real nature of the threat posed by the Luftwaffe in a combined arms offensive. The interwar preoccupation of the air staff with a bombing policy and imperial policing reduced the likelihood of such weaknesses receiving a timely solution, but the fundamental reason for the Allied catastrophe was German superiority in strategy, applied technology, and doctrine.

The Air Ministry, subsequently, issued a "Restricted" study of operations in France. It naturally emphasized the contribution made by the RAF but implicitly identified many weaknesses. There were two command chains and two independent missions. They were under the nominal command of Air Marshal A. S. Barratt, but he had no command authority over the bomber force. There was no Allied coordination, and there were differences over bombing policy and the French order forbidding attack until German forces had taken the initiative. The BEF had no reliable ground-air communications. There were few reconnaissance aircraft and initially no intelligence-coordinating cell. Airfields were too few and vulnerable. Until December 1939, there was no coordinating maintenance authority. There was insufficient mechanized transport and no early warning for air defense. The Allied air forces were outnumbered in theater and, with the exception of the Hurricanes, outclassed.[104]

Within three months, a joint RAF-army report recommended the establishment of an air support control and communications system, and the attachment of army air liaison officers (ALOs) to squadrons.[105] In December 1940, an Army Cooperation Command was established, but had no operational responsibility. RAF-army cooperation would be forged by experience in North Africa rather than from conclusions from France.[106]

NORTH AFRICA

Italy declared war on Britain in June 1940 and invaded Egypt that September. The combat in North Africa never reached the scale of the Russian front or Northwest Europe in 1944–45, but the issues at stake—the Suez Canal and the Middle East oilfields—were crucial to the outcome of the war.

After Italian forces were expelled from Libya, German forces under Rommel, in February 1941, forced the British Army to retreat. RAF-army cooperation was ineffective. Army and RAF headquarters were separate, communications were slow and unreliable, air-land and situational awareness were difficult, and there were no identification procedures.[107]

In June 1941 Air Marshal Arthur Tedder became air officer commander in chief, Middle East Command. In the following month, Air Vice-Marshal Arthur Coningham took over 204 Group, which comprised fighter, light bomber, and reconnaissance squadrons based in Egypt, and swiftly began to redress the problems. On September 30 an air support directive was issued that "detailed the conceptual principles that informed cooperation . . . for the remainder of the war."[108] The directive included the merging of RAF and army headquarters, the principles of direct and indirect support, the primary requirement for air superiority, and air support control (ASC) headquarters that would "meet, modify or reject requests for support . . . so that the maximum effort is obtained from the available aircraft." Guidelines included target selection, bomb lines, communications, and recognition signals. Subsequently, "forward air support links" of ALOs were attached to army units that could pass requests to the ASC for tasking and, subsequently, update pilots directly on target location.[109]

While the war in the desert ebbed and flowed, the RAF, with South African and Australian support, protected the Eighth Army. Although it took some time to implement all of the September directive, partly because of a lack of forward airfields, cooperation between the army, reconnaissance, fighters, fighter bombers, and bombers progressively improved while the Luftwaffe was weakened by attrition and the diversion of German priorities to the eastern front.

Rommel's radio traffic was intercepted by the Ultra organization in the United Kingdom and by the RAF's own "Y" service. Awareness of German movements and resupply, including the Alamein battle plans, enabled the RAF to centralize command and control, allocating aircraft across its combat radius to the greatest effect. The RAF's destruction of Rommel's air and sea transport undermined the Afrika Corps' ability to sustain an offensive and hastened its defeat.

Tedder, Coningham, and General Montgomery demonstrated while in North Africa the importance of personal relations at high level. Command was unified, creating a mini-air force. There was no divisive bomber doctrine because there were no strategic targets and, therefore, no conflicting priorities. The desert provided few hiding places for ground forces and few civilian communities to inhibit air attack. The sole task was to defeat the enemy ground forces.

After a hesitant start, the Middle East principles of army cooperation were adapted by the joint U.S.-UK forces that invaded Northwest Africa in Operation Torch in November 1942. Initially, national army commanders wished to retain control of "their" aircraft, but combat experience quickly revealed the weaknesses in the system.[110] Coningham relayed the remedial principles to senior U.S. and British officers. Army and air force commanders must act together on a combined plan, under the overall command of the army. Air power command and control must be centralized under an airman for maximum flexibility and not distributed in "penny packets." Air superiority was essential to permitting friendly concentration of forces and supply while denying them to the enemy. The battlefield had to be isolated by destroying enemy access to it: the process of interdiction.[111]

Coningham's principles and example led directly to the reconstitution in the United Kingdom of Army Cooperation Command as the Second Tactical Air Force (2TAF). The U.S. War Department incorporated the principles in a new field manual. According to Vincent Orange, "The doctrine made possible one of the most effective collaborations known to military history."[112]

The lessons of ground force cooperation in North Africa were transferred to Northwest Europe after the Allied invasion of June 1944. From the outset, the combination of Allied air superiority and army cooperation reversed the 1940 experience of the BEF. German defenses were attacked, counterattacks were disrupted, and retreats turned into routs. Everywhere, fighters established air supremacy, enabling consistent ground attacks that "imposed severe constraints on the German strategy for fighting an aggressive defence."[113] A last effort by the Luftwaffe to regain tactical superiority in January 1945 failed, and for the remainder of the war, Allied aircraft provided close air support and interdiction of transport at will, with the only significant opposition coming from ground fire. Permanent principles of air-land cooperation had been established for the RAF.

THE BATTLE OF BRITAIN

On June 18, 1940, Prime Minister Churchill announced to the House of Commons, "The battle of France is over. I expect that the battle of Britain is about to begin." It changed the direction of history. The RAF denied the Luftwaffe the air superiority necessary for an invasion of Britain. Britain survived, the war continued, the base was secured for the Allied invasion of Europe, and occupied nations retained hope of liberation. The heroism and sacrifice of "the few" of

Fighter Command are justifiably commemorated annually in the United Kingdom, but there were several reasons why the RAF narrowly defeated the Luftwaffe, prompting Hitler on September 17 to postpone the invasion indefinitely.

German plans for an air war against Britain did not include invasion. The RAF had exaggerated the size and effectiveness of the Luftwaffe bomber force and assumed that it would face an immediate knock-out blow against London and other major cities. Instead, the Luftwaffe initially focused on Britain's economic vulnerability. Shipping, London, and other ports would be the major targets. Civilians would not be attacked but "should the enemy resort to 'terror' measures, for example to attacks on towns in Western Germany, similar operations could be carried out by the Luftwaffe with even greater effect, due to the greater density of population of London and big industrial centers."[114]

The RAF was well prepared. Headquarters Fighter Command received inputs from the radar stations, the Royal Observer Corps, the RAF "Y" service, and Ultra. Information flowed through four Groups to Sector Operations Centers that tasked, informed, and controlled the fighter squadrons. Antiaircraft and Balloon Commands completed CinC Dowding's forces. The three RAF commanders—Dowding, Park at No. 11 Group, and Leigh-Mallory at No. 12— had their differences, but were professional airmen with considerable experience and judgment. The only tactical RAF weakness was an early failure to amend the three-aircraft Vic formations, designed to intercept bombers but vulnerable to the Luftwaffe four-fighter *Schwarm*.

A number of inconsistent war directives from Hitler weakened German strategy. In November 1939 he identified the conquest of Britain as a prerequisite for victory and tasked the Luftwaffe with independent missions in May 1940. In June he directed the strategy of economic strangulation by air and sea alone, with no preparation for invasion. Luftwaffe CinC Hermann Goering subsequently ordered the Luftwaffe to attack industry and air force targets, by individuals or groups, with every effort to avoid unnecessary civilian loss of life. Invasion was still not mentioned. The Luftwaffe attacked shipping and coastal targets, seeking to lure RAF fighters into a war of attrition. Yet, at the same time, the general staff was contemplating an invasion as a last resort, "only after Germany had gained control of the air."[115]

On July 16, while still working on a peace treaty, Hitler ordered the preparation of an invasion of England, in directive no. 16, code-named "Sea Lion." As England was refusing to make peace, "I have decided to prepare and, if necessary, to carry out, a landing operation against her . . . to remove a base from which the war could be continued. Preparations for the entire operation must be completed by the middle of August."[116] The directive included the instruction: "The task of the Air Force will be to prevent interference by the enemy air force." Just three days later, he made a final public appeal to Britain to make peace, which was rejected.

On August 1, Hitler directed the Luftwaffe to destroy the RAF while conserving its strength to "take part in full force in Operation Sea Lion."[117] On August 2, Goering ordered "Eagle Attack" to establish air superiority and target British naval strength, ports, and factories. The following day, General Jeschonnek, chief of the Luftwaffe General Staff, ordered the Second and Third Air Fleets, based in France and Belgium, to "eliminate British radar stations as a prelude to the intensified air offensive."[118] On August 12, radar stations along England's south coast were struck. Most were operational again after a few hours, but Ventnor was knocked out and a gap created in the early warning chain for three days before a mobile unit replaced it. The apparent ease with which the damage to the radars was repaired discouraged the Luftwaffe from repeating such attacks.

In Eagle Attack, on August 13, the Luftwaffe enjoyed overall numerical superiority of three to one, which frequently rose to six or eight to one in combat. The three Luftwaffe fleets contained 2,288 serviceable combat aircraft: 998 medium bombers, 261 Stuka dive-bombers, 224 long-range Me-110 fighters, and 805 Me-109 fighters.[119] Aircraft from Bomber, Coastal, and Fighter Commands would contest the battle, but air combat by day would be waged by some seven hundred Spitfires and Hurricanes, with a further one hundred Blenheims deployed ineffectively as night fighters.[120]

Luftwaffe intelligence seriously underestimated its opposition.[121] Exaggerated Luftwaffe "kills" were unquestioned. The impact of heavy attacks on RAF bases and aircraft factories was weakened by poor coordination and concentration, and by the vulnerability of the Me-110s and especially the Ju-87s:

> The original objective of the Operations staff, the destruction of the Royal Air Force by means of an intensive lightning blow, was to have been carried out in conjunction with the German plan to carry out a large scale landing on the South coast of England with combined forces of the Army, Navy and Air Force. . . . As our aims were not achieved in three to four days, the orders for further attacks had to be repeated at certain intervals.[122]

Nevertheless, Fighter Command lost 231 aircrew in two weeks, including one-third of the flight commanders and one-fifth of the squadron commanders. Novice replacements frequently failed to survive their first forty-eight hours. All the squadrons fell 25–30 percent below strength. Sector stations and other airfields were severely damaged. The Luftwaffe was also suffering heavy losses, but German numerical superiority weighed heavily and induced combat fatigue among the Fighter Command survivors. Dowding was unable to continue rotating squadrons in and out of No. 11 Group. On September 6, Fighter Command was threatened with defeat.

Then, on September 7, the Luftwaffe switched its attacks away from the seriously wounded Fighter Command to London, immediately relieving the pressure on airfields and offering a single massed target to the fighters. The attacks continued for eight more days until Sunday, September 15, when the battle reached its climax as the Luftwaffe flew more than one thousand sorties against the capital. All of Dowding's reserves were committed, but Luftwaffe cumulative losses proved unsustainable: "These operations, carried out with relatively high losses, did not achieve the required success, because German air superiority was not achieved over Southern England."[123]

The disastrous decision to attack London may have been prompted by Hitler's fury at Bomber Command raids on Berlin. A Luftwaffe study in 1944 acknowledged that "Hitler ordered concentrated reprisal attacks on London to commence on 7th September," but added, "In this way a blow would be struck at the political and economic centre of the British Empire and at the morale of London's civilian population."[124] Ironically, when Luftwaffe strategy had finally become focused on destroying Fighter Command, the attack on civilian morale, in the way expected by the Air Ministry in 1939, was a costly failure.

Meanwhile, RAF Bomber Command and the navy had been attacking invasion barges in North Sea and Channel ports. On September 17, the German Navy reported further very heavy losses to the barges, noting "the RAF are still by no means defeated . . . and the Fuhrer therefore decides to postpone 'Sea Lion' indefinitely."[125]

While German errors contributed to their defeat, the RAF success was firmly based in Britain. The British objective was unambiguous: survival. The RAF was prepared and equipped to defend Britain. British tactical intelligence based on radar was invaluable. Fighter Command accurately identified bombers as the enemy's center of gravity. The British aircraft industry produced more fighters than were lost while German production decreased during the battle despite the Luftwaffe losing twice as many aircraft.[126] Fighter Command leadership was experienced, intelligent, and in Dowding, Park, and other Group commanders, inspirational. The Luftwaffe suffered from the personality of Goering and from lack of experience, operational awareness, and coordination among the Luftflotte commanders.[127] The Luftwaffe continued its attacks by night on London and other cities, in a futile attempt to break civilian morale, but as Hitler's attention turned eastward, the battles intensified over Germany and in the Atlantic.

BOMBER COMMAND AT WAR

In September 1939 Bomber Command had 576 medium and light bombers that were deficient in range, self defense, navigation aids, and bomb sights, for the mandate of attacking industrial targets in Germany. By December 1939 a number of daylight attacks simply on coastal targets had incurred unacceptable losses. In the same period, ineffective leaflet raids had confirmed the difficulties of night navigation but revealed the comparative ineffectiveness of German air defenses at night.

In 1939 the Air Ministry, realizing that parity with the Luftwaffe had not been achieved, sought to avoid provoking the expected knock-out blow before Bomber Command had removed weaknesses and reequipped with heavy bombers. It imposed target restrictions on Bomber Command, which were repeated in 1940: "purely military objectives in all senses of the word" and "intentional bombardment of civilians as such is illegal."[128]

Nevertheless, when Air Marshal Portal became CinC Bomber Command in April 1940, he reported his bombers' inability to damage designated oil targets and argued that scattered attacks would increase their impact on morale over a wider area.[129] Churchill concurred and indeed identified the need for a "devastating exterminating attack on the . . . Nazi homeland."[130] The combined chiefs of staff confirmed that "massive bombing would be relied on to destroy the German economy and morale."[131] After Portal's elevation to chief of the Air Staff in October 1940, the Air Ministry's preference for the disruption of Germany's oil industry was subordinated to attacking German morale with no concern about reprisals.[132]

Aircrew debriefs claimed widespread destruction and dislocation of German industry. In May the retired Lord Trenchard reiterated his arguments for increased bomber production to sustain an unremitting offensive directed at

German civilian morale, which, he asserted, was much more vulnerable than the British.[133] In July Bomber Command was directed to attack transportation targets, which would usually have indirect effects on surrounding built-up areas.

In September 1941 the RAF received a devastating shock. Professor R. V. Jones instigated a review of aerial photographs to verify Bomber Command's effectiveness.[134] Mr. D. M. Butt of the War Cabinet Secretariat examined 650 bomb-release night photographs taken in June and July 1941 from 6,100 sorties. The study revealed that, on average, only one in five aircraft came within five miles of the target. In full moonlight, two out of five achieved that, but it dropped to one in fifteen in thick haze, and in the Ruhr, the ratio was one in ten. Butt observed that a five-mile radius would often include open areas so not all bombs even within the ring would do any damage.[135]

The impact of the Butt Report was swift. Precision bombing was abandoned on February 14, when Bomber Command was directed to attack area targets with the aim of "undermining the morale of the enemy civilian population and, in particular, of the industrial workers," with the observation that this was a good time to derive the greatest impact from "concentrated incendiary attacks."[136] Eight days later, Air Marshal Arthur Harris replaced Air Marshal Pierce at Bomber Command. After twenty-three years of advocacy, Trenchard's insistence on attacking morale was formally adopted as Bomber Command strategy. The logic, or more accurately the faith, remained the same as in 1919. The capacity to inflict heavy destruction by air attack in 1942 was slight; only the diffuse, undefined target of German civilian morale appeared vulnerable.

Air Marshal Harris inherited a front line of less than four hundred serviceable, mainly twin-engined, relatively short-range aircraft. The commander in chief was aware of his force's shortcomings and of uncertainty in high places about the bombing strategy itself.[137] However, with the United States entering the war, he looked forward to cooperation with the United States Army Air Forces (USAAF). Meanwhile, the first squadrons of "heavy" Stirlings, Halifaxes, and Lancasters were forming, new navigation aids were on trial, and an operational research section (ORS) had been established.

Radio navigation aids Gee and Oboe began to improve bomber effectiveness during 1942, facilitated by ORS recommendations to use Oboe with aircraft designated as target markers and to concentrate bombers in time and space over targets.[138] Harris quickly applied ORS advice, mustering every serviceable bomber in the command to launch more than one thousand against Cologne on the night of May 30, destroying large areas of the city. In the following month, Essen and Bremen received similar attacks.

After a string of military disasters, the thousand-bomber raids were important for their impact on British morale and the future of the bomber offensive, rather than their effectiveness. Public reaction was immediate, Harris was knighted, and doubts about Bomber Command's strategy and achievements were silenced. But effectiveness had been diminished by cloud cover, while casualties over Bremen had risen to 4.9 percent with significantly higher losses among the training crews pressed into the operations. Navigation training was under pressure. ORS recommended the deployment of target-marking aircraft, and a specialist Path Finder Force (PFF) was established in August 1942. Initially impeded by the German jamming of Gee, the PFF, by the end of 1942, was improving bombing accuracy, which was further enhanced by the introduction of H2S, a ground-mapping radar.

The USAAF brought its own bombing strategy of daylight attacks on specific targets by large, self-defending formations and declined invitations to be incorporated in Bomber Command operations. The Casablanca Conference of January 1943 directed the primary Allied objective to be "the progressive destruction and dislocation of the German military, industrial and economic system and the undermining of the morale of the German people to the point where their capacity for armed resistance is fatally weakened."[139] In May 1943 the combined chiefs of staff issued a combined bombing offensive directive, "Pointblank," which listed submarine construction yards and bases, aircraft industry, ball bearings, oil, synthetic rubber and tires, and military transport vehicles.[140] The directive identified the German fighter force as the first priority to enable the offensive to be extended and the necessary air superiority established for a subsequent Allied invasion.

Harris simply incorporated the specified targets in his strategy of attacking cities. In July, four raids devastated Hamburg when the first use of "window" (chaff) to confuse German radar plus the concentration of the force resulted in low casualties of 2.8 percent. In August, a V1 and V2 missile research and development installation at Peenemünde on the Baltic was severely damaged. In November 1943, Harris proposed to attack Berlin and nine other cities, confidently informing the Air Ministry that Germany would be forced to surrender by the destruction of between 40 and 50 percent of its principal towns, which could be achieved by April 1944 if Lancaster production was maintained, navigation aids received, and priority given to the bomber offensive.[141]

Meanwhile, USAAF daylight strategy foundered in the face of strengthened German air defenses. Unsustainable casualties were suffered in daylight raids on Schweinfurt and Regensburg in August and October, while clouds impeded precision bombing. Gen. Ira Eaker fell back on the morale argument to defend his position, but was replaced by Gen. Carl Spaatz in the following month.[142]

During 1943 ORS enhanced Bomber Command's effectiveness by producing detailed studies of losses by aircraft types and groups, employment of the new equipment, training needs, German fighter tactics and control, bomb loads, route planning, concentration, diversions, and the integration of the PFF.[143] Unfortunately, German fighter strength had doubled, ground defenses extended, and tactical and electronic countermeasures introduced. Bad weather and darkness could conceal the targets but not the bombers.

Between November 1943 and March 1944, Bomber Command flew 9,111 sorties against Berlin and 11,000 elsewhere. Loss rates to Berlin rose from 4.85 in December to 6.1 in January and, in the final attack, to 9.1 percent. The total loss rate during the winter of 1943–44 exceeded the strength of the command. Industrial and political targets in Berlin were spread over a wide area and were less vulnerable to incendiary attack. There was no discernible impact on German morale.

Many years later, an obdurate Harris recalled that his squadrons had fought a three-year campaign involving nightly battles, that the overall casualty rate was slightly over 3 percent, and that in the following months the command launched equally heavy attacks on other German targets. He dismissed the critical verdict of the official historians on the Berlin raids with a terse and typically forthright "some defeat!"[144]

Fortuitously, in April 1944 the U.S. and RAF bomber forces were directed to support preparations for Operation Overlord, the Allied invasion of Europe under Allied Deputy Supreme Commander Sir Arthur Tedder, who directed priority to attacks on transportation and the gaining of air superiority. In February 1944 heavy attacks on German fighter production and attritional air battles marked a turning point in the air war. The USAAF's new long-range Merlin-engined Mustang inflicted heavy casualties on the German fighters. Their loss of over one hundred experienced pilots proved irreplaceable. Meanwhile, Bomber Command was largely attacking closer targets in France and the Low Countries, which were not defended as strongly as Berlin.

After the invasion, Allied numerical and technical superiority, the reduction of German air defense depth, the destruction of Luftwaffe fuel supplies, and the

cumulative impact of fighter-pilot losses created a much more favorable environment for Bomber Command. Nonetheless, cloud cover over the target still impeded precision on most raids, resulting in little difference between the "area"
bombing of the RAF and the "precision" bombing of the USAAF.

In late 1944 Harris objected to Portal's demand that Bomber Command
could and should now hit the petroleum industry.[145] Recent research suggests
that Harris's case was stronger than has often been recorded,[146] while an erstwhile critic in the Air Ministry in 1944 observed in 1959 that "the difference
of view between the Air Staff and the C-in-C was not as great as the amount of
paper absorbed in its discussion would indicate."[147]

Unfortunately, the end of Bomber Command's campaign was marked
by controversy over the destruction of Dresden in February 1945. The Soviets asked for Anglo-American bombing to aid the advance of Soviet armies.
Churchill suggested Berlin and other eastern German cities could now be at-
tractive targets. The air staff directed Harris to attack Leipzig, Chemnitz, and
Dresden to impede German troop movement. Harris recalled carrying out the
directive without compunction, placing it in the context of the impending Yalta
Conference and the "mare's nest" of contemporary intelligence reports about
the withdrawal of German troops southward to establish a last redoubt at Ber-
chtesgaden. Dresden was located on both north-south and east-west routes.[148]
In perfect conditions for bombing, with no opposition, the RAF and USAAF
heavily attacked Dresden on February 13 and 14. The baroque city center was
leveled, a firestorm ensued, and civilian casualties were later estimated at twenty-
five thousand.

THE BOMBER BALANCE SHEET

The war against Germany was not decided by an unremitting bomber offensive, but on land, especially on the eastern front. German civilian morale neither
broke nor contributed to the collapse of Hitler's regime. German war production almost trebled between 1941 and 1944. Air superiority had to precede an
effective bomber offensive rather than be achieved by it and was established
largely by the defeat of the Luftwaffe in the air in early 1944, facilitated by the
late acquisition of a long-range escort fighter. Bombing remained a relatively
blunt instrument, with accuracy being measured in hundreds of yards.

The real measure of Bomber Command's effectiveness was not against prewar dogma, but in its actual contribution to the defeat of Germany.[149] Bombing

had a severe impact on ground campaigns. During 1944 synthetic oil production dropped by 86 percent, explosives by 42 percent, aircraft by 30 percent, and
tanks by 35 percent. In January 1944, 68 percent of German fighters were defending the Reich, 81 percent by October. Less than two hundred faced twelve
thousand Allied aircraft over Normandy. German armies on all fronts were bereft
of air support and faced largely unchallenged Allied air power. Resources diverted to antiaircraft defenses included eight hundred thousand people, one-
third of all gun production, 20 percent of ammunition, half the production of
the electrotechnical industry, and one-third of all optical equipment. Production
from dispersed and protected industries demanded scarce engineering resources
and became vulnerable to transportation attacks. In sum, "bombing took the
strategic initiative away from German forces. . . . With more men, more heavily
armed, an intact transport system and an uninterrupted flow of industrial resources, Germany might well have kept the Allies at bay in 1945."[150]

Bomber Command lost fifty-five thousand aircrew. Less than one in three
completed the mandatory thirty missions. They paid a heavy price sustaining
Trenchard's offensive dictum, but they undoubtedly shortened the war and
saved thousands of lives on battlefields. Their failure to receive a campaign medal
remains a stain on Churchill's reputation.

COASTAL COMMAND

Preparations for maritime operations at the outbreak of World War II were seriously impeded by resource constraints, other procurement priorities, and lack
of cooperation between the two services. In 1939 Coastal Command had 180
frontline aircraft, which, at the behest of the admiralty, were to concentrate on
the German surface fleet on the assumption that the submarine threat could be
countered by aircraft carriers and surface attack. As a result, the aircraft were neither equipped nor trained for antisubmarine warfare and their maximum combat
radius was 250 miles.[151] With the fall of Norway and France in 1940, U-boats
deployed from French Atlantic ports. Air attacks on U-boat pens were ineffective, initially because of bomber vulnerability by day and the inability to find
and hit precise targets at night, and, subsequently, when the installations proved
impervious to bombing, the boats had to be found and destroyed at sea.

From 1940, radar supplemented the visual detection of U-boats on the
surface. In 1941 air-launched depth charges replaced ineffective bombs. An
operational research section was established in Coastal Command that quickly

increased depth-charge success rates. From the spring of 1941, British analysts began to break German Enigma ciphers, enabling convoys to be diverted from known "wolf pack" ambushes and aircraft to refine their search-and-attack patterns. In 1942, however, the U-boats regained the initiative with devastating consequences when the Kriegsmarine introduced a more complex Enigma cipher and broke the Royal Navy's (RN) own convoy cipher. Heavy losses were inflicted off the East Coast of the United States, on the north Atlantic convoys, in the South Atlantic, and the Indian Ocean. U-boat numbers increased from 100 to 150, and the rate of Allied merchant ships being sunk threatened to outstrip construction. Until 1943, 90 percent of maritime patrol aircraft failed to locate the convoys they were detailed to protect.[152] Bomber Command retained priority for navigation and radar equipment, while Air Marshal Harris, supported by Churchill, opposed the diversion of aircraft from his offensive. Although the entire war could be lost in the Atlantic, the allocation of very-long-range aircraft to the U-boat war was a very low priority until March 1943, when the gap in air cover in the north Atlantic was closed by RAF Liberators and USAAF long-range aircraft. Surface passage by night across the Bay of Biscay was disrupted by the Leigh airborne searchlight exploiting radar location. Coastal Command's strength increased to 450, the new German cipher was broken, and decrypts revealed increasing concern among U-boat crews about air attack.[153] The presence of any aircraft was always a deterrent, usually compelling a submarine to submerge and lose contact with a target convoy.[154]

In May 1943 Adm. Karl Doenitz withdrew his boats out of reach of Allied aircraft. In November he abandoned pack attacks and withdrew most boats from the North Atlantic while continuing to suffer heavy losses. The introduction of the Schnorkel breathing tube was too late to influence the outcome as the submerged boat became vulnerable to underwater radar from sonobuoys. As a result, the transport of U.S. troops to the United Kingdom and the Allied invasion in June 1944 proceeded virtually unhindered. Finally, the RAF disrupted the entry into service of a Type XXI oceangoing high-speed boat, which, with Schnorkel, could cruise submerged, carrying new attack weapons, detection systems, and unbreakable Enigma. Bomber Command attacked its shipyards and mined its intended trials area. Consequently, 120 were completed but only one sailed.

Victory in the Atlantic was achieved by close cooperation between air and naval surface forces, heavily reliant on Ultra intelligence. By 1945 Coastal Command deployed almost a thousand well-equipped aircraft with effective weapons and highly trained crews. Aircraft destroyed just under half of the 784 U-boats lost, the great majority by Coastal Command.[155]

THE NUCLEAR TRANSFORMATION

By 1945 air power primacy had passed to the United States, which deployed more than thirty-one thousand modern combat aircraft and was the sole possessor of nuclear weapons. In 1946 the RAF was reduced to one thousand front-line aircraft and fifty-five thousand people. Its wartime successes, with leadership in the fields of jet propulsion and electronic warfare, faded. The government was financially exhausted. Relations with the Soviet Union deteriorated quickly into a cold war across the Iron Curtain that divided Europe. Whereas Britain and the United States had swiftly demobilized in 1945, the Soviet Union maintained some three million men under arms, representing considerable numerical superiority.[156]

The atomic attacks on Hiroshima and Nagasaki in August 1945 seemed at last to vindicate strategic-bombing advocates. A scientific subcommittee of the Chiefs of Staff Committee foresaw the devastating potential of atomic weapons and urged the production of jet bombers to carry them because retaliation was the only answer that might deter an aggressor.[157] Prime Minister Clement Attlee reinforced the sentiment by recalling in 1945: "We recognize, as some of us did before this war, that bombing would only be answered by counter bombing. We were right. . . . The answer to an atomic bomb on London is an atomic bomb on another great city."[158]

The production of atomic energy for industrial and military use was authorized. The Cabinet Defence Committee accepted the recommendations for the delivery of atomic weapons by long-range, high-performance aircraft against a civilian population to preclude a clash between military forces or their buildup or the exercise of sea power. On August 1, 1946, President Truman signed the McMahon Act, which ended nuclear cooperation between the United States and the United Kingdom. On August 9, the RAF issued its first operational requirement for an atomic bomb weighing ten thousand pounds to be carried by a new generation of high-performance, long-range aircraft.

On January 7, 1947, the air staff issued operational requirement 229 for a four-engined jet bomber capable of carrying a ten-thousand-pound bomb or a multiple bomb load of twenty thousand pounds at five hundred knots at heights up to fifty thousand feet for fifteen hundred nautical miles. On January 8, the prime minister authorized the research and development of atomic weapons after a meeting in which Foreign Secretary Ernest Bevin, supported by the Minister of Defence, had argued that Britain could not afford to acquiesce in an American monopoly, that other countries might develop atomic weapons, and that, unless an effective international system could be established to prohibit production and use, Britain must develop them. The Chiefs of Staff were anxious lest other countries possessed the weapon and Britain did not.[159]

Driving the arguments for a nuclear-capable bomber was the former chief of the Air Staff, now Lord Charles Portal, first as chairman of the Chiefs of Staff and then as controller of production of atomic energy. Chief of the Air Staff Sir Arthur Tedder argued that dependence on the United States for a nuclear strike force would "involve complete subservience to United States policy and would render Britain completely impotent in negotiations with Russia or any other nation."[160]

In 1947 the procurement and development of three bombers were prepared: Vickers Valiant, Avro Vulcan, and Handley Page Victor, collectively known as the V-force. Bomber Command was given the primary responsibility for the defense of the United Kingdom, by deterrence and, if necessary, retaliation. In 1939 bomber policy had been based, however intuitively, on RAF warfighting concepts. In 1947 it was a political decision. The possession of nuclear weapons was considered essential to sustain Britain's freedom of international action, to avoid dependence on the United States, to deter a nuclear-armed enemy, and to sustain British international status. Now there really was no defense against the *nuclear* bomber; fighters would be ineffective, cities would become hostages, and conventional forces would become irrelevant.

Before 1939, RAF procurement for the front line had been largely determined by imperial policing. After 1945, the Cold War and the North Atlantic Treaty Organization (NATO) alliance would dominate. Yet before the new bombers entered service, the RAF was engaged in an emergency in Malaya from 1948 to 1960, against a communist insurgency fighting the British colonial government. More than 40,000 Commonwealth troops, 60,000 police, and 300,000 Home Guard were deployed. The RAF bombed insurgent camps, cultivated clearings, and suspected insurgent movements but with little discernible impact. Maps and reference points were often unreliable, and the opposition quickly learned to exploit jungle concealment.[161] Conversely, RAF airlift by aircraft and, for the first time, by helicopters was invaluable. Insurgent mobility was countered, the duration of foot patrols could be doubled or trebled, fortified outposts deep in the jungle could be sustained, and casualties could be swiftly recovered.

Similar assistance was given to ground forces in Kenya against an uprising between 1952 and 1955. Once again, air attack harassed insurgents without inflicting heavy casualties but air mobility was a considerable force multiplier. In both Kenya and Malaya, the opposition was far less vulnerable to air attack than the tribesmen and villagers of the Northwest Frontier of India and the Middle East because of their mobility and concealment.

Nor were RAF bombers required for the Berlin Airlift in June 1948, when the Soviet Union closed all surface traffic from the Western zones of Germany into Berlin. The following day, British Foreign Secretary Ernest Bevin ordered the Chiefs of Staff to prepare the largest available transport force for the supply of the city. Within twenty-four hours, RAF transports began the airlift, alongside USAAF, French, and Commonwealth air forces. In April 1949 the blockade was lifted, and in September the airlift was officially ended. The RAF delivered 541,937 tons of supplies, out of a total of 1.78 million, at a cost of 41 lives. The Berlin Airlift changed the direction of the Cold War and was arguably the most important single contribution of air power to it.

In April 1949 Britain became a founding member of NATO, which envisaged "strategic bombing by all types of weapons" being the primary responsibility of the United States, with the European allies providing the bulk of tactical air support and air defense.[162] Nevertheless, in 1952 the British government authorized the production of the V-bombers charged with delivery of a British nuclear weapon.

A fourth jet bomber, the shorter range Canberra, designed for high-speed, high-altitude "blind" bombing, entered RAF service in 1951. Canberra squadrons deployed to 2TAF in Germany in 1954 and subsequently to Cyprus. In 1958 they were given a nuclear role against Warsaw Pact ground forces and airfields. Conventional tactical air power for close air support, interdiction, or counterair operations was no longer required.[163] This was just as well because all the RAF jet bombers were configured for high-level visual bombing. Initially, they lacked blind, all-weather bombing equipment and depended for night and all-weather navigation on the guidance of Gee ground stations, which were range limited and confined to the European theater.

The RAF played little part in Korea between 1950 and 1954. Chief of the Air Staff Sir John Slessor saw the Russian threat to Western Europe as paramount and would not contemplate any weakening of it.[164] As a result, the RAF only deployed a handful of Sunderland flying boats to supplement maritime reconnaissance and volunteer pilots to serve with U.S. and Royal Australian Air Force Fighter squadrons.

In November 1956, however, the new bomber force saw combat in Operation Musketeer: the abortive attempt by Britain, France, and Israel to coerce President Gamal Abdel Nasser into accepting Western control of the Suez Canal and the presence of allied forces in the canal zone. Political objectives were vague. Training was impeded by ceremonial commitments. So much secrecy was imposed by Prime Minister Anthony Eden that until final briefings some crews did not know which country they were to attack.

In the first phase, the Egyptian Air Force was to be neutralized on the ground by Bomber Command Valiants and Canberras from bases in Malta and Cyprus. Fighter bombers from RN and French aircraft carriers and bases in Cyprus would also attack airfields and support invading ground forces. A parallel objective was to break the Egyptian will to resist by means of an extended air bombardment of "infrastructure" targets.[165]

The air attacks, which began on October 31, were virtually unopposed but the bombers had little impact on their targets.[166] The frank and necessary postconflict Musketeer Report noted that Bomber Command was geared to a "radar" war in Western Europe and was neither prepared nor constituted nor organized to undertake such an operation.[167]

Conversely, attacks by Fleet Air Arm and RAF fighter bombers on airfields were very successful, before ensuring that the allied invasion was closely supported by "cab rank" attacks on Egyptian ground forces and unopposed by Egyptian aircraft. Military success was, however, curtailed after seven days when the fighting was ended with a cease-fire at midnight on November 6 after increasing international pressure, with Egyptian morale unscathed.

RETRENCHMENT

Between 1950 and 1954, resources were strained by Cold War commitments and imperial residue, as the defense budget doubled and the RAF expanded from 50,000 to 270,000. The British Defence White Paper of 1955, therefore, asserted that deterrence would rest primarily on strategic air power with nuclear weapons to counter the massive preponderance of Soviet conventional forces in Europe. Thereby, a forward strategy would preclude the need for another "grim process of liberation."[168] The decision prefaced a reduction in British military strength, which, punctuated by occasional recovery, would continue into the twenty-first century.

In 1956 the cabinet approved a memorandum titled "The Future of the United Kingdom in World Affairs" from Cabinet Secretary Norman Brook.[169] It argued that a thorough review of policy was essential because the strategic situation had been transformed and conventional forces had become relatively less important. A substantial reduction in government expenditure was necessary. Defense demanded nuclear deterrence with conventional forces reduced. The feasibility of the air defense of Britain against nuclear attack was questioned. The Air Council independently forecast considerable reductions in Fighter Command, 2TAF, Coastal Command, and even the V-force itself, with only a token presence left in the Far East.

The 1957 Defence White Paper confirmed the policy of nuclear deterrence.[170] Britain's defense forces were now beyond the nation's capacity to fund. Military planning had to prevent war rather than to prepare for it.[171] It confirmed fighter defense of the bomber airfields, but in view of the good progress perceived to be made in surface-to-air missile (SAM) development, "the RAF are unlikely to have a requirement for fighter aircraft of types more advanced than the supersonic P1 and work on such projects will stop." The V-force would be supplemented by ballistic rockets, and considering the likely progress of ballistic rockets and missile defense, a supersonic manned bomber would not be developed.[172] National Service conscription would end in 1960. Canberra squadrons in Germany and Cyprus would receive nuclear weapons. The transport force would be expanded to provide army reinforcement beyond Europe.

In 1957 the RAF confirmed that nuclear weapons delivered from the air had become the great deterrent, the primary function of air power, and the responsibility of the RAF.[173] Because of the destructive power of nuclear weapons, the opening phase in a global war was likely to be decisive. Therefore, preparedness became all important, rather than the spiritual and material resources of a nation.[174] The primary RAF functions were to attack the enemy's military, industrial, and administrative centers essential to his power of resistance and to defend the United Kingdom against air attack. Secondary RAF functions were to provide intelligence, contribute to defeating threats to sea communications, establish a favorable air situation in the land battle, and exploit it by interdiction and paralysis.[175] Air Publication (AP) 1300, *The Royal Air Force War Manual*, echoed the 1922 Operations Manual and would remain the "bible" of RAF doctrine until after the end of the Cold War.

On August 2, 1957, the Cabinet Defence Committee reduced the V-force to 144 aircraft from the 184 recommended by the Minister of Defence. The figure was determined by the severe economic strains on the force and because "the UK would never in practice be expected to challenge the Soviet Union alone."[176]

Nevertheless, in October 1957, the Chiefs of Staff planned for both integrated and unilateral contingencies. In the former, target policy for Bomber Command would be determined by allied operational circumstances. In the joint UK-U.S. nuclear strike plan of 1958, it was allocated 106 targets, comprising 69 cities of government or other military significance, 17 long-range airfields associated with the nuclear threat, and 20 air defense installations.[177] If, however, "the UK should be forced to take unilateral action against the USSR, the target policy should be to attack the Soviet centres of administration and population," which would comprise ninety-eight cities whose population exceeded one hundred thousand, located within twenty-one hundred nautical miles of the United Kingdom. The cities were prioritized by population, administrative and economic importance, and transportation. The list would be periodically updated.[178]

The RAF bombers could reach their targets from airfields in the United Kingdom ahead of the Strategic Air Command (SAC) strike forces, prompting the chief of the Air Staff of the day to assert that the V-force would be "the spearhead of, and largest contributor to, the first and possibly decisive Anglo-US retaliatory strike."[179] Bomber Command demonstrated its credibility during the next decade by well-rehearsed dispersals to airfields in the United Kingdom, by deployments to overseas bases, by maintaining a proportion of the force on fifteen-minute quick reaction alert status from 1962, and by being able to launch a high proportion of the force within the four-minute warning expected of a Soviet nuclear strike.

From 1959 to 1963, the bombers were supplemented from the United States by sixty Thor liquid-fueled, static, intermediate-range ballistic missiles (IRBMs), which could deliver a megaton warhead fifteen hundred nautical miles. The warheads were controlled by United States Air Force (USAF) officers, the missile launch by RAF.

The 1960s dealt severe blows to Bomber Command. Improved high-altitude Soviet air defenses, together with the deployment of Soviet IRBMs against British targets, placed a high premium on mobility and low-level penetration to sustain deterrence credibility. The development of a British IRBM was canceled in March 1960 because of its perceived vulnerability and costs. Later in the year,

Britain sought to extend the V-force's standoff ability by purchasing the U.S. Skybolt, which was expected to have a range of one thousand nautical miles and an accuracy of less than two miles, to replace the shorter range Blue Steel.

Skybolt was canceled by the United States in December 1962. With no RAF alternative, Britain accepted a U.S. offer of Polaris submarine missiles to be deployed with British warheads. Meanwhile, the V-force was downgraded to low-level attacks, mounted ground alert from thirty-six airfields, procured new navigation and electronic countermeasure equipment, and modified Blue Steel for low-level delivery. The whole of the force was allocated to NATO subject to diversion to British national interests beyond NATO if necessary.

The RAF had finally acquired the strategic potential of air power envisaged by Trenchard. Targets explicitly included the enemy population. The costs and casualties of a protracted conventional war of attrition would indeed have been avoided, but only on Faustian terms. Nuclear weapons took bombing theories into a surreal world of potential devastation, unimaginable even by the standards of two world wars. Yet even the V-force lost its deterrent credibility because of enemy air defenses and resource constraints.

During the 1960s, further defense reviews reduced the size of the V-force and canceled several aircraft programs, including a Canberra replacement, an intended purchase of U.S. F-111As, as well as a new aircraft carrier. All future RAF frontline aircraft would be internationally collaborative. Meanwhile, Britain withdrew from the Far East. In June 1969, although the RAF retained a nuclear capability in the V-force and other aircraft, the RN assumed responsibility for the British nuclear deterrent.

Retrenchment continued in the 1970s, alongside the need to provide conventional forces after 1967 to sustain NATO's revised strategy of "Flexible and Appropriate Response." Defense expenditure was reduced to 5.5 percent of gross domestic product (GDP) and the RAF cut to ninety thousand. Combat squadrons in Germany were further reduced. Greater RAF priority was given to the protection of the Atlantic and North Sea by the deployment of small numbers of Nimrods and Buccaneers carrying conventional and nuclear weapons. As the RAF withdrew from the Middle East and Malta, virtually all combat and combat-support aircraft were allocated to NATO. The remark by a distinguished retired official, "In the 1960s, decisions about foreign policy and defence were dominated by financial considerations," could equally have been applied to any subsequent decade.[180]

By 1976 the Air Force Board was concerned about the direction and public image of the RAF after the loss of the deterrent, overseas withdrawals, and compulsory redundancy. The AP 1300 of 1957 had been withdrawn in 1970. In June 1976, Chief of the Air Staff Air Marshal Sir Neil Cameron provoked debate in Parliament by publicly expressing his concern at the disproportionate reductions imposed on the RAF since 1957 and his fears of further cuts. He had also asserted, "Somehow we have grown shy of talking about air power . . . about the critical importance of the Air Force in our defences."[181] The government emphasized the importance of the RAF but would not rule out further cuts. The underlying problem was that the RAF had become synonymous with the hugely publicized V-force. The air defense of Britain was apparently now only associated with the defense of the deterrent. If both had gone, what was the remaining rationale for an independent service?

Later that year, Neil Cameron established a new post, the Director of Defence Studies (RAF), with a remit to "revive and maintain an interest in the study of the present and future use of air power in all its various military operations."[182] A new definition of air power was approved that would transcend the confines of strategic bombing and be progressively refined and incorporated into new issues of RAF doctrine into the twenty-first century:

> Air power is the ability to project military force by or from a platform in the third dimension above the surface of the earth, . . . which is exploited to advantage . . . for example for manoeuvre, deployment, concealment or surprise. . . . Air power may apply force directly, as in bombing or air to air combat, or distribute it by providing air mobility or resupply, or amplify it by reconnaissance or surveillance.[183]

The spirit of the offensive would be maintained, but the RAF acquired a comprehensive doctrinal base, which would support it in a new era of widely differing operations and enable it to avoid the resurrection of bomber dogma that infected the 1990s elsewhere.

AN EXPEDITIONARY AIR FORCE

On April 2, 1982, Argentina invaded the Falkland Islands in the South Atlantic, eight thousand miles from London. Prime Minister Margaret Thatcher immediately ordered the armed forces to expel the invaders.

Although there were no contingency plans, three days later a task force, led by the aircraft carriers *Illustrious* and *Hermes* with twenty Sea Harriers (SHARs), set sail from the United Kingdom while preparations began for combined service operations. There was an intermediate USAF base on Ascension Island but aircraft had to cover two legs of approximately four thousand miles each, making inflight refueling (IFR) essential. Uniquely, the Alert Measures Committee authorized new equipment and weapons without seeking financial authority. Thatcher was reported to have observed, "We will not count the cost, but we will keep account of the cost."[184]

Twenty Victors, previously converted for IFR, supported the operation, including three immediately trained for maritime reconnaissance. RAF Harrier GR3s landed on aircraft carriers for the first time, Nimrods and C-130s were fitted with IFR equipment, and crews were authorized in three weeks. Five Vulcan bombers were given additional navigation and electronic countermeasure equipment and reinstated IFR currency; pilots were trained in conventional bombing tactics. The United States provided bulk fuel and prefabricated accommodation at Wideawake Airfield on Ascension Island, AIM 9 air-to-air Sidewinders for Harriers and Nimrods, and antiradiation Shrike missiles for the Vulcans.

On May 1, while the task force was ninety-five miles northeast of Port Stanley, a single Vulcan, refueled seven times by twelve tankers from Ascension, hit Port Stanley runway with a single bomb. The airfield remained open, but its vulnerability precluded its use by Argentine jets deployed from the mainland. The Argentine government reacted to the threat by withdrawing the majority of its Mirage fighters to defend the mainland.

Harriers and Special Forces attacked the islands' airfields and gave close air support to the ground forces following landings on May 21. Despite interdiction, the Argentine troops fought bitterly until June 15. In the last engagement of the war, three Harriers "turned the tide" at Goose Green by taking out Argentine artillery that had pinned down exposed British infantry.[185]

Meanwhile, Argentine aircraft, despite heavy losses, continuously attacked the task force at low level in clear weather, destroying six ships, damaging six, and hitting four more with bombs that failed to explode because of inadequate fusing. The transport *Atlantic Conveyor*, carrying Chinook helicopters intended for ground force support, was sunk, while the landing ship *Sir Galahad* was hit with heavy loss of life. Super Étendards with Exocet missiles compelled the carriers to remain out of range, thereby reducing the time on station of the Harriers. Command of the air was never achieved because of the lack of airborne early warning (AEW) and extended combat air patrols. RAF offensive air power was a major contributor to victory but defensive weaknesses could have been disastrous. Had the carriers or the troop-carrying *Canberra* been hit, or indeed if "they had got the fusing of their bombs right, they would have virtually wiped out the fleet."[186]

The unlikely success over a distance of eight thousand miles was due to strong political leadership, a clear strategic objective, considerable ingenuity, good joint service cooperation, technological superiority, freedom from budgetary restraints, and extensive U.S. military support. Deficiencies were exposed in AEW, reconnaissance, battle damage assessment, and air-land cooperation. The impact of the loss of three helicopters was considerable.

While the RAF maintained a detachment in the Falklands, the main focus returned to NATO for another nine years. The Vulcan squadrons were disbanded in 1982, and the multinational strike/attack Tornado GR1 replaced Phantom and Buccaneer squadrons in Germany. In 1989 the fall of the Berlin Wall marked the end of the Cold War, but within twelve months the RAF's attention was once again turned to the Middle East when Iraq invaded Kuwait in August 1990.

THE FIRST GULF WAR AND THE TWENTY-FIRST CENTURY

Twelve Tornado F3 fighters were dispatched to Saudi Arabia from an existing detachment in Cyprus and immediately mounted combat air patrols. Their rapid deployment, followed by a further six F3s, was a signal of British intent and politically acceptable as a defensive measure.[187] The F3 was not an air superiority fighter and, therefore, was tasked throughout the Gulf War with defending rear areas against any Iraqi aircraft that managed to penetrate the forward defenses of U.S. F-15s, F-18s, and SAMs, freeing the U.S. fighters to operate further forward.

This was the first occasion on which the RAF had flown in coalition with the U.S. air forces since World War II, with the marginal exception of the United Nations (UN) operations in Korea. Now, the RAF was a very junior partner. The air campaign in Desert Storm was compiled entirely by the USAF, and more than 90 percent was executed by U.S. air forces. Unlike the close personal relationships established between Allied commanders in 1942–45, USAF Air Commander Gen. Charles Horner had not previously met Air Vice-Marshal William Wratten, the RAF commander in theater, although a good relationship quickly evolved. Conversely, for many years RAF crews had trained alongside their U.S.

allies in Red Flag and other exercises, sharing common operating procedures harmonized within the NATO structure. Consequently USAF Lt. Col. David Deptula had no hesitation in allocating RAF aircraft to the vitally important offensive counterair role in his prewar mission planning.[188]

The daily Air Tasking Order (ATO) integrated RAF aircraft while reserving national control. It was produced by USAF staff, with one RAF representative from the Tornado force. Before Desert Storm began in January 1991, there was difficulty in aligning the rules of engagement (ROE) between the two allies. General Horner coolly advised the visiting chief of the UK Defence Staff that if the UK could not resolve the problem, "If you are not ready, we will go without you."[189] In NATO operations ROE were well established, but in Desert Storm fighter interception ROE were tighter for the RAF F3s than for U.S. interceptors. Problems of coalition ROE would persist in the new strategic environment, where the United States would dominate air operations but for political reasons had to take into account different ROE among coalition partners.

The RAF contributed 4 percent of the total coalition effort: 92 combat aircraft, 17 tankers, a number of Nimrods, and air transport of British ground forces. The United States deployed 1,200 combat aircraft, supported by 240 tankers.[190]

RAF operations in the first Gulf War were determined by its configuration for a war against the Warsaw Pact. Jaguars and Tornado GR1s had trained to operate in Europe at low level: the Jaguars delivering one-thousand-pound bombs and BL-755 cluster munitions, the Tornados attacking airfields with JP233 runway-cratering bombs at night in all weather. With many large bases with hardened shelters, and equipped with some modern Soviet-sourced combat aircraft capable of delivering chemical weapons, the Iraqi Air Force threatened coalition operations. Consequently, airfield denial was initially a major coalition objective. The GR1s began airfield attacks from the outset of the war. Typically, eight Tornados were supported by U.S. Navy (USN) A-7s and 3EA-6Bs suppressing enemy defenses, F/A-18s attacking airfield installations, and F-14s providing top cover. The whole package was supported by KA-6 tankers with a USAF EF-111 acting as a standoff jammer.[191]

After four days, it became obvious that the threat from the Iraqi Air Force was negligible. Moreover, the Iraqi air bases were far bigger than those of the Warsaw Pact, and there were insufficient Tornado/JP 233 systems to close them down. Consequently, Wratten, after a difference of opinion with the Ministry of Defence (MOD) in London, authorized medium-level attacks to avoid the low-level SAMs and guns.[192] Subsequently, twelve Buccaneers were deployed to provide laser-target marking, and the force was further enhanced by the accelerated deployment of a Thermal Imaging Airborne Laser Designator pod. Only the reconnaissance GR1A operated throughout the war as it had trained in Europe, flying at high speed at low level. In a campaign where reconnaissance assets were always in high demand, it emerged unscathed from Scud searches, support to Special Forces, preattack reconnaissance, and battle damage assessments.[193]

By 1990 a central staff of civil servants "put a new level of challenge and response into the financial part of the system" that contained "a recipe for conflict in the future."[194] In the next two decades, the gap between strategic aspirations and resource provision continued to widen as the United Kingdom and several NATO allies sought a "peace dividend" from the ending of the Cold War, while, at the same time, embarking on more operations. In 1991 the RAF front line was reduced by 25 percent even before the government had completed an appraisal of the procurement implications of Desert Storm.

Just one year later, the RAF began to support UN humanitarian intervention in the civil war in the disintegrating federation of Yugoslavia.[195] The conflict that developed in the Federal Republic of Yugoslavia (FRY) in 1991 was completely different from that in the Gulf. Instead of a confrontation between states, it would become a civil war between three former members of the Yugoslav Federation, which in turn stimulated the need for extensive humanitarian intervention. Instead of desert sands, the Balkans was an area of mountains and forests. Instead of comparatively good weather, low clouds could be expected for 75 percent of any year. Instead of lines of armor conventionally deployed in largely uninhabited desert, ground forces in Bosnia would frequently operate among the civilian population. The different environment would have been sufficient on its own to prompt a reconsideration of the best way to apply air power in Bosnia. In the event, other differences would present even more difficult problems to air and ground commanders. UN action was hesitant, reflecting international uncertainty about whether or how to intervene in the dispute over territory between Croatia, Bosnia and Herzegovina, and Serbia. Other than the constraints of international law, there had been no humanitarian complications in Desert Storm, although the coalition had been sensitive to casualties.

From 1992, RAF Hercules began flying supplies to the beleaguered Bosnian city of Sarajevo. In April 1993 the North Atlantic Council (NAC) agreed to NATO aircraft enforcing United Nations Security Council Resolution (UN-SCR) 816, which imposed a ban on fixed-wing flights over Bosnia, where Serbian and FRY aircraft had come to influence ground operations. The RAF contributed eight Tornado F-3s to combat air patrols and two E-3D Sentry Airborne Warning and Control System (AWACS) aircraft.

There was, however, no international consensus about using offensive air power. Troops of the United Nations Protection Force (UNPROFOR), deployed as peacekeepers in July 1992, were seen to be vulnerable to Serb retaliation. Strict ROE confined air attacks to military units, such as individual tanks or guns, that had been identified as initiating an attack. The need to reduce collateral damage and civilian casualties hastened the employment and procurement of precision-guided munitions (PGMs). Out of the U.S. and NATO munitions, 96 and 28 percent were PGMs, respectively, compared to an allied total of 9 percent in Desert Storm.[196]

In August 1993 the NAC initiated NATO operational planning for air strikes. It included command and control coordination between the UN and NATO, which became known as "the dual key" and comprised the principle of proportionality of force and sensitivity to collateral damage. "The discussion on air strike options noted that the selection of targets needed to take into consideration proportionality as well as the importance of showing resolve and capability; the selection should also discourage retaliation."[197]

The RAF contribution was determined by contemporary UK political perspectives, which were affected by the uncertainties of a new strategic environment. Fear that air strikes would provoke retaliation on British and other UN ground forces, as well as the disruption of humanitarian missions, would prove enduring and inhibiting concerns. UK defense policy categorized humanitarian operations only as "Defence Role Three," after the "Provision of an Effective Strategic and Sub-Strategic Nuclear Capability" and commitment to NATO.[198] In January 1993, British defence secretary Malcolm Rifkind told a press conference that the role of British forces in Bosnia would continue to be "the provision of humanitarian aid. . . . Our primary responsibility as the United Kingdom is obviously the safety of our own forces."[199] Foreign Secretary Douglas Hurd announced that sixteen NATO countries had agreed to prepare plans for air strikes "subject to arrangements for the UN to be involved so that our troops are not at risk."[200] Lest there be any remaining doubt, Mr. Rifkind observed the following day from Bosnia, "It remains essential to ensure that any new initiative by the UN is fully compatible with the physical safety of British and UN forces in Bosnia."[201]

Only in August 1995 were ROE relaxed, after Serb recalcitrance remained unshaken, a single NATO chain of command had been established, and clear strategic objectives identified.[202] In twenty-two days, 3,515 sorties were flown in Operation Deliberate Force, against Serbian command and control centers, air defense units, fielded forces, and supplies. Coupled with international diplomatic pressure and advancing Croatian and Bosnian ground forces, the air offensive coerced the government of President Slobodan Milosevic to accept the terms of the Dayton Peace Accords. RAF Harriers and Jaguars flew 9 percent of the sorties from the Italian base of Gioia del Colle. The fact that one squadron of Harriers and one squadron of Jaguars represented one-third of each fleet illustrated the extent of the cuts already imposed on the RAF's front line after the Gulf War. Tankers and reconnaissance Nimrods and Canberras flew from other Italian bases.

Nevertheless, in 1995 the RAF front line was cut again by 40 percent, under the banner "Front Line First," together with the removal of nuclear weapons and a reduction in manpower from 72,000 to 52,000, to be achieved by redundancy and civilizing a number of support activities. The policy was bitterly debated in Parliament at a time of increasing overseas commitments with reduced base facilities.[203] The government was unmoved.

Meanwhile, the RAF had sustained a presence with the USAF over Iraq from bases in Turkey, Saudi Arabia, and Kuwait, contributing in two no-fly zones to the protection of Kurdish minorities in northern Iraq and Marsh Arabs in the south. Tornados, Harriers, and Jaguars rotated in providing armed reconnaissance, frequently responding to Iraqi air defense installations that sought to engage them. In 1998, twelve GR1s and eight Harriers flying from HMS *Invincible* joined U.S. forces in Operation Desert Fox, designed to compel Saddam Hussein to accept UN inspectors in the country. In the 1990s, the RAF operated in the Middle East, the Balkans, the Falklands, Belize, Cyprus, Northern Ireland, and Sierra Leone. While operations expanded, the service was reduced by almost 50 percent.

In 1999 the RAF saw combat once more in the Balkans. The Yugoslavian province of Kosovo had not been included in the Dayton Peace Accords of 1995. Britain was among several countries that became increasingly concerned by Serbian oppression in Kosovo. In April 1998 the UK's Permanent

Joint Headquarters began informal contingency planning for a contribution to a NATO effort to intervene in Kosovo to contain the increasing violence.[204] UNSCR 1199 of September 23, 1998, called upon the FRY government to cease all repressive action in Kosovo, facilitate the safe return of refugees to their homes, and allow unimpeded movement by humanitarian organizations.[205] The following day, the NAC approved an Activation Warning Order for a limited air operation, subsequently followed by the approval of a plan for air strikes against Serbia.[206] In October UNSCR 1203 welcomed conciliatory statements from the FRY government but, "[a]cting under Chapter 7 of the Charter of the United Nations," demanded that it "comply fully and swiftly" with UNSCR 1199.[207] At Rambouillet in January 1999, attempts to broker an agreement on the October terms between FRY president Milosevic and a Kosovo delegation collapsed. On March 24, 1999, the United States and NATO allies began air strikes against the FRY to reduce the military capability being used against the civilian population of Kosovo.[208] It was widely and mistakenly believed that coercive air attacks would quickly succeed.

RAF participation in Operation Allied Force reflected the evolution and problems of the overall air campaign. NATO was a defensive alliance. It had no contingency plans for offensive action and no experience of humanitarian intervention. Limited air operations was the only action acceptable to all nineteen alliance members. Political differences and rigid ROE prevented high-tempo, continuous, overwhelming attacks as used in Desert Storm. President Milosevic's unexpected resilience prompted an expansion of the NATO master target list from 169 to 976. Each required a review and approval by NATO and national authorities. Each of the thirteen participating nations could veto any target.[209] The RAF contribution was progressively increased from eight Harrier GR-7s, two tankers, two E-3D AWACS aircraft, and one reconnaissance Nimrod to sixteen Harriers, twelve Tornado GR1s, and eight tankers. All operated from Italian bases, except for the Tornados, which flew from their home base at Bruggen in Germany, necessitating additional IFR, before deploying to Solenzara Air Base in Corsica. The House of Commons Defence Committee subsequently identified the failure to deploy the Tornados to the theater as the product of the United Kingdom's overoptimistic assumptions about the duration and intensity of the air campaign and the impact of overstretch on the RAF's ability to deploy outside home bases.[210]

A total of 1,618 sorties were flown by UK aircraft, including 102 combat air patrols by seven RN Sea Harriers from HMS *Illustrious*. The RAF released

1,011 munitions, of which 244 were PGMs, 230 were gravity bombs, and 531 were cluster weapons.[211] Persistent bad weather in the area, together with strict ROE designed to minimize civilian casualties and collateral damage, frequently induced abortive sorties and rendered the RAF's laser-guided Paveway II and III munitions inoperable against small, mobile targets. Strategic arguments over targeting of infrastructure and fielded forces, allied friction over ATOs, the heavy constraints of ROE, and the need to sustain alliance cohesion continued until June 9, when President Milosevic finally accepted the Rambouillet Agreement and the air campaign ceased.

The campaign exposed RAF deficiencies in all-weather PGMs; battle damage assessment; secure communications; the ability to find, target, and engage mobile forces; and the ability to deploy units from main operating bases. Despite sustained resource constraints, all would be addressed before the next major commitment in 2003.

In 2001, after the Twin Towers attack in September, the RAF began to support NATO operations in Afghanistan. Between October 2001 and January 2002, Canberras and Nimrod R1s provided reconnaissance, Nimrod MR2s offered maritime reconnaissance, E3D Sentries flew one-third of battle management sorties, C17 and C130 Hercules provided strategic and tactical airlift, and Hercules and Chinook helicopters supported special forces operations. TriStar and VC10 tankers supported RAF E3Ds and USN fast jets. RAF personnel were involved from the outset in the planning and conduct of the air campaign both in theater and at HQ U.S. Central Command in Florida.[212] At this stage, no RAF combat aircraft were involved.

In 2002, U.S. and allied attention switched from Afghanistan to Iraq. RAF representatives joined U.S. planning staff in Florida in the summer of 2002 as a result of the close cooperation between the two air forces in the no-fly zone operations over Iraq during the previous decade.[213] The declared British overriding objective was to disarm Saddam Hussein of his weapons of mass destruction, which threatened his neighbors and his people.[214] UNSCR 1441, passed in November 2002, declared Iraq to be in breach of previous resolutions, provided Iraq with a final opportunity to comply with its disarmament obligations, and threatened serious consequences in the event of further Iraqi noncooperation. The United Kingdom, deciding that noncompliance with UNSCR 1441 provided legitimate authority for military action against Saddam, joined with Australia and Canada in the U.S.-led coalition that attacked Iraq on March 24, 2003.

The 1998 UK Strategic Defence Review had identified the need for Britain's armed forces to be configured for joint service rapid deployments to operate in a single battlespace with speed and precision. A Joint Helicopter Command was established, and "Joint Force 2000" would initiate joint RAF and RN Harrier operations at sea and from land. Priority was given to enhancing strategic air-lift; standoff PGMs; intelligence, surveillance, target acquisition, and reconnaissance (ISTAR); and improved command, control, and communications.[215] The review noted that since 1990 defense expenditure had fallen in real terms by 23 percent and the armed forces reduced by almost a third. Consequently, "there are personnel shortages in important areas and unsustainable pressure on many of our people."[216] The review guidelines were quickly overtaken by events. Programs were underfunded, commitments increased, and manpower was stretched still further, but by 2003 the RAF was armed and configured to make a substantial contribution to Operation Iraqi Freedom—known as Telic in the United Kingdom.

The RAF was totally integrated with U.S. forces in the planning and execution of the air campaign,[217] contributing 112 fixed-wing aircraft and 27 helicopters. Although only approximately 6 percent of the coalition total, they included attack, air defense, surveillance, reconnaissance, airlift, IFR, and AEW. Of the munitions dropped on hardened Iraqi installations, 85 percent were PGMs, including Storm Shadow missiles. RAF tankers off-loaded 25 percent of their fuel to U.S. aircraft. Although much smaller than the force deployed in Desert Storm, the RAF's contribution was powerfully enhanced by the reequipment of the intervening decade.

Lessons of Kosovo were applied. ROE were tight, protecting civilians and reducing the risk of collateral damage, but clearance procedures were carefully planned and executed, reducing the need for high-level authorization to a minimum and enabling the high-tempo warfare desired by the United States and identified by the UK Defence Review. Unlike the preparatory air phase of Desert Storm, Iraqi Freedom began with a heavy simultaneous air and land offensive. Darkness, bad weather, and sand storms offered no protection to the Iraqi ground troops against incessant air attacks. The conventional phase of the war was over in three weeks.

Air Vice-Marshal Glenn Torpy gave a typically forthright briefing to the House of Commons Defence Committee. Admitting satisfaction with the contribution of the RAF, he nevertheless observed that, for the first time in very many years, the RAF had been called upon to provide close air support to ground forces in a very fluid environment. Both parties needed to improve air-land integration: they needed more combined service training, better communications, and a much more comprehensive forward air control presence and structure.[218]

One source of friction in Kosovo was, however, not totally resolved in 2003. The secure U.S. command and control system, Secure Internet Protocol Router Network, was not accessible to the allies, resulting in frequent delays in the transfer of intelligence and other information to the RAF, although good personal relations were seeking to minimize the problem.[219] Even if the United Kingdom could catch up with U.S. communications technology, the problem of national sensitivity would remain and the more coalition partners, the greater the problem.

The conventional war was quickly won by the allies, but six more years of insurgency elapsed before the RAF withdrew from Iraq. Meanwhile, by the time allied attention returned to Afghanistan in 2005, the Taliban had taken advantage of the diversion to consolidate and regain much of their position in the south of the country.

The RAF was called upon to fight a very different kind of war in both countries. In Iraq a residue of Saddam supporters and other dissidents waged an insurgency against coalition forces, complicated by bitter internal struggles between Sunni and Shi'ite Muslims. In Afghanistan, the Taliban sought to regain political control, expel coalition forces, and overthrow the Afghan government. In both theaters, the insurgents were indistinguishable from civilian noncombatants, and used them for concealment and protection when fighting. They relied on terror to sustain civilian support. They practiced traditional guerrilla tactics of ambush, increasingly relying on homemade mines, known as improvised explosive devices, to disrupt coalition movement and isolate coalition units. They used civilian cars, trucks, and motorbikes for mobility. They manipulated the media to engender local and international sympathy and support whenever coalition action caused civilian deaths and collateral damage. They exploited hostility to the "occupying" powers. Wherever possible, they avoided confrontations with coalition ground forces, unless they perceived considerable local superiority. In short, they too had learned progressive lessons from the wars of the previous decade. There were no traditional front lines, giving rise to the expression "war amongst the people." "Asymmetric warfare" was now fully fledged.

In such an environment, air power became the coalition's own asymmetric advantage. Once again, the RAF made a major contribution to the U.S.

and NATO forces. Over five years, RAF Harriers flew 8,557 sorties, including close air support, convoy escort, and demonstrations of "presence." Enhanced Paveway II PGMs enabled attacks on small groups of insurgents without risk of civilian or friendly force casualties. Fast jet speed enabled rapid response across the theater to beleaguered ground forces, which were now usually accompanied by a ground force joint tactical air controller (JTAC). He could summon close air support and direct the pilots onto Taliban who were frequently very close to his own position. RAF intervention often saved the lives of soldiers.[220] Access to such precise, proportionate, and responsive firepower reduced considerably the number of combat ground troops required. In 2009 the Harriers were replaced by Tornado GR IVs, delivering light-weight, dual-mode laser and millimetric radar seeker all-weather Brimstone and Paveway IV PGMs, which reduced the area of impact still further.

C130 Hercules and C17s provided strategic and tactical airlift. Chinook and Merlin helicopters reduced the volume of vulnerable surface transport, achieved surprise by placing ground forces in all weather in proximity to Taliban positions, and evacuated casualties. Air mobility was an essential counter to the insurgents' own mobility and unpredictability. Air base security was provided by the RAF Regiment.

The operational environment demanded swift and precise intelligence, followed where relevant by swift and precise attack. The Tornado carried a RAPTOR (Reconnaissance Airborne Pod for Tornado), a day-night, all-weather reconnaissance pod, with real-time down-links as well as Litening targeting and reconnaissance pods. The Reaper reconnaissance unmanned aerial vehicle carried PGMs, while the Sentinel R1 airborne standoff radar provided long-range surveillance and reconnaissance. All systems were netted within the command and control structure.

In 2006 the crew of a reconnaissance Nimrod were killed over Afghanistan when their aircraft caught fire after IFR. A judicial inquiry confirmed the technical failure that caused the accident but delivered a scathing verdict on civilian contractors, consultants, and especially on the UK MOD. Previous RAF identification of the specific weakness, reminders of the age of the Nimrod, and requests for more maintenance resources had all been ignored by the MOD.

The MOD suffered a sustained period of deep organizational trauma between 1998 and 2006, beginning with the 1998 Strategic Defence Review. Financial pressures and cuts drove a cascade of multifarious organizational changes, which led to a dilution of the airworthiness regime and culture within the MOD, and a distraction from safety and airworthiness issues as the top priority. There was a shift in culture and priorities toward business and financial targets, at the expense of functional values, such as safety and airworthiness.[221]

The report induced a public apology from the defense minister, but the roots of the tragedy lay in the repeated reductions in RAF resources, accompanied by unremitting and increased commitments. Peacetime cost-effectiveness endangered wartime operational effectiveness.

In February 2010 CAS Air Chief Marshal Sir Stephen Dalton reaffirmed the inherent flexibility of air power and the RAF's readiness to face the challenges of Afghanistan and beyond. He emphasized the asymmetric advantage of, and enduring requirement for, "control" of the air in every kind of operation. He explained the ability of air power to shape the context of operations on land and sea. He described the RAF's agility and adaptability in achieving a core competence of "combat ISTAR": the fusion of dominant intelligence with real time response. He observed that sometimes air power could achieve the desired political or strategic effects on its own, but his primary focus was joint service operations.[222]

In October 2010 the coalition government sought to reduce a £38 billion Defence Budget overspend, while seeking to sustain operations in Afghanistan. The Joint RAF/RN Harrier Force was disbanded, Tornado squadrons reduced, Nimrod replacement canceled and purchases of the Lockheed Joint Strike fighter and Eurofighter Typhoon II cut. The RAF was reduced to 33,500.[223]

The RAF was a fraction of the size inherited by Trenchard, but could dominate a new spectrum of warfare, in alliance or coalition, to an extent that would have satisfied the early visionaries. Their forecast of a massive air strike determining the outcome of a war had been replaced by a doctrine that sustained the fundamental importance of command of the air but emphasized the inherent flexibility of air power and its indispensable contribution to all operations.[224] Indeterminate "strategic bombing" had been replaced by achieving offensive effects that could have either a direct or cumulative strategic impact, now delivered only by the RAF. Air power was succinctly defined not as hitherto by means, but in terms of ends: "the ability to project power from the air and space to influence the behaviour of people or the course of events."[225]

Trenchard's spirit of the offensive continued to imbue the service, but dogma was long gone. The RAF had embarked soundly on its second century of independence.

2

U.S. AIR POWER

Richard P. Hallion

"For all but the resolutely sightless, it is now obvious that air com-
bat determines the outcome in modern war."
—Gen. Merrill A. "Tony" McPeak, Chief of Staff,
USAF (Ret.), 2003

While military aviation began shortly after the invention of the balloon, during
the French revolutionary wars,[1] in the United States, the balloon's introduction
into military service did not take place until the Civil War. In the opening years
of the war, the observational balloon proved its great value to the Union army
as an intelligence and surveillance system. Yet after nearly two years of largely
exemplary service—including shipboard service—command disinterest led to its
withering and eventual withdrawal. Thus, at Gettysburg, the decisive battle of
the war, both the Union and Confederate armies fought without the benefit of
aerial observation, something available to the world's leading armies since the
time of the French Revolution. It is intriguing to contemplate what an observation balloon might have accomplished for either side at Gettysburg, particularly
for the Confederacy.[2]

American interest in airborne observation reawakened at the time of the
Spanish-American War, coincident with increasing European interest in heavier-
than-air flight. Already, in 1892, the French war ministry had contracted inven-
tor and air power theorist Clément Ader to develop an aerial destroyer, this
becoming the bat-like steam-powered *Avion III*, which unsuccessfully attempted

flight in 1897.³ In America, Smithsonian Institution secretary Samuel P. Langley, an astrophysicist, pursued the development of large flying models he termed
Aerodromes. In 1898, at the behest of Assistant Secretary of the Navy (and
future U.S. president) Theodore Roosevelt, a joint Army-Navy board reviewed
Langley's work, and from this sprang a War Department contract for a man-
carrying observational aircraft. The result was the Langley Aerodrome, a colossal
failure. Following two accidents (the second of which nearly killed its pilot), the
program collapsed in 1903, having, as with many joint-service aircraft programs
since, failed to meet requirements, schedule, and cost goals.⁴

The Langley failure undoubtedly retarded the efforts of the Wright brothers
to win swift military support for their aircraft, likely delaying the introduction
of the first military aircraft into service by approximately two years. Their per-
sistence, and rising foreign interest, led to the U.S. Army's issuing the brothers
a development contract in February 1908 for an airplane. Its subsequent flight
testing at Fort Myer, Virginia, and the demonstrations of another Wright Flyer
in France that same year, so galvanized public and official opinion that even the
crash of the military airplane on September 17, 1908 (an accident that claimed
the life of Army Lt. Thomas Selfridge), failed to dim official interest. In 1909 the
army acquired its first aircraft, and the Wrights opened a military aviation school
at College Park, Maryland. From this act began an evolutionary process that,
slightly less than four decades later, led to the United States Air Force (USAF).

Navy interest in winged flight came later, owing to the official view of Navy
secretary Victor Metcalf that the airplane "held no promise" for naval service.⁵
Undaunted by such disdain, Capt. Washington Irving Chambers, a prescient
and dedicated engineering officer, asked Wilbur Wright if he would consider
making an experimental flight from a ship. He refused, but Eugene Ely, a test pilot for the Wrights' archrival Glenn Curtiss, swiftly agreed to make the attempt.
On November 14, 1910, he launched from the cruiser *Birmingham*, landing
on shore at Willoughby Spit, Virginia. Then, on January 18, 1911, taking off
from a field outside San Francisco, he landed aboard the cruiser *Pennsylvania*,
anchored in San Francisco Bay.⁶

ESTABLISHING IDENTITY AND ORGANIZATIONAL FORM

Despite these notable accomplishments, the development of America's military
air power—and American aviation in general—surprisingly stagnated and even

reversed in the years after 1909. The combination of a debilitating patent fight between the Wrights and Glenn Curtiss, the lack of government support for aeronautical science and technology, and no clear military threat accelerating defense investment worked to limit the innovation, design progression, basic research, and incorporation of aircraft in military service.[7] By the time of the First World War, American aviators were increasingly flying imported European aircraft designs and using European-rooted aeronautical science and technology. Despite Americans' having invented the airplane just over a decade previously, when the European nations went to war in August 1914, their air services numbered, respectively, 244 Russian, 232 German, 162 French, and 113 British aircraft. The United States possessed only twenty-three.[8]

America's subsequent record in the Great War is well known. Uncle Sam's aviators proved themselves in the skies over Europe, but did so in foreign, not American, designs, for the nation's optimistic hopes of building a huge combat air fleet collapsed amid mismanagement, corruption, and industrial chaos. Tellingly, Gen. John Pershing remarked in his final report on America's involvement in the war, "In aviation we were entirely dependent upon our Allies."[9]

That included the development of military doctrine, operational concepts, and tactics, as well as the structuring of military-industrial and government relations. The best-known example of this influence was that of Britain's Maj. Gen. Sir Hugh Trenchard, chief of the Royal Flying Corps (RFC) and first chief of the Air Staff of the Royal Air Force (RAF), on Brig. Gen. William "Billy" Mitchell. A relative latecomer to aviation, Mitchell had been shocked at the losses incurred by conventional battles, such as France's failed Aisne offensive in 1917, and perceived air power as an alternative form of less costly and more productive attack. His meeting with Trenchard and later exposure to British (and, to a lesser degree, French) methods further convinced him he was on the right track.[10] His subsequent command and employment of Allied air power forces in the Battle of St. Mihiel effectively made him the prototype for the modern Joint and Combined Force Air Component Commander (JFACC/CFACC). His order, written in the young adulthood of the wood-and-fabric open-cockpit biplane, that airmen would be responsible for "destroying the enemy's air service, attacking his troops on the ground, and protecting our own air and ground troops," explicated, with admirable succinctness, the function of joint and combined aerospace forces operating in a combat theater even in the present day of transonic stealthy strike aircraft, cruise missiles, and unmanned aerial vehicles (UAVs).[11]

Exposure to European aeronautics and aeronautical practice during the last years of the First World War exacerbated the already-sensitive feelings of American military and industrial authorities regarding the abysmal showing of their own wartime aviation effort. A postwar European study mission led by Assistant Secretary of War Benedict Crowell returned the finding that Europe's aviation structure merited copying, recommending a British Air Ministry–like "Department of Aeronautics" to oversee civil and military aviation; unified national defense; an RAF-like independent air force; an aviation academy equivalent to West Point; a Federal airways network with weather reporting, communications, and searchlight beacons; "flying fields in strategic locations"; training facilities; subsidies for carriers who "[kept] their facilities available for use in time of war"; and a "well-defined and continuing" military production program that would support, sustain, and nourish America's aviation industry until commercial demand proved sufficient to "form an effective nucleus upon which can be built a war-time production in case of need."[12]

The Crowell Mission's recommendations were rejected virtually immediately, though they continued to stimulate debate and discussion over the next several years on how American aviation should be structured. Secretary of War Newton Baker had little regard for aviators, testifying they were "prima donna[s]" possessed of "a feeling of superiority," and both the navy and army were fiercely opposed to any type of centralized defense leadership, particularly an air ministry–like establishment overseeing both an independent air force and a civil aviation authority.[13] Ironically, all of the Crowell Mission's recommendations eventually came to pass over the next four decades, with the formation of the Department of Defense (DoD), the USAF, the Air Force Academy, and the Federal Aviation Administration (FAA), among others.

The Crowell report thus defined the future structure of America's national military (and civil) aviation effort. It also cast the terms of the subsequent debate between traditionalist surface warfare adherents and advocates of the new form of air power force projection that would play out throughout the interwar years and into the post–World War II era as well. Conveniently, if misleadingly, encapsulated in Mitchell's bombing campaign against the battleship, the debate actually involved a range of issues affecting both military and civil aviation, indeed, virtually the entire structure and future of America's aeronautical industrial enterprise.[14] The military issues included the nature of both land- and sea-based air power; the efficacy of air power as a means of strategic, operational, and tactical attack; the relationship of air power to land armies and surface navies; and the relationship of airmen to their surface warfare colleagues, including promotion, benefits, and recognition.[15]

An unfortunate aspect of this debate was the increasingly heated salvoes exchanged between various partisans, particularly the acrimony that developed between Mitchell and his acolytes on the one hand, and the army's and navy's senior leadership on the other. [16] Ironically, William "Billy" Mitchell and his principal antagonist, the navy's William "Billy" Moffett, were, in many ways, far more alike than different. Both functioned within services that remained primarily focused on surface operations, and that regarded aviation—and aviators—as largely noisy upstarts. Both supported the development of aircraft carriers. Both recognized that aircraft rendered conventional concepts of surface warfare (especially naval operations) obsolete. Both were exceedingly and persistently firm minded, even obstinate. Mitchell's led him to a court-martial and the death of his career; Moffett's led him (and seventy-three others aboard the navy dirigible *Akron*) to a watery grave one stormy night off New Jersey. Thus, it was perhaps inevitable that neither showed the least inclination to work with the other; their antipathy was so extreme and undisguised that at least one diplomatic attaché reported on the "jealousy between Admiral Moffett and General Mitchell" and how it was having a deleterious impact on their two services.[17]

The hostility between the Mitchell and Moffett camps underscored a fundamental difference in the early shaping of British and American air power. In Britain, airmen of the RFC and the Royal Naval Air Service (RNAS) had chafed at their enforced second-class status. Thus, they stood together as airmen, and out of their united discomfiture, and with the catalysts of German air raids and the vigorous advocacy of Jan Smuts, came the creation of the independent RAF.[18] It was different in America, however. The U.S. Navy (USN), in general, treated its airmen better than the U.S. Army treated its own aviators. Indeed, throughout the 1920s, the navy was more generous in supporting naval aviation than the army was in supporting its air service.[19] Thus, calls by Mitchell and other advocates (such as Crowell, earlier) for an independent American air service never enjoyed the support among Uncle Sam's naval airmen as the calls for an independent RAF among Britannia's flying sailors; the American naval airmen did not suffer the same discontent or feeling of inferior status as did their Royal Navy counterparts.[20]

Mitchell's court-martial further estranged the army's airmen from their leadership, and even their own more temperate air commanders. Indeed, to a generation of influential air leaders, including Henry H. "Hap" Arnold, Carl Spaatz, Ira Eaker, and Robert Olds, it would be Mitchell, not Mason Patrick or Benjamin Foulois, who became the exemplar and embodiment of the American airman, rather than these others who were, rightly or wrongly, perceived as more conciliatory and malleable.[21] As a consequence, it would take more than a quarter-century—and the experience of another war—before the air force would gain its independence with the formation of the USAF in 1947, and even then arguably only because of the strong endorsement and concurrence of an air-minded army chief of staff, the venerated Gen. Dwight D. Eisenhower.[22]

Played out openly in the media, congressional hearings, professional journals, and various commissions and boards, and conducted more stealthily behind the scenes, the multifaceted debate over the future of American aeronautics involved a larger issue: rebuilding American civil and military air power and the national aviation industrial and research base, which had largely lain fallow since the enervating prewar patent battle between the Wrights and Curtiss.

In 1923 an army investigation board found the air service's condition "alarming," with the service "practically demobilized": 80 percent of its not quite two thousand aircraft were obsolescent or inadequate; as well, the board found the nation's aircraft industry "entirely inadequate" and predicted it "will soon practically disappear."[23] Two years later, in 1925, a congressional committee, formed by Florian Lampert, documented that American aviation was in desperate shape, with "no uniformity of Army and Navy policy" on basic military roles and missions. At best, coordination had been only "sporadic and occasional," with "a wide divergence of opinion" on issues involving coastal defense, antiaircraft effectiveness, and the vulnerability of ships to air attack. The aircraft industry had "dwindled" from "lack of continuity" in orders, a "failure to recognize design rights," and "a destructive system of competitive bidding" required of all airplane production contracts, which greatly endangered struggling private firms.[24] That same year, French designer and industrialist Louis Breguet, president of France's *Chambre Syndicale des Industries Aéronautiques,* dismissed America's aviation industry as "not having evolved as rapidly as ours."[25] In 1925—as Billy Mitchell went to his court-martial—the Army Air Service had only twenty-six frontline fighters, roughly one for every two states, and the navy a paltry forty-two.[26] Nor was civil aviation any better: two years afterward, at the beginning of 1927, one airmail operator still pronounced American aviation "not only miles behind Europe in airplane passenger service, but not even in the running."[27]

Out of this tumult came a virtual restructuring of American military and civil aviation in 1926. The Army Air Service became an Army Air Corps. Assistant secretaries for aeronautics were established in both the War and Navy Departments; five-year acquisition plans (one thousand aircraft for the navy and eighteen hundred for the air corps) stabilized a tottering aircraft industry and imposed some rationalization on military aircraft procurement; the professional status, pay, and promotion of military aviators were affirmed and rationalized; and civil aviation was strengthened (via the Air Commerce Act of 1926) by establishing a bureau regulating pilots, mechanics, and aircraft—the beginnings of the FAA.[28]

THE MATURATION OF AMERICAN AIR POWER

By the early 1930s, America had been restored to the front rank of aeronautical nations owing to the effect of the two five-year plans; the establishment of assistant secretaries for aeronautics in the War, Navy, and Commerce Departments; the Air Commerce Act; and the reshaping of the American aeronautical education and research organization via the activities of the National Advisory Committee for Aeronautics and the private philanthropy of the Daniel Guggenheim Fund for the Promotion of Aeronautics. This turnaround was exemplified by the introduction into service of the Martin B-10 bomber, global progenitor of the streamlined twin-engine all-metal monoplane bomber; the Boeing Model 247 airliner, which accomplished the same for the civil airliner; and the more refined Douglas DC-2, which formed the basis for the Douglas DC-3, the most productive and remarkable airliner of the pre-jet era. Additionally, the drive to develop efficient long-range flying boats for commercial airline service led directly to the USN's extensive interwar investment and development of long-range maritime patrol seaplanes culminating, at the end of the 1930s, in the remarkable Consolidated PBY Catalina.[29]

To a degree far surpassing that of other nations, American aviation was characterized by an emphasis on long range, both for civil and military purposes. Civil air advocates saw air transport as a means of speeding communication across a vast nation, while military air advocates saw aircraft as a means of protecting far-flung outposts, attacking from a distance, and controlling the sea frontier. Though the range claims American firms, such as Boeing, Douglas, Sikorsky, Martin, and Consolidated, made for their aircraft were initially greeted with skepticism, bordering on derision, by 1935 their accuracy was acknowledged, evidenced by surprising events such as the 1934 MacRobertson Air Race from England to Australia in which American Boeing 247 and Douglas DC-2 airliners had successfully competed with the best of the specialized high-performance streamlined racers.[30] The result of the rapid expansion of America's aeronautical competency was the development of a dual-use long-range aircraft industrial base that furnished streamlined monoplane transports in the 1930s, small numbers of streamlined advanced bombers and seaplanes, and then, at the end of the decade, increasingly large numbers of military long-range aircraft, such as the Boeing B-17, the Consolidated PBY, and the Consolidated B-24. Indeed, this dual-use base sometimes resulted in the direct transfer of identical technology from the civil to the military side, and vice versa.[31] For example, Northrop's family of low-wing, all-metal, single-engine, express transports led directly to the Northrop A-17 attack bomber, which itself influenced the design of the navy's Midway-winning Douglas SBD dive-bomber. The wing design of the Boeing XB-15 transferred directly to Boeing's Model 314 flying boat. The B-17 furnished the wings and tail surfaces for the Boeing 307 Stratoliner. The Douglas DC-2 transferred its wing and tail section to the Douglas B-18, and the bigger DC-3 did likewise to the Douglas B-23. The DC-3, of course, put on its own war paint to become the ubiquitous C-47 of Second World War fame, as did the Lockheed Model 14 Super-Electra, which became the Hudson maritime patrol bomber.

The complex interplay of foreign-rooted air power theory, steadily growing technological acumen, lessons drawn from interwar combat experience, and per-ceived military threats and needs (particularly that of the rise of Japan) strongly influenced Army Air Corps and naval aviation air doctrine. Even so, formal doc-trinal development proceeded at a pace far slower than the technological chang-es taking place within aviation itself. In 1919 Lt. Col. William Sherman, Chief of Staff of the First Army Air Service, issued the first "tentative" doctrinal guide for American air power employment, thereby setting forth the subordinate re-lationship the army's airmen would endure until 1943 and the issuance of Field Manual 100-20 by the army chief of staff, Gen. George Marshall. Sherman's manual started almost apologetically, noting, "The name Air Service is, to some extent, a misnomer." It then stressed:

> It is a fundamental of human nature for man to fear man more than the
> chance action of steel and lead. Therefore, in the future, as in the past, the

final decision in war must be made by man on the ground, willing to come hand to hand with the enemy. When the infantry loses, the Army loses. It is, therefore, the role of the Air Service, as well as that of the other arms, to aid the chief combatant: the infantry.[32]

Postwar battles between air and land partisans resulted in a nearly decade-long doctrinal stalemate. In 1926 the War Department issued a more comprehensive explication in the form of a training regulation, TR 440-15, one still strongly reflecting the predominant surface warfare culture of the service, emphasizing yet again that air power forces existed to support their surface warfare brethren. Once established, TR 440-15 remained largely unchanged for the next fourteen years, until the emerging realities of air-land war as reflected in European and Asian conflict prompted a search for more appropriate doctrine.[33]

Doctrinal development proceeded more smoothly within the USN, where the lines between air power partisans and surface-minded traditionalists were not so sharply and bitterly drawn. Still, there were differences and, consequently, difficulties. Within the surface warfare community, aviation was seen in terms very similar to how traditionalist army officers viewed it: as an adjunct to surface maneuver, providing reconnaissance and artillery fire control, some limited form of air attack, and with a fighter force required primarily to ensure that an enemy would not enjoy similar advantages. Just as army doctrine emphasized air power's role as a supporting element of surface warfare, until Pearl Harbor, naval doctrine emphasized the subordinate role of the carrier and land-based naval aviation forces as a means to help the battle fleet reach the culminating point of naval war, defined as a decisive encounter between opposing big-gun battle fleets.[34]

Despite such constraints, both Army Air Corps and naval aviation advocates went well beyond the established doctrinal thinking embodied within their respective doctrinal manuals. Within the Army Air Corps, instructors at the Air Corps Tactical School, influenced by foreign and domestic strategic air power advocates, enunciated what historian Robert Finney termed the "primacy of bombardment," looking to various applications, including coastal defense.[35] Air Corps airmen assigned to other army professional study centers, such as the Command and General Staff School, spread this thinking to the rest of the service.[36] While this line of argument ran counter to the thinking of most battlefield-focused army traditionalists, it did coincide with a growing realization within the War Department that the service's air corps had to be organized, trained, and equipped to undertake a range of missions, including "pursuit" (air superiority), battlefield attack, strategic bombardment, and coastal defense. This recognition, stimulated as well by the short-lived airmail debacle of 1934 (which revealed woeful deficiencies in the air corps' training and equipment), was transformed into official policy by the findings of the Drum Board of 1933. Its conclusions, reaffirmed by the Baker Board a year later, set the air corps on the path to being an air arm with both tactical and strategic functions, the latter encouraged by the formation of the so-called GHQ Air Force, the technological advancement of long-range aviation, and the personal interest in long-range military aviation expressed by President Franklin Roosevelt.[37]

That shift in policy encouraged the exploitation of America's growing dual-use aircraft industrial base to pursue the development of new all-metal monoplane attack and, especially, long-range bombardment aircraft equipped with more advanced bombsights. These initiatives (exemplified by the prototype Boeing Model 299, progenitor of the Second World War's B-17 Flying Fortress, and the navy-developed and air corps–adopted Norden bombsight) caught the attention of President Roosevelt, who supported the air corps' interest in strategic attack with ever-increasing enthusiasm, particularly after Japanese air attacks savaged Chinese cities, and France and Britain caved in to Nazi pressure at Munich. Roosevelt's support of long-range air power significantly eased the task of partisans within the air corps. His call for drastically increased aircraft production provided the opportunity—indeed, almost furnished an excuse—for the formation of an air corps study team that crafted Air War Plans Division–1 (AWPD-1), America's first strategic air war plan, and one strongly reflecting the ambitions and beliefs of strategic bombing advocates within the U.S. Army Air Forces (USAAF).[38] Approved in the high summer of 1941, AWPD-1 conceptualized a campaign of strikes against Nazi power, transportation, synthetic fuel production, and air defense forces (including the aircraft industry). It formed the basis for the USAAF's subsequent strategic air offensive against the Third Reich. Indeed, the same mix of targets has generally formed the strategic core of the nation's air campaign plans to the present day.[39] Within the navy, a small coterie of key partisans, Moffett chief among them, steadily pressed for the development of a carrier force. He carefully integrated it within contemporary notions of fleet employment, but also exploited the opportunity of fleet exercises and war games to illustrate the growing (and soon formidable) power of carriers operating in concert with other carriers, thereby creating the carrier task force that would become the normative naval striking force at sea for the Pacific war. (Nor, of course, were they alone in this: so, too, did the Japanese, as evidenced by their six-carrier attack on Pearl Harbor in December 1941.)[40] Although carriers figured with increasing prominence in the navy's interwar contingency plans for operations against Japan, even the most ardent enthusiast of naval aviation could hardly have imagined the vast nature of naval air forces prosecuting the air war against Japan in 1945, when the navy had over ninety carriers in service, many others nearing completion, and a large land-based and seaplane force as well.[41]

From 1934 onward, European and Japanese fighter aircraft development progressed rapidly, reflecting both growing tensions and the recognition that the likelihood of future war demanded the development of large and robust fighter forces. Not surprisingly, this emphasis resulted in the emergence of the high-performance monoplane fighter, typified by the Soviet Polikarpov I-16; the French Morane-Saulnier M.S. 406, Dewoitine D.520, and Bloch M.B. 150; the British Hurricane and Spitfire; the German Messerschmitt 109; and the Japanese Nakajima Ki-43 and Mitsubishi A6M Type 0, some of which saw combat in Spanish, Chinese, and Manchurian skies prior to the outset of the Second World War in September 1939. While critics have alleged that American fighter aircraft development lagged behind these nations, and while some deficiencies did exist, the basic direction of American fighter aircraft and engine development was, in fact, both logical and technically sound, and formed the technical base for the war-winning aircraft, such as the F6F, F4U, P-38, P-47, and P-51, that followed.[42] The relative excellence of America's aircraft industry in general was evidenced by the large numbers of aircraft ordered by foreign nations, particularly France and Britain. By the end of August 1940, as the Battle of Britain entered its full fury, the Air Ministry and Admiralty had already ordered twenty thousand American aircraft and forty-two thousand engines. Not including lesser support types, these included twenty American combat aircraft on order, and, indeed, out of these orders would spring what was arguably the finest all-around piston-engine air superiority fighter of the Second World War: the North American P-51 Mustang.[43]

AMERICAN AIR POWER IN GLOBAL WAR

The outbreak of the Second World War resulted in an immediate effort by American military officials to gain access to information on foreign developments and technology, and a series of technical missions were dispatched overseas. The Battle of Britain drew particular attention, as Spain had several years earlier. Ironically, in both cases, the significance of key aspects was misinterpreted, distorted, or missed entirely by many observers, including those of the United States. In the case of Spain, the valid lessons learned were the continuing necessity of developing high-performance fighter aircraft able to secure control of the air, the ability of aircraft to sink ships at sea under combat conditions, the value of air transport (indeed, Franco's Spanish revolt could not have succeeded without it), and the ability of battlefield air attackers to control ground movement, pin forces in place, and demoralize them. One attaché report noted "the morale effect was unbelievably severe," particularly "the lowering of the will to resist," which led to numerous requests for friendly aircraft "to appear over such-and-such sector for no other reason than to reassure the ground forces."[44]

But in one very important respect, observers misinterpreted the results of bomber operations. The new, fast twin-engine monoplane bombers—aircraft such as the Tupolev SB and Heinkel He 111—inspired and influenced by America's development of the fast, streamlined twin-engine monoplane transport and bomber in 1933–34, had little difficulty evading older biplane fighters, such as the Italian Fiat C.R. 32, the Polikarpov I-15, the Heinkel He 51, and the Hawker Fury. Since this combat experience seemed to confirm prewar notions in Europe and America of the bomber always "getting through," bomber advocates missed that the Spanish case was simply an anomaly of two technologies, one waxing, the other waning, coming into conflict. When unescorted modern monoplane bombers clashed with modern monoplane fighters, the outcome—as Britain, France, and Germany found out in 1939–40, Japan in 1942, and America in 1943—was vastly different.

For the United States, the ability to send military experts to London during the Battle of Britain in 1940 afforded an opportunity to dramatically revise thinking on the survivability aspects of long-range bomber operations. As in Spain, while some significant lessons were learned (for example, the need to equip fighters with powerful machine gun or cannon armament, which drove the USAAF and the USN to standardize on four, six, or even eight 0.50 caliber machine guns; the need to prioritize agility and power in new fighter aircraft; and the value of radar control of air defense forces), others were missed. These included the value of using range-extending jettisonable fuel tanks to increase the combat radius and duration of fighter aircraft, even though the Army Air

Corps and navy had employed them in the interwar years, and Japan had routinely used them on its naval and army aircraft in its China operations. Most significantly, the weakness of the unescorted or weakly escorted bomber in the face of even modest fighter forces was missed. As Maj. Gen. Haywood S. Hansell, the principal air corps architect of AWPD-1, recalled:

> For all its ferocity the Battle of Britain could not duplicate the sort of air battle the American air planners had in mind. As a result, concrete "lessons" simply did not materialize. True, both German and British bombers proved vulnerable to fighters, but then they were medium bombers, poorly armed and flying at relatively low altitude. . . . The experience seemed inconclusive.[45]

In other words, because the *combat experience* in the skies over Britain did not match *theoretical expectation*, USAAF planners rejected reality in favor of theory.[46] This constituted a grave misreading of what happened in British skies, and thousands of Eighth Air Force bomber crewmen would pay for that error, many with their lives, in the various air raids and battles of 1943.

The misreading of the long-range unescorted bombers' vulnerability was matched by an equal misreading of the value of battlefield attack aircraft, both of these misperceptions reflecting that, as one study concluded a half-century later, American air doctrine on the eve of the war was "not very well entrenched, and employment concepts were in a state of confusion."[47] Strategic bomber enthusiasts overconfidently believed new four-engine, turbo-supercharged, high-altitude bombers were essentially invulnerable to enemy defenses. Ground attack proponents remained convinced that ever-larger twin-engine ground attack airplanes—some the size of medium bombers—and slow and limited-value dive-bombers could survive in an increasingly dense air defense environment characterized by the proliferation of rapid-firing, mobile, multibarrel 20mm, 37mm, and 40mm light antiaircraft cannon.

Tactically, exposure by U.S. airmen to British close air support (CAS) and battlefield air interdiction operations (then termed "trench strafing" and "ground strafing," respectively) during the First World War had led to the U.S. Army Air Service's extensive pursuit of "attack aviation" in the interwar era, including a brief flirtation with dive-bombing that influenced the U.S. Marine Corps' development and exploitation of dive-bombing (and CAS) during its interwar Nicaraguan campaign.[48] Throughout the interwar years, the United States, Germany, Italy, France, Russia, and Japan had vigorously pursued the development of specialized attack aircraft, guided by wartime experience (for example, Germany's *Schlachtflieger* of the First World War) and encouraged by advocates, such as Italy's Amedeo Mecozzi, a First World War fighter ace turned strong advocate for so-called *aviazione d'assalto*, or "assault aviation."[49] Such aircraft flew in various interwar conflicts, achieving one notable success: at Guadalajara, in 1937, concentrated attacks by Soviet Polikarpov R-Z attack biplanes played a decisive role in routing an Italian mechanized assault aimed at Madrid.[50] Generally, air corps officers regarded attack aviation as a required branch and operational competence within the corps, necessitating the development of specialized ground attack aircraft, and, in contrast to much myth that has grown up since, were generally eager to use aviation as a means of supporting ground forces engaged in close combat with a foe.[51]

But the air defense environment was transforming rapidly, and by the end of 1940, airmen, if not soldiers, were recognizing they had seriously overestimated the value and survivability of specialized ground attack aircraft. In the First World War, the performance disparity between specialized support aircraft and fighters had been essentially nonexistent, and in some cases, the specialized support aircraft actually had better overall performance. But the technological paths of the two had diverged so sharply that by 1940 the fighter had significant advantages over the attack airplane, and the attack airplane was so big and slow that it was very vulnerable to light antiaircraft defenses. British and French attack aircraft were destroyed in large numbers during the battles of France and the Low Countries, and the specialized dive-bomber—which attracted much favorable comment from the embattled Allies—was proven a chimera when the Stuka encountered the Spitfire and the Hurricane during the Battle of Britain. The future, except in the most favorable of circumstances, clearly belonged to the bomb-carrying fighter, what the Allies would call a fighter-bomber, and what the Germans, in similar fashion, would call a *Jagdbomber* (*Jabo*).[52]

However, to the German, British, Soviet, and American armies, the special-purpose attack aircraft, and particularly the dive-bomber, held continued appeal, even as its allure tarnished in the eyes of most airmen. In the United States, War Department doctrinal manuals issued in 1940 largely ignored the special circumstances that had to be in place—particularly overwhelming air superiority, even air supremacy—for attack aircraft to operate with impunity. A veritable cult of the dive-bomber sprang up, forcing the air corps and RAF to acquire large numbers of aircraft that ultimately proved of essentially no value to either service after their acquisition.[53] Indeed, for the U.S. Army, the appeal of the battlefield attack aircraft would linger until the swing-role fighter-bomber forcibly demonstrated its far-greater value and survivability in Normandy. The employment of attack aircraft was a central tenet of prewar army air support doctrine, as enunciated in two (and somewhat contradictory) Air Corps Field Manuals, FM 1-5 and FM 1-10, as was a general assumption that fighter aircraft would be of minimal, if any, value. While the former cautioned against using air attack against both "well-intrenched [*sic*]" and dispersed forces, the latter envisioned a wide range of possible operations across the battle area, including close support and interdiction.[54] Based on its "small wars" experience, the U.S. Marine Corps took a far broader view, holding that, as a deployed landing force operating in conjunction with sea-based power, "[t]he primary mission of combat aviation in a small war is the direct support of the ground forces," and adding significantly, "This implies generally that all combat aviation will be used for ground attack."[55]

The shocking efficiency of the Wehrmacht air-land combat team, evidenced by the swift overrunning of most of Continental Europe, caused the army to readdress its air-land thinking, and coming out of this, in April 1942, was Field Manual FM 31-35.[56] Issued just months after Air Vice-Marshal Arthur Coningham had established a ground support system furnishing swift and efficient delivery of air support for the British Eighth Army by RAF fighters and light bombers, FM 31-35 constituted an attempt to develop a similar process. The key to Coningham's system was the recognition of the basic coequality of air and land forces, the collocation and close liaison of air and land headquarters, streamlined communication, and rigorous training. In contrast, FM 31-35 was overly cumbersome and complex, reflecting the desire of Corps and Division commanders to have absolute control over the air assets available to them. At Kasserine in February 1943, in the face of an intensive assault by remnants of the Afrika Korps and the Luftwaffe, its weaknesses were glaringly evident, and FM 31-35 was swiftly replaced by FM 100-20, which, issued five months after the debacle in the Tunisian desert, governed subsequent Army Ground Forces–Army Air Forces relations through the end of the war. FM 100-20, which also replaced FM 1-5, drew heavily from the experience, thinking, and doctrinal inspiration of Coningham.[57] It stated (somewhat hysterically) that "LAND POWER AND AIR POWER ARE CO-EQUAL AND INTERDEPENDENT

FORCES, NEITHER IS AN AUXILIARY OF THE OTHER," a red-flag-to-a-
bull phrase that led some subsequently to call it the Army Air Forces' Declaration of Independence—which, in effect, it was.[58]

FM 100-20 prioritized three phases of theater-wide air operations: first, the
securing of air superiority; second, interdiction "into or within the theater of
operations"; and third, battlefield air support via combined air-land attack "to
gain objectives on the immediate front of the ground forces."[59] If viewed with
the most acute suspicion by many traditionalists, who thought the third priority meant that army ground forces would, at best, receive grudging assistance,
FM 100-20 proved a most appropriate doctrine. At the end of the war, the Air
Effects Committee of Gen. Omar Bradley's Twelfth Army Group interrogated
a number of German commanders in support of the United States Strategic
Bombing Survey (USSBS), concluding, "From the high command to the soldier in the field, German opinion has been agreed that air power was the most
striking aspect of allied superiority."[60] At a meeting of air and ground commanders called by Eisenhower at Bradley's headquarters in May 1945, participants
"unanimously reaffirmed centralized control of air power as prescribed by FM
100-20," an outcome strongly endorsed at the meeting by the colorful and outspoken commander of the Third Army, Gen. George Patton. He had already,
during the Battle of the Bulge, ordered his chaplain to write a prayer for clear
weather—so that fighter-bombers could attack German forces.[61]

If U.S. Army commanders overly focused on the control of tactical air
power had their comeuppance amid the wastes of Kasserine in February 1943,
USAAF strategic bomber advocates had their own in the skies over Schweinfurt
in October that same year.[62] In May 1943, attendees at the Trident Conference had approved the plan for a combined strategic bomber offensive against
Germany. But when it was implemented, particularly by bombers relying primarily on mutual interlocking and overlapping fields of defensive machine-gun
fire, the intensity of radar-cued German fighter attacks and surprisingly accurate
antiaircraft fire generated heavy losses (in one case more than 20 percent of
the force dispatched, and routinely exceeding 10 percent), effectively forcing a
reconsideration of the strategic bomber offensive, and tactical and operational
changes upon subsequent American bomber operations. The initial emphasis on
attacking key centers of gravity shifted to the more immediate goal of destroying
German fighter forces, particularly because doing so was crucially important to
any planned invasion of Europe.

This was evident in Hap Arnold's urgent directive—essentially both a warning and a desperate plea—to the commanders of the Eighth and Fifteenth Air Forces that "Overlord and Anvil will not be possible unless the German Air Force is destroyed. Therefore, my personal message to you—this is a MUST—is to '*Destroy the Enemy Air Force wherever you find them, in the air, on the ground, and in the factories.*'"[63] When the already-legendary James H. "Jimmy" Doolittle took over as commander of the Eighth Air Force, he refined this guidance further, recognizing immediately the need to free the fighters from defensive close escort to offensive fighter sweep. He directed that a sign in the Eighth AF fighter headquarters reading "THE FIRST DUTY OF THE EIGHTH AIR FORCE FIGHTERS IS TO BRING THE BOMBERS BACK ALIVE" be torn down and replaced by one reading "THE FIRST DUTY OF THE EIGHTH AIR FORCE FIGHTERS IS TO DESTROY GERMAN FIGHTERS." "Your first priority," Doolittle emphatically told the commander of the Eighth's fighter forces, "is to take the offensive."[64] Doolittle's directive, plus the advent of long-range drop-tank-equipped fighters (typified by the P-51 Mustang), the freeing up of fighters from close-escort duties to fighter-sweep operations instead, and the emphasis on eliminating the German Air Force as a serious threat to Allied strategic operations turned the daylight bombing offensive from a costly failure to a reasonable success. By the late spring of 1944, the Luftwaffe was on the skids, and the Third Reich was open to round-the-clock aerial assault, with attackers flying from Britain and Italy. Though a sometimes-bitter targeting debate broke out between advocates of an "oil campaign" or a "transportation campaign," both, in retrospect, ensured the collapse of German fortunes.[65]

By early January 1945, Germany's war economy was shattered, its production disrupted, its oil supplies vastly reduced, and the clock on the Thousand-Year Reich rapidly running down. Afterward, the independent USSBS concluded:

> Allied air power was decisive in the war in Western Europe. Hindsight inevitably suggests that it might have been employed differently or better in some respects. Nevertheless, it was decisive. . . . The German experience suggests that even a first class military power—rugged and resilient as Germany was—cannot live long under full-scale and free exploitation of air weapons over the heart of its territory. By the beginning of 1945, before the invasion of the homeland itself, Germany was reaching a state of helplessness.

Her armament production was falling irretrievably, orderliness in effort was disappearing, and total disruption and disintegration were well along. Her armies were still in the field. But with the impending collapse of the supporting economy, the indications are convincing that they would have had to cease fighting—any effecting fighting—within a few months. Germany was mortally wounded.[66]

If Taranto and Pearl Harbor had marked the decline of the battleship and two-dimensional surface combat as the linchpin, the fulcrum of naval strategy and tactics, the widespread employment of the submarine and the aircraft carrier marked the fulfillment of the dreams of submariners and naval aviation advocates who had envisioned a three-dimensional model of warfare in which surface forces would be attacked and controlled from above and below the sea's surface. In 1942 the exigency of the Pacific war had resulted in several early attempts at joint air operations. The Doolittle raid in April imaginatively blended the aircraft carrier and the land-based medium bomber in a mission that profoundly shocked Japan's military leadership, achieving strategic effects out of proportion to the force involved and any damage it caused. Midway, in June, was a crucial American victory, but a failure as a joint operation, reflecting prewar failures to develop and train to appropriate doctrine. A far more successful example came subsequently, with the landing on Guadalcanal, and the formation of the so-called Cactus Air Force, which blended joint Army–Navy–Marine Corps fighter and strike forces that eventually wrested control of the air and secured control of the seas around the Solomons.[67]

By the summer of 1945, Japan was doomed, its once-seemingly invincible fleet destroyed, its aviation forces attacked over the Home Islands, and its industry shattered. It was essentially defenseless against increasingly severe air attacks delivered by the USAAF and carrier raids, and ultimately incapable of preventing both its atomic bombing, or, had such proven necessary, an invasion. In the steady reduction of Japan since April 1942, as Eliot Cohen has written, "[g]round and purely naval forces had served mainly to seize and hold forward bases for the projection of air power." [68] The increasing primacy of USAAF air attack so worried one naval planner that he had warned, "The danger is obvious of our amphibious campaign being turned into one that is auxiliary support to permit the USAAF to get into a position to win the war."[69] To airmen, the subsequent dropping of two atomic bombs, and the Japanese collapse that followed, seemingly fulfilled such prophecy; in late 1945, USAAF Lieutenant General Doolittle reputedly stated, "The Navy had the transport to make the invasion of Japan possible; the Ground Forces had the power to make it successful; and the B-29 made it unnecessary."[70] In fact, of course, it had been the Allies' joint-service advantage in three-dimensional power—the power of the submarine and sea- and land-based air warfare—that had thwarted Japanese aggression and destroyed its imperial ambitions.

CONFRONTING THE POSTWAR WORLD

The end of the Second World War marked the ascendency of the three-dimensional warrior—airmen and submariners—into the highest levels of America's national military command. This rendered interwar arguments over whether the United States should have an independent air force essentially moot. During the war, General Arnold of the USAAF had possessed authority and influence more akin to those of a chief of staff of an independent service rather than a subordinate commander operating within a service. Indeed, as Chase Mooney wrote in 1946, from the outset of the war:

> Air power, along with defensive air supremacy, was a *sine qua non* for the successful conduct of any kind of major operation. Thus, with the creation of the Joint Chiefs of Staff in early 1942 the head of the air arm acquired a status equal to that of the Chief of Naval Operations and Chief of Staff of the Army, and his membership on the Combined Chiefs of Staff (American and British) gave to him an even more important role in the determination of world-wide strategy.[71]

It was evident in how other service chieftains of the Allied nations treated Arnold at various conferences and meetings, and the wide-ranging latitude given by the Combined Chiefs of Staff to strategic bomber forces both in the war against Germany and Italy, and the war against Japan, which granted them virtually the status of an independent military branch.

The backdrop behind the virtually inevitable establishment of the USAF in the immediate postwar years was the Normandy invasion. Recognition of air power's central role in the liberation of the Continent, especially its criticality for the invasion at Normandy, which Arnold had forcefully alluded to in his address to the commanders of the USAAF's bomber forces in December 1944, played at least as an important role as did his post-Hiroshima desires for an independent air force as America's primary deterrent force. In May 1944, Chief of Staff Gen. George C. Marshall had written to Secretary of War Henry Stimson, "We are about to invade the Continent and have staked our success on our air superiority."[72] At Normandy, just weeks after D-Day, General Eisenhower had told his son, "If I didn't have air supremacy, I wouldn't be here."[73] Afterward, testifying before Congress in November 1945, Eisenhower reminded legislators that

> [t]he Normandy invasion was based on a deep-seated faith in the power of the Air Forces in overwhelming numbers to intervene in the land battle. . . . The air commander and his staff are an organization coordinate with and coequal to the land forces and the Navy . . . that seems to me to be so logical from all our experiences in this war, such an inescapable conclusion that I, for one, can't even entertain any longer any doubt as to its wisdom.[74]

Clearly, then, the remembrance of Normandy played a major role in ensuring the swift independence of the postwar air force. Conversely, had air power failed at Normandy and afterward, the debate over an independent air force certainly would have continued for some years, perhaps a decade until Sputnik provided a goad for separation, as it did for the creation of the National Aeronautics and Space Administration (NASA). In a message to Congress in December 1945, President Harry S. Truman strongly advocated the unification of national defense within a single department, "a goal Truman had been striving for since taking office."[75] In words echoing Eisenhower's testimony before the Senate Committee on Military Affairs, he stated, "Air power has been developed to a point where its responsibilities are equal to land and sea power, and its contribution to our strategic planning is as great"; it was the first time a president had publicly endorsed, however veiled the words, the establishment of an independent air force.[76]

The National Defense Act of 1947 fulfilled the recommendations of the Crowell Mission, the vision of Billy Mitchell, and the aspirations of other advocates of both an independent air force and a unified DoD. But the establishment of the USAF as a separate and coequal service in September 1947, and its pivotal role in preserving Berlin as a free city in the face of a Soviet blockade just months thereafter, did little to reduce debates over air doctrine, roles, and missions. Indeed, such debates took on a special vehemence, with advocates of various positions arguing with special ardor, given the helter-skelter postwar drawdown of military forces and the acute struggle for resources in an environment of extremely constrained defense spending. How significant was air power to postwar America? What was to be the relationship of this new service to its traditionalist brethren? What would its primary function be? What would the relative prioritization be among the services for the acquisition of new weapons and capabilities?

For the Truman administration, at the national political level, there was little doubt of the significance and importance of air power. In June 1947, President Truman had appointed an air policy commission chaired by Thomas K. Finletter. Its report, based on detailed hearings and testimony from military and government officials, industry executives, and other knowledgeable authorities, was issued at the end of the year. It left little doubt that a new and dangerous age had dawned. Consequently, the commission unanimously recommended:

> [America's] military establishment must be built around the air arm. Of course an adequate Navy and ground forces must be maintained. But it is the Air Force and naval aviation on which we must mainly rely. *Our military security must be based on air power.*[77]

They also warned:

> The Air Force as presently composed is inadequate. It is inadequate not only at the present time when we are relatively free of the danger of sustained attack on our homeland, but is hopelessly wanting in respect of the future.[78]

The essential structure of the postwar air force had been set forth in January 1946 by Eisenhower and General Spaatz, who envisioned a balanced force built around air power's core functions: strategic air command, tactical air command, air defense command, and air transport command, with supporting commands, services, and organizations, including an air university charged with responsibility for writing air doctrine.[79] Within the air force, however, there was little doubt that the service's primary functions would be furnishing strategic deterrence, long-range global striking power, and continental air defense, using the power of the atomic bomb and the capabilities furnished by new developments

in aerodynamics, propulsion, and electronic sensors and communications, typified by the swept and delta wing, the jet engine, the long-range rocket, and radar. Indeed, an essential technological determinism came to dominate postwar
American air power thought, particularly that of the air force, stimulated by the
advanced state of Nazi technical development. The introduction of German jet-
and rocket-powered fighters, and the "robot weapon" campaign that began in
1944 with the V-1 cruise missile and (subsequently) the V-2 ballistic missile, had
shocked Allied air leaders confronted with the specter of technological inferiority. In early November 1944, Arnold established an advisory body to formulate
a long-range research and development program for the USAAF, warning his
staff two months later that the next war could start with a pilotless attack on
Washington and other American targets.[80]

When, immediately after the war, American technical intelligence investigators had an opportunity to assess German advances in aviation science and meet
with key investigators and researchers, their reports and assessments further underscored the dramatic transformations in high-speed flight and weapons technology taking place at mid-century, particularly the danger of potential atomic
attack by transonic and supersonic aircraft, "robot" cruise missiles, and ballistic
rockets.[81] The technical disparity between Germany and the United States appeared so grave that authorities arranged for the swift importation of researchers
and the transfer of great quantities of technical material and reports.[82] The impact of German research was immediately felt, demonstrated by the swift onset
of advanced programs for swept-wing jet-propelled aircraft, tactical surface- and
air-launched missiles, and long-range ballistic missile study efforts in both the air
force and navy.[83] Accompanying this was the building, by both services, of new,
specialized research and development centers and facilities, and command structures for them.[84] This technological emphasis stimulated preparations for strategic attack and continental air defense. In 1949 the Soviet Union's detonation of
its own atomic bomb further secured the dominance of these two mission areas
within air force planning and triggered changes in force structure and command
leadership, epitomized by the growing influence of commanders (especially the
gruff, no-nonsense yet charismatic Gen. Curtis LeMay) who had cut their teeth
in strategic bombers and been bloodied in the great bomber offensives in European and Pacific skies.[85]

As the "Grand Alliance" collapsed into postwar drift and uncertainty, followed by the early onset of the Cold War, external events and various potential

threats, dangers, and weaknesses made their own mark on air force and navy doctrine, planning, and operations. For example, the startling loss in operational competency manifested by Strategic Air Command (SAC) bomber crews in a bombing competition in June 1948 (coincident with the onset of the Berlin blockade) prompted the swift elevation of LeMay as SAC commander, who subsequently instituted and oversaw a top-to-bottom restructuring of the troubled command. [86] By the early 1950s, the imposition of standard procedures, rigorous training, continuous evaluation, and the intolerance of less-than-fully satisfactory performance had transformed the image of SAC into that of a supremely trained, confident, and determined force of professional airmen ready to go to war quite literally at a moment's notice.

In March 1948 Secretary of Defense James Forrestal called a meeting of the service chiefs at Key West, Florida. Out of this came the Key West Agreement, approved by President Truman the next month, the setting for the paper "Functions of the Armed Forces and the Joint Chiefs of Staff." Generally speaking, the agreement gave each service primacy over its natural operational environment: the army on land, the navy and Marine Corps on sea and the littoral environment, and the air force in the air. The air force was given responsibility for continental air defense; gaining and maintaining air supremacy; undertaking strategic air attack; establishing forces to support joint warfare, including amphibious assault, airborne operations, and battlefield air support; and furnishing transport and logistical support to joint forces.[87] But the agreement failed to dampen the fires of interservice rivalry then smoldering into open flame between the air force, army, and navy.[88]

Much as the RAF had struggled under Air Minister Sir Samuel Hoare and Chief of the Air Staff Sir Hugh Trenchard during the 1920s to survive various assaults on it from the Royal Navy and the British Army, the air force under Secretary Stuart Symington and Gen. Hoyt Vandenberg was involved from the outset in a punishing roles-and-missions battle with the navy in the last years of the 1940s.[89] The battle was encapsulated in an appropriations fight pitting land-based air power, symbolized by the Convair B-36, a massive six- (later ten-) engine intercontinental atomic bomber, and sea-based naval aviation, symbolized by a projected supercarrier, the USS *United States*, capable of carrying its own long-range atomic bombers. Supported strongly by Louis Johnson, who had replaced Forrestal as defense secretary in 1949, the air force came out on top, with the B-36 (which could carry the heaviest of nuclear weapons then contemplated) going forward into production, while the proposed carrier did not. The battle, which resulted in a so-called Revolt of the Admirals, inflamed old passions among partisans dating to the days of the Mitchell-Moffett rivalry, furnishing weeks of dramatic Washington political theater. The air force had the strong support of Gen. Omar Bradley, newly appointed chairman of the Joint Chiefs of Staff (JCS)—again a reflection of how USAAF air power from Normandy onward had transformed his thinking—and this backing, coupled with pressure from Defense Secretary Louis Johnson, proved decisive.[90]

Ironically, the B-36 was far from an ideal strategic bomber. Piston powered and straight winged, it was a throwback to the pre-air-refueled era where planners feared that bombers might have to operate from a "Fortress America" and its possessions against enemies far across the oceans. Consequently, it was massive, with thick high-aspect-ratio wings that gave it an ability to reach well over forty thousand feet, but limited its maximum speed, making it vulnerable despite claims of immunity to interception. It was so slow that the air force subsequently added four jet engines in wing pods to boost its over-the-target speed to 435 mph. While this was impressive when compared to a B-17 or B-29, it was still slower than late–World War II propeller-driven fighters. It was thus destined to serve as an interim weapon until replaced in 1955 by the far more suitable jet-powered B-52. For their part, extreme navy partisans argued that the service could do a more efficient job delivering atomic power against foes, a claim even more suspect than the claims put forward for the B-36; they had, at this time, no long-range aircraft other than the new land-based Lockheed P2V, and if modified to serve aboard ship, this maritime patrol bomber could not possibly have survived a hostile air defense environment long enough to deliver a nuclear weapon. The outcome of this brouhaha was a corrosive suspicion between the air force and navy (and, to a somewhat lesser degree, the navy and the army) in the years ahead.[91]

Interestingly, neither the navy nor the air force subsequently suffered the decline in significance and resource share that each had feared at the time of the roles-and-missions debate. The rapid deterioration of international relations signaled by the Greek insurgency, the Berlin blockade, the detonation of the Soviet atomic bomb, the fall of China, and the onset of the Korean War stimulated their rapid expansion. The air force pushed ahead with long-range bomber development, progressing from the era of the propeller-driven B-29, B-50, and B-36 to the air-refueled all-jet B-47 and B-52. Fascination with exotic technological possibilities led the service to the B-58, eventually to the XB-70, and then to an attempt to develop an all-purpose single-stage-to-orbit Aerospaceplane. The navy eventually received its large carriers, beginning with the appropriately named USS *Forrestal*, progenitor of the supercarrier, and (after the development of early tactical nuclear weapons) new generations of atomic-armed jet-propelled fighter and attack aircraft to operate from them. By the end of the 1950s, for the navy as well as the air force, the piston-powered, propeller-driven aircraft was largely an anachronism, saved from extinction only by its suitability, a few years later, for counterinsurgency and air rescue operations in Southeast Asia, where its long-loiter and high-payload capacity made it (briefly) invaluable.

KOREA: THE UNANTICIPATED CHALLENGE

The Korean War had a profound impact on the perception of air power and the roles and missions of air power forces, particularly in the area of battlefield air support, which had, in the postwar years, once again flared as a point of contention between airmen and soldiers. In September 1944, following the great success of the air-land Normandy campaign and breakout, Twelfth Army Group Commander Lieutenant General Bradley, a staunch air critic for much of the war, wrote Arnold to praise the "very close cooperation" between air and ground forces, adding, "In my opinion, our close cooperation is better than the Germans ever had in their best days."[92] By VE (Victory in Europe) Day, the AAF's performance had relieved any anxieties generated by the issuance of FM 100-20 that the USAAF was uninterested in supporting army ground forces in combat. Unfortunately, the period 1945–50 witnessed the disestablishment of American tactical air power in both the navy and the air force. In a postwar world dominated by big-ticket fights over the future of strategic bombers and aircraft carriers, there was little interest in the kind of swing-role air-to-air and air-to-ground tactical fighter-bomber missions that had proven so valuable in 1944–45. When Lt. Gen. Elwood R. Quesada, the air force's master of tactical air support, retired from the service, he did so in part because of what he considered the breaking of a 1945 USAAF promise to Gen. Dwight D. Eisenhower that the army would always be able to call on strong tactical air support assets even after the USAAF became a separate air force.[93] After the creation of the USAF, its Tactical Air Command (TAC) was deliberately placed at Langley Air Force Base (AFB), next to the headquarters of Army Field Forces at Fort Monroe, Virginia. But then the service realigned both the TAC and Air Defense

Command under the Continental Air Command. Fearing that TAC's subordination once again indicated a lack of interest in furnishing tactical air support, the army pressed for a new joint doctrine on air support operations. The army's concerns coincided with doctrinal discord with the navy over joint amphibious warfare doctrine. Neither was resolved before the outbreak of the Korean War, nor was the larger issue of joint-service air power doctrine and command relationships across all the services, something that would plague interservice relationships into the post–Cold War era.[94]

The sudden onset of the Korean War in June 1950 shocked America's military and political establishment and revealed the woeful unpreparedness of American forces in theater and at home, a result of the headlong plunge into willy-nilly demobilization that had characterized the years after 1945. Within weeks, an army and Marine Corps, having been overwhelmingly victorious in 1945, were at risk of being pushed off the Korean Peninsula by an army that had not even existed five years previously. The army, in desperate need of tanks, even took some that had been on concrete pedestals at Fort Knox and sent them to Pusan.[95] For their part, the air force, navy, and Marine Corps were struggling to deploy sufficient aviation forces in time to prevent catastrophe from descending into disaster.

For all its violence, Korean combat was characterized by limits and constraints placed on the freedom of theater commanders and their application of air, land, and naval power, to the consternation of many, particularly a public conditioned by the no-holds-barred combat of World War II.[96] It reflected the prevailing international strategic context and an understandable desire not to spread the conflict into a general global war. Like any war, Korea offered a mixed collection of lessons and experiences, none more mixed than those applying to air power. In contrast to criticisms offered during the roles-and-missions debate, carrier air power immediately proved of great value, particularly as, in this last great non-air-refueled air war, Communist forces had rapidly overrun South Korea's airfields, forcing land-based operations from Japan, and reducing combat endurance to mere minutes once strike aircraft reached their targets. After the war, both the air force and the navy prioritized introducing air refueling for tactical as well as strategic aircraft, making Korea the last American air war fought without routine aerial tanking.

But on balance, despite the chaos, many things had actually gone surprisingly well. In the crisis of the summer and fall of 1950, the application of air force strategic and tactical air power was so crucial that Lt. Gen. Walton Walker, commander of U.S. ground forces, affirmed, "If it had not been for the air support that we received from the Fifth Air Force we would not have been able to stay in Korea."[97] Then Douglas MacArthur's brilliant Inchon invasion set the stage for the rout of invading forces.[98] When he then overextended his advance to the Yalu, and China intervened in the war, the application of massive air power and the logistical support and transport capabilities of air force combat cargo enabled a reasonably orderly withdrawal and stabilization of the front, something confirmed by captured documents and by MacArthur himself, who noted, in his characteristically grandiloquent fashion, that "the support that our tactical air has given to our ground troops in Korea has perhaps never been equaled in the history of modern war."[99]

Key to the successful use of air power at the front was control of the air over Communist-held territory. So quickly did the air force secure air dominance over the front that, after the summer of 1950, the threat of Communist air intervention was never a factor influencing the plans or scheme of the United Nations (UN) ground maneuver: UN forces operated in the open, with complete freedom of action from air attack, at all times.[100] Conversely, Communist forces were at all times subjected to battlefield observation and on-call battlefield air support that inflicted overwhelming casualties, restricted supply movement largely to night, and forced elaborate deception and concealment strategies.[101] While interdiction operations proved of less value than those of tactical bombing at the front, these reflected the "steady state" conditions of fighting a limited war of deliberately chosen stalemate in the face of fears of "growing" the war into a global conflict between the West and the Communist bloc. Under these conditions, with an absence of materiel-draining offensives, interdiction (in the pre-precision era), however costly to a foe, could not prevent a steady trickle of supplies sufficient to replenish entrenched forces at the front.[102]

Strategic bombing swiftly destroyed the few industrial targets existing in North Korea, and thereafter turned to other targets. But SAC's "legacy" bomber force (propeller-driven B-29s) was seriously threatened and constrained by the MiG-15, a modern and heavily armed jet fighter often flown by Soviet pilots that forced Far East Air Forces (FEAF) to emulate RAF and Luftwaffe practice early in the Second World War, with night operations superseding daylight ones. Joint-service screens of (primarily) Marine and Navy Douglas F3D nightfighters, together with Air Force Lockheed F-94s, significantly reduced but did not totally eliminate MiG depredations against B-29 bomber streams. Here, too, was an aspect of joint air operations that had not been foreseen in prewar years, but that emerged as an urgent need from combat experience over the North.[103]

Joint-service air power could not win Korea on its own, but in its absence Korea would clearly have been lost. While historians will continue to debate the precise value of air power in Korea, it is fair to conclude that joint coalition air power ensured that the Communists could not fulfill their war aims. Certainly, the air force's own corporate interpretation of its performance immediately after the war was one of prideful accomplishment and confidence, even exuberance, in the service's future prospects.[104]

But the application of air power in Korea was hardly free of controversy. As RAND's James A. Winnefeld and Dana J. Johnson have written, it was

> a painful lesson on the clash of doctrine with combat realities, the down-stream costs of interservice conflict, the expense in blood of "savings" extracted from peacetime defense budgets, and the failure of peacetime and wartime command alike to deal adequately with the requirements for truly effective joint operations.[105]

Despite the favorable judgments of combat commanders, such as Generals Walker and MacArthur, no air issue was more controversial than the control and execution of battlefield air support. Ostensibly, as a result of post–Second World War doctrinal coordination of the army and air force, they agreed on their joint air support doctrine, which had been codified by the issuance of a new edition of FM 31-35 in August 1946 reflecting the aftermath of FM 100-20 and the combat experience of the Twelfth Army Group working with the USAAF's Ninth Air Force.[106] Army artillery would control the battlespace from the forward line of troops to the effective reach of the available cannon (the artillery line), with the air force, in coordination with the army, attacking from the artillery line to a point, designated the bomb line, beyond which it could conduct air operations without the necessity of coordination with army ground forces.

However, when army forces exposed to Marine CAS during the fighting around the Pusan perimeter in the opening weeks of the war saw strike aircraft strafing, bombing, rocketing, and napalming to within a few meters of embattled troops, many expressed a strong preference for the Marine-Navy model of CAS, despite the air force's application of CAS conforming to previously agreed-on joint Army–Air Force battlefield air support doctrine. While the combination of both systems—the Army–Air Force system operating beyond the artillery line, and the Navy-Marine system operating from the forward line of troops to the artillery line—certainly worked to effectively "smother" the entire battlefield and its approaches (and, as the Naktong fighting showed, thus slowing, stopping, and attriting Communist mechanized and infantry forces), serious doctrinal differences (exacerbated, in some cases, by bad personal relationships among senior service representatives) remained a source of contention over the length of the war and fueled continuing controversy afterward.[107]

The Korean War demonstrated that competencies previously thought secure could easily be lost within just a few years. The smoothly functioning tactical air coordination processes of the late Second World War period were not restored until well into the second year of the war, and then only after doctrinal rigidity in all services gave way to a more flexible in-theater practicality. Basic problems in communication highlighted inadequate equipment and prewar planning, further worsening relations between airmen and soldiers. The relationship between the air force and naval aviation forces in prosecuting both battlefield support operations and deep attacks north of the thirty-eighth parallel was contentious, resolved only when Far East Command determined that coordination authority would be delegated to the commander of FEAF. Korean experience shaped the subsequent division of command responsibilities in Vietnam, though issues of control of battlefield air support, the relationship with naval and land-based aviation forces, and the relationship of the theater air commander with the Marine Corps' aviation component continued to trigger both popular and official debate.[108]

The air war in Korea had amply revealed that the opportunity to derive and develop a genuine joint air operations doctrine in the time between VJ (Victory over Japan) Day and the onset of the Korean invasion had been squandered in mutual suspicion and the punishing roles-and-missions debate, a cautionary and sobering tale. As the war raged in Korea, the air force and army undertook a number of joint training exercises in the United States, with very mixed results. Air support and other missions (such as troop and cargo delivery) went generally well, as did service-specific missions, such as air superiority, strategic and tactical bombing, and simulated nuclear weapon delivery. But joint planning was only occasionally satisfactory. The very different ways the air force and army perceived air command and control over the battle area resulted in serious disconnects, particularly in larger and more complex exercises, leading one analyst to bemoan "this discouraging tendency to continue making the same mistakes."[109]

The exercise results further signaled the need for doctrinal reform, already under way within the air force. As the Korean War entered the talks phase and then wound down toward armistice, the Air University was preparing a series of doctrinal manuals reflecting its most recent combat experience. In early 1953, following an Air Staff review, the service issued Air Force Manual (AFM) 1-2, a basic doctrine for the air force, which effectively replaced the decade-old USAAF-legacy FM 100-20; next, in September, came AFM 1-3 on theater air operations, AFM 1-4 on air defense operations, and AFM 1-5 on air operations supporting amphibious operations. Debate over theater issues and strategic air power required more work until, finally, during 1954, the service issued AFM 1-7 on theater air forces in counterair, interdiction, and CAS; AFM 1-8 on strategic air operations; AFM 1-9 on theater airlift operations; and AFM 1-11 on theater air reconnaissance. Differences remained even within the air force on the finer points of these doctrinal manuals, and on larger issues such as the degree of control—if any—accorded to ground commanders over air power forces. But if far from perfect, taken as a whole this body of doctrinal expression established a far more thought-out and supportive rationale for the service and its air power employment as it entered the post-Korean era.[110]

TOWARD A NUCLEAR-CENTRIC AIR AND SPACE FORCE

After the armistice, the air force and navy responded differently to their Korean experience, each reflecting their own special circumstances. For the navy, Korea had highlighted serious deficiencies and shortcomings in naval aviation that had to be addressed. Most importantly, the service needed to redesign the aircraft carrier to meet the demands of larger, faster, and heavier jet aircraft. It subsequently did so by adopting the British angled deck, mirror landing system, and steam-catapult, adding these to existing carriers, and placing the first supercarrier, the *Forrestal,* in service by mid-decade. It needed better aircraft, which drove the navy toward the transonic force structure that it eventually operated in Southeast Asia and beyond.[111]

The air force, having generally had a "good" war, concluded that while Korea had reaffirmed the necessity of maintaining a strong competency in tactical air power, it had to prioritize, addressing the offensive and defensive requirements associated with the nuclear warfare threat. All new tactical strike aircraft and swing-role fighters, such as the later F-86F and F-86H, the F-84F, and the F-100, were equipped to employ newly available tactical nuclear weapons, and their crews practiced delivery tactics for such weapons.[112] Defending against the threat of Soviet atomic bombers became a major air force priority after the first Soviet atomic test, driving the development of an integrated air defense network that comprised new missile-armed jet interceptors, such as the F-101, F-102, and (ultimately) F-106; the semi-automatic ground environment system, a sophisticated command and control system; early warning radars (some on towers in the Atlantic Ocean); the so-called distant early warning line running across northern Alaska, Canada, and Greenland; and long-endurance airborne early warning airplanes (the genesis of what, a quarter-century later, would evolve into the E-3 airborne warning and control system [AWACS]).[113] It was the designated service for long-range air transport and logistical support. As well, it had growing responsibilities for the emerging field of tactical and intercontinental atomic-armed missiles, both winged jet-propelled cruise missiles, such as the Matador and later Mace, and rocket-propelled exoatmospheric ballistic missiles, such as Atlas, Thor, Titan, and Minuteman.[114] Missile programs naturally led to speculation over the service's future role in "the coming of astro power," with one optimist, writing in 1958, predicting that the air force would have upward of ten thousand rated space-vehicle crew members by the early 1980s, routinely making spaceflights "from a few hours to perhaps a day or two."[115] In order to support these developments, the air force would need to undertake comprehensive research across a variety of technical disciplines, and do so at a time when the technology of flight (both within the atmosphere and out of it) was transforming rapidly.[116]

The air force of the post-Korean era had to confront (then as now) a wide range of possible contingencies across the spectrum of conflict, from low-intensity counterinsurgency operations to global and theater nuclear war. Nuclear warfighting preparedness was a key area of concern—witness the air force's lead role in developing the Single Integrated Operational Plan (SIOP) for the control and execution of the joint-service nuclear warfighting force—but was far from the service's only concern.[117] Spurred by what was happening in Africa, Asia, and the Middle East, air force analysts at the Air University were evaluating the use of air power in limited war contingencies and what became euphemistically known as wars of national liberation, particularly the Greek, Indochinese, Malayan, and Algerian experiences.[118] Thanks to the advent of aerial refueling for fighters and bombers in general, the service devoted considerable attention to the challenge of expeditionary air power, evidenced subsequently by its deployment of Composite Air Strike Forces (CASFs) to both the Lebanon and Taiwan Straits crises of the late 1950s, and another to Europe in response to the onset of a second Berlin crisis in 1961.[119] It developed a robust military-industrial-scientific advisory infrastructure that built a family of ballistic missiles and derivative launch vehicles that, today, still stands at the heart of America's strategic nuclear deterrent forces and civilian space enterprise. (Indeed, in retrospect, it was the air force's Gen. Bernard Schriever, with his vision of careful, incremental projects tied to practical needs, not NASA's Wernher von Braun and his vision of successive large projects, who contributed most to the building of America's practical military and civil space power.)[120]

One of the major responsibilities the air force had after Korea was supporting the growing North Atlantic Treaty Organization (NATO) alliance. Created in 1949, NATO and the air force—especially the deployed air force in Europe, which became the United States Air Forces in Europe (USAFE)—effectively grew up together. Air power was crucial to maintaining a free Europe, evident from the earliest days of the Berlin blockade, and its use was implicit within NATO's evolving security vision. It quickly became the veritable linchpin of NATO strategic thought, built around the steadily expanding capabilities of USAFE and backed by USAF forces in the Continental United States (CONUS) and deployed naval aviation forces. NATO's first strategic planning concept, DC 6/1 (issued in 1950), stated that the alliance's first undertaking in war would be to "insure the ability to carry out strategic bombing *by all means possible* with all types of weapons, without exception."[121] After Joseph Stalin's death in 1953, the Eisenhower administration, seeking a reduction in the size and expense of American forces both overseas and at home, adopted a "New Look" defense doctrine emphasizing a strategic and tactical nuclear response to Soviet aggression, known by the shorthand of "massive retaliation." Set forth in National Security Council document NSC 162/2, it stated that "in the event of hostilities the United States will consider nuclear weapons to be as available for use as other weapons."[122] In 1954 NATO followed with MC 48, which emphasized that NATO would use atomic weapons from the outset of a war with the Soviet Union—whether or not the Soviets did. MC 48 gave the Supreme Allied Commander Europe (SACEUR) the same prerogatives for the automatic use of nuclear weapons as existed for the SAC's commander in chief, firmly setting NATO in line with the policy of massive retaliation. In 1957 NATO issued revised versions of two earlier posture statements, MC 14/2 and 48/2. MC 48/2 bluntly stated that if conflict with the Soviets escalated beyond a local border incident or infiltration that might stem from the unauthorized actions of a Soviet commander, "the situation would call for the utilization of all weapons at NATO's disposal, since in no case is there a concept of limited war with the Soviets," claiming such weapons would be used "from the onset" of conflict.[123] Under this scenario, NATO's conventional ground forces were, in effect, a "trip-wire" to trigger an overwhelming nuclear response from SAC, USAFE, RAF Bomber (later Strike) Command, and other air partners.

In the wake of Sputnik, the rapid development of Soviet offensive capabilities (which eroded American nuclear primacy), and the advent of the Kennedy administration in 1961, the doctrine of massive retaliation gradually gave way to one of "flexible response." Flexible response received its first test in the Berlin crisis of 1961. Faced with a second attempt by the Soviets to isolate and possibly seize Berlin, President John F. Kennedy authorized the largest deployment of aircraft since the Second World War. Here the doctrinal, organizational, and technological capabilities investment since Korea paid off. The TAC quickly deployed a 210-plane CASF, consisting of F-100 and F-104 fighters, and RF-101 and RB-66 reconnaissance aircraft. The Air National Guard subsequently deployed another 180 F-84, F-86, and F-104 fighters. Berlin simmered for some months until the focus of international tension shifted to Cuba, where America faced its severest Cold War test, the Cuban Missile Crisis of October 1962.[124]

Aside from the air force flying strategic and tactical reconnaissance missions (an air force U-2 flown by an air force pilot detected the presence of Soviet SS-4 missiles in Cuba), the service's other roles in that seminal crisis have received scant attention.[125] At the U.S. Navy's request, SAC located and identified numerous Soviet ships beyond the range of naval maritime patrol air and sea forces, using its tankers and bombers for sea surveillance. Out of nearly four hundred sightings, two cases particularly stood out. In the first, the navy asked SAC for help in locating the freighter *Odessa*, believed to be carrying nuclear warheads for Soviet SS-4 missiles in Cuba. When a roving B-52 subsequently located and overflew it (echoing the famous *Rex* mission of the 1930s), the *Odessa* stopped, turned about, and returned to the Soviet Union.[126] In the second, SAC responded to another urgent navy request to locate and photograph the Soviet tanker *Grozny*, believed to be transporting missile fuel. An RB-47 located, photographed, and tracked this ship until a navy destroyer could intercept it. *Grozny* also turned back (though it carried liquid ammonia, not rocket fuel after all).[127] SAC crews had once again demonstrated the enduring value of using long-range air force aircraft for maritime patrol.[128]

THE VIETNAM ORDEAL

By the early 1960s the air force, navy, army, and Marine Corps were all services in transition. Each had progressed from the propeller to the jet era, and into the era of the rocket. Seven astronauts representing the air force, navy, and Marine Corps stood poised to become the first Americans boosted into earth orbit. The air force and navy possessed sophisticated intercontinental ballistic missile systems, as well as the Mach 2+ aircraft, and the army and Marine Corps operated increasingly capable vertical-lift helicopters as well, enabling a transformation of envelopment strategies. These capabilities engendered continuous discussion over roles and missions, typified by the revisiting of Key West arrangements with regard to army aviation forces. The new administration's defense chief, Robert S. McNamara, would take particular interest in aircraft and missile acquisition, and seek to advance joint-service commonality and interoperability. These developments occurred while the flames of insurgency smoldered and then erupted into open fire in the lush terrain of Southeast Asia. Much as the launch of Sputnik in 1957 became the defining moment for American science and technology after 1945, the ten-year involvement in Southeast Asia constitutes another watershed, marking both the nadir of America's military fortunes and the source of a doctrinal and planning renewal that paved the way for the success of Desert Storm near the end of the century.

Korea taught the price of reckless military drawdown; the challenges inherent in fighting a politically constrained conflict; the difficulty of fixing doctrinal, training, and equipment deficiencies while "in the fight"; and the danger of overly focusing on atomic conflict at the expense and neglect of conventional forces. Unfortunately, these lessons learned were forgotten in less than a decade. All were inherent in equal or greater measure in America's Vietnam experience, which, additionally, taught its own hard lessons. These included waging a war without clearly defined goals; no coherent definition of what constituted "victory" (what a later generation, sobered by the occupational realities of Iraq and Afghanistan, would term an "exit strategy"); no comprehensive and consistent strategy; no enduring commitment from the national political leadership to pursue what was in the best interests of the nation's deployed combat forces; and misuse of air power to send diplomatic signals rather than achieve decisive combat effects.[129] "We were really unhappy about how that war was conducted," one Vietnam airman recalled decades later, adding, "We bitched about it in the bar at night and got drunk, but we sobered up and went flying the next day. We were as good as we could be."[130]

For the USAF, Vietnam constituted a turning point, illustrating a myriad of doctrinal, organizational, force-structure, and training deficiencies. Over much of the 1950s, accelerated by the impact of Sputnik and the consequent development of bipolar strategic nuclear missile forces, the sharp focus on conventional warfare brought by Korea and the "low-intensity" counterinsurgency wars of the 1950s was blurred, replaced by a far stronger emphasis on the danger of nuclear attack. Slightly more than a year after the Korean War ended, the National Security Council had drafted a new statement of national security policy, NSC 5501, which the Eisenhower administration issued at the beginning of 1955. It defined deterrence and defense as the two critical challenges facing the American military:

> Our first objective must therefore be to maintain the capability to deter an
> enemy from attack and to blunt that attack if it comes—by a combination
> of effective retaliatory power and a continental defense system of stead-
> ily increasing effectiveness. *These two tasks logically demand priority in all
> planning.*[131]

Given these two demands, the air force prioritized the development of strategic bomber and missile forces. After 1953, air force fighter development, having followed a pattern of swing-role employment embodied in the precepts of FM 100-20 from the F-80 to the F-84, F-86, and on through the development of the F-100, now diverged sharply from the needs of conventional war toward nuclear warfighting. Subsequent air force fighters were either bomber-destroying interceptors (F-101B, F-102, F-104A, F-106) or themselves nuclear-strike aircraft (F-101A/C, F-104C, F-105). Indeed, of the entire 5,525 so-called Century series of fighters (those beginning with the F-100 onward), less than a quarter—only 1,274—were fully mature fighter-bombers "designed from the ground up" for swing-role conventional air-to-air and air-to-ground operations, but capable as well for tactical nuclear warfare needs.[132]

By the time of the Kennedy inauguration, the result was an increasing disconnect between the force structure of the air force and the kinds of aircraft it needed for the combat environments of the 1960s. Secretary of Defense Robert S. McNamara has justifiably come under a great deal of criticism for his stewardship of the DoD during the Kennedy and Johnson administrations.[133] McNamara reviewed and canceled several major aircraft, missile, and space programs, including the North American B-70, Douglas Skybolt, Nike-Zeus, Lockheed YF-12 Blackbird, and Boeing X-20. He possessed an unjustified and zealous faith in the "do-everything" airplane, resulting in the mediocre North American Rockwell OV-10 and (most memorably) the misbegotten Tactical Fighter Experimental (TFX) program that spawned the troublesome F-111.[134] The product of a forced marriage of two widely differing navy and air force requirements, the TFX cost the navy a decade of fighter development time, took years before becoming a useful air force asset, and featured in repeated congressional hearings occasionally rivaling the debate on the B-36 versus the carrier roles-and-missions, for their acrimony and searing testimony.[135]

But if his decisions on the TFX, B-70, and so forth remain controversial, his decisions on two other acquisition programs, the McDonnell F-110 (which became the F-4) and the LTV A-7, redressed serious air force mission capability shortfalls. In 1961, to make up for the shortage of tactical air force aircraft suitable for conventional and nuclear air warfare, he directed the air force to acquire the navy-sponsored F-4 multipurpose fighter, enabling the expansion of air force tactical fighter wings from eighteen to twenty-four. Shortly thereafter, in part to allay army fears (yet again) that the air force was not serious about the CAS mission, he directed acquisition of the navy A-7, itself effectively a stubby subsonic stripped-down ground attack version of the supersonic F-8 fighter. It was a good replacement for the aging low-payload F-100, then the air force's primary air-to-ground CAS airplane. Though neither was a completely satisfactory aircraft, and each required significant modifications and upgrading over time, both the F-4 and A-7 went on to complete three decades of sterling service, from Southeast Asia through the Cold War and on to Desert Storm of 1991.[136]

Operational competency, however, withered. Writing in 1968, one noted fighter-leader stated:

Between 1954 and 1962, the USAF training curriculum for fighter pilots included little, if any, air-to-air combat. *This omission was partly a result of doctrine, which then regarded tactical fighters primarily as a means for delivering nuclear ordnance. . . .* As late as October 1963, it was reported that only four of 30 pilots in one fighter squadron had ever shot aerial gunnery.[137]

The air force paid heavily for such failings in the skies over North Vietnam. In Korea, its pilots shot down an average of ten Communist aircraft for every friendly aircraft shot down. But, in Vietnam, that ratio fell to 2.4:1, necessitating changes in tactics, training, and acquisition.[138]

The war in Southeast Asia involved multiple air wars, over North Vietnam, over Laos and Cambodia, and over South Vietnam. It was both joint and coalition warfare, and some partners—Australia, and particularly South Vietnam, for example—contributed in ways that have not generally been recognized or acknowledged to the degree that they should.[139] The "in-country" air war (and that over Laos and Cambodia) was not without risk for air attackers but (until the introduction of the SA-7 man-portable surface-to-air missile [SAM] in 1972) was generally benign enough that otherwise-vulnerable aircraft, such as attack helicopters, Korean-vintage propeller-driven attack airplanes, fixed-wing transports modified as gunships, and light jet trainers modified as counterinsurgency aircraft, could survive.

The war "up north" was very different; virtually from its outset, attackers faced a growing missile and fighter threat that increased steadily over time. The air war over North Vietnam differed dramatically from the time of Rolling Thunder under Lyndon Johnson and Linebacker II under Richard Nixon. Under Johnson and McNamara, targeting and tactics were controlled directly from the White House, governed by infamous Tuesday luncheons at which targets and rules of engagement were determined. But it was not only they: through their remarkable docility, the JCS bear significant culpability for the dismal strategy of the Johnson years. The staunchest opponent of gradualism, the fiery Curtis LeMay, retired as chief of staff at the beginning of 1965. Thereafter, the chiefs meekly acquiesced, so much so that historian and Army Brig. Gen. H. R. McMaster has damningly concluded that Vietnam "was lost in Washington, D.C., even before Americans assumed sole responsibility for the fighting in 1965 and before they realized the country was at war."[140] In contrast, under the Nixon administration the military was given greater latitude in planning, targeting, determining rules of engagement, and execution. That, plus changes in technology (such as the maturation of Wild Weasel anti-SAM operations, increasingly so-phisticated radar jamming and communications intelligence, and the employ-ment of first-generation laser-guided precision munitions) made Linebacker the success that Rolling Thunder was not.[141]

Although not seen so at the time (and unfortunately little acknowledged afterward), the record of joint-service air power (and air force air power in par-ticular) was, in general, commendable. At particular points—the siege of Khe Sanh, Tet, the 1972 spring invasion, Linebacker II—air power proved of crucial, even decisive, significance. Indeed, since 1975 former Viet Cong and North Vietnamese commanders and veterans have increasingly acknowledged the se-vere impact air power had on their operations (particularly after 1968), in words echoing those of Axis commanders after the Second World War.[142]

Vietnam witnessed a continuance of the interservice struggles for control of air power, and also a continuance of the seemingly endless debates over the apportionment of air effort and the intricacies of battlefield air support. Indeed, the authors of the most comprehensive and sweeping study of joint air opera-tions concluded that Vietnam marked a "regression" in joint operations, noting that "the services and the joint commanders had learned little from the Korean experience to enhance joint air operations."[143] Certainly throughout the John-son-McNamara era, that is, up through 1968, the combination of political influ-ence, service single-mindedness, and a frequently disrupted and unclear chain of command worked to hinder effective joint air operations. It fatally damaged the Rolling Thunder campaign over the North and worked to hinder air support and operations elsewhere throughout Southeast Asia, though, fortunately, the amount of air support available to support ground forces fighting in the South was always more than sufficient to meet their needs.[144]

Lessons on the effectiveness of battlefield air attack lost since the Second World War were rediscovered: as one army commander commented, "I learned after a while that my casualties were tremendously decreased if I used the air power and air strikes and used [them] properly. And it was there to use."[145] One of the most valuable battlefield support and interdiction systems used was the B-52, echoing the use of strategic bombers for battlefield support in the Second World War (exemplified by Operation Cobra in 1944, which set the stage for the breakout across France). Conceived in the heyday of strategic nuclear thought, the B-52 was modified with pylon adaptors to carry conventional "iron bombs" externally as well as within its cavernous bomb bays. B-52 strikes impressed friend and foe alike: Army Gen. John R. McGiffert considered the B-52 "the most effective weapon we have been able to muster," while assorted Viet Cong and North Vietnamese Army commanders recalled its attacks as leaving troops "demoralized," "shaking so badly it looked as if they had gone crazy," and "terrifying."[146] (A generation later, similar comments were offered to coalition interrogators from captured Iraqi prisoners after the first Gulf War.) Given this record, it is hardly surprising that, despite problems with joint operations, a postwar army survey revealed that 60 percent of respondents considered Army–Air Force cooperation "excellent," with only 2 percent rating it "unsatisfactory."[147]

By 1972 the experience of the 1960s had shaped a far more capable joint-service force, strengthened still further by the introduction of new precision munitions and (for their time) advanced electronic warfare and radar-homing missiles that made the suppression of enemy air defenses more practicable and achievable than at any previous time. It was that force that shattered the ambitions of the Hanoi regime when it launched a massive conventional air-armor assault into South Vietnam in the spring of 1972, and that force that, later in the year, used the punishing air power of Linebacker II to bring its representatives back to the negotiating table.[148] Arguably, air power salvaged all the previous bad decision making of the preceding decade, turning stalemate into settlement—a settlement later thrown away by a fatal, even immoral, lack of political will within the U.S. Congress and executive branch in 1974–75.[149]

REACTION AND REBUILDING

The bitter lessons of Vietnam, coupled with the disturbing experience of the Israeli Air Force in the 1973 Arab-Israeli War immediately afterward, provided the springboard for the United States to rebuild its air power and redirect systems acquisition and weapon systems development that would shape the aerial victory of Desert Storm almost two decades later. In 1973 the linkage of radars, SAMs, interceptor aircraft, and the proliferation of mobile radar-directed gun systems and man-portable air defense missiles (exemplified by the Soviet ZSU-23-4 *Shilka* quad–23mm gun carriage and the SA-7 *Strela* man-portable air defense systems) had led to extraordinarily high losses, largely obviating the Israeli Air Force's ability to operate with freedom either for deep attack or over the battlefield. An emergency USAF airlift, subsequently acknowledged by Prime Minister Golda Meir as having "saved Israel," not only furnished critically needed ammunition, weapons, and supplies, but also illustrated the crucial need (later addressed) for equipping American long-range transports for aerial refueling.[150] In addition to disturbing losses of B-52 bombers to SAMs during Linebacker II, the Israeli experience affirmed the growing power of integrated air defense systems, synergistically aided by the concomitant rise of increasingly small yet sophisticated computers.[151]

Out of this came a far greater investment in applying advanced technology to meet the challenge of evolving air defense threats, typified by greater investment in electronic warfare and, most notably, the development of so-called low-observable stealth aircraft and missile systems. The air combat lessons of Vietnam, as well as the acquisition failure of the TFX's air combat fighter variant, the F-111B, stimulated the development of a new "fourth generation" of fighter aircraft characterized by high thrust-to-weight ratios, exceptional agility, and refined handling characteristics. Purchased in a "high-low" mix of F-14s and F-15s, and F-16s and F/A-18s, these supplemented, and then eventually replaced, the post-Korean legacy fighters exemplified by the workhorse joint-service F-4.[152]

If the air force had returned to, and then prioritized, a nuclear warfighting strategic focus after Korea, the same was not true after Vietnam. Bomber programs went ahead—notably the advanced manned strategic aircraft, which became the B-1 of the late 1970s, followed by the advanced technology bomber, which became the B-2 of the late 1980s—as did advanced weapons for them, such as the short-range attack missile, the air-launched cruise missile, and then the AGM-129 stealthy cruise missile. Missile programs continued as well, with the retirement of the Titan force and the introduction of Peacekeeper, stimulated by the steady introduction of new missile systems into the Soviet arsenal and the increasing capabilities of those systems to effect a "first-strike" knockout of American forces. But the air force, having addressed these needs and its obligations to the SIOP construct, placed great emphasis on tactical air warfare and (since the Soviets had reaffirmed their support for so-called wars of national liberation) to that of irregular warfare and counterinsurgency as well.

Vietnam had largely been a fighter-and-attack-aircraft war, and afterward it gave rise to tactical fighter pilots to command responsibility. Those air force leaders whose formative experiences had been in the great European and Pacific bomber offensives of the Second World War were moving on, and soon to disappear entirely. In their place stepped a new leadership tempered in fighter cockpits in Korean and Southeast Asian skies.[153] By the time of Desert Storm, the air force's chief of staff and vice chief of staff (Merrill McPeak and Michael Loh), its greatest theorist of air power (John Warden), the Central Command theater air commander (Charles Horner), and his key planning staff (for example, Buster Glosson, Larry Henry, and David Deptula) would all be from the tactical fighter, not strategic bomber, community; even the chiefs of SAC would be former fighter pilots and fighter commanders, something unthinkable in the LeMay era.[154]

Further encouraging this transformation to a more tactically oriented warfighting force was the ominous world after 1968, when five thousand Soviet tanks and two hundred thousand Soviet troops swiftly crushed the bright promise of the "Prague Spring." The reminder of the Soviet threat in Europe imparted extreme urgency to addressing the problem of defeating a rolling, unfolding Warsaw Pact armored and mechanized assault through the Fulda Gap and into the heart of Western Europe, one that emphasized "tacair" application. The Soviet Union's renewed emphasis on operational maneuver groups (revealed in Soviet literature and then confirmed in subsequent Warsaw Pact exercises) had triggered a variety of NATO and member-state interagency programs and study efforts to break the power of such assaults.[155] Existing programs to develop counterinsurgency capabilities and systems—for example, the Air Force A-X program, which became the antiarmor A-10, and the army's various helicopter gunship programs—were reevaluated and reconfigured to meet the threat of Soviet armored forces.[156]

In 1970 NATO had begun its search for a new air support doctrine applicable to resurgent Soviet power, spawning doctrines on both offensive air support and tactical air usage, differentiating between CAS and battlefield air interdiction. Complicating this quest for unified doctrine were not only the traditional differences in how air power was regarded by the air and land communities (differences that showed a not-so-surprising unanimity of viewpoint within the airmen and the ground forces of the various NATO nations), but also the always-contentious position of Germany as a divided and, indeed, occupied power torn between the West and East and with special commitments to each side.[157] These doctrines might well have remained little more than ambitious position papers were it not for the development of new guidance and weapons capabilities—typified by the laser-guided bomb, but also innate in such transformational technologies as satellite-based communications and navigation—that made them practical, reasonable responses to the Soviet threat.[158]

The Soviet challenge and NATO's needs renewed and reinvigorated the spirit of cooperation between the USAF and the U.S. Army, which was also stimulated by the linkage (and close physical proximity) of the air force's TAC and the army's Training and Doctrine Command (TRADOC), and the personal interest and encouragement of the two service chiefs of staff, Air Force Gen. George Brown and Army Gen. Creighton Abrams. In June 1975, the commanders of TAC and TRADOC created the AirLand Forces Application (ALFA) Agency that vigorously pursued the development of joint air-land doctrine and the emphasis on the attack of follow-on forces. Effective tactical air power was crucial to NATO's surviving any Warsaw Pact encounter. A secret 1981 quality-versus-quantity analysis by Air Force Studies and Analysis projecting NATO force capabilities required by a 1985 Central European campaign scenario concluded:

> NATO's air and ground forces will have to fight outnumbered in the foreseeable future. . . . When the programmed air forces are pitted against Warsaw Pact air threat, NATO's air forces are marginally capable of meeting requirements in the air war. However, this capability is delicately balanced and could be negated by threat-force growth or simply by the Soviets introducing additional tactical air assets from other regions. . . . NATO's ground forces depend directly upon tactical airpower to maintain a successful defense. This dependency lays a requirement upon NATO's air forces that the Soviet/Warsaw Pact air armies do not have to the same magnitude.[159]

By 1984 TAC-TRADOC-ALFA had developed a series of thirty-one key initiatives covering a variety of air-land needs, requirements, and operations, ranging from confronting tactical missiles to the joint suppression of air defenses and CAS.[160] Out of these initiatives sprang a variety of programs to furnish and enhance requisite capabilities to combat forces, particularly new sensor and precision weapon programs (the greatest challenge NATO tactical air power proponents faced was air support at night and in adverse weather), command and control architectures, joint tactics, the development of more sophisticated laser-guided munitions, and the acceleration of the space-based Global Positioning System (GPS) navigational satellite constellation program. Thus began arguably the most harmonious and fruitful period in Army–Air Force relations since the creation of the air force, largely because the air force willingly prioritized so many army needs over its own.[161] (Even so, by the end of the 1980s, arguments over Corps or JFACC control of theater air were reemerging, triggering senior officers to intervene in the debate.)[162]

The Falklands, Bekaa Valley, and Afghanistan experiences illustrated both the value of high technology and the deadly necessity of keeping an opponent possessing it outside striking distance of one's own forces. The Falklands offered a case study in the risks of not projecting overwhelming force. Having retired the last of its large Phantom-equipped carriers, Britain's expeditionary task force triumphed only after enduring a debilitating war of attrition forced by its lack of robust long-range maritime air power.[163] Over the Bekaa, American-built F-15 and F-16 fighter aircraft dominated contemporary Soviet-built (and other) designs. Additionally, the first manifestations of the post-Vietnam investment in electronic warfare and the suppression of air defenses could be seen in the swift destruction of Syria's air defense network and command and control structure that left Syrian pilots flying essentially blind.[164] In Afghanistan, man-portable SAMs (most notably the American Stinger, but also the British Blowpipe and Chinese Hun-Ying 5) hindered and ultimately negated the Soviet aviation forces' ability to assist ground forces in close combat, setting the stage for the subsequent Soviet withdrawal from that country.[165]

For the United States directly, a series of occasional engagements and episodes ranging from the Grenada and Panama incursions to actions against Libya and Iran enabled the measurement of American military effectiveness and its steady refinement, taking it, figuratively, from the debacle of Desert One to the eve of Desert Storm.

GLOBAL REACH, GLOBAL POWER

The swift collapse of the Soviet bloc in 1989 caught America's military services by surprise. Perhaps, in retrospect, it is not surprising that a national debate began virtually immediately about the future of American military power, how it might be reduced in size and scope, and a healthy (and four-decades-overdue) "peace dividend" be made available to the American people. Ironically, in less than a year, the United States would be positioning its air, sea, and land forces to confront an opponent operating a force structure of Warsaw Pact systems. Indeed, it may be said that the high-tech, high-tempo mechanized war America expected to fight in Europe was fought, instead, in Kuwait and Iraq.

In the years between Vietnam and the Gulf, the air force approach to doctrine had been at best inconsistent, and the end of the Cold War found the service in a period of doctrinal drift, disarray, and indecision. For many in the air force, doctrine was perceived as having value in inverse relationship to its complexity and length. What substituted for formal doctrine was a practical "work-around" wisdom distilled from historical experience and expressed in Bacon-like squadron aphorisms, shorthand, and quotes from noted airmen: "fly, fight, and win," "anything else is rubbish," etc.[166] To many, it all came down to technology, and, in a service intrinsically built around technology, having the right technology and the right training, bolstered by robust investment in science, technology, and research facilities, was crucially important. To do otherwise was to risk "nothing less than disaster for the way of life that America has come to represent to all the free peoples of the world."[167] Doctrine advocates went to some length to build upon such common wisdom, emphasizing the interrelatedness of doctrine, technology, and air power thought to the larger purposes and missions of the service, in hopes of broadening its acceptance.[168] As one noted:

> Doctrine is more than a theoretical luxury of value only in the classroom. It must instead be the binder, the adhesive, *justifying* our future technological research and development, *rationalizing* our planned acquisition strategy and *governing* our present employment of forces.[169]

Certainly, brevity is the soul of good doctrine, and it must be focused on attaining practical, measurable, and achievable "real world" effects and ends, lest it degenerate into jargon-laden, overspecialized introspection and self-absorption, the worst features of medieval scholasticism.[170] A 1979 edition of AFM 1-1, on the functions and basic doctrine of the USAF, drew more comment for its design layout—heavy on drawings and visual images—than for its content, which was largely unchanged from previous editions. A 1984 revision stressed, for the first time, the interplay of air and space power. Neither spoke adequately (though the 1984 revision commendably tried) to the broad, overarching purposes of air and space power, or how either could contribute across the spectrums of conflict and of operations presented by various levels of conflict.[171] In part, it was such failures of formal air force doctrine to address issues at the operational level of war that impelled Col. John A. Warden III, as a student at the National Defense University, to study air campaign planning. The result was the most influential, nonofficial expression of air power thought since the days of Mitchell and Alexander de Seversky. Warden's book, positing effectively a generic "AWPD-1" for the world of the late 1980s, was, much like Mitchell's and Seversky's writings before, very controversial and inherently counter to the implicit "air-power-in-support" dominant mind-set of most TAC-TRADOC-ALFA initiatives.[172] Thus, when the Cold War came to an abrupt and unanticipated end, the service was in a period of doctrinal drift and indecision, with AFM 1-1 then undergoing yet another revision, and with little official consensus on what direction it should take, except that it should be better sourced and more authoritative.[173]

It was in this climate that the air force ended the Cold War and began its journey into the uncertain world that followed. Even before the fall of the Berlin Wall, defense observers had debated the nature and future of American military power. Central to their often-heated dialogue was the role of the air force and its relationship to other services.[174] As the Soviet bloc self-destructed, then–secretary of the air force Donald B. Rice grew increasingly concerned over perceptions that the end of the Cold War had, somehow, left the service without a purpose or a mission. As with many such debates, at root was the dividing of resources and the sharing of the now-shrinking defense budget. Rice launched an internal study whereby senior air force military and civilian officials were asked what they believed to be the function and purpose of their service. Surprisingly, virtually no unanimity of viewpoint existed. The service in its major command construct—TAC, SAC, Military Airlift Command (MAC), and Space, followed by "everything else"—was so "stovepiped" that few members had a sense of the purpose and working of the air force as a whole. Instead of seeing themselves as airmen in the sense that all army personnel consider themselves soldiers, and all navy personnel consider themselves sailors, they saw themselves as representatives of an occupational specialty, such as pilots, engineers, physicians, and so forth. While they knew a great deal about their own function, they had little sense of where they fit within the larger air force institutional construct. Thus, when confronted with questions about the service, they tended to champion their own specialty area or major command, or speak in, at best, vague and general terms. Accompanying this was a surprising lack of awareness of air force heritage and history.[175]

So whither the air force? Rice determined that the service needed a strong position statement that could function as a unifying and cohesive statement of core service precepts, indeed even as a strategic planning framework. From this objective came the most incisive and important White Paper issued by the service since its creation in 1947: *The Air Force and U.S. National Security: Global Reach–Global Power*.[176] Written in 1989, reviewed by a small number of trusted associates, and then issued just weeks before Saddam Hussein invaded Kuwait, it was authored principally by then–lieutenant colonel Deptula, a doctrine-minded fighter pilot working within SAF/OSX, the Secretary's Staff Group at the Pentagon.[177] *Global Reach–Global Power* enumerated the "inherent characteristics" of the air force—speed, range, flexibility, precision, and lethality—and related how these, working in concert, enabled a planning framework supporting America's national defense strategy of sustaining deterrence (through nuclear forces), providing versatile combat force (for theater operations and long-range power projection), supplying rapid global mobility (via airlift using the tanker "air bridge"), controlling the high ground (via command, control, communications, and intelligence systems in orbit or in the atmosphere), and building American global influence (via strengthening security partners and relationships). It concluded:

> The Air Force is building a force with agile and responsive capabilities tailored for the world we see unfolding before us. We will continue developing these capabilities—planning the "pieces" of our Air Force to complement each other, complement the capabilities of the Army, Navy, and Marines, and create optimum power to underwrite our national security strategy. . . . With the Air Force's range and rapid reaction, we are prepared to meet the challenges of the future: to provide *Global Reach–Global Power*.[178]

Global Reach–Global Power was—amazingly—the first senior-level statement of purpose and the air force's "first principles" ever issued at the secretariat or chief of staff level. As such, in retrospect, it constituted the most important and influential position paper and strategic planning framework issued over the previous forty-plus years of the air force, even more than Theodore von Kármán's *Toward New Horizons* study of the immediate postwar era. While the latter had expressed the service's commitment to a future based on the high technology of jets, rockets, supersonic aerodynamics, advanced electronics, and nuclear power, the former spoke to more significant foundational issues—what the air force brought as an independent service to American defense policy and national military power.

In one of those ironies scattered throughout history, no sooner was *Global Reach–Global Power* "on the street" than a crisis erupted that would test its validity: Saddam Hussein's invasion of Kuwait. The response to that invasion, from the Desert Shield airlift through the combat operations of Desert Storm, confirmed the essential wisdom underlying the paper. It illustrated as well three critical aspects of air force operations: the essential expeditionary nature of the service's aerospace force that permitted the rapid packaging of strike forces and their deployment from CONUS, the increasingly artificial distinction between "strategic" and "tactical," and the emergence of effects-based operations.

Within thirty-eight hours of receiving permission from the Saudi government to deploy to the Arabian Peninsula, twenty-four F-15s had deployed from Langley AFB and were flying combat air patrols in the Gulf. Within two weeks, sufficient force was deployed to ensure the safety of Saudi Arabia. Within a month more than twelve hundered aircraft were based in theater, capable of executing both strategic and tactical missions. Over the previous thirty-three years of the air force, the two words had become synonymous with particular sorts of aircraft, commands, and particular missions: strategic meant SAC's large bombers and nuclear war, while tactical meant TAC's fighter and attack aircraft and conventional war. This unfortunate typecasting was highlighted when, in the early phase of campaign planning, a senior air force officer took exception to the use of the word "strategic" to describe the air campaign being planned by General Horner's Black Hole team. "You can't say strategic," he informed one planner, "because we are not contemplating using nuclear weapons."[179] Horner's planners pressed ahead, using an effects-based targeting approach, and focused on achieving decisive effects through simultaneous, parallel attacks against key nodes across the theater, rather than sequential, linear attacks focused on a step-by-step working through of a series of target categories.

Planning and executing the war were not without controversy; in the early part of the crisis, Secretary of Defense Dick Cheney relieved Air Force Chief of Staff Gen. Michael Dugan ostensibly for intemperate remarks about American intentions, but seemingly more for his bold advocacy of an air-centric, as opposed to ground-centric, campaign.[180] Thereafter, for much of the fall, the air force labored under a disadvantage in presenting its options for attack, until a briefing before President George H. W. Bush, where the new chief, General McPeak, succeeded in convincing the president and his senior advisers that air could play the dominant role in undoing the Saddam regime. The results, once the air campaign began, swiftly spoke for themselves, as Saddam's military infrastructure and his deployed forces were disrupted and destroyed in a whirlwind of attacks.[181]

During the war, the theater commander, Gen. Norman Schwarzkopf, had used his joint-power coalition forces to good effect. As a result, the triumph of coalition arms was complete—a country liberated, and another country's military humbled, at remarkably little cost in lives and materiel. The air attacks were so successful and overwhelming that they overturned decades of conventional wisdom that air attack did little more than rearrange a landscape.[182] Almost as profound as the physical destruction of the Iraqi military infrastructure was the psychological effect of air attack on Iraqi forces in the Kuwaiti theater of operations. A postwar RAND study noted that

[j]udging by interviews with enemy prisoners of war (EPWs), the net effect of the air campaign was a story of increasing hardship, a sense of helplessness, and a loss of the will to fight. By the time the ground war began, EPW reports indicate that half the Iraqi forces had deserted. EPWs frequently commented that aircraft were always overhead. This meant that they were constantly under threat of attack. Compounding the psychological pressure was the fact that they believed that they were defenseless against air attacks. . . . Debates will probably rage for a long time over whether airpower won the war, and the extent to which ground forces are necessary to seize and hold territory. However, in this conflict new conditions have emerged. In the past, air has been a support element that contributed around 10 to 20 percent to the outcome of the land battle. In this conflict, coalition air forces were responsible for 50 percent or more of the enemy ground force's losses. This represents a significant shift in the contribution of the respective forces on the outcome of the air-land battle.[183]

For airmen, Desert Storm confirmed a revolution in the application of air power. Indeed, all that has happened in air power development since has reflected the impact of that revolution. It demonstrated that in the intensely integrated command, control, communications, computers, intelligence, surveillance, and reconnaissance era, with space-based assets, electronically cued and controlled flight (and flight control systems), and electronic sensors and architectures that could be applied to offensive air power systems (aircraft, missiles, and platforms), air power could so unhinge an opponent as to essentially "deliver" a prostrate nation to the mercy of occupation forces. This new model of war emphasized the interplay of precision long-range engagement forces using aerospace power (whether aircraft, helicopters, surface-air-subsurface-launched missiles, etc.), partnering with special operating forces. Thus, the need for punishing force-on-force mass encounters was rendered obsolete, for the underlying strategy of such encounters was itself rendered largely so: from a mass-driven Clausewitz model to a Sun Tzu model emphasizing the use of deceptive attack, "nodal" effects-based war, and the destruction of capabilities, not people—in short, if not blood-free, then at least less bloody.[184] Desert Storm accomplished this in the relative early era of the revolution—when only a small percentage of attackers could use smart munitions, and when the numbers of the munitions themselves were quite small, less than 10 percent of the whole. It signaled to the global aerospace community that the stealth revolution was more than hype, launching a rapid proliferation of interest in stealth that has continued to the present. It encouraged the UAV revolution that occurred in the early to mid-1990s and was first evidenced over the Balkans and later in the global war on terror.[185]

Tellingly, President Bush chose to preside over the Air Force Academy graduation ceremony held in May 1991. There he extolled air power, confessing that when he first heard McPeak explain the plan at Camp David, he wondered if air power could indeed accomplish what McPeak believed it could, and pointedly noting, "Gulf Lesson One is the value of air power." In that moment, without question, the air force was at the apogee of its existence.

THE STORM AFTER THE STORM

The 1990s have been excoriated as a "holiday from history," in which "every major challenge to America was deferred" and "every can was kicked down the road."[186] However true otherwise, the 1990s constituted quite the opposite for the air force: a time of intense work, characterized both by great accomplishment and great frustration, a nonstop series of more than fifty international crises and interventions, a variety of initiatives, and rapid changes in key leadership that destabilized and distracted the service, damaging its institutional cohesiveness and focus. To give a brief and incomplete review, over the length of the 1990s, the air force repositioned itself to confront the disorder of the "New World Order"; participated in numerous contingencies; enforced UN sanctions and imposed a policy of air denial and control over Iraq via Operations Northern and Southern Watch and joint combat operations, such as Desert Fox; flew humanitarian relief and combat air patrol missions in the Balkans with Provide Promise and Deny Flight; undertook numerous other humanitarian and presence operations;

established and deployed air expeditionary forces; and went to war twice, with Operations Deliberate Force and Allied Force, the latter coinciding with the fiftieth anniversary of the formation of the NATO alliance.[187]

In retrospect, the air force was most fortunate with the leadership it possessed as it began the decade. Secretary Rice and General McPeak, though very different in personal styles and temperament, constituted a highly effective team, certainly the finest executive pairing the service enjoyed since the fruitful and mutually supportive partnership of Secretary Symington and Chief of Staff Gen. Vandenberg at the birth of the service in 1947. Uniquely among air force civil and military leaders, they each left a published corpus of key papers and documents that shed light not only on their thinking and key decision making, but also on their reaction to the circumstances surrounding the air force and its operation in the years surrounding the end of the Cold War and the onset of the post–Cold War period.[188] Over what was subsequently dubbed A Thousand Days of Change, each worked strenuously and resolutely to extend the principles of *Global Reach–Global Power*, and reshape the air force, Rice until he left office in January 1993, and McPeak until his retirement from service in late 1994.[189]

The successes the air force has continued to enjoy, and the competencies that it routinely displays, owe more to this partnership than to any other single combination of secretary and chief. Less than a year after the conclusion of Desert Storm, Rice and McPeak oversaw a dramatic restructuring of the air force. Gone were TAC, SAC, MAC, and Air Force Systems and Air Force Logistics Commands. In their place were Air Mobility Command (the "global reach"), Air Force Combat Command (the "global power"), and AF Materiel Command.[190] Rice's grasp of technical issues, strategy, and doctrine was virtually unique, and he used it in compelling testimony and forceful essays. McPeak was supremely organized and farsighted, a careful organizer and rigorous thinker, always contemplating several moves ahead. He was intolerant of bloat—physical and organizational alike—and slashed headquarters and unnecessary bureaucracy vigorously, streamlining and realigning the air force structure to eliminate stovepiping and redundancies. Though he favored revolutionary rather than evolutionary long-range planning, he insisted it be tied to clearly identified requirements and realistic funding estimates, and brought in a widely experienced and brilliant planner, Maj. Gen. Robert E. Linhard, to serve as director of planning. He appointed Maj. Gen. Charles D. Link as a special assistant to act as point man for arguing the air force position on roles and missions, a task

Link subsequently undertook with vigorous and memorable dedication. At his direction, the Air University undertook an influential space-power forecasting study effort, *Spacecast 2020*. As a result of this study, which highlighted the contribution of a space-based view and awareness to national security, McPeak expanded on the original *Global Reach–Global Power* formulation to incorporate the concept of Global Presence, an initiative that would bear some fruit under his successor.[191]

All this occurred against a backdrop of surprising contentiousness over the nature and future of American military power, played out, ironically, even as other nations moved rapidly to emulate the very aspects of American air power now under traditionalist attack in the land of their origin.[192] The difficulties the service—and air power generally—were to face were signaled within weeks of the conclusion of Desert Storm by a series of seemingly small events.

Critics who had, just months previously, forecast the possibility of tens of thousands of coalition casualties now swung wildly in the other direction. Air power's very success, some now argued, constituted proof Iraq had not been as robust as they had previously thought.[193] Others argued that the four-day ground incursion had accomplished what thirty-nine days of air attack had not, coining the expression "Hundred-Hour War." That most Iraqis—more than eighty thousand, in fact—advanced toward incurring ground forces with their hands in the air, that most Iraqi prisoners of war credited air attack with breaking their ability to fight, and that no single patch of ground existed in Iraq where one could raise a monument (as at Waterloo, Gettysburg, Stalingrad, El Alamein, etc.) signaling the "culminating point" of the conflict where the power of Iraqi force was broken, were virtually ignored. Instead, with a disturbing loss of historical perspective, mechanized infantry and armor actions that, in the context of the Second World War or the various Arab-Israeli wars, would have been regarded as small-scale routine operations were now hailed as veritable El Alameins and Kursks.

During the preparation of the so-called Title V interim report on the Gulf War to the U.S. Congress, Joint Staff partisans questioned whether the authors of the section referring to air operations could use the expression "air campaign," arguing that there was only one campaign, that of the theater commander. It was a small but telling indication that what had been readily visible daily for six weeks to global audiences was, in the arcane world of the Pentagon, subject to the filtering of a joint-force political correctness that persisted in seeing air as operating only "in support of" a surface commander's wishes.[194]

In November 1992 the army and navy held the first meeting of the Army-Navy Joint Board to take place since the creation of the air force in 1947 in order to discuss joint planning. The air force was not invited to participate, eerily echoing the head-in-the-sand response to revolutionary air power exhibited at an earlier Army-Navy Joint Board meeting more than seventy years ago. Then, participants had studiously ignored the results of Billy Mitchell's recent bombing of the *Ostfriesland* and continued to champion the battleship's primacy.[195] Their prewar criticisms having been utterly discredited by the wartime performance of America's high-tech joint-service forces, members of the 1980s defense reform movement grimly fought to retain some shreds of credibility and relevance. Not content to accept Schwarzkopf's and Horner's ascension among the great captains of war, they advanced increasingly fanciful reasons for the success of Desert Storm, anointing various figures as the architects of victory. Among the military cognoscenti, despite clear evidence of how technological, organizational, and operational excellence had come together to produce victory, a debate broke out over whether Desert Storm actually marked a "revolution in military affairs." Played out over several years, this debate admittedly triggered some thoughtful and useful scholarship, though at times partisans debated with the fervor and intensity of medieval theologians arguing whether angels could dance on pins. It reached no real consensus and ended at best tepidly and inconclusively.[196]

Certainly there were logic and reason enough to examine the nature and use of combat force in light of the changes in military power and capabilities that had taken place since the Vietnam era, the Desert Storm experience, and the lessons learned from both. The air force's Tony McPeak, already in the midst of transforming the service to fulfill the vision of *Global Reach–Global Power*, was up for it. McPeak believed both theater air defense and battlefield air support needed better combat integration, and that each could be done best by a service having ownership of the issue.[197] Since history taught that air defense required a "system-of-systems" approach to be successful, it made little sense, he believed, to artificially divide fighters and SAMs between two different services, with all the attendant command, control, and communications problems that entailed. As well, the army had, for years, sought more control over battlefield air support and had, at various times, experimented with reintroducing its own fixed-wing air support, an implicit recognition of the limitations of helicopters for the mission. After the war, in his cut-to-the-heart fashion, he met privately with Gen. Gordon Sullivan, the army chief of staff, proposing a swap: the air force would absorb the army's Patriot SAM force and air defense mission, and give the army its A-10s for organic ground support. "Sullivan gave me the cold shoulder," McPeak recalled a decade later, "claiming the Army could not afford to operate the A-10. Of course the real problem was not cost but an inability to imagine how our combat effectiveness might be dramatically improved by bold, non-incremental change."[198] Alas, in 2003's Operation Iraqi Freedom, the air defense disconnect noted by McPeak proved more chasm than seam: despite any credible Iraqi air threat, itchy-fingered U.S. Army Patriot crews shot down two friendly coalition aircraft, killing those on board. Likewise, concerns for the survivability of army helicopters proved equally well founded: on one operation, all thirty-two gunships of an attack helicopter battalion were hit by enemy fire, with many badly damaged and one lost, as they attempted to prosecute a deep night attack against an Iraqi division.[199]

In the wake of Desert Storm, a variety of panels, private initiatives, and congressionally mandated commissions examined roles and missions, joint-force relationships, acquisition plans, and similar issues. Since, at heart, all of these involved the apportionment of increasingly scarce defense resources, discussions developed an intensity mirroring that of the roles-and-missions debate of the late 1940s. They rarely manifested a similar intemperate outspokenness, but were, nevertheless, conducted with an earnestness worthy of those earlier times. Simultaneously, an increasing number of generally thoughtful assessments on the nature of air power and its applicability to future warfighting appeared, many by authors who were themselves civilian or military air force members, or who had long-established professional ties to the service, and who expanded on the principles and concepts set forth in *Global Reach–Global Power*.[200]

The most influential of formal reviews was the so-called White Commission, the congressionally mandated Commission on Roles and Missions of the Armed Forces, chaired by John P. White. Its report, issued in May 1995, emphasized the necessity of improving the way the DoD conducted joint unified operations and identified four mission areas that "demand[ed] immediate attention": combating the proliferation of weapons of mass destruction, information warfare, peace operations, and operations other than war. Played out behind the scenes among the service support staffs was a struggle to ensure that each service presented its own position and capabilities in the best possible light.[201]

From the fallout of the White Commission came a shaping conceptual template for the post-Gulf era, *Joint Vision 2010*, prepared at the instigation of JCS

Chairman Gen. John Shalikashvili. *Joint Vision 2010* developed four operational concepts for future war, shared among the various services, though to different degrees: dominant maneuver, precision engagement, full-dimensional protection, and focused logistics. In turn, it energized each service to develop its own strategic plan for fulfilling the objectives of *Joint Vision 2010*.[202] The air force response, called *Global Engagement*, was released in late 1996. It unapologetically pronounced, in words that would have done credit to Rice and McPeak, that in the twenty-first century, "the strategic instrument of choice will be air and space power"; it explained that the synergistic interaction of six core competencies— air and space superiority, global attack, rapid global mobility, precision engagement, information superiority, and agile combat support—linked by space-based global awareness and command and control, provided the critical air and space power that any joint-force team would need to employ in future war, thus contributing to the chairman's vision of joint warfare in 2010. As anticipated, these have, in fact, remained at the heart of air force thinking throughout the first decade of the new century.[203]

"WE DON'T NEED ANY BILLY MITCHELLS": THE USAF ENTERS TROUBLED SKIES

Yet as air force planners addressed the White Commission and *Joint Vision 2010*, all was not well; indeed, in mid-1995 a comprehensive Air University survey of the air force's future prospects concluded, "Five years after its greatest triumph, the USAF finds itself besieged by orchestrated attacks from the other services."[204] And it wasn't just that. A rapid turnover in senior DoD leadership influenced the priorities and progress of the service. Secretary of Defense Dick Cheney, who had sacked Dugan and canceled the navy's troubled A-12 stealth aircraft program, left office with the change in administrations. In his place stepped Leslie "Les" Aspin. But despite being a Capitol Hill veteran with a deserved reputation as a defense expert and reformer, Aspin proved surprisingly ineffectual. In 1993 Aspin ordered a bottom-up review to assess national strategy and defense investment, predicated on having forces sufficiently robust to fight two regional wars: prioritizing and winning one, "holding" the other, and, after having won the first, then applying American combat power to win the second.[205] His role in the flawed decision making that led to the disastrous Black Hawk Down tragedy of Mogadishu, and the terribly cumbersome command and control structure overseeing that characterized the failed attempt by Army Rangers to seize key associates of the warlord Mohamed Farrah Aidid, permanently tarnished his tenure.[206] Increasing ill health forced him from office in early 1994, and he died the next year. He was replaced by William Perry, a DoD veteran from the Carter years, who proved a superb secretary. Though a former army officer, he strongly believed in air power (having been one of the fathers of the stealth revolution in the Carter era), once stating, "Desert Storm taught us something about air dominance. We had it, we liked it, and we're going to keep it."[207]

Shortly after taking office, Perry had his own encounter with the coercive and deterrent value of air power. On October 6, 1994, as senior leaders gathered for their Corona Fall meeting at the Air Force Academy, satellites and U-2s spotted Iraqi forces massing once again on the Kuwait border. The response, Operation Vigilant Warrior, involved stepped-up patrols over the Northern and Southern no-fly zones, aircraft and carrier deployments, and mock bombing raids by B-52s and other aircraft, within sight of Iraqi forces. Four days later, Iraq ended its provocation, withdrawing its forces from the border. "The Air Force has really deterred a war," Perry remarked afterward. "When we deployed F-15s, F-16s, and A-10s in large numbers, I think they got the message very quickly."[208] Unfortunately for joint-service air power (and the air force in particular), Perry remained in office only until early 1997, succeeded by William Cohen, a career legislator appointed in the interests of bipartisanship. Cohen subsequently presided over the preparation of the Quadrennial Defense Review (QDR) and the slashing of air force and navy aircraft and ship-building programs, remaining in office until the end of the Clinton era.

By the time Cohen took office, the strong partnership of Rice and McPeak was history, having dissolved with the election of President William Jefferson Clinton, which resulted in Rice's leaving office in January 1993. The critical teaming of secretary and chief, so necessary for effective dealings with the executive and legislative branches, was broken—and precisely at the time when the post–Gulf roles-and-missions debate entered full fury. Incoming Secretary Sheila Widnall energetically plunged into air force decision making, but her great forte was science and technology, not force structure, operations, and joint-force relations. Her launching of a path-breaking multidiscipline forecasting survey, *New World Vistas*, revitalized the Air Force Scientific Advisory Board and constituted both a major accomplishment and a fitting legacy.[209]

The expression *Global Reach–Global Power* had served as a unifying slogan for the air force, an easily remembered reminder and tagline of what the service was about, and one that resonated well with outside audiences. In the early to mid-1990s, it had tremendous currency and clear lasting power, in direct relation to its proximity to the success of Desert Storm, various ongoing humanitarian contingencies, and rapid reaction responses, such as Vigilant Warrior. But with McPeak's departure near the end of 1994, the tempo and rhythm of *Global Reach–Global Power* were fatally disrupted, ironically by the unintended consequences of one who embraced its tenets and worked to further them.

The incoming chief of staff, Gen. Ronald Fogleman, was a fighter pilot and military historian whose appointment was warmly welcomed by air and space advocates. Fogleman saw *Global Reach–Global Power* as the "guiding construct" of the air force, though he, like McPeak, believed it needed to address the presence issue, and so endorsed his predecessor's intent to refine it, but with more of a joint focus. *Global Reach–Global Power* thus began a slow but accelerating fade, tarnished with the false perception that it was in need of revision. Following the untimely death of Major General Linhard, Fogleman appointed an equally gifted successor to serve as special assistant for long-range planning, Maj. Gen. John Gordon. Fogleman launched *Air Force 2025* (another Air University study effort following on from McPeak's *Spacecast 2020*), created a Chief of Staff professional reading list, established a new doctrine center, encouraged development of remotely piloted aircraft systems, and proved a forceful advocate within and outside the DoD community for air force modernization. As well, he took the first steps toward institutionalizing an expeditionary operational construct, deploying a prototype Air Expeditionary Force (AEF) to the Persian Gulf. All these were very welcome initiatives.[210]

Establishing the doctrine center constituted a signal accomplishment in the life of the air force and built upon previous interest in doing so. In 1988 then–Chief of Staff Gen. Larry Welch had directed the establishment of a new specialist school at the Air University, and this institution, the School of Advanced Airpower Studies (SAAS), seated its first class in the summer of 1991. Thereafter, SAAS stimulated a renewed interest in doctrine and air power application, evident in two comprehensive examinations of air and space thought prepared by its faculty and published over the next eight years.[211] Over this same period, a number of SAAS and other Air University students examined particular aspects of air doctrine, air operations, and the history of air warfare, thereby strengthening both their own and the service's intellectual underpinnings. There was no attempt to enforce an organizational or consensus view; indeed, skepticism was encouraged, and many SAAS faculty themselves were hardly among the ranks of air power "zealots." The intellectual ferment resulting from this and other air force professional military education activities was unprecedented, echoed the prewar days of the legendary Air Corps Tactical School, and, like then, was carried back into the real world of planning and operations.[212]

Thus emboldened, air force members increasingly took on long-standing shibboleths (such as the notion that air forces existed only for "strategic" bombing, and that strategic bombing, except for the use of nuclear weapons, was, essentially, a waste of time), now arguing with insight matching their instincts.[213] In 1996, Fogleman directed the consolidation of the service's doctrine development within a new Air Force Doctrine Center, located first at Langley AFB, and then relocated to Maxwell AFB the following year. Center commander Maj. Gen. Ronald Keys and his successors oversaw a long-overdue revitalization of the crafting and publication of doctrine, beginning with the issuance of Air Force Doctrine Document 1 (AFDD-1), *Air Force Basic Doctrine,* in September 1997, coincidentally the fiftieth anniversary of the air force as an independent service. Since then, the center (merged in 2007 with the College for Aerospace Doctrine, Research and Education and renamed the LeMay Center for Doctrine Development and Education) has prepared a series of doctrine documents covering the range of air force operations; coordinated them with the corresponding Joint Publications on doctrine; and served both as the service's "single face" on doctrinal matters within the air force, and as the service's doctrinal authority representing the air force doctrinal position when presenting (and, in some cases, "deconflicting") it in the joint- and combined-force environment.[214] The latter was particularly crucial, given that the long-standing (and likely never-to-be-resolved) struggle over whether an army ground commander or an air commander should exercise control of theater air power still roiled the joint doctrine arena.[215]

If in on AFDD-1's conception, General Fogleman, alas, was not present for its birth. In July 1997, two months prior to its release, he abruptly retired amid rising controversy over several contentious issues. Growing disagreements with the secretary of defense and the Joint Staff over the QDR (one of whose errand boys had told him, "We don't need any Billy Mitchells"); intemperate cuts reducing the planned F-22 buy from 448 to 339; a messy and well-publicized disciplinary case involving a wayward and duplicitous female B-52 pilot; and differences with Secretary of Defense William Cohen over Brig. Gen. Terryl

Schwalier, a fine officer whose promotion to major general was unjustly canceled by the secretary in the wake of the Khobar Towers terrorist attack (even though Schwalier had been found blameless for the disaster) had all influenced his decision, as had a reading of H. R. McMaster's *Dereliction of Duty* on the failure of the Johnson-McNamara–era JCS to live up to the standards of professional integrity expected of military officers. And so, rather than catalyze further conflict and divisiveness, he placed the good of the service and the DoD above his own privilege, and thus chose to retire early.[216]

Fogleman's retirement, coming almost on the eve of the air force's fiftieth anniversary, rocked the service. In retrospect, he could not have done otherwise.[217] He had entered office believing "that the country had a tremendous number of internal needs, that the external threats were lower than we had faced in half a century," and that service members had "allowed ourselves . . . to lose sight of our values."[218] His subsequent quest to "restore the soul of the Air Force" was encapsulated by a new service slogan emphasizing a renewed commitment to ethics, service, and performance: "Integrity First, Service before Self, Excellence in All We Do." His steadily growing differences with the DoD leadership and its decision making were so readily apparent that remaining much longer would have necessitated either boldly confronting the secretary of defense or meekly submitting before him. The latter course, signaling a "do as I say not as I do" moral relativism, would have smacked of the rankest hypocrisy, destroying any credibility he possessed and breaking faith with the men and women under his command.

There were, unfortunately, other unintended consequences of the new slogan. *Global Reach–Global Power* had succinctly summed up the air force's essential operational character. "Integrity First, Service before Self, Excellence in All We Do" now summed up what was expected of its people. But however commendable the motive behind it, the new slogan inadvertently helped derail the air- and space-power express Rice and McPeak had crafted (and, as well, much of Fogleman's own program). It transformed the public persona of the service from being one that reached out and won wars to being one made up of people whose moral compasses were so adrift that they needed to be constantly reminded of their responsibilities as public servants and guardians of the nation, even by having the slogan emblazoned on all institutional documents, letterheads, and briefing slides.[219] For almost three years—as the roles-and-missions debate intensified, as other services articulated competing strategic visions, as the air force underwent a complex QDR, and at a time when the air force leadership needed to focus public attention on the case for force restructuring—the service was effectively taken "off-message."[220]

In this environment, the McPeak-rooted attempt to rework *Global Reach–Global Power* to include presence came to naught. Instead, presence appeared as its own White Paper, *Global Presence*, issued in early 1995, but then overtaken by the subsequent release of the Commission on Roles and Missions of the Armed Forces report. Unlike with the "one-force, one-source, one-voice" solidarity of *Global Reach–Global Power*, little consensus existed as to what the air force's next "vision vector" should be, or even who should "own" its crafting process. As one Air University student noted at the time, "Even at the lower level, between Air Staff offices, the most fundamental of questions of purpose, intent, and message were never answered."[221] Given these circumstances, the issuance of *Joint Vision 2010* at least served to furnish a common "top-down" departure point to drive an air force response, which, of course, became *Global Engagement*. But *Global Engagement* certainly wasn't *Global Reach–Global Power:* less than five years after its release, that easily recognized and remembered framework—at once so simple, so appropriate, and so profound—now lay, for the moment, in dismembered dormancy.[222]

The baton now passed to Gen. Michael E. Ryan. Richly experienced, Mike Ryan was past commander of USAFE and the son of a former air force chief of staff. In August–September 1995, as Sixteenth Air Force commander and commander Allied Air Forces Southern Europe, he had overseen Operation Deliberate Force, an immensely successful Bosnian air campaign that ended Serbian threats to Sarajevo and refugee safe areas. Bosnia was a morass, the victims of Serbian aggression little aided by a diffident, largely ineffective (and thus oxymoronically named) UN Protection Force (UNPROFOR); its leadership was so fearful lest air power antagonize the Serbs, widen the conflict, and cause unnecessary casualties, that its attitude might rightly be considered "air phobic." Such an attitude went down badly with NATO, and particularly American, airmen. When UNPROFOR commander Gen. Sir Michael Rose complained to Adm. Leighton "Snuffy" Smith about his pilots firing antiradar missiles at Serbian acquisition radars that had locked onto them, Smith bluntly informed Rose that "it was no good asking him to provide air cover for the UN and then trying to stop him from doing his job."[223]

In July 1995, Serbian forces overran two safe areas, Srebrenica (whose inhabitants they murdered) and Zepa, and the UN abandoned a third at Gorazde, tacitly acknowledging before the world that its Balkans policy had failed. Later that month, representatives of Great Britain, the United States, and France met in London, determined to prevent further savagery. Under pressure from these nations and the NATO leadership, UN Secretary-General Boutros Boutros-Ghali finally agreed to looser restrictions on air strikes.[224] Ryan had supervised the preparation of an air campaign plan to protect UN safe areas and strike at Serbia's military infrastructure. A brutal shelling of Sarajevo on August 28 triggered retaliatory action: on August 30, Deliberate Force got under way. Virtually as soon as bombs started falling, Serbian president Slobodan Milosevic requested a bombing halt, which was granted on September 1. But on September 5, after several fruitless days of delay and intransigence, Deliberate Force resumed. NATO airmen targeted Serbian air defenses, command and control, depots, ammunition dumps, and communications. Slightly over a week later, Bosnian Serb commander Ratko Mladic sued for peace, and on September 20, following a six-day bombing halt, Deliberate Force came to a triumphant end.[225]

Deliberate Force confirmed that the experiences with precision engagement in Iraq were universal, confounding skeptics who had, after Desert Storm, suggested that while air power might "do deserts" it could not replicate such success elsewhere. In Desert Storm, precision weapons had constituted just 9 percent of expended munitions; in Deliberate Force, they constituted more than 98 percent of those employed by American airmen and 28 percent of those dispensed by non-U.S. NATO attackers. NATO military and political objectives previously unattainable were met, including removing threats to safe areas, opening roads and airport access to Sarajevo, removing heavy weapons from designated areas, and securing a path to a peace agreement, subsequently sealed with the so-called Dayton Accords (with a dinner, tellingly, held under the warplanes of the National Museum of the United States Air Force). Afterward, American diplomat Richard Holbrooke noted, "One of the great things that people should have learned from this is that there are times when air power—not backed up by ground troops—can make a difference."[226]

It was with this background that Ryan, who served as chief of staff from October 1997 to just before 9/11, entered office. But he was in an unenviable position. Of the last four chiefs of staff, Welch had been unjustly rebuked by Cheney over a peacekeeper missile issue, Dugan had been removed, McPeak had been caricatured by his critics to marginalize his effectiveness, and Fogleman had stepped down.[227] It fell to Ryan to restore a command stability sorely tested over the previous decade, and at a time when the service was in flux, confronting the turmoil engendered by the senior-level changes in the DoD, the QDR (which made deep force-structure cuts in the air force, but left the army and navy largely untouched), the demands of continuing combat operations in the Gulf and Balkans, and the routine challenge of watchful preparedness around the globe, in anticipation that it might have to fight and win not merely contingencies, but two nearly simultaneous major theater wars (MTWs). In all of this he was fortunate to have the top cover of an excellent service secretary, F. Whitten "Whit" Peters. A former naval officer who demonstrated a keen interest in air and space power, Peters had a willingness to engage the service's critics and an openness to novel ideas and concepts. Peters served as secretary in the latter years of the Clinton presidency and after leaving office continued his strong support of American aerospace by cochairing a commission on the U.S. aerospace industry's future. If neither Peters nor Ryan was as publicly outspoken on air power issues as Rice and McPeak before them, they were, nevertheless, equally committed to the primacy of air and space power, and the air force as a "full-service" air and space power provider. Further, both were politically astute and conciliatory by nature. When a controversy arose over the location of a memorial to the air force that, in the view of Marine Corps supporters, threatened to diminish the stature of their iconic Iwo Jima memorial, Ryan and Peters worked quietly with congressional and memorial officials to relocate the memorial to a more suitable site, thereby defusing what had been festering into an ugly interservice spat.[228]

Meeting the two-MTW construct was a major burden, for, even as demands on the air force increased, it was shrinking, having already lost fully a third of its personnel and force structure since 1991. "The mismatch between resources and requirements," Ryan recalled later, "was forcing the men and women of the USAF into a lifestyle characterized by high personnel tempo at the expense of family life."[229] It was a situation that could not go on without breaking the force. In response, over the mid- to late 1990s, successive air force leaders had championed returning to an expeditionary force structure echoing the CASFs of the 1950s, but with more robust and balanced power as a means to furnish theater commanders "rapid, responsive, and reliable" air power tailored to meet their needs.[230] In October 1995 the air force deployed an AEF to Bahrain, filling a two-month void created by differences in the scheduling of departing and arriving aircraft carriers. Its eighteen F-16s, configured either as low-altitude navigation and targeting infrared for night–equipped precision bombers or as high-speed antiradiation missile–firing air defense suppressors, were flying sorties twelve hours after arriving, validating using land-based aircraft as a quick, flexible, responsive, and effective means to project global presence and influence. Thus encouraged, the air force sent two more, with nearly double the aircraft, to Jordan and Qatar, where they operated for three months in support of Operation Southern Watch. From this experience sprang a formalized Expeditionary Aerospace Force (EAF) construct that then–acting air force secretary F. Whitten Peters and Ryan announced to the world in August 1998, and which, fourteen months later, became an operational reality with the rollout of ten EAFs. For the air force, "expeditionary warfare" was now institutionalized as a way of life.[231]

The two greatest crises Ryan faced during his tenure were Operation Allied Force in 1999, and a congressional "drive-by shooting" that nearly ended the F-22 program. Peters and Ryan formed a senior-level "Raptor Recovery Group," and then Brig. Gen. T. Michael Moseley, director of Air Force Congressional Liaison, skillfully marshaled support that kept the F-22 alive.[232] While this battle to save American air dominance raged, the air force was once again engaged in a full-fledged Balkan air war, this one vastly different from Deliberate Force.

Operation Allied Force, conducted in the spring of 1999, constituted not a serious challenge merely to the United States, but also to the integrity of the NATO alliance, and to the relationship and use of air power as a force for coercion and ejection.[233] Undertaken with far greater political constraints and interventionist meddling than Desert Storm and Deliberate Force, it proceeded at a plodding and indecisive rate that threatened to make its acronym—OAF—all too appropriate. UN authorities were profoundly unwilling to use the leverage of air power (its representatives, on occasion, even going to bizarre lengths to avoid meeting with military authorities), and the then–SACEUR Gen. Wesley Clark repeatedly micromanaged and interfered with targeting and target selection.[234] Plans (later abandoned) to cavalierly use helicopters over Kosovo threatened to repeat the loss rates from dense antiaircraft fire experienced by airmobile forces during the Nixon-era Cambodian incursion. Task Force Hawk, the army's movement of fifty-five helicopters from Germany to the Balkans, took weeks of planning, twenty-two days of execution, and required disproportionate use of C-17 transports urgently needed for other purposes. It tripled in size, to 5,100 personnel—including Bradley Fighting Vehicles and Abrams tanks.[235] Afterward, an Army War College study concluded:

> The Army's deployment during the Kosovo crisis illustrated that force pro-
> jection remains its Achilles heel and underlines the fact that the remnants of
> the Cold War infrastructure, thereby, and doctrine remain. The Army has
> yet to make deployment a core competency as the Navy, Marine Corps, and
> Air Force have already done.[236]

As the crisis deepened, the stakes swiftly reversed: from whether Milosevic's hold on Kosovo and his regime would survive, to whether the NATO alliance would prevail and survive as other than a toothless tiger, an outdated Cold War relic. The growing crisis built up as NATO ministers gathered in Washington for the alliance's fiftieth anniversary, and it was only resolved when the NATO military leadership collectively persuaded its political overlords to invest far larger numbers of aircraft and loosen the ties hindering effective military action. Thus bolstered, NATO air power, spearheaded by the U.S. Air Force, subsequently shattered the military capabilities of the Milosevic machine, employing the full range of capabilities shown previously in Desert Storm and Deliberate Force, including the use of new weapons, such as the Joint Direct Attack Munition (JDAM [a GPS-guided bomb]), and the Predator remotely piloted aircraft. Under NATO air attack, Serbian forces in Kosovo were constrained, threatened, and intimidated. As one stated later, revealing the exaggeration of perspective commonly found among those exposed to the psychological effects of air attack:

> They knew everything about us. There wasn't anything they didn't know.
> If we lit a cigarette they could see it. God knows what they were dropping
> on us, all sorts of bombs. We didn't expect that intensity. We couldn't fight
> planes with mortars. And our antiaircraft guys couldn't do anything.[237]

In early June Milosevic folded. Kosovo gained its independence, and NATO's civil and military leadership breathed easier. Noted military historian John Keegan proclaimed:

> There are certain dates in the history of warfare that mark real turning
> points. Now there is a new turning point to fix on the calendar: June 3,

1999, when the capitulation of President Milosevic proved that a war can
be won by air power alone. . . . The air forces have won a triumph, are
entitled to every plaudit they will receive and can look forward to enjoying
a transformed status in the strategic community, one they have earned by
their single-handed efforts. All this can be said without reservation, and
should be conceded by the doubters, of whom I was one, with generosity.
. . . This was a victory through air power.[238]

But if it ended well, its Rolling Thunder–like beginning—targets personally
approved within the Oval Office, aping the worst aspects of Johnson-McNama-
ra-Bundy Vietnam-era micromanagement—profoundly disturbed professional
airmen, and even the land-centric Clark. Adm. Leighton Smith, commander
of all NATO forces during Deliberate Force, judged Allied Force "possibly the
worst way we employed our military forces in history."[239] The tortuous relations
among coalition partners and the UN made it, in the words of Benjamin Lambeth, "an object lesson in the limitations of alliance warfare."[240] Operational and
intelligence mistakes led to the loss of an F-117 stealth fighter from poor mission
planning and the bombing of the Chinese Embassy in Belgrade from a Central
Intelligence Agency error in target identification. Reflecting on the surprisingly
challenging campaign against Serbian air defenses, USAFE Commander Gen.
John Jumper stated that his greatest fear had been that "Milosevic would find a
way to float an SA-10 or SA-12 up the Danube River, put it together and bring
it to bear as a part of this conflict. If that had happened it would have profoundly
changed the balance of the threat and our ability to maintain air superiority."[241]
"I saw mistakes made in Kosovo in 1999 that I saw made in Southeast Asia and
North Vietnam in 1967," Lt. Gen. Michael Short, commander of Sixteenth Air
Force and Allied Air Forces Southern Europe, reflected two years afterward:
"[In Desert Storm] we used air power appropriately from the very first night. . .
. We did it entirely differently in Kosovo and Serbia, and I think we did it wrong.
I call this 'victory by happenstance' as opposed to 'victory by design.'"[242]

In the wake of Allied Force, emboldened by how air power (whatever the
problems in execution) had literally saved the credibility of the NATO alliance,
Ryan and Peters oversaw another reshaping of the air force vision, coinciding
with an update and extension of *Joint Vision 2010* undertaken by the J5 Strategy
Division on the Joint Staff. In June 2000, Joint Chiefs Chairman Gen. Henry
H. Shelton issued the revision, titled *Joint Vision 2020,* which stressed the need

for American forces to become "faster, lethal, and more precise."[243] Again, each service had its own responses, and that of the air force, carefully crafted by Maj. Gen. John L. Barry, whom Ryan had appointed director of strategic planning in February 2000, was *Vision 2020*. It stressed the air force mission ("To defend the United States and protect its interests through aerospace power") and emphasized the service's dependency on quality people "who live our core values," buttressed by enumerating the now-commonplace mantra of "Integrity First, Service before Self, Excellence in All We Do." But significantly this was relegated to the interior of the document. Instead, on its cover, it prominently featured a new tagline, *Global Vigilance, Reach, & Power*.[244] The grafting of *Global Vigilance* (defined subsequently as "the persistent, world-wide capability to keep an unblinking eye on any entity—to provide warning on capabilities and intentions, as well as identify needs and opportunities," clearly an allusion to the use of the space domain) onto the now-decade-old *Global Reach–Global Power* restored to the air force some of the unity of vision, directional vector, mission focus, and momentum it had lost with the passing of *Global Reach–Global Power*.[245] Simple, easily remembered, and comprehensive—like its predecessor—it remained a unifying vision for the service from the latter Peters-Ryan era, through the post-9/11 James Roche–Jumper era, and into the Wynne-Moseley era that followed, with some change in terminology, but none in spirit or underlying philosophy.[246]

FROM 9/11 TO?

On September 11, 2001, jihadist terrorists ascribing to an extremist form of Islamic fundamentalism seized four commercial airliners, flew two into the World Trade Center, one into the Pentagon, and crashed another into a farmer's field in Pennsylvania, thereby killing themselves and murdering three thousand people of various backgrounds, nationalities, faiths, and creeds who were doing nothing more offensive than trying to live productive lives. Lacking the means of creating or acquiring cruise missiles and strike aircraft, the terrorists turned America's vaunted commercial air transport system to its own purpose, transforming civil aircraft into deadly missiles and frustrating conventional notions of air power and air defense. The attack came on the first duty day of new Air Force Chief of Staff Gen. John Jumper and was heard by his predecessor, Mike Ryan, who was enjoying his first morning of retirement just blocks away. As the Pentagon shook and people died, a new and challenging era opened for America and its air force.[247]

Over the next two years, the air force engaged in three major counterterror actions—Operations Noble Eagle, Enduring Freedom, and Iraqi Freedom—and numerous smaller ones that sorely stretched its expeditionary abilities and saw its airmen deployed and fighting around the world. Noble Eagle, which began literally before the last airliner had crashed on 9/11, exemplified the "Total Force" contributions of the U.S. Air Force, Air National Guard, and Air Force Reserve to protecting the homeland. Enduring Freedom liberated Afghanistan from the Taliban and al Qaeda. Iraqi Freedom destroyed the Hussein regime in Iraq. All three continue.

The first focus for American military action was Afghanistan, with planning beginning the morning after the 9/11 attack.[248] As with Desert Storm in 1991, at the onset of post–9/11 operations, little consensus existed on what value air power might have in operations against the Taliban. Their leader, Mullah Mohammad Omar, warned the Bush administration against any military action. Comfortingly, when America did take action, it had the strong support of the international community. NATO, the Organization of American States, and the Australia, New Zealand, United States partnership invoked their treaty obligations to assist the United States, and 142 nations froze the assets of terrorists and their organizations. Of these, 136 offered military assistance, with 89 granting overflight or landing rights, and 16 eventually deploying forces in, above, or around Afghanistan.[249]

Conventional wisdom held that the United States would have to execute a long build-up phase, insert up to one hundred thousand troops, and then undertake a ground offensive to destroy the Afghan government in the spring of 2002, with air power playing at best a supporting and relatively minor role, and at worst proving a disappointing failure.[250] But the reality was considerably different. Operation Enduring Freedom opened on October 7. Mazar-e-Sharif fell on October 9, Kabul followed on November 13, and the Taliban surrendered their stronghold of Kandahar on December 7. The hard core retreated to the caves of Tora Bora, near the Pakistan border. Intensive bombing followed, and by December 18 this wild region was under the control of anti-Taliban and anti–al Qaeda forces. The focus of the campaign shifted from seizing control of the country to searching out the fleeing and disheartened remnants of the Taliban. Certainly much fighting remained, but the focus, thanks to air and space power and their synergistic effect on other military operations, had been shifted from the forcible entry of a hostile nation and the defeat of its forces to one of working with local authorities to ensure that stability could be restored to a now-liberated country. Afghanistan had been a "proving ground," President George W. Bush remarked on December 11, where American forces had shown that the combination of "real-time intelligence, local allied forces, special forces, and precision airpower" had "taught us more about the future of our military than a decade of blue ribbon panels and think-tank symposiums."[251]

Air power had once again confounded its critics, again largely through air force air power application. Air force combat controllers and special operations forces (SOF), deployed with local insurgents and equipped with modern communications, navigation, and designation systems, routinely called down GPS-cued JDAM bombs or other weapons from on-call orbiting strike aircraft, including B-52s. As in Vietnam and the Gulf War, the B-52 generated powerful psychological effects. "We call it 'grandma,'" Northern Alliance commander Abdullah Mohammed said, "because it's very big and dangerous."[252] B-52 strikes profoundly demoralized Taliban and al Qaeda forces; one journalist even noted how survivors "shuffled into view like zombies, living dead."[253] Within days of being exposed to air attack, Taliban and al Qaeda forces were "sensitized" so that they broke and ran for cover whenever aircraft appeared. In mid-November, the Taliban attempted to retake Tarin Kowt, a town that had fallen to anti-Taliban forces, by a mass assault of at least a thousand Taliban soldiers riding in pickup trucks. Special forces on the ground cued air support that destroyed numerous vehicles, leading an Afghani provincial governor to remark later, "When the Americans destroyed their vehicles and also killed their troops, we knew we could take Kandahar soon. It was easy. They fled."[254] When the Taliban fled Kandahar, one sympathizer confided to a reporter, "There were no Americans to fight. You only bombed. It would have been lovely if you came on the ground."[255] "It's time to face facts," Fareed Zakaria presciently noted, "American air power today is an amazing weapon of war. The combination of the information revolution and precision munitions has produced a quantum leap in lethality."[256]

The rapid collapse of overt Taliban–al Qaeda power in Afghanistan had exemplified something broadly hinted at by Desert Storm over a decade before: the power of precision air attack coupled with special forces action. Instead of having to rely on a traditional mass invasion of Afghanistan, the United States was able to topple the Taliban regime via the use of small numbers of SOF, coupled with joint-service air power forces incorporating the latest in technical advances, from satellite-based navigation, communications, and observation, to conventional and remotely piloted aircraft, working with indigenous forces, such as the Northern Alliance.[257] Summing up this broader lesson, Richard Cohen wrote:

> It is now absolutely clear that air power works. The evidence has been accumulating in recent years—the Gulf War, Kosovo—but it has taken the war against the Taliban to show just what can be done from the wild blue yonder. The use of air power coupled with proxy fighter—the Northern Alliance—has meant that American casualties have been minimal. We all have our fingers crossed on that one, but even when that changes—and it will—the zero casualty rate that stood for some weeks will still have been a major accomplishment.[258]

Such success masked the continuing need for good joint planning and execution to ensure the policing of the Afghan state. This need was dramatically illustrated by Operation Anaconda in early 2002, where a combined American-Afghani task force under the command of Maj. Gen. Franklin "Buster" Hagenbeck intemperately commenced a mountain assault against dug-in Taliban remnants without proper intelligence, planning, and coordination with supporting air elements.[259] Very quickly it faced potential disaster: enemy forces were five times stronger than estimated, and their air defenses were hindering the ability of helicopters to operate over the battle area, even helicopter gunships. The deteriorating ground situation eerily recalled Maj. Gen. Lloyd R. Fredendall's failure to employ air power properly not quite sixty years previously at Kasserine. Fortunately, Anaconda was saved by the heroism and improvisation of those who fought it on the ground and in the air. Though incorporated in emergency "after-the-fact" fashion, air power's ability to swiftly bring massive power to bear—"the greatest number of precision munitions dropped into the smallest geographic space in the history of air warfare"—prevented a replication of the disaster that had befallen II Corps at Kasserine.[260] It was—yet again—a reminder of the necessity of ground commanders' incorporating air into their planning and their "conops" before, not after, they commenced ground operations.

Operation Iraqi Freedom in 2003 removed the Hussein regime from power in Iraq. Although frequently portrayed in the media as an air campaign, and an example of a "shock-and-awe" strategy, in fact, it represented a very different approach than that followed by Norman Schwarzkopf over a decade previously. Gen. Tommy Franks, the U.S. Central Command commander, took more of a "land-centric" than "air-centric" planning approach. Even so, air proved of crucial importance. Iraqi Freedom benefited greatly from the Northern and Southern Watch operations that had constrained Saddam Hussein's ability to reconstitute his air defenses, from the intensive investment in precision weapons development over the decade since Desert Storm, and from the lessons (re-)learned in battlefield command and control—particularly the near-debacle of Operation Anaconda—from Operation Enduring Freedom. But there were some problems. Iraqi ground fire proved so intense that Army Apache attack helicopters had to suspend their operations; the army had placed the Fire Support Coordination Line so far ahead of friendly forces that it constrained the freedom of the air force to strike Iraqi forces; and weather conditions (in the form of a massive sandstorm) limited the ability of coalition forces to press their attacks. Here was where prewar investment in sensors, air- and space-based ISR platforms, and remotely piloted aircraft, such as the transport-sized Global Hawk, paid off. Despite the sandstorm, B-1s and B-52s armed with sensor-fuzed munitions destroyed multiple Iraqi T-72 tanks and other mechanized vehicles. As in Desert Storm a decade earlier, the use of air attack obviated the need for the kind of punishing force-on-force ground combat characteristic of classically mass-based Clausewitzian war.[261]

Operation Enduring Freedom and Operation Iraqi Freedom reaffirmed earlier lessons from Desert Storm, Deliberate Force, and Allied Force that, in the precision weapon era, air power forces, using cuing from air- and space-based platforms, or from SOF on the surface, are able to destroy an enemy's conventional military strength, rendering opposing forces helpless to prevent the introduction of friendly occupation forces. Those forces, which in older times had to fight for the "right" to enter enemy territory, are now, thanks to the leverage of air and space power, introduced intact into a nation to undertake "winning the peace." As the tortuous history of Afghanistan and Iraq since 2003 amply illustrates, the challenges for armies are no less significant, and the role for armies no less significant as well, than in previous times. Dealing with the problems of occupation, policing, and working with local forces to establish stability and cohesion in fractured societies requires different abilities and "skill sets" than the traditional ones expected of an army in classic conflict. It is a lesson some armies—the British Army, for example—learned years ago, in places such as Malaya, Cyprus, and Northern Ireland. It is a lesson that the U.S. Army is still learning today, in the demanding environment of Southwest Asia and the Persian Gulf.

Just over a century after the first military airplane appeared, air power, joined with the power of space-based systems, has repeatedly demonstrated its dominance in modern war. Frustrated in earlier eras by a lack of precision and ability to "see" the foe, it now benefits from advanced sensing and electronic technologies so that its accuracy and awareness are remarkable. It is not omnipotent, and it is not omniscient—but it is more so than any other form of military power and likely will remain so into the future.

In the real world of combat, air and space power have proven their value. But in the popular mind they are too often seen as support elements for traditional surface warfare. This view reflects, in large measure, the retained sense of loss and sacrifice that accompanied the First and Second World Wars. To many Americans, warfare will always be inextricably tied to images of punishing force-on-force land combat in the Second World War. Popular culture has reinforced this through land-centric films and television programs, such as *Patton, Saving Private Ryan, Band of Brothers, Enemy at the Gates, Wind Talkers, Flags of Our Fathers*, and (most recently) *The Hurt Locker* and *The Pacific*. As a consequence, that kind of warfare—brutal, costly, inconclusive, consistent with traditionalist assumptions and pre-1914 military experience—is seen as normative war. That such conflict involved victories bought at loss rates that a modern society is ill prepared to accept—fewer American servicemen and servicewomen's lives have been lost in combat since 9/11 than in the many island invasions and small battles in the Second World War—tends to be missed. That difference in casualties, of course, is the result of the leverage that modern air power brings. When Saddam Hussein moved confidently into Kuwait in 1990, he did so believing that he could inflict heavy casualties on any conventional force and that this expectation would deter attacks against him. Indeed, in the congressional debate that preceded the Gulf War, the question of casualties loomed large. No less than a former U.S. Army Chief of Staff—Gen. Edward C. "Shy" Meyer—estimated that it would take between ten and thirty thousand American casualties to dislodge Iraq from Kuwait, as did many lesser commentators and analysts.[262] And, without the air power advantage, he would likely have been proven correct.

Unfortunately, this cultural emphasis on "boots on the ground" has been accompanied by another problem, namely, the tendency to typecast the wars in

Iraq and Afghanistan as a virtual template for future war. Just as it would have been incorrect to use Vietnam or Desert Storm as a benchmark to forecast the nature of future war, so, too, is it incorrect to use the policing campaigns in Iraq and Afghanistan, and the war against terror generally, as a template through which to evaluate future conflict. While it is likely true that expeditionary conflicts, such as Iraq and Afghanistan, will proliferate in the future, nations have to be able to fight across the spectrum of conflict, from low-intensity conflicts through high-intensity ones, and everything in between. Prudently, even as America's military forces study counterinsurgency campaigns of the past, rediscovering forgotten truths from older doctrinal manuals, such as the Marine Corps' *Small Wars Manual* of 1940, those services study the requirements of high-tempo advanced technology war, whether it be army studies of theater ballistic missile defense, navy studies for new generations of high-speed vessels and aircraft carriers, or air force studies for reshaping nuclear forces, and future survivability requirements. Given the pace of force modernization in many nations, it is clear that this continued interest in high-tempo, high-technology war is justified. Certainly, as even a casual perusal of defense trade shows and of professional journals indicates, the international security environment is becoming more, not less, sophisticated. Systems that were revolutionary just a few years ago—for example, remotely pilot aircraft or precision weapons—are increasingly the common currency of big and small militaries alike.[263]

Several factors work against the fullest acceptance of the equality of air power forces, of which the following are but a sampling. First is the tendency to accept mass-on-mass brute-force war as "normative" war (accepting the individual soldier as a figure of little leverage), and the related tendency to "template" post–9/11 conflict as the form in which such conflict will occur. Second is what might be termed the appeal to classic military theory—Clausewitz in particular—which argues the primacy of the close fight and the mass clash of arms, accompanied by an impenetrable "fog and friction" of war that leads to "chaotic" and "unpredictable" conditions.[264] Third are the SOF and tactical air mobility elements of the air forces themselves, who are so much in daily service and support to the land-warfare community that they function little differently from their predecessors within the Army Air Forces in the Second World War. They do not, as a rule, share the passion for deep attack and air domination that characterize the fighter and bomber communities. Fourth is the investment issue. With most casualties occurring to ground warriors, it is only right that investment and equipment decisions be prioritized to address their needs. At the same time, basic questions about whether such forces should be put at risk—should be put in close proximity to the foe, where they effectively solve the enemy's "deep-attack" problem, not their own—remain too often unanswered or even unasked. As a consequence, it is easier to argue for and justify improvements to, say, armor and mechanized vehicles than to improve air-to-surface missile systems or new initiatives for advanced aircraft.

Regarding the investment issue, undoubtedly, air power projection forces (whether air forces, or naval aviation forces, or land-force air components) *are* expensive to develop and maintain. Each service, over the last three decades, has had problems meeting basic acquisition goals, schedules, and costs with its most treasured programs, some of which have suffered cancellation as a result. A "joint-service" list of failed, stretched-out, truncated, or costly aviation programs would read like a roster of modern military acquisition efforts. To list just four: the RAH-66 Comanche (Army), the A-12 Avenger (Navy), the F-22 Raptor (Air Force), and the CV-22 Osprey (Marines). As this is written, another program is entering the baleful light of critical scrutiny: the F-35 Lightning II Joint Strike Fighter. And these are just the American ones. The experience with foreign programs, such as the EADS A-400 airlifter program, shows that the American experience is not, alas, unique.

The acquisition process is one badly needing reform; there has been, over the last six decades, a tenfold increase in fighter aircraft development times. But, as well, these aircraft (and other mission types, such as bombers, ISR platforms, etc.) have the potentiality of decades, not years, of service. Upgraded periodically with advanced avionics, sensors, and weapons that keep pace with the (still-constant) electronic revolution typified by Moore's Law, there is no reason not to expect that they will be as relevant to future conflict as they are to present. A good example is the B-52, which has been, successively, a high-altitude nuclear bomber, a low-altitude nuclear bomber equipped with short-range attack missiles, a high-altitude cruise missile launch platform, and now a high-altitude bomber equipped with satellite-cued munitions—all over more than a half-century since its first flight. Unfortunately, given the long development times of some of these systems, the perception that they are "legacies" of outdated Cold War–era strategies and needs works to hurt their procurement, as does rising costs.

The cost issue is a frustrating one, for much of the cost escalation reflects the trimming of program size by decision makers (typically legislators seeking short-term fixes) who seemingly do not realize that much of a program's costs are "up front" in the purchase of the industrial base infrastructure, and the science and technology investment, that goes into the aircraft. For example, when the B-2 was conceived, production plans called for 132 aircraft with a final flyaway cost less than that of a 747 jet airliner. But when these numbers were cut to twenty (with one added subsequently) the cost per aircraft went to over a billion dollars. The F-22 has been the subject of continuous cuts and, consequently, rising individual aircraft costs. Assuredly, the F-35 will be as well.

One of the major downsides of acquisition controversies is the ill will they engender between the senior DoD leadership and the leadership of the service promoting the program. Within the United States, questions surrounding the ground rules of an air tanker competition led to fiery exchanges between Congress and the air force, whose secretary (James Roche) and chief of staff (Gen. John Jumper) had served the nation both honorably and well, and who each deserved great credit for air force performance in the war on terror. Roche, a retired navy captain, had even established a program, *Horned Owl,* that demonstrated in combat the viability of using a small RC-12 ISR platform to locate improvised explosive devices and frustrate those who planted them.[265] After first Roche, and then Jumper, left; continued disagreements between their successors and Secretary of Defense Gates over the tanker program and several others, coupled with some well-publicized problems with internal air force processes relating to the accounting and transfer of nuclear components and weapons, and the role of the chief in supporting public outreach, led to the removal of the next secretary, Michael Wynne, and the early retirement of Chief of Staff General Moseley—the latter a key architect of victory in the second Gulf War—even though they had identified the problems and put solutions in place. To airmen, their removal seemed to have far more to do with their outspoken advocacy of the F-22 and other advanced systems than with any accountability issues, and thus was widely perceived as reflecting a generalized bias in senior DoD circles against the air force (their perceptions were reinforced when Gates and his staff promptly set about dismantling and canceling a variety of air force acquisition programs and initiatives).[266]

In conclusion, it is impossible to predict the future of the USAF, other than to say that whether or not the service itself continues in its present form, the need for and desirability of an independent air force certainly will. It is interesting to note that, one hundred years after the first military airplane, the battle on the role, control, and value of air power and air power forces remains unde-cided. Even as combat experience has repeatedly taught the value and power of the independent air force, in various nations—Britain and America both being convenient examples—calls arise periodically to reverse nearly seventy-five years of practical experience and place air power within the hands of surface warfare traditionalists. Accompanying this is an often-cavalier minimization of the value of air power, which results in its too-easy dismissal. Too often, as in Mitchell and Trenchard's day, airmen simply have a problem getting a seat at the table of military decision making, and when they do get to the table, they have to watch what they say. In the modern joint world, alas, to advocate for air marks one in many eyes as a zealot. It is somewhat ironic that this is so, because one never has to seek an opinion on the value of air power from an airman—one only has to go to those who have been savaged by it, who know its strengths all too well. That is, in its own way, perhaps the greatest tribute one can pay to American airmen, and those of other nations, who took the genius of the early pioneers of flight and built upon it to redefine war in the twentieth and now twenty-first centuries. But it is, likewise, a lesson we need to teach within the joint-warfare environment, lest it be appreciated more by our foes than by ourselves, to our consequent (and mutual) misfortune.

3

ISRAELI AIR POWER

Itai Brun

The current form of the Israel Defense Forces (IDF) was largely determined in the 1950s and the 1960s. During these two decades, the nascent state of Israel articulated its security strategy and established the principles underlying the IDF's offensive operational doctrine. The vision of Israel's air power was shaped as well. According to the principles set forth at the time, air power is concentrated entirely in the Israeli Air Force; unlike other nations' military services, the IDF has no aviation capabilities in its ground forces or navy. The air force has responsibility for operating and equipping the force. Its assets include fighter aircraft, cargo aircraft, helicopters, and unmanned aerial vehicles (UAVs). The command and control over its units is centralized. However, the Israeli Air Force is not an independent service as the air force is in some other countries: the air force commander is subordinate to the chief of the IDF's general staff and the air force itself falls under IDF General Headquarters.

This chapter discusses the role of Israel's air power in the nation's overall military profile. It demonstrates that the Israeli Air Force has always been a central pillar of Israel's military power. Nevertheless, the chapter also explains why, in accordance with Israel's military strategy, the air force was perceived until the last two decades as an entity whose task was mainly to defend the nation's airspace and support the ground forces and the navy in their missions. Even so, the uses to which the political and military leadership put the air force throughout that time were daring, imaginative, innovative, at times even virtuosic, and often deviated from Israel's overall military doctrine. The chapter goes on to describe

137

how this occurred both during and between Israel's wars, and how the air force dealt with various challenges throughout the years. The chapter's overall thesis is that the Israeli Air Force's development was linked mainly to those challenges, and that its current form, as demonstrated during the conflicts of the past decade, derives from this continuing evolutionary process.

THE WAR OF INDEPENDENCE (1948–49)[1]

The structure of the Israeli Air Force and its role in Israel's overall military strategy resulted from a series of decisions the political and military senior leadership made during the first two years following the establishment of the state of Israel and the IDF. The Israeli Air Force was founded during the War of Independence, on May 16, 1948, two days after the state of Israel itself and one day after the first of a series of attacks carried out by Egyptian Dakota and Spitfire fighter planes against targets in the heart of the Jewish state.

The Israeli Air Force grew out of the small-scale air service that operated in the framework of the Haganah organization during the rule of the British mandate in Palestine and that to a large extent constituted the nascent Jewish state's air force. The air service had no fighter aircraft, and only a few days after its foundation did it receive any such aircraft: the Avia S-199, a Czechoslovakian version of the German Messerschmitt Bf 109. Their assembly within a few days enabled the Israeli Air Force to perform its first combat mission, which occurred at the end of the month. On June 3, an Israeli fighter pilot, Mordechai "Modi" Alon, shot down two Egyptian Dakota aircraft that had attacked targets in the Tel Aviv area—the IDF's first shootdowns of enemy aircraft.

Israel's War of Independence was prolonged and resulted in extensive loss of human life (approximately 6,000 casualties out of a population of 650,000). The conventional wisdom is that the inexperience and small size of the Israeli Air Force prevented it from being a significant factor during this war. Nevertheless, the air force did achieve a few successes. For example, except for the first month of the invasion by Arab armies following Israel's declaration of independence, the skies of the Jewish state were relatively protected. Even though Arab aircraft infiltrated Israel's airspace and attacked targets in its territory, they did not do so systematically and their influence on the outcome of the war was negligible. In some major operations, the Israeli Air Force even assisted the ground forces in their missions.

The Israeli Air Force contributed more significantly through cargo and transport missions. It carried munitions to Israel as well as supplies and weapons to forces that were cut off from the center of the country. In one unique operation, the air force succeeded in leading, through the air, a battle-weary brigade and replacing it with a fresh one.

These achievements did not impress the senior military leadership, who attributed only a small role in the outcome of the war to the air force and were mostly skeptical of its capability. Indeed, once the War of Independence was over, the Israeli Air Force's condition worsened. More than seven hundred foreign volunteers (pilots, technicians, and other professionals—Jews and non-Jews alike), who had constituted its main operational force during the war, left the air force. It lacked an operational doctrine, and the command and control system was inefficient.

The Israeli Air Force's organizational structure resulted from the requirements of the war and was based on a large variety of aircraft, as purchases during the war reflected availability rather than deliberate planning. An attempt to restructure the force encountered a slew of problems, the main one being the limitations that the superpowers imposed on the sales of munitions to the new Jewish state. In July of 1950, about a year after the war ended, the Israeli Air Force's inventory included 173 different types of aircraft, the majority of which were unusable. The force had only fifteen combat-ready fighter-bomber aircraft in addition to forty-five aircraft that were used for instructional purposes.[2]

Under these circumstances, it is no wonder that the Israeli Air Force's short-lived struggle to gain independence ended in failure and led to the resignation of its second commander, Aharon Remez, a former pilot in the Royal Air Force (RAF) who had served since July 1948 (the first commander, Israel Amir, held office for about two and a half months). The new commander, Shlomo Shamir, was an infantry soldier who had a civilian pilot's license but no experience in military aviation. His successor, Haim Laskov, was a famous ground force officer but lacked aerial experience. Both Shamir and Laskov were appointed in order to subordinate the Israeli Air Force to the IDF General Staff and to thwart the air force's ambition to become conceptually and organizationally independent.

In 1953 Dan Tolkovsky, who had been an RAF fighter pilot during the Second World War and held a number of Israeli Air Force staff positions, was appointed the force's commander. Tolkovsky served as commander until 1958, and the relationship between the air force and the IDF's general staff was shaped during his five-year tenure. While the model developed and formalized during those years provided the Israeli Air Force with a large degree of independence as well as direct access to policymakers, the air force remained subordinate to the general staff and to the overall military strategy.

In retrospect, it is easy to discern the structural logic that governed the Israeli Air Force and its adaptation to the IDF's operational doctrine that evolved during its first decades. This doctrine viewed ground force maneuver as the main component of a military decision (*Hachra'a*) and emphasized the idea of moving the battlefield to the enemy's territory. This emphasis stemmed from the IDF's lack of strategic depth and the need for a fast, visible, and clear victory in every war. The central position that ground force maneuver occupied in this doctrine also stemmed from the perception that dynamic maneuver battle, in which conditions change constantly, would enable the IDF to take advantage of the high quality of its commanders and offset the quantitative advantage of its enemies.[3]

The importance of maneuver and of ground forces did not contradict the priority given to developing the Israeli Air Force. Since its inception, the force has been considered a key component of Israel's military power and received a generous amount of resources. The concept of combat-ready, flexible air force with a regular service, capable of assisting ground forces in defensive as well as offensive operations, and achieving air superiority were among the most basic building blocks of the Israeli defense strategy. This imperative originated in the ideas of David Ben-Gurion, the founding father of the Jewish state and the main designer of its national security strategy. From the beginning the leadership recognized that the Israeli Air Force also had a part to play at a strategic level. This role was mainly identified with the capability to attack national, military, and civil infrastructure in enemy territory. Along with the implications of such a capability in relation to Israel's deterrence potential, the young Israeli Air Force had even tried to turn this potential into reality in five attack missions on Arab capitals (Amman, Damascus, and Cairo) in June and July 1948, but these attacks produced no significant results.

However, David Ben-Gurion explicitly rejected the possibility that wars could be decided from the air, and during the debate concerning the air force's independent status he favored the opinions expressed by the chief of the general staff, Yigael Yadin. Nevertheless, in hindsight, it seems that his own insights helped create a situation in which the problematic status of the Israeli Air Force changed significantly in the following years.

THE SINAI WAR (1956)[4]

In the first years after the War of Independence, the Israeli Air Force still suffered from technological inferiority in relation to the air forces of the Arab nations, which already possessed supersonic jet fighters. The supersonic age only began for the Israeli Air Force in 1953, when Britain agreed to sell Israel fifteen Gloster Meteor jet fighters. The most important developments occurred in 1955, the year preceding the Sinai War. Cooperation with France became closer and ushered in the "French era" of the Israeli Air Force, which lasted until after the Six Day War. France included Israel in the development of the most advanced military equipment and allowed it to make limitless purchases as long as it could pay the price. The jewel in the crown was the Israeli Air Force's purchase of sixty-one Dassault Mystère IV fighter planes as well as thirty Dassault Ouragan jet fighter-bombers. In addition, new radar stations, as well as munitions and additional weapons, were purchased. On the eve of the Sinai War, the Israeli Air Force order of battle consisted of 221 fighter-bombers, of which 115 were supersonic jet fighters. Of this total, 117 aircraft were usable for combat purposes and half were new supersonic jet fighters.

The Sinai War, secretly coordinated between Britain, France, and Israel, was the first significant challenge the IDF had to face after the War of Independence. Israel entered this war because of fears regarding the change in the Middle East's balance of arms, particularly in light of information concerning Egypt's plans to purchase a large number of munitions from Czechoslovakia and a wish to remove the limitations Egypt imposed on Israel's freedom of navigation. Israel's leaders also clearly recognized that the retaliatory actions Israel had carried out in the first half of the 1950s had failed to reduce terrorist activities at the nation's border. The military and political leadership in Israel also had underlying concerns about the possible collapse of the balanced deterrence created at the end of the War of Independence, the growing power of Egypt's charismatic leader Gamal Abdel Nasser, and the political constellation enabled following the nationalization of the Suez Canal in July 1956. The Sinai War was also the first significant operational challenge to the new Israeli Air Force in the supersonic age. By that time, the air force was led mostly by officers who had risen through its ranks.

The war started on October 29, 1956, when a brigade of Israeli paratroopers parachuted into the heart of Egyptian territory in Sinai from sixteen Dakota aircraft protected by different jet fighters (Meteor, Ouragan, and Mystère). Preceding this major operation, the Israeli Air Force had carried out two special aerial operations. The first was an attempt to shoot down an Ilyushin Il-14 aircraft in which the Egyptian chief of the general staff and other senior officers were about to fly, using a Meteor night-interceptor aircraft. In fact, the Israeli Air Force succeeded in intercepting and shooting down the aircraft, but it turned out that the general was not on board. The second and more successful operation, carried out by P-51 Mustang aircraft, disconnected the telephone lines in Sinai by tearing the upper cables.

This was not how the Israeli Air Force wished to start the war. The doctrine that had evolved in the air force before the war in Sinai was based on the concept of a preliminary surprise strike on the enemy's airfields, even though the force had not yet developed the full capability needed to execute such a strike. This operational plan was even distributed to bases and squadrons, and the air force was prepared to implement it. However, its role was limited in this war as well, since political coordination with Britain and France resulted in a situation where French squadrons protected the Israeli skies and the force concentrated mainly on CAS and interdiction missions. This situation reflected, first and foremost, the priority given to political considerations (coordination with Britain and France) over military decisions. It also expressed the militaries' and political leaders' considerable mistrust of the Israeli Air Force.

The Sinai War had a complex outcome. During the war, the IDF achieved great military success and conquered the Sinai Peninsula in just four days. The Egyptian forces retreated hurriedly, leaving behind thousands of soldiers, who became prisoners of war, as well as large amounts of military equipment. But, at the end of the war, Britain, France, and Israel had to accept the conditions imposed on them by the superpowers, the United States and the Soviet Union, which forced a cease-fire and a withdrawal from the occupied territories. Thus, the war did not lead to a strategic change in the policy of the Arab countries, which continued to call for the destruction of the state of Israel. Nevertheless, the IDF's success in the Sinai War helped to increase Israel's deterrence and resulted in a relatively quiet decade for its borders.

The Sinai War increased the trust of the senior military leadership in the Israeli Air Force despite the limited role the force had played in protecting the Israeli skies. After the air force shot down seven enemy aircraft during the war the fears concerning its air-to-air capability were dispelled to a certain extent. Success in CAS missions shed light on the role and the importance it would have in overall military doctrine; during the war the Israeli Air Force carried out 489 attack missions in which 308 vehicles and 22 tanks were hit. The air force's success in parachute and supply operations and in the evacuation of wounded soldiers also contributed to the overall perception regarding its role.

Indeed, following the Sinai War, we can see the high priority accorded to the Israeli Air Force in the allocation of resources. Between 1957 and 1967, the overall acquisition budget (excluding domestic development) reached $581.1 million. Of this sum, $385.7 million (approximately 66 percent) was allocated to the Israeli Air Force, compared to the budget allocated to the ground and naval forces, which, on average, reached 28.5 percent and 5.3 percent, respectively.

FROM THE SINAI WAR TO THE SIX DAY WAR[5]

The IDF's offensive doctrine was formed in the decade following the Sinai War and, to a large extent, in light of the lessons learned from that war. Abstract concepts, especially using ground force maneuver and taking the war to the enemy's territory, that emerged prior to the war were practiced during the war. To ensure maneuverability, the IDF was structured as a mobile and mechanized army that could operate quickly and uninterruptedly, drawing on appropriate logistical support. After the Sinai War, the armored corps bore the burden of taking the battle to enemy territory. The central role of ground force maneuver and the dominance of the armored corps also had a direct influence on the composition of the senior military leadership, which consisted mostly of ground force officers.

Approximately two years after the Sinai War, Ezer Weizman replaced Dan Tolkovsky. Like Tolkovsky, Weizman was a fighter pilot who had graduated from an RAF flying course and had held a number of positions in the Israeli Air Force. Weizman remained at the head of the force throughout the decade preceding the Six Day War—a decade during which the air force's capabilities improved dramatically. The most conspicuous manifestations of progress were the ongoing improvements in its aircraft, in terms of speed, flight range, armament-carrying ability, versatility, and day- and night-time capability under relatively harsh conditions. These improvements were accompanied by advances in the aircrafts' equipment: air-to-air missiles, electronic weapon systems, and communication. During the Sinai War, the Mystère was the frontline aircraft in the order of battle, yet a large proportion of the order of battle still relied on piston aircraft. At the end of the war, the Israeli Air Force completed the process of converting to exclusively supersonic jet fighters. This process continued for the rest of the 1950s; by 1961, the air force order of battle consisted only of French-made supersonic jet fighters. In April 1962 the first two Dassault Mirage IIIC aircraft landed in Israel and brought in the age of Mach-2 flight, with airborne radar systems as standard military equipment. Through the end of 1964, France provided Israel with seventy-two aircraft of this type.

This decade also marked the emergence of the Israeli Air Force as the heart of the evolving military doctrine. Decision makers were fascinated by the availability and flexibility of air power. Indeed, a basic principle of the air force was its reliance on regular active duty forces not only as combatant units of the aircrews, but also in all other units. This principle was diametrically opposed to the practices governing the structure of the IDF, which relied on reserve forces. During these years, doctrine regarding the combat readiness of the air force evolved and was reflected in a corresponding state of alertness. When Egypt surprisingly moved its forces to Sinai in February 1960, known as the Rotem alert, the Israeli Air Force stood out as the only military force that could take action on short notice.

Even so, the importance accorded to the Israeli Air Force was, first and foremost, related to the fear that an Arab aerial attack on the Israeli home front would cause serious damage and interfere with mobilization of the reserve soldiers who had become the IDF's main offensive force. The increase in the Arab air forces' offensive capabilities intensified these fears. At the end of the Sinai War, the Egyptian Air Force underwent a reorganization during which many airfields were built and the aerial order of battle came to include the best Soviet weapons available—MiG-17, MiG-19, and MiG-21 jet fighters. The number of MiG-21 jet fighters grew, and on the eve of the Six Day War, Egypt had approximately one hundred jet fighters equipped with AA-2 (Atoll) Soviet air-to-air missiles. At the beginning of the 1960s, Egypt acquired its first Tu-16 jet fighters, which could carry 10-ton bombs. In 1963 the Soviet Union began supplying Egypt with SA-2 surface-to-air missiles (SAMs); by the time of the Six Day War, Egypt had approximately thirty batteries of such missiles that protected strategic facilities and airfields. The Syrian Air Force grew stronger as well, though to a lesser extent. On the eve of the Six Day War, the Syrian order of battle comprised sixty MiG-21 jet fighters. The Jordanian Air Force remained small, however, and was based on the British Hunter aircraft.

These developments led Israel to formulate a doctrine that identified air superiority as the most significant contribution that its air force could make to achieving a military decision. Regarding how to gain air superiority, the Israeli Air Force adopted a conspicuously offensive approach. At its heart was the concept of destroying enemy air forces on the ground. Attack on airfields, therefore, became the air force's top-priority mission, and nearly all issues that did not touch on this mission were neglected or given a low priority (except air-to-air capability). At the beginning of the 1960s, the IDF's general staff agreed with this basic doctrine, which stated that it was essential to open every military campaign with a preliminary surprise attack on the enemy's airfields.

The operational plans for attacking the enemy's airfields changed throughout the decade that preceded the Six Day War. In 1957 the Israeli Air Force decided to attack the runways at the airfields—not only the aircraft. This decision necessitated fundamental changes in both munitions and operational doctrine, and put great emphasis on shortening the interval between a jet fighter's landing and subsequent takeoff for another mission. This was viewed as the main way to reduce the overall time needed for achieving air superiority. Additional changes in this plan were required in light of developments in the Arab countries' orders of battle and the deployment of their forces. In this context, the Israeli Air Force had to consider particularly the distance to some Egyptian and Syrian bases, as well as the antiaircraft artillery and SAMs protecting the bases.

Until 1964 the Israeli Air Force was not involved in significant operational activities. Israel was engaged in border skirmishes, but avoided using its air force believing it could lead to an undesired escalation. After a heated debate, the Israeli government decided to deviate from this policy in March 1962, but realization of this policy was postponed until November 1964, when aircraft were used during a confrontation between Israel and Syria; Israel had paved a road on its side of the border with Syria, which responded by launching artillery attacks on northern settlements. The use of aircraft was perceived to be very successful, both in the tactical and strategic aspects. Contrary to initial fears, it did not lead to escalation, and did not harm Israel's international status. Following these events, the Syrians stopped attacking northern settlements and instead focused on hampering the regular activities of the National Water Carrier of Israel next to the border and assisting terrorist activities inside Israeli territory.

In April 1966, after eight years of heading the force, Ezer Weizman was replaced by Mordechai (Moti) Hod, the first Israeli Air Force commander who had received his pilot's wings as a member (he had taken a short course to supplement a civil aviation course he had completed prior to the establishment of the state of Israel). His first year in office was especially tumultuous. In July 1966, when the IDF failed in its attempts to attack and destroy Syrian equipment aimed at diverting the water of the river Jordan, the air force was called into action. During this operation one of its Mirage jet fighters shot down a MiG-21 jet fighter for the first time. Aerial operations continued sporadically until May 1967, which improved the air force's skills in surface-to-air as well as air-to-air activities; in the year preceding the Six Day War, the force shot down eleven enemy aircraft. In August 1966, a MiG-21, flown by an Iraqi pilot, was forced down in Israel as a result of a joint operation between Mossad and the Israeli Air Force, allowing the latter's pilots and planners to discover its secrets. These developments dramatically improved the force's capability leading up to the Six Day War.[6]

THE SIX DAY WAR (1967)[7]

Israel's sense of isolation increased significantly in 1966–67, eventually resulting in the Six Day War. The actual escalation began in the middle of May 1967, when Egypt's ruler, Nasser, informed the United Nations (UN) forces that they had to leave the stations they had occupied since the end of the Sinai War. During this period, Egypt once more blockaded the Straits of Tiran (whose opening was one of the few achievements of the Sinai War) and moved its forces to Sinai. These actions resulted in a three-week period of heightened alert and waiting during which the IDF prepared itself for war. Israel initiated hostilities on June 5, 1967.

At the beginning of the war, the Israeli Air Force order of battle consisted of 211 jet fighters organized into 10 squadrons: 65 Mirage 3 jet fighters, 35 Super Mystère jet fighters, 19 Vautour jet fighters, 51 Ouragan jet fighters, and 41 Fouga Magister aircraft used for instructional purposes and converted into combat fighters. The air force took advantage of the period of high alert to update its operational plans, train its aircrews, and recruit reserves (pilots and maintenance personnel). This preparatory period enabled it to utilize most of the order of battle, and during the first wave of attacks the majority of the air force's jet fighters (85 percent) took part, including the old Fouga jet fighters.

During the Six Day War, the IDF's original operational doctrine was put into practice and proved a great success. The importance of the Israeli Air Force was highlighted by its preliminary air strike on the first day of the war (Operation Moked), in which it attacked enemy aircraft on the ground. The takeoffs were timed so that all the attacking jet fighters would arrive at their destinations at exactly the same time. To maintain the element of surprise, the aircraft flew at low altitude to avoid being tracked by the radar systems. They also maintained complete radio silence. At 0745, the air force jet fighters arrived simultaneously at eleven Egyptian airfields, attacking them using the Israeli-developed antirunway bombs. Egyptian aircraft that could not take off were destroyed on the ground. In the first wave of attack, 189 aircraft were destroyed and the Egyptian Air Force was severely damaged. Toward 1200 on the first day, the Jordanian and Syrian air forces joined the war and attacked targets in Israel. A short time following the first wave of attack, the Israeli Air Force struck these air forces as well. By 1200, the Israeli Air Force had gained absolute air superiority in the skies of the Middle East.

From the beginning, Operation Moked was intended to achieve air superiority, and it reflected, first and foremost, Israel's considerable fear of the Arab air forces' offensive abilities. Nevertheless, in retrospect, the operation had much broader implications. The chief of the general staff, Yitzhak Rabin, later stated that the outcome of Operation Moked determined, to a large extent, the outcome of the entire war. The operation resulted in Israel's total air superiority in its own skies and the entire airspace for the duration of the war, which allowed the Israeli Air Force to direct many of its resources to assisting the ground forces in carrying out their maneuvers. By contrast, the military and political leadership in the Arab countries was in a state of confusion.

During the six days of fighting, the Israeli Air Force destroyed 451 aircraft belonging to the Egyptian, Syrian, Iraqi, Jordanian, and Lebanese air forces out of 600 jet fighter–bombers and bombers at their disposal on the eve of the war. Eighteen additional aircraft were destroyed by accidents and antiaircraft weapons. Of this total, 391 aircraft were destroyed on the ground (376 of them during the first day of the war) in an attack on 28 airfields, and 60 aircraft were shot down in air-to-air combat. The Egyptian Air Force incurred the greatest losses—approximately 80 percent of its aircraft.

Over the six days, the Israeli Air Force carried out 2,400 air strikes on Egyptian, Syrian, and Jordanian ground forces. During these operations, approximately 850 tanks and infantry fighting vehicles were hit, and 144 cannons, 1,670 vehicles of different types, and 162 bases and air force facilities were destroyed. Israel's ground forces could move unimpeded. The Israeli Air Force paid a heavy price for these achievements. Twenty-seven pilots were killed (fifteen during the first day of the war), seven were captured, and eighteen were wounded. Sixty-nine jet fighters (approximately 25 percent of the overall jet fighter order of battle) were hit during the war.

The Israeli victory was quick and decisive. Within six days, the Egyptian Army was crippled and had almost ceased to exist as a fighting force. Nasser, whose inflammatory rhetoric and political decisions contributed significantly to the ongoing escalation before the war, was a broken man. He resigned his position and his minister of war committed suicide. The Jordanian Army suffered similar damage, and Jordan lost Jerusalem and the West Bank. The Syrian Army, whose provocative actions toward the IDF in the years preceding the war had also contributed to the escalation, was severely damaged and Syria lost the Golan Heights, which it had regularly used to threaten settlements in Israel for nineteen years. Israel now occupied an area three times the size of its territory on the eve of the war and controlled the entire Sinai Peninsula up to the Suez Canal.

The Six Day War was a formative event regarding the Israeli Air Force's array of helicopters. At the time of the Sinai War, the force possessed only four helicopters (Healer 360s and Sikorsky S-55s). During the decade between the Sinai War and the Six Day War, the helicopter order of battle grew significantly and reached forty-five helicopters (S-55s, S-58s, and Aérospatiale Super Frelons). This increase was influenced by the way the French forces had used helicopters during the Sinai War and the conflict in Algiers, which produced lessons learned by a joint delegation of the paratrooper corps and the Israeli Air Force. By the end of the 1950s, the general staff had decided that the air force would operate the helicopters, but its operational doctrine would be developed together with the paratrooper corps. Special relationships formed between the helicopter squadron and the ground forces, especially with the elite unit established at that time (*Sayeret Matkal*). These relationships contributed to the development of an airborne commando theory that developed and matured in a series of operations before the war. For example, a drill performed in 1966 demonstrated the helicopters' ability to land a team of paratroopers deep in enemy territory. During the war itself, a series of helicopter operations took place; the most significant was the raid at Umm Qatf in Sinai, which was headed by the division commander, Ariel Sharon.

The Israeli Air Force's achievements in the Six Day War marked, to a large extent, the successful conclusion of the process that had formed and structured it since its inception. The air force, which could not demonstrate its capabilities during the War of Independence, had grown in power and turned into a state-of-the-art fighting machine. The roots of its development can be found in the vision of its founding fathers, who understood the importance of air power and allocated the proper resources to structure the air force. This vision enabled its transformation from a group of light aircraft and a team of brave and brash volunteers into an efficient and modern air force.

The air force's success in the war made it a very significant factor in Israel's overall military power. Faith in the force's ability to carry out its missions grew significantly among the political and military leadership as well as the wider public. The war strengthened the status of the Israeli Air Force and the armored corps, which were recognized as the forces that produced the final military outcome. Naturally, the largest proportion of resources (approximately 48.3 percent of the security budget) was allocated to the air force.

At the same time, a more profound examination shows that the Israeli Air Force's operational success did not change the original doctrine that viewed ground force maneuver as the dominant factor leading to a military decision. The extent of Israel's victory also brought about a situation that marked a key turning point in the threats that the air force would face in the coming years.

After the war, France placed an embargo on the provision of munitions to Israel, and the United States became Israel's sole supplier of munitions. At the end of 1967, the Israeli Air Force received the first Douglas A-4 Skyhawk and Bell-205 helicopters, but the most important breakthrough in its capabilities was made with the acquisition of McDonnell Douglas F-4 Phantom jet fighters in 1969. For the first time, the air force had an advanced, two-seater, twin-engine, multipurpose jet fighter that could carry large loads of weapons, had an efficient radar, and possessed missile-guidance systems. The Israeli Air Force's helicopter units were also boosted with the arrival of the Sikorsky CH-53 helicopters. The cargo units received the Boeing 377 Stratocruiser aircraft and the first Lockheed C-130 Hercules aircraft. During this time, the Israeli Air Force removed the Vautour, Ouragan, and Mystère aircraft from its order of battle.

THE WAR OF ATTRITION (1969–70)[8]

The defeat in the Six Day War caused a military, political, and social upheaval in the Arab countries. Yet, contrary to Israel's expectations, the extent of the Arab defeat did not lead to a cessation of belligerent activities and border violence. In all areas, especially on the border with Egypt along the Suez Canal, a genuine war took place throughout Israeli territory: the War of Attrition. The war actually started in September 1968 with massive artillery bombings of Israeli forces stationed by the Suez Canal. The number of bombings increased gradually in the following months. On March 3, 1969, Egyptian president Gamal Abdel Nasser officially declared that the cease-fire with Israel had ended, which marked the official beginning of the War of Attrition. In April the Egyptians deployed two SA-2 rocket systems to the Suez Canal area, limiting the Israeli Air Force's activities there.

The Arab rationale underlying the War of Attrition reflected a deep understanding of the limitations of Israel's forces to handle a protracted war with many casualties. The Egyptian Army adopted a new offensive doctrine that included commando raids, artillery bombing, and sniper attacks. This type of combat forced the IDF into a static mode of combat, which was inherently different from its original operational plans. The Egyptian Army had an advantage in the number of artillery systems, and the IDF, whose major firepower was traditionally concentrated in the Israeli Air Force, lacked an adequate response to this threat.

The increase in the power of the Egyptian and Syrian air forces manifested itself mainly in defensive moves. The number of Arab airfields grew significantly and a few hundred aircraft shelters were built. These elements of the force design stemmed directly from the events of the Six Day War and led to a situation in which Israel's ability to destroy aircraft on the ground was completely lost. At the same time, the Egyptian and Syrian air forces shifted the emphasis from jet fighters to SAMs.

In July 1969 the Israeli Air Force entered a period of offensive activity along the Suez Canal in order to neutralize the Egyptian artillery, reduce the number of Israeli casualties, and send a message regarding Israel's intention to escalate the situation in a manner that would demonstrate the perils of war to Egyptian leadership. Although this aerial activity inflicted damage and losses on the Egyptians, it could not bring about a cease-fire. This state of affairs led to nineteen Israeli operations that took place deep in Egyptian territory. The intent behind these attacks was to reveal the Egyptian military's helplessness and cause a rift between the Egyptian people and their leaders, resulting in a weakened leadership and regime change, and bringing an end to the war. The targets attacked included military bases, military factories, warehouses, radar stations, and SAM systems.

These attacks exposed the Egyptians' weakness, but did not achieve Israel's objective. In fact, they only led to an increase in Soviet assistance to Egypt and the deployment of a protective Soviet air defense system that included interception jet fighters, such as MiG-21, as well as SAM systems, such as the SAM-2 and SAM-3. The arrival of the Soviet SAMs resulted in a fundamental change, since the missiles shot down several jet fighters, including five of the new Phantom jets, which caused the Israeli Air Force to doubt its equipment and strategy. The SAMs forced Israel to change the targets of its air strikes from large cities to the Nile delta, and then to stop these attacks altogether. This series of events provoked a deep crisis in the Israeli Air Force and expedited the end of the war.

At the same time, an ongoing combat took place between Israeli and Egyptian (and later Soviet) jet fighters, as well as a series of airborne commando operations deep in Egyptian territory. The Israeli Air Force completely dominated these air battles and reached an unprecedented ratio of shooting down seventy-three enemy jet fighters at the cost of three of its jet fighters (24:1). The force achieved these results through meticulous planning and by taking advantage of real-time intelligence and control systems that evolved throughout the War of Attrition. These systems enabled the Israeli Air Force to plan interception operations and forced Egyptian fighter pilots to use tactics that enabled Israeli dominance to prevail. On July 30, the air combat reached its peak in operations that included a well-planned ambush and ended in five Soviet jet fighters being shot down; the Israeli Air Force did not incur any losses.

The special operations deep in Egyptian territory were basically intended to serve a purpose similar to that of the in-depth air strikes: to instill fear, confusion, and uncertainty among armed forces and civilians alike. When Israel acquired CH-53 helicopters in October 1969, the flight range as well as the carrying capacity of the Israeli Air Force's helicopter order of battle increased significantly and new areas for military operations became available. Overall, helicopters made approximately six hundred sorties in all types of missions. In one of the special operations, a special force, flown into Egypt in a Super Frelon helicopter, took over an Egyptian radar station; the soldiers took down the P-12 radar and it was airlifted by a CH-53 helicopter to Israel.

The war eventually ended on August 7, 1970, following diplomatic pressure from the United States. All in all, the Israeli Air Force had made 10,520 jet fighter sorties, of which 8,200 were for offensive purposes (more than 50,000 bombs were dropped during these operations) and 1,500 were intercept sorties. The allocation of jet fighter sorties to in-depth strikes was very limited, totaling only 120 sorties. After the war, and in violation of the cease-fire agreement, the Egyptians deployed SAM systems to the Suez Canal.

At a cursory glance, the way in which Israel used its air power, and especially the emphasis on in-depth strikes, can be seen as the first indication of the concept of turning the air force into a main force for achieving a military decision. Nevertheless, a more thorough examination shows that these strikes did not rest on a coherent alternative to the doctrine stating that military decisions would be achieved by ground force maneuver. Instead, they were primarily a manifestation of the uncomfortable situation in which Israel found itself when for the first time it had to deal with a confrontation conducted according to principles that did not tally with the IDF's original operational doctrine.

THE YOM KIPPUR WAR (1973)[9]

The War of Attrition ended before the Israeli Air Force could develop a solution to the "SAM problem." This situation had worsened by the time the Yom Kippur War broke out. In the years preceding the war, the air force focused its attention and efforts on countering SAMs, yet it could not identify their weaknesses. At the beginning of the war, the Israeli Air Force faced closely spaced SAM batteries, which included approximately 180 mobile and fixed batteries at the Egyptian and Syrian fronts, and was fundamentally wrong in its assessment of their specifications and operational doctrine. It was to pay a heavy price for this misguided assessment and these information gaps.

In May 1973, Mordechai Hod, who headed the Israeli Air Force during the Six Day War and the War of Attrition, was replaced by Binyamin (Benny) Peled, the first commander to have taken the force's complete pilot's course. A trained aeronautical engineer, Peled had commanded Israeli Air Force squadrons and bases, but in the years preceding his appointment, he dealt with the development of new weapon systems.

On the eve of the Yom Kippur War, the Israeli Air Force order of battle consisted of 380 jet fighters: 125 Phantoms, 185 Skyhawks, 45 Mirage jet fighters, and about 25 improved Super Mystère jet fighters. The helicopter order of battle included seventy-six Sikorsky CH-53 helicopters as well as Super Frelons and Bell-205s. The aircraft-strength combat readiness percentage stood at eighty.

The Yom Kippur War constitutes a classic case study of strategic surprise. Until the outbreak of hostilities at 1350 on October 6, 1973, the most senior military and political decision makers in Israel estimated that the probability of a war between Israel and the Arabs was very low. This assessment was based on an intelligence concept that had two components. First was a prevalent belief that Syria would not initiate a war on its own, but would only do so together with Egypt. Second was the belief that Egypt would not launch a war until it had achieved a significant ability to attack the Israeli heartland. The assessment remained unchanged even after intelligence revealed several activities that demonstrated a heightened military alert in Egypt. In light of the overall estimate, these military moves were explained as being related to a major exercise under way in Egypt and a possible Syrian response to a large air battle on September 13, 1973, in which Israel had shot down twelve Syrian jet fighters.

This intelligence assessment, which would probably have been accurate at the time of Nasser, completely failed to detect the fundamental change that the Egyptian strategy had undergone upon the rise to power of Nasser's successor, Anwar el-Sadat. Sadat articulated a different strategic concept based on the idea of waging war in order to achieve limited military objectives that corresponded to the abilities of the Egyptian Army and that would lead to diplomatic results.[10] The Egyptian generals translated this doctrine into an operational idea that stressed the need to cross the Suez Canal and hold the area beyond it as a bargaining chip for future negotiations. Egypt's coordination with Syria was meant to split the limited Israeli resources and make the conduct of war more difficult.

At the beginning of the war, the Egyptian and Syrian forces crossed the Israeli border and enjoyed great success. The Egyptians crossed the Suez Canal, destroyed or defeated all but one of the strongholds along the canal, and deployed a military force, almost one field army, in the Sinai, where the Egyptians enjoyed effective aerial defense. The Syrians were blocked at the northern part of the Golan Heights, yet succeeded in reaching the center of the area. In the south of the Golan Heights, they were only seven kilometers from Lake Kinneret. The IDF, whose forces were only partially deployed, was caught by surprise and found it difficult to execute its traditional doctrine, which called for it to conduct ground force maneuver in order to transfer the war to enemy territory within a short period of time.

At this point, Israel activated its reserve forces. In the south, the IDF was unable to defeat the Third Egyptian Field Army from Sinai, but still managed to surround it and maintain a significant presence on Egyptian territory, on the western bank of the Suez Canal. The Syrians were pushed back, and within a few days Israel had reoccupied the entire area that they had captured and two Israeli divisions crossed the Syrian border. At the end of the war, Israeli forces were approximately 35 kilometers from Damascus, the Syrian capital, and 101 kilometers from Cairo, the Egyptian capital.

The Israeli Air Force was less surprised by the outbreak of the war than the IDF ground forces, yet the conflict demonstrated the great change in the balance of power since the end of the Six Day War. The air force had been preparing for war since Friday, October 5, and had already briefed its senior officers and recruited some of its reserve units. Even so, despite these preparations, the first day of the war was characterized by great turmoil. The air force was called to provide assistance on both the northern and southern fronts. The missions assigned to its bases and squadrons changed frequently, which created confusion and mistrust among some of the squadrons. Eventually, it was decided to concentrate aerial strikes on the northern front, against Syria.

On October 7, 1973, the second day of the Yom Kippur War, the Israeli Air Force switched to Operation Dugman-5 to attack SAM systems in the Syrian Golan Heights. The operation, based on the air force's extensive planning efforts, failed in every respect. During the operation, six jet fighters were shot down, two aircrew members were killed, and eight were captured by the Syrian Army. No SAM batteries were found in the locations identified by Israeli Air Force intelligence; the strikes only destroyed one fixed battery and badly damaged another, but no mobile battery was hit. The trauma of Dugman-5 caused the air force to avoid launching any significant strikes against missiles during the war itself and led Ezer Weizman to write a few years later that "the missile bent the aircraft's wing."[11]

The difficulty of dealing with the SAM batteries was clear and tangible, and the magnitude of this problem, alongside other factors, obscured the realization that the main failure related to the Israeli Air Force's operations during the Yom Kippur War was, in fact, a failure at the strategic-military level. More than anything else the war revealed the lack of a coherent operational doctrine that would explain the air force's actions at the beginning of an Arab-initiated offensive, when the mobilization of the IDF reserve units had not even begun and the regular ground forces were inferior in relation to the opponent's.

The offensive doctrine, whose centerpiece was the concept of a preemptive strike by the air force, proved irrelevant to the unique circumstances of this war. In the early morning hours, when it became clear that a war was about to break out, the chief of the general staff, David Elazar, instructed the air force commander, Binyamin Peled, to prepare for a preemptive strike against Syria. Yet, later that morning, Prime Minister Golda Meir announced that she had not approved such a strike. Some experts believe that if the preemptive strike had taken place, the war would have taken a completely different turn. Even though this hypothesis cannot be ruled out entirely, it is doubtful whether such a strike would have achieved actual military successes. In any event, there is a broad consensus that the strike would not have prevented the war.

On the third day of the war, the Israeli Air Force was used for the first time to strike Syrian infrastructure, including both civilian systems (electricity, fuel, and bridges) and military infrastructures, such as the general staff and the air force headquarters in Damascus. These strikes were approved during a dramatic morning when it seemed that Israel was facing a heavy defeat on both fronts. The main reason for approving the strikes on these strategic targets was the desire to remove Syria from the war and to deter Jordan and Iraq from opening another front against Israel. On the night before the strikes were approved, Syria had fired a number of FROG-7 missiles at the Israeli Air Force base in Ramat David, which made it easier for the political leadership to approve the attack. But again, the strikes did not achieve their goal. Syria remained a combatant throughout the war and even continued a war of attrition against Israel a few months following the cease-fire. Moreover, after the war the Syrians quickly repaired the damage caused by the aerial attacks.

All in all, the Israeli Air Force carried out 11,300 sorties, of which 6,730 were air-to-ground missions. The force lost 102 jet fighters and 7 helicopters during the war. Fifty-three aircrew members were killed and forty-four were captured. The Israeli Air Force achieved its greatest success in defending Israel's skies. The Egyptian and Syrian air forces collectively carried out 12,600 sorties during the war, of which only about 1,000 were for offensive purposes, especially in battlefront areas. The two Arab air forces lost 433 aircraft, 277 in air-to-air battles. Their activities were limited and their effectiveness minimal.

The Israeli Air Force's failure to achieve air superiority led to many difficulties in providing assistance to the ground forces, which caused these forces as well as the military leadership to lose faith in the air force's CAS capability. During the Six Day War, air superiority had enabled the Israeli Air Force to fly freely and safely throughout the battlespace—air and sea—and independently locate targets to attack. This situation, as well as the unique manner in which air power was used during the War of Attrition, to a large extent led the air force to neglect development of a coherent CAS doctrine. Operational doctrine rested on the idea that the air force would obtain air superiority at the beginning of the war and would then shift to CAS. However, in the Yom Kippur War, the air force had to engage in CAS before achieving air superiority. The Egyptian and Syrian ground forces penetrated beyond the IDF's lines, and the air force, which was supposed to assist in blocking efforts, was not able to do so.

The Israeli Air Force had made plans for such a situation, but its basic premise was that it would receive detailed information regarding the location of the enemy's forces. The existing doctrine stated that intelligence regarding the attack would accompany the ground forces' demand for air support. During the War of Attrition, the air force had received such intelligence, which was based on the continuous effort and planning of intelligence and operational units. The Israeli Air Force's pinpoint strikes in the Suez Canal and in the heart of Egyptian territory were carried out on the basis of precise and timely intelligence. However, at the beginning of the 1970s, prior to the Yom Kippur War, fundamental changes took place in the air force's CAS array, as a series of highly capable officers were appointed to head these arrays. During the war itself, it turned out that these changes were counterproductive and that relevant intelligence regarding the enemy's position and forces did not reach the air force.

After the war, Israeli Air Force commander Binyamin Peled stated that if the force had received the information—which he claimed did exist—its achievements in this area would have been completely different. Peled concluded that the force's responsibility for CAS should be matched with the authority to gather the required intelligence for this mission. His demands were not fully accepted, but they brought about changes in the structure and organization of Israeli Air Force intelligence that improved the link with intelligence units in the ground forces.

The doctrine for coping with SAMs necessitated a different solution, since it was obvious that Israel had no intelligence regarding this issue and that such intelligence had to be produced. Indeed, in retrospect, it was clear that the failure of Operation Dugman-5 provided the major learning experience regarding the challenge posed by SAMs. After the Yom Kippur War, General Peled led the Israeli Air Force's process of absorbing and implementing the lessons learned from the conflict. This noticeably impressive learning process took place from 1973 to 1978 and resulted in a number of significant doctrinal and organizational changes. The second half of the 1970s was filled with drills and experiments that were designed to examine and improve the applicability of those changes. Partly for this reason, General Peled is regarded as one of the most influential commanders in its history and the one responsible, to a large extent, for the creation of the current form of the Israeli Air Force.

AFTERMATH OF THE YOM KIPPUR WAR[12]

The Yom Kippur War did not lead to a revision of the IDF's overall operational doctrine, since many considered that the outcome of the war had strongly borne out the original doctrine, especially the importance of ground force maneuver. This interpretation resulted in a substantial increase in the ground force order of battle, which was considered to be a necessity for the realization of the original operational doctrine. The number of tanks grew by 50 percent, artillery by 100 percent, and armored personnel carriers by 800 percent.

The jet fighter order of battle also grew noticeably—by approximately 30 percent—which was enabled, among other things, by the development of the Kfir jet fighter by Israel Aviation Industries (IAI). The decision to build an Israeli jet fighter was made in light of the French embargo that followed the Six Day War. At the end of the 1960s, the IAI had introduced an Israeli version of the Mirage 5 jet fighter, which was called *Nesher* (Eagle). This aircraft achieved great success during the Yom Kippur War, shooting down more than one hundred enemy aircraft. The Kfir jet fighter entered service in 1975 and, for a short period, was the leading combat aircraft in the Israeli Air Force. In 1977 the Kfir was used for the first time during an offensive strike in Lebanon; in 1979 it shot down an aircraft for the first and last time. At the end of the 1970s, the Israeli Air Force already had enough Kfir jet fighters to replace the Mirage and Nesher aircraft.

At the beginning of the 1980s, the IAI began developing the Lavi multipurpose combat aircraft, which was intended to become Israel's front-line jet fighter and one of the most advanced combat aircraft in the world at the beginning of the twenty-first century. However, the Lavi project came to a halt in 1987 following pressure from different actors, especially the U.S. government. The Kfir aircraft were taken out of service during the 1990s.

In the meantime, the Israeli Air Force gradually regained the trust of the political and military leadership and of the Israeli society. A milestone in this process was the night of July 3–4, 1976, when the Israeli Air Force's cargo unit played a central role in one of the most daring operations ever carried out by the IDF—the release of 105 Jewish and Israeli hostages, in addition to aircrew members, who were kidnapped during an Air France flight from Israel to France. The kidnappers had landed the airplane at the airport in Entebbe, Uganda. Four C-130 aircraft flew combatants from special and other IDF units to Uganda, while two Boeing 707 aircraft served as airborne command posts and an airborne hospital.

In the following years, the Israeli Air Force order of battle underwent major changes. In December 1976 the force acquired its first McDonnell Douglas F-15 Eagle jet fighters. Israel was supposed to receive the first Lockheed Martin F-16 Fighting Falcon jet fighters in mid-1981, but because the Shah's regime was deposed in 1979, a major agreement to sell F-16s to Iran was canceled and Israel received the first of these fighters (type A) in July 1980. Within a few years, the F-15 and F-16 jet fighters became the Israeli Air Force's major air-to-air aircraft, while the Phantom aircraft were used mainly for air strikes and CAS missions.

In October 1977, prior to these developments, Peled was replaced by David Ivry, a fighter pilot with a diverse background as a fighter and planner. A month later, Egyptian president Anwar el-Sadat surprised the world when he decided to travel to Israel and deliver a speech in the Knesset. This visit paved the way for the peace accord signed between Israel and Egypt in March 1979. Following the peace accord, Israel retreated from the Sinai Peninsula, and Ivry was required during his first years as commander to manage the evacuation of airfields and Israeli Air Force bases in Sinai (Refidim, El-Arish, Eitam, and Etzion) and the relocation of these assets to the Negev. The peace accord removed Egypt from the circle of nations at war with Israel and made Syria the main enemy on which the IDF's military power was focused.

An aerial conflict in 1979–81 between Israel and Syria in the skies over Lebanon enabled the F-15 and F-16 jet fighters to confront Soviet MiGs for the first time, resulting in the first shootdowns of these aircraft in history. On June 27, 1979, an Israeli F-15 pilot shot down a Syrian MiG-21; on February 13, 1981, another F-15 pilot shot down a Syrian MiG-25. On April 28, 1981, an F-16 recorded its first kills, when Israeli pilots shot down two Syrian helicopters in Lebanon. In response, the Syrians for the first time deployed SAM batteries on Lebanese territory. On July 14 of that same year, an F-16 pilot for the first time shot down a Syrian MiG-21 aircraft.

During that period, the F-15 and F-16 aircraft also took part in a much more significant operation. On Sunday, June 7, 1981, at 1735 (Israeli time), eight F-16 Israeli Air Force aircraft, accompanied by six F-15s, attacked the Iraqi nuclear power plant on the banks of the river Tigris next to the town of Al Tuwaitha, approximately seventeen kilometers west of Baghdad. The flight range from Israel was about one thousand kilometers. Within a few minutes, the plant's main building and the reactor core were destroyed. The strike showcased the precise and long-range capability of Israeli air power and demonstrated Israel's determination to employ its military power when confronted by a threat to its national interests.

The IDF's original operational doctrine did not ignore considerations related to Israeli civil society, but until the Yom Kippur War these considerations had only a limited influence on the use of military power. The IDF's success in the War of Independence, the Sinai War, and the Six Day War had strengthened the confidence that Israeli citizens had in their army, its commanders, and its doctrine of operations. In that sense the Yom Kippur War marked a sharp break in Israeli public opinion. The war caused an unprecedented wave of protest that focused primarily on the political leadership and undoubtedly made a contribution to the dramatic political changes that took place in 1977.

The combination of the loss of confidence in the military and political leadership and the acknowledgment of the high price paid for the war led to a more fundamental change in the attitude of Israeli society toward war itself. This change was reinforced following the peace accords with Egypt. The accords reflected a vision that until the middle of the 1970s was shared by only a negligible minority in Israeli society: that there was a genuine alternative to the state of war perceived to be imposed on Israeli society. These developments were manifested in the public debates about the first Lebanon war.

THE FIRST LEBANON WAR (1982)[13]

Israel entered the first Lebanon war in June 1982 to eliminate the threat posed by Palestinian organizations that had turned Lebanon into their main base of operations after they were banned from Jordan in 1970. A more ambitious strategic goal lay behind Israel's decision: to produce a "new order" in Lebanese politics. The operation was designed to take forty-eight hours and penetrate forty kilometers deep into Lebanese territory; in fact, the IDF went much deeper, beyond the government's declared purposes, and finally reached Beirut.

The way the IDF functioned during the first Lebanon war reflected, to a large extent, adherence to the original operational doctrine. Israel invaded Lebanon using a large ground force and engaged in relatively successful battles with the Palestine Liberation Organization (PLO) forces in the western and central areas, and with Syrian forces in the east. Nevertheless, this war clearly showed the emergence of a preference for using standoff firepower. The ground force maneuvers certainly resulted in considerable achievements, yet an analysis of the battle showed a change in the way commanders made use of ground force maneuver. This change was related directly to the loss of life resulting indirectly from the social and political constraints that first appeared after the Yom Kippur War.

The recognition that success in attacking SAM batteries depended, first and foremost, on intelligence resulted in the higher priority being given to Israeli Air Force intelligence and highlighted the need to significantly shorten the time between locating the SAMs and attacking them. For example, Dugman-5 had been executed according to an outdated intelligence picture produced by a reconnaissance sortie carried out forty-eight hours before the operation began. This realization strengthened cooperation between intelligence officers and operational planners.

During the same period, the Syrian air defense system remained unchanged. It still operated under the assumption that the Yom Kippur War had reflected the basic superiority of SAMs over jet fighters. This assumption even grew stronger when a combination of political decisions and weather considerations forced Israel to avoid attacking the Syrian batteries positioned in Lebanon during the "missile crisis" in April 1981. During Operation Artzav-19, the new doctrine developed by the Israeli Air Force led to an unprecedented achievement.

This operation, which destroyed the Syrian SAM batteries on the fourth day of the war, marked the high point of the force's activities in this conflict. The operation was named Artzav-19 for the nineteen Syrian batteries targeted. Thanks to accurate and precise intelligence, the Israeli Air Force destroyed the vast majority of the batteries during the operation. Further, Israel's success in shooting down more than eighty Syrian aircraft demonstrated the air force's supremacy. Yet this success also led to the next stage in the evolution of the challenge confronting the air force. The precision-attack capabilities manifested during this operation led Israel's adversaries to place a stronger emphasis on factors such as concealment, camouflage, and deception in all systems that could become targets of air strikes, especially SAM systems. This created increasing difficulties in locating and attacking the targets. In the area of force design, awareness of the Israeli Air Force's absolute supremacy led Israel's opponents to rely on surface-to-surface missiles (SSMs) as the main means to penetrate Israel's airspace.

The first Lebanon war also marked Israel's first massive use of attack helicopters and UAVs. The Israeli Air Force first showed an interest in attack helicopters following the Six Day War and the U.S. military's use of attack helicopters in Vietnam. The Yom Kippur War strengthened this interest, since one of the lessons learned was that attack helicopters could have enabled the air force to block the Syrian and Egyptian armies in a much shorter period of time and at a significantly lower cost. In light of these conclusions, Israel decided to purchase AH-1G Cobra helicopters, which were incorporated into the air force in 1975. In 1979, improved AH-1S Cobra helicopters began to arrive. Since the U.S. government had limited the sale of Cobra helicopters to twelve, the Israeli Air Force also decided to purchase about thirty MD 500 Defender helicopters, which were incorporated into the arsenal in November 1979.

The air force added its first UAVs at the beginning of the 1970s as a result of demands that originated in the War of Attrition. The earliest vehicles were Ryan Firebees, which the United States used in Vietnam. For the Israeli Air Force, these vehicles served as platforms for taking photographs, including in the Yom Kippur War and the first Lebanon war. Additional UAVs made by Ryan served as decoys against the enemy's SAM systems.

A major advance in the area of UAVs occurred at the beginning of the 1980s when the Israeli Air Force acquired the first remotely piloted vehicle, the Scout. The Scout, built by the IAI, carried a payload and TV cameras, and could transmit real-time images. During the 1990s, the Israeli Air Force equipped itself with more advanced models, such as the Searcher and the Searcher 2; the Scout went out of service in 2004.

The first Lebanon war resulted in the expulsion of the PLO leaders from Lebanon and eventually brought about a fundamental change in the PLO's position regarding the armed conflict with Israel. The war temporarily limited Syrian hegemony in Lebanon, but it could not reinstate the moderate Lebanese regime that had collapsed during the civil war of 1975–76. This state of affairs created the vacuum that enabled Iran to establish Hezbollah in 1982. Hezbollah became a central player in the Lebanese arena toward the end of 1983, sponsoring a series of terrorist assaults that resulted in the withdrawal of foreign forces from Lebanon in the spring of 1984 and later caused Israel to withdraw from Lebanon to the international border in the summer of 1985. During the 1980s, Hezbollah was still fighting another Shi'ite organization, AMAL, but a confluence of Syrian and Iranian interests resulted in a situation that since the early 1990s has seen Hezbollah become the primary entity initiating attacks against Israel. Hezbollah continued attacks on the security strip in southern Lebanon until the final withdrawal of the IDF from Lebanon in 2000.

A few months after the end of the fighting in the first Lebanon war, Amos Lapidot became the new Israeli Air Force commander, replacing David Ivry. Lapidot's tenure was characterized by massive budget cutbacks owing to the crippling inflation in Israel during the mid-1980s. During Lapidot's term in office, the air force received F-16 Cs and Ds and advanced munitions, such as the Popeye missile, from the United States. In October 1985 the air force launched long-distance attacks on PLO camps in Tunisia; it then shot down two Syrian MiG-23 aircraft in November.

In September 1987, Lapidot was replaced by Avihu Ben-Nun, one of the most senior Israeli Air Force pilots, who had shot down four enemy jet fighters. During his tenure, the air force received its first Apache AH-64 attack helicopters. Ben-Nun headed the air force during the first Palestinian Intifada and the first Gulf War. Both events reflected the wide gap between the IDF's original operational doctrine and the security challenges that evolved during these times.

THE FIRST INTIFADA (1987–93) AND THE FIRST GULF WAR (1991)[14]

The IDF found itself facing the Intifada without any preparation and without a conceptual framework that would enable it to cope with the events in the field. Generally speaking, the Intifada significantly decreased the IDF's focus on the type of warfare that it had practiced up to that point and marked the beginning of a process that, after two decades, dealt a blow to the ground forces' ability to execute maneuver warfare. The Intifada necessitated a constant and growing ground force presence in the West Bank and the Gaza Strip, as well as the establishment of a new headquarters. The IDF began performing tasks normally reserved for police forces in other countries; the Israeli Air Force was not used during the first Intifada.

The Intifada had a major influence on Israel's attitude toward the use of military power. It demonstrated to the Israeli public the limits of military power and the difficulties inherent in occupying territory and controlling a hostile population over a long period of time. The Intifada also revealed to the political and military elite the problems involved in using large-scale military units for politically controversial missions.

The invasion of Kuwait by Saddam Hussein, the ruler of Iraq, in August 1990 led to the first Gulf War in January 1991. At the beginning of the war, Iraq fired SSMs at targets in Israel and Saudi Arabia, and continued firing missiles almost until the end of the war. For the first time, Israel chose not to employ its military power when it was attacked. Many reasons underlay this decision, but there was also the recognition that Israel had no proper response to the challenge it faced at the time. The war clearly demonstrated that the original operational doctrine reflected a specific reality that had characterized Israel's strategic environment in the first few decades. The changed environment created new threats and undermined this doctrine's overall relevance.

In a more concrete way, Iraq's missile attacks demonstrated that the age of "clear skies" that embodied the Israeli Air Force's long-term air superiority had passed. The lessons of the war revealed the need to prepare—in terms of doctrine, organization, and military equipment—for conflicts inherently different from the ones that had shaped the original doctrine. The establishment of the Home Front Command (1992) and Israel's investment in the Arrow antiballistic missile during the 1990s reflected the increasing importance of defense in the Israeli doctrine.

Ben-Nun ended his tenure as the Israeli Air Force commander in January 1992 and was replaced by Herzl Bodinger, who held this position until July 1996. Bodinger headed the air force during intensive activities in the Lebanese arena. During his tenure, the air force received its first UH-60 Black Hawk helicopters and signed a deal to purchase F-15I jet fighters.

Bodinger's tenure coincided, to a large extent, with the beginning of Israel's preference for using air power in operations that, in the past, had relied on ground forces. This preference was manifested in Operation Accountability in 1993 and Operation Grapes of Wrath in 1996, carried out to deal with Hezbollah's firing of Katyusha rockets at northern Israeli cities and towns. In both cases, Israel chose to employ its firepower, using the Israeli Air Force and artillery, and avoided using its ground forces. This was the first obvious deviation from the original operational doctrine, and it constituted a total contrast to the way Israel had handled the Lebanese threat in Operation Litani (1978) and the first Lebanon war (1982).

One school of thought claims that the temporal closeness of these operations to the first Gulf War was not coincidental and that the preference for using firepower originated in the lessons learned from that conflict. Indeed, Israeli military officers studied and wrote about the first Gulf War, and it appears to have had some influence on the military thinking behind the operations in Lebanon in the 1990s. Yet it is doubtful whether the influence was direct, as this viewpoint suggests.

The attempt to link the military operations of the 1990s only to the lessons of the first Gulf War ignores the influence of the social and political processes occurring in Israeli society. The preference for the use of firepower reflected the effect of those processes on the army's ability to execute its original operational doctrine.[15] By then it had already become evident that a fundamental change in the relationship between the IDF and Israeli society was under way and that the collectivist model of a "nation in uniform," which was created in the 1950s and had persisted until the 1980s, had diminished in power.

In July 1996, Bodinger was replaced by Eitan Ben-Eliyahu, an ace fighter pilot who had shot down four enemy aircraft. During Ben-Eliyahu's tenure, the Israeli Air Force acquired its first F-15I jet fighters—a version of the F-15E Strike Eagle that was especially manufactured for Israel. The considerable carrying capacity of these aircraft, combined with their advanced systems, enabled them to perform in-depth strikes with more weapons and against more distant targets (approximately 4,450 kilometers, and further with aerial refueling) at low altitudes, around the clock, and in all weather conditions. During Ben-Eliyahu's tenure, the Israeli Air Force also incorporated the Arrow air defense system operationally and the first Israeli astronaut, Col. Ilan Ramon, began his training program with the National Aeronautics and Space Administration.

During Ben-Eliyahu's tenure, the possibility of conflicts other than force-on-force confrontations with an army of a country bordering Israel (a model that pertained only to Syria) clearly necessitated an update of the IDF's original operational doctrine. One type of conflict was a limited war with a non-state entity (such as the ongoing confrontations with the Palestinians and Hezbollah), while the other was war with a nation-state that had no common border with Israel (for example, Iraq or Iran). During these years, the political and military leadership concluded that the probability of a war with Syria was very low, and additional resources should therefore be allocated to the forces that could achieve a military decision in the other types of conflict.

These new conflict scenarios challenged the IDF's original operational doctrine, with its central idea of achieving a military decision by ground force maneuver. A limited war must have legitimacy both at home and abroad—something to which the maneuver doctrine was ill suited. Ground force maneuver was equally inappropriate in the case of a confrontation with a state that shared no border with Israel. Thus, Israel's military doctrine needed a new framework that would formulate strategies for obtaining a military victory and achieving the objectives of the war when ground forces either could not operate or were under strict limitations. As a result, Ben-Eliyahu concentrated on conducting operations in distant locations, thus diverting the emphasis from close operations to SSMs.

In April 2000, Ben-Eliyahu was replaced by Dan (Dani) Halutz, an experienced fighter pilot who had fought during the War of Attrition, the Yom Kippur War (during which he had shot down three enemy jet fighters), and the first Lebanon war. One month later, in May 2000, Israel unilaterally withdrew from the security zone in southern Lebanon, trying to end eighteen years of frustrating military presence in the country.

Halutz headed the Israeli Air Force during the main part of the second Intifada and prepared it for possible involvement in the second Gulf War (2003). During his tenure, the air force acquired its first F-16I jet fighters in 2004. These combat aircraft are capable of carrying Python 4 and 5 infrared-guided, air-to-air missiles; radar-guided, advanced medium-range air-to-air missiles (AMRAAMs);

Litening and Low-Altitude Navigation and Targeting Infrared for Night navigation pods; and joint direct attack munition bombs. After serving as Israeli Air
Force commander, Halutz was appointed as deputy chief of the general staff,
and a year later, in June 2005, he became the first chief of the general staff to
come from the air force. Halutz commanded the IDF during the second Lebanon war and resigned following the public criticism of the IDF's performance
in this war.

THE SECOND INTIFADA (2000–04)[16]

Since the signing of the Oslo Accords with the PLO in 1993, the Israeli political
system had operated under the assumption that the long-standing conflict with
the Palestinians would lead to negotiation and be resolved by diplomacy, with
both sides making concessions. Israeli political leaders did not discount the conflict's core issues, and the underlying assumption was that violence would break
out by 2000. Nevertheless, the dominant viewpoint among Israel's leaders was
that the Palestinians, too, were interested in resolving the conflict via dialogue
and negotiation. Therefore, Israel believed that even if violent incidents were
to occur, they would be concentrated and short lived and that, in the end, the
parties would meet at the negotiating table. Israel did not foresee that it was
heading toward a massive terrorist offensive, at the heart of which would be the
Palestinian attempt to strike at the centers of Israeli civilian life with the intention of undermining and threatening the society's powers of endurance.

The violent clashes with the Palestinians erupted in late September 2000.
The lynching of two IDF soldiers by Palestinians on October 12 in Ramallah
led to the first air strike by attack helicopters on targets in the areas held by the
Palestinian Authority. Following this attack, the use of the Israeli Air Force,
especially for "targeted-killing" operations, became one of the conflict's unique
aspects.

The Israeli Air Force was employed throughout the conflict, but its missions
varied according to shifts in the conflict and in Israel's operational policy. The
number of operations targeting Palestinian infrastructure rose from six in 2000
to thirty-six in 2001 and reached its peak in 2002, a year in which sixty-four air
strikes were carried out (fifty-three of them in the first three months of the year,
until the beginning of Operation Defensive Shield). Thereafter, the number of
air strikes against infrastructure decreased significantly, and in 2003 only eleven
air strikes were carried out. In 2004 the number of air strikes rose once more,

and the Israeli Air Force carried out thirty-two air strikes aimed at infrastructure, most of them targeting the rocket-production facilities located in the Gaza Strip.

During the conflict, the number of targeted-killing operations rose gradually. In 2000 only one such operation was carried out; in 2001 there were fifteen, and in 2002, fourteen. The use of this method increased significantly in 2003 and 2004: twenty-seven targeted-killing operations were carried out in each of these years.

The number of Israeli casualties decreased dramatically after 2002. Israel's success in disrupting the wave of Palestinian terrorism and reducing it is a commendable achievement—and, to a large extent, a unique one, as will be subsequently addressed.

Examining the Israeli Air Force's role in this achievement is complicated, since the use of the air force was only one component of a long series of moves—military and other— that Israel made against the Palestinian Authority throughout most of the conflict. Nevertheless, the air force undoubtedly played a distinctive role during the Second Intifada. From the first months of the conflict until April 2001, air power was the only type of military force used in attack missions against the Palestinian Authority, since Israel avoided using its ground forces in the "A" areas (those under total Palestinian control). In the later months of 2001, ground forces were used in these areas, but their activities were specific and of short duration. At this time and until the beginning of 2002, the Israeli Air Force still served as Israel's main tool to force the Palestinian Authority, and especially Yasser Arafat, to take action against terrorist entities. After that, Israel used its ground forces in the "A" areas to a greater extent, and the role of the air force was proportionally reduced. The air force also played the leading part in Israel's offensive against the Hamas leadership during 2003 and 2004.

All in all, Israel made an efficient use of its air power during the conflict and turned it into a significant tool in a battlespace inherently different from the one for which the Israeli Air Force was originally designated. Given the unusual circumstances of the Second Intifada and Israel's view of the uprising as a "limited conflict," the air force proved an appropriate tool that assisted Israel in achieving its goals. The attacks on infrastructure constituted an effective tactic that enabled Israel to retaliate against the assaults waged by its adversaries, and to gain internal and international legitimacy, which it perceived as a sine qua non for undertaking ground maneuvers that assisted in reducing the wave of terrorism. Further, the targeted-killing operations created the pressure that was so vital in order to restrain Hamas.

Halutz was succeeded as Israeli Air Force commander by Eliezer Shkedi, a prominent representative of the generation of pilots who enlisted following the end of the Yom Kippur War. During most of his service he had flown F-16s. In 1980, Shkedi took part in the first conversion course for F-16 aircraft that took place in Israel, and later on he commanded an F-16 C/D squadron. Shkedi headed the air force during the conflict with the Palestinians and during the second Lebanon war.

THE SECOND LEBANON WAR (2006)[17]

The second Lebanon war is the official name the Israeli government gave, in retrospect, to the military confrontation that took place between the state of Israel and the Lebanese Hezbollah from July 12 to August 14, 2006.[18] The war occurred within the framework of the ongoing conflict between Israel and Hezbollah that had existed since Hezbollah's founding in 1982. The war began after Hezbollah attacked an IDF patrol moving along the border between Israel and Lebanon, killing three soldiers and kidnapping two. It ended after both sides agreed to accept UN resolution 1701, which called for a cease-fire and the strengthening of the UN force in Lebanon. At the same time, Israel managed to launch a significant military offensive against Hamas and other terrorist organizations in the Gaza Strip, following another kidnapping that took place in the area.

As known, in this war Israel preferred to exercise its firepower using the Israeli Air Force and its artillery, and was very hesitant to use its ground force maneuver capability. While Israel undertook some limited raids, no major ground force maneuver was carried out until the late stages of the war. When the government finally decided to authorize such a maneuver, the action was carried out only partially and stopped before it could achieve its goals.

During the war, the Israeli Air Force made approximately nineteen thousand jet fighter and helicopter sorties. Of those, approximately twelve thousand were jet fighter sorties, most carried out for offensive and CAS missions (an average of three hundred sorties a day). Approximately forty-two hundred additional sorties were made by attack and transport helicopters. Most of the aerial operations were aimed directly at Hezbollah. The Israeli Air Force attacked approximately seven thousand targets during the war, dropping nineteen thousand bombs as well as firing twenty-two hundred missiles. Approximately 35 percent of the munitions used during the war were precision guided.

The targets attacked fell into two main categories. The first included mainly buildings that served as headquarters, warehouses, and hideouts. The second included locations from which rockets were fired and targets that were suspected of containing rocket launchers. In addition, the Israeli Air Force struck communication lines, gas stations, bridges, coastal radar stations, groups of terrorists, vehicles, rocket launchers, and blockaded areas. Alongside the kinetic offensive operations, the air force also operated in the field of information warfare as aircraft dropped approximately 17.3 million leaflets.

Syrian and Iranian support had enabled Hezbollah to become a terrorist organization with significant military capabilities. At the beginning of the war, it had a large number of long-range rockets (approximately one thousand with a range of 250 kilometers), an even larger number of short-range rockets (thirteen thousand), air power (UAVs for offensive missions), and naval power (anti-ship missiles), as well as considerable ground forces (approximately ten thousand combatants) that operated like guerrilla forces, armed with advanced personnel antitank and antiaircraft rockets. Hezbollah established formations in southern Lebanon that were armed with antitank weapons; the organization also built underground bunkers and prepared a logistics supply line to support continuous battle. These facilities were partly located on the outskirts of Shi'ite villages.

Hezbollah was no doubt surprised that the kidnappings led to war, but it had prepared in advance for a confrontation that would follow patterns similar to those of previous conflicts. The concept underlying Hezbollah's force design and the organization's preparation for a possible confrontation with Israel was one known as winning by not losing. This variation of the notion of attrition was based on a deep-seated understanding of Israel's core doctrines on security and military issues. Hezbollah recognized Israel's need for clear-cut and unequivocal victory within a short period of time. By contrast, Hezbollah simply had to survive and, especially, demonstrate its survivability. The organization thereby continuously fired missiles into Israel's territory.

Apart from deep feelings of disappointment shared by many Israelis, the strategic outcome of the war remains a topic of dispute in Israel. One view is that Israel failed in the war and that its deterrence abilities were severely damaged. Another is that the war strengthened Israel's deterrence abilities, proven in the years following the war when Hezbollah did not initiate any attacks against Israel, a stark contrast to the years preceding the war.[19] This school of thought claims that, contrary to the exaggerated expectations of the Israeli public, the more realistic objectives of the war were in fact achieved, mainly by the Israeli Air Force.

As a direct result of the second Lebanon war, Israel has enjoyed more than three years of exceptional peace along its northern border. However, public opinion in Israel holds the war as a missed opportunity, not because of the number of Israeli casualties (120 soldiers and 42 civilians), but rather because of the wide gap between expectations at the beginning of the war and its final outcome. The IDF, with its advanced capabilities, could not prevent the continuous firing of rockets into Israel's northern region and had failed to defeat Hezbollah by the time the war had ended after thirty-four days. To many Israelis, the four thousand rockets that landed in Israel demonstrated the limitations of their country's military might and particularly air power.

CONCLUSION

Six decades after the Israeli Air Force's establishment, the second Lebanon war marked a significant change in the balance of power between air and ground forces, with a clear deviation from the IDF's original operational doctrine. Indeed, since the war, some experts have expressed the opinion that the IDF abandoned its original doctrine in the years preceding the conflict and adopted a different operational doctrine based on the concept of attaining military decision by using air power. This opinion has usually attributed the abandonment of the old doctrine to the profound impression that advanced technologies made on military decision makers, and to the success that U.S. forces achieved in applying those technologies to conflicts in Iraq, Kosovo, and Afghanistan. Further, this new operational doctrine was conceived and implemented by the first chief of the general staff who had come from the Israeli Air Force, and it failed miserably.[20]

Even though this school of thought provides a somewhat simplistic explanation of a complicated issue, the second Lebanon war undoubtedly signaled a change in the attitude of the political and military leadership toward the importance of air power in achieving military decisions. Yet the priority given to the use of the Israeli Air Force in the second Lebanon war did not stem exclusively from senior IDF commanders' fascination with advanced technology or with the Americans' use of firepower in conflicts after the first Gulf War. Instead, as this chapter has shown, the high priority resulted from a process that lasted approximately three decades, during which the traditional ratio between ground

force maneuver and air power, captured in the IDF's original operational doctrine, was continually modified. The shift in the balance reflected the increase
in threats to ground force maneuver as manifested during and after the Yom
Kippur War. Yet, more than anything else, it resulted from social and political
changes that characterized Israeli society as well as other Western liberal democracies.[21] These changes significantly increased the restraints on the use of military
power in general and highlighted the importance of the Israeli Air Force.

The air force's capabilities have undergone a fundamental change over the
last decades. Thanks to these developments, the Israeli Air Force is now able to
conduct massive attacks on a large number of targets, fixed and mobile, at short
and long ranges, in bad weather conditions, in poor visibility conditions, and
in areas presenting complex threats. The interesting question is whether such a
massive attack would still be relevant in the current strategic environment.

There is no easy answer to this question. The second Lebanon war, as well
as the second Intifada, directed Israel's attention to the new types of military
conflicts typical of the current era: low-intensity, usually protracted, conflicts
conducted in an urban environment against asymmetric threats and non-state
adversaries. They differ fundamentally from the wars that the Israeli Air Force
had been designed to deal with since its foundation. These new types of conflicts
create an increasing gap between impressive advanced capabilities and the ability
to apply them to a significant extent in military conflicts.

It seems that the Israeli Air Force has coped quite well with this challenge.
In Israel's two most recent armed confrontations, the air force played a major
part, offering decision makers an appropriate military tool. Alongside its capability to initiate a massive attack, which remains relevant in the case of a confrontation with Syria or another nation-state, the air force did develop a variety of
capabilities that are highly applicable to the new types of military conflicts. Some
of them were displayed recently when, in late 2008 and at the beginning of
2009, the Israeli Air Force was used in Operation Cast Lead in Gaza. One of the
central features of this operation was the massive use of precision-guided munitions (around 80 percent of the total), reflecting the emerging linkage between
accuracy and legitimacy.

Contrary to accepted opinion, the IDF waged the second Lebanon war
without a coherent doctrine. The original doctrine, which had been formulated
in the first years of the IDF's existence and had been successful until the 1970s,
was considered irrelevant. The collection of understandings developed then did

not represent an alternative doctrine that could have served as a solid basis for responding to the challenge presented by Hezbollah.

The lack of a relevant doctrine and concept directly damaged Israel's ability to use its air power efficiently. Because of this absence, the Israeli Air Force operated in separate missions that did not form a comprehensive conceptual framework. The need to develop such a doctrine is thus the main lesson learned from the second Lebanon war.

This issue is connected to the shortcomings of the decision-making process in Israel at both the political and senior military levels, and to a more fundamental problem related to the mind-set of air forces. The second Lebanon war therefore also demonstrated the need for strategic and operational thinking among air forces and for an appropriate planning process that would best translate thinking into action. The Israeli Air Force executed its assigned missions very efficiently, but the logic behind these missions has drawn the majority of criticism in regard to the way Israel employed military force during the war. The air force has always concentrated on maintaining its tactical excellence. The second Lebanon war and the importance of employing air power in this and other conflicts compel the Israeli Air Force and other air forces to develop operational and strategic excellence as well.

PART II

The following three chapters take a closer look at a few emerging and reemerging global players: Russia, India, and China. The fundamental debates on doctrine, organization, force structures, joint-force relations, acquisition, roles, and missions are the same as those that occurred within the Royal Air Force (RAF), the U.S. Air Force, and the Israeli Air Force, but the conclusions drawn by each country vary, sometimes significantly.

SOVIET-RUSSIAN AIR POWER
Dr. Sanu Kainikara describes the development of Soviet-Russian air power: growing from 244 aircraft at the outset of the First World War into the massive air arm of a superpower. Currently, after two decades of considerable transformation, "the Russian Air Force in its new incarnation is a force to be reckoned with at all levels." Chapter 4 demonstrates that while the Russian Air Force is a result of several defense reforms, doctrinal reassessments, and technological upgrades, most profoundly it reflects the legacy of the Soviet Air Force, which was imbued with a Marxist-Leninist ideology and a Communist Party philosophy. Throughout its existence, the Soviet Union impressed a defensive and land-centric orientation on its military forces. Consequently, the air force was accorded only secondary status; its sole purpose was to support ground campaigns.

The Second World War was a decisive experience not only for the British and the Americans, but also for the Russians. The German attacks that commenced on June 22, 1941, were so devastating that the Soviet Air Force lost more than twelve hundred aircraft on the first day of the war. Despite this loss, the Soviets were able to carry out seventy-three thousand sorties during the following month alone. As the war proceeded, the Red Air Force became ever more capable and ultimately applied air power very successfully against the Germans when integrated with the ground forces. The Red Air Force conducted both defensive and offensive operations, especially through massive air strikes close to the battlefront and in support of the land offensive. Typically, these operations first sought to obtain air superiority over a corridor in the direction of the Red Army's proposed main thrust; next, air forces would supply either close air support or deep air interdiction focusing on the enemy rear and its command and control centers, transportation hubs, and communication nodes. The principles of mass, concentration of firepower, and centralized control always lay at the heart of these operations, but the Soviet Air Force never developed a doctrine or concept for conventional strategic bombardment. To some extent, the leadership believed that reliance on nuclear weapons made the concept of obtaining control of the air and strategic attack irrelevant, but Operation Desert Storm served as a wake-up call.

Kainikara also provides insight into the Soviet Air Force in its proxy wars during the Cold War and the ways in which nuclear weapons influenced the development of air power. He then explains the strengths and weaknesses of the Russian Air Force, how it has tried to adapt to new challenges, and how over the last two decades it has achieved some success in changing from a quantitatively massive force to a smaller, but more agile, air power: one that continues to reorganize and improve its operational procedures.

INDIAN AIR POWER

Air Commodore (Ret.) Jasjit Singh examines the birth, maturation, and possible future of the Indian Air Force. Modeled on the RAF in terms of organization and training, early formations were used for air policing and close air support together with the British; however, a strong element of "Indianization" was always involved. The Indian Air Force was declared an independent service in 1932 and worked alongside the RAF during the Second World War. After the war, independent India faced downsizing its military forces due to the country's partition in 1947, and self-reliance in defense policy became a natural part of the policy and strategy of nonalignment. As a consequence, India kept investment in the military to a minimum, and the forces had a defensive orientation that fit within the framework of deterrence.

Nevertheless, the Indian military and its air forces were put to the test on several occasions, mostly against Pakistan and China. Singh examines the use of air power in various conflicts from 1947 to 1999. In all cases, air power was employed as a political instrument with restraints and constraints. Typical missions were airlift and close air support, and in the 1965 war, the air force was explicitly ordered not to take offensive action in East Pakistan. The Indian Air Force also took part in United Nations peacekeeping and peace support operations, for example in the Congo in 1961–62 and Sri Lanka in the late 1980s, besides disaster relief and air maintenance in the Himalayas.

The Indian Air Force had been created as a tactical force, but its first mission in the Second World War was strikes against Japanese bomber bases in Thailand. Until India's independence in 1947 the British retained strategic missions for themselves, with the Indians concentrating on tactical army support, artillery spotting, and reconnaissance. Within the limitations of aircraft capabilities, and as a non-nuclear power until 1998, India's strategic role was confined to conventional deterrence and air superiority. After 1962 India increasingly relied on Soviet aircraft built with license in the country, and until the late 1980s, Soviet fighters had a short range and carried a small payload. Indian acquisitions resulted in the indigenous development of different versions of the original aircraft that were better suited to national requirements; the air force's agenda never included employment of multirole combat aircraft and conventional bombing by light bombers.

In assessing the Indian Air Force's future role, Singh discusses perceived threat scenarios that India may confront. In his view, China is growing stronger and more ambitious, and its increased strength in air power makes it a considerable global player; Pakistan is a major power in its own right, with a strong interest in gaining control of Kashmir, Jammu, and the Punjab. What the author is concerned with is the China-Pakistan strategic nexus. In light of these threat scenarios, Singh recommends that the Indian Air Force undertake a force modernization that enhances jointness, takes advantage of India's economic growth, and reduces its dependence on Soviet-Russian weapons and equipment. The air force's way forward, according to Singh, is to transform itself from a predominantly tactical strike force into one with strategic reach and effect, whose rationale is to provide conventional and nuclear deterrence, and if deterrence fails, to win the nation's wars rather than merely carry out territorial air defense of India.

CHINESE AIR POWER

The Chinese Air Force is the largest air power in Asia and the third largest in the world. Dr. Xiaoming Zhang examines how China has sought to increase its air power over the last sixty years, but argues that only recently can the People's Liberation Army Air Force (PLAAF) be considered capable of offensive air and space power.

Zhang provides insight into the development of the Chinese Air Force and of the PLAAF's organization, equipment, and doctrine by examining how Soviet influence, guidance from the Chinese Communist Party, and China's strategic culture strengthened the force in certain areas and weakened it in others. He concludes that lack of progress over the years resulted from a backward military philosophy and lack of access to modern industry and technology. The Chinese culture of favoring man over machine, the theories of Mao Zedong on guerrilla warfare, and the insistence on copying the Soviet system (sometimes blindly and mechanically) resulted in less-than-optimal warfighting concepts. Given its experiences from the Korean War and homeland air defense operations during the Cold War, the PLAAF focused on air defense and the support of ground operations; the priority accorded to the latter meant that army officers were chosen to command the air force.

From the beginning, the air force's development included elements of the "Chinaization" of Soviet aircraft, training, and doctrine. The 1950s witnessed some questioning of Soviet influence, and senior officers began to write new operational concepts and regulations. Still, it was not until the 1990s that the Chinese seriously contemplated air power as an offensive force, and that quality rather than quantity became a driving factor. Mao's thinking on war and warfare still remains relevant to many PLAAF senior officers, but gradually—and heavily influenced by Deng Xiaoping—the Chinese have adopted a strategic vision that includes a substantial air force with long-range strike capabilities and third- and fourth-generation fighter aircraft, combined with Chinese-made air surveillance aircraft for integrated command, control, and communication systems. China's rise as a global economic and political power in recent years has enabled the PLAAF to significantly transform itself from a territorial air defense force to one capable of both offensive and defensive operations.

The Chinese Air Force has undergone an impressive modernization over the last twenty years. Even so, it will take some time for the PLAAF to nurture an institutional culture that embraces offensive and defensive capability and especially, according to its own ambition, to "bear the brunt of, and play a sustained, independent role" in modern warfare. Like the Russian Air Force and the Indian Air Force, the Chinese Air Force has gone through downsizing, but is a force to be reckoned with in the future.

4

SOVIET-RUSSIAN AIR POWER

Sanu Kainikara

The fountainhead of military doctrine has always been a force's ability to accept unfettered and free-ranging intellectual development and yet encompass that development within the bounds of a pragmatic set of conditions. While the two requirements could be viewed as opposites, they complement each other in the development of strategic military doctrine. The absence of either intellectual freedom or pragmatism would unbalance the doctrine development process. Incorrectly aligned doctrine will cause the suboptimal employment of the force, almost certainly leading to defeat.

A nation's strategic culture plays an important role in the formulation of air power doctrine. The overwhelming influence of doctrine on all areas of a force is perhaps most obvious in the Soviet, and later Russian, Air Force. Over the years, the Russian Air Force has demonstrated the inherent flexibility of the concept of air power itself. The transformations and transitions that the Russian Air Force has undertaken in its long history bear out the truism that air power is the most versatile military capability.

Although a great deal of study has been undertaken to understand doctrinal development in the Soviet-Russian forces,[1] the foundations of Soviet-Russian military thought have not always been evident. However, political ideology has clearly been a critical determinant in the formulation of military policy and development of force structure. For most of the twentieth century, the nation's threat perception was permeated by the concepts of Marxism-Leninism. Since the erstwhile Union of Soviet Socialist Republics (USSR) was mainly landlocked and the government's primary objective was the spread of Communism—initially to the neighboring region and subsequently globally—it is not surprising that the USSR developed a land-centric military posture. Soviet ideology considered the army the basic arm of defense, with the air force an essential but subservient component. The overriding concern was the availability of adequate air support for the ground campaign. This secondary status of the air force had a direct and detrimental effect on the Soviet Air Force's doctrinal and strategic development, resulting in a desultory doctrine development process guided almost completely by ideology.

The following analysis of doctrine development in the Soviet-Russian Air Force has two underlying themes. First, it provides historical insight into the developments that took place within the Soviet Air Force in response to political ideology. Second, it examines the evolution of air power doctrine and the relationship of that process to the ideological, sociopolitical, and economic forces influencing the nation. In doing so, the chapter analyzes the wars in which the Soviet-Russian Air Force has so far been involved and also explores the more recent security perceptions of the Russian nation. The narrative follows a chronological order in order to enable the reader to understand the dual strands of history and doctrinal development within the Soviet-Russian Air Force over the past hundred years.

THE CONVERGENCE OF HISTORY, IDEOLOGY, AND DOCTRINE

The story of humankind is woven around successive waves of people moving in search of an environment—fertile soil, good water, and more congenial climatic conditions—that favors a settled life. These population movements are worldwide phenomena, and the clash of civilizations described by Professor Samuel Huntington has always existed in history. The well-documented conflict between Christianity and Islam in the Middle Ages; the European fear of threats from the East, embodied in the subliminal expression "the Yellow Peril"; the European colonial conquests that almost wiped out the Aztecs, Incas, Mayans, and Maoris; the marginalization of the native populations of North America and Australia through forced displacement by European immigrants; and the impact of decolonization that is tearing apart Africa are all manifestations of the clash of civilizations.[2] Security, stability, and development are symbiotically related and must be addressed in a holistic manner to ensure a viable environment for progress. Further, international bilateral and multilateral relationships have different connotations for different nations and are shaped by history, culture, demography, commerce, and security perceptions.[3]

Russia as an entity can be traced to the ninth century, when Norse rulers established Novgorod in the north and Kiev in the south as garrisons to defend the newly united areas. However, by the twelfth century, the region's defenses were weak, and Russia, still in its nascent state, was subjected to a number of invasions. Between AD 1055 and 1462 the nation was attacked 245 times.[4] The impact of these successive invasions on the collective national psyche is difficult to comprehend fully, but some vestiges of the paranoia that stemmed from these constant invasions are still visible in Russian strategic thought.

The most important conquest was that of the Mongols, who, by 1242, had overrun the country. They not only established an autocratic Asiatic rule that would last for several decades, but they also dominated political and economic life throughout Eastern Europe.[5] This period was an unmitigated disaster for Russia. By denying access to the seaports, the Mongols effectively isolated Russia from the rest of Europe, leading to relative technological and cultural backwardness. The Grand Duchy of Muscovy (Moscow) defeated the Mongols in 1380 and, by the middle of the fifteenth century, controlled the entire nation, establishing a despotic tradition that continued for the next three centuries. The rest of Russian history consists of a prolonged and continuous effort to attain parity and reestablish contact on an even footing with the West.

From the beginning, Russia was ruled by force, meaning that political objectives were traditionally achieved by military might. Of the 550 years prior to 1900, Russia was at war for 310, clearly underlining the preeminence of the military in matters of state.

Historically, Russia has often been on the defensive, seeking to protect its national boundaries. This stance has had two major long-term consequences for the development of the Russian military ethos. First, the outcome of every battle was critical to the nation's safety and security. This stands in contrast to a majority of colonial Western powers, which could afford to lose battles, and even wars, in far-flung parts of the world without the home countries being even indirectly threatened. However, Russia's territory is so enormous that the nation has, on a number of occasions, sacrificed some of its land so that it could buy time and avoid defeat. Second, because Russia is a landlocked country, the army has had the primary responsibility for defending the nation and therefore absolute primacy in all matters of military thinking and organization.

The lack of adequate port facilities resulted in the absence of a seagoing tradition and minimal development of the naval forces. Because of this, the Soviet Navy and indirectly the Soviet air forces were later placed under army control, not only in the organization of the force, but also in the development of operational concepts.

Since the military has always been the primary tool of national survival, it is not surprising that it had an inordinately powerful influence on foreign policy until the collapse of the Soviet Union. The national psyche of extreme insecurity, resulting from Russia's long history of suffering under foreign invaders, supported the propensity to build and maintain large military forces. Thus, Russia might well have built a vast military machine regardless of the kind of government that came to power in the early twentieth century.

Although the revolution in 1917 brought an end to tsarist rule, it was only in 1922 that the Bolsheviks gained total control of Russia and established the USSR. The foreign policy of the newly formed USSR was based on Lenin's interpretations of Marxist doctrine. The cornerstone of this doctrine was the imperative to spread Communism throughout the world. The Bolsheviks combined the principles of international diplomacy with political expediency, treating war and the use of military force as complementary, and necessary, tools to fulfill this overriding mission. However, until the outbreak of World War II, the Soviets used the military only in indirect and subtle ways to influence its neighboring areas. This mainly took the form of supporting incipient uprisings by providing weapons and at times personnel in the guise of advisers.

EARLY HISTORY OF RUSSIAN MILITARY AVIATION

Although airplanes can at present still not make very long flights or rise to any great heights, and in general they are not suitable for military purposes, in the future they will nevertheless play a tremendous role in military affairs and so will undoubtedly be introduced into the armament of the army.[6]

The history of Russian aviation runs parallel to the country's socioeconomic development and mirrors its changes during the nineteenth and twentieth centuries.[7] Fascination with flight was as evident in Russia as it was in other parts of the world from the time hot-air balloons made lighter-than-air flights feasible. Russian interest in military aviation can be traced back to 1812, when

Napoleon's invasion of the country prompted some consideration of building a balloon, although it was only in 1831 that the first recorded balloon flight took place.[8] Alexander F. Mozhaiski, a captain in the Imperial Russian Navy, designed an airplane in 1875, patented it in 1881, and completed construction by 1883. He is reported to have successfully tested it the following year in a short flight at Krasnoye Selo, near St. Petersburg.[9] The Russians claim this as the first powered flight in history. However, reports conflict, and a number of authorities refute this claim.[10] Nevertheless, Mozhaiski must be given credit for his ingenuity and enterprise in building and testing what must rank as the first full-size powered airplane. His achievement clearly indicates that technological innovations were taking place in Russia at the same time as in the rest of the world.

In 1869 the war minister, Gen. D. A. Muliutin, set up the Commission on the Use of Aeronautics for Military Purposes, and in 1895 Russia established the Officers' Aeronautical School.[11] By 1909 the Imperial Russian War Ministry had concluded that the airplane had great military potential. More importantly, the officer corps unanimously acknowledged the superiority of airplanes over cavalry in the reconnaissance role. Although not officially endorsed or articulated, this recognition marked the first entry of the airplane—and air power—into the perceptions governing Russian military doctrine. The acceptance of air power as a military capability is confirmed by the Russian military's efforts to encourage the officer cadre to become "air-minded" through pilot training activities.

In 1909 Igor Sikorsky built the first helicopter, marking the beginning of an illustrious career. Around the same time, another designer, Aleksandr Porokhovschikov, built a military aircraft with an armor-plated cabin, which could be considered the ancestor of the famous World War II ground attack aircraft, the Sturmovik.

Although the intelligentsia expressed some concern that the government was indifferent to aviation and unaware of its strategic implications, the progress of military aviation in Russia in the first decades of the twentieth century was comparable to that of any other nation. The War Ministry clearly understood the long-term need for a domestic industry that could create a modern and well-equipped frontline air service, and by 1914 Russia's aviation industry had orders for up to one thousand planes to be built in the next three years. Further progress was hampered by the outbreak of World War I and the subsequent turn of events in Russia.

WORLD WAR I

When war began in 1914, Russia had 244 aircraft, compared to Germany's 232 and France's 138. However, these planes were comparatively obsolete, Russia's rate of production was far inferior to that of other countries, and maintenance facilities were not geared for prolonged operations.[12] Within a month, the number of Russian frontline aircraft was reduced to 145.

Following the declaration of war, the Russian government made great efforts to expand and modernize military aviation through increased domestic production and the purchase of aircraft from France. Both of these initiatives had far-reaching consequences for Russian military aviation. First, Russia's aviation industry failed to produce aircraft whose performance and quality matched those of Western models. This failure clearly demonstrated that nations must have a broad-based, sound scientific educational system if they are to design, develop, and manufacture state-of-the-art aircraft domestically. The Soviet Union would later put in place extensive nationwide technical education systems to overcome this drawback. Second, the French, though Russia's allies, supplied only aircraft whose inferior performance made them unsuited to their own air force requirements and provided fewer than demanded. The Russian high command fully learned the disadvantages of being dependent on foreign industry for critical military capabilities. This historically obscure fact contributed to the Soviet emphasis on self-sufficiency in all military matters, especially in the 1920s when the USSR was almost a pariah nation.

Perhaps more importantly, the Russian air services during World War I were organized according to the French model, with air units—designated field companies or squadrons—assigned to the army at the corps level and, at times, to higher echelons, but always under a single area commander.[13] This laid the foundation for the army-centric organizational development of the Soviet air forces.

As long as it remained a belligerent, Russia made efforts to improve the performance of the Imperial Air Force through reorganization and an increase in the number of squadrons available. However, it never matched the performance of the enemy. The Germans retained control of the air throughout, even though their efforts were concentrated on the western front. Russia withdrew from the war in 1917, having learned two main lessons: that military aviation required a robust domestic industry to support it and that such an industrial base required strong economic, political, and cultural foundations within the nation.[14]

THE IMPACT OF THE BOLSHEVIK REVOLUTION

The Bolshevik Revolution directly affected Russian military aviation in terms of operations and industrial development. Only about one-third of Russia's pilots joined the Reds; the others either joined anti-Bolshevik groups or sought asylum in the West. The result was an acute shortage of qualified fliers in the Red Air Force. The impact on the aircraft industry was equally disastrous. Some outstanding scientists and aircraft designers were killed, and the majority of the survivors left the country. Further, a number of aviation factories were completely destroyed during the revolution. This made the recovery of the industry, already suffering from a "brain drain," a long and arduous process.

On February 23, 1918, the Red Army was officially founded, encompassing the fledgling air element as an integral part. The Revolutionary Military Committee, under the chairmanship of Leon Trotsky, also controlled the air element—referred to as the Red Air Fleet—although it was not represented in the committee for several more years.

Most air units were placed under the command of officers with no professional appreciation of military aviation. The organization of the air fleet mirrored that of the army. Two or more squadrons formed a group, which was the basic tactical unit; two or three groups combined to form a division. Several divisions operated under the command of a single headquarters.[15] This organization was deceptively simple, but when the army command structure was superimposed on it, the air element was deprived of all independence. It took the Soviet-Russian Air Force a very long time to rid itself of this rigidity in command.

Even though its effectiveness was minimal, the Red Air Fleet was employed in all operations during the civil war. As a result of these campaigns, fought at different levels of intensity and involvement, military aviation leaders gained considerable insights into air power doctrine and tactics, command and organization, and training and equipment requirements. They learned two major doctrinal lessons from these campaigns. First, it seemed desirable to subordinate some air units directly to ground formations to increase the effectiveness of the air-land combination. Second, central control of the tactical air elements was important to ensure adequate and concentrated support to ground operations. While these perspectives were new, they still represented an army-centric view and meant that the air force remained fully under the control of the land commanders.

Subordinating the air forces to the land forces had many consequences. From this point on, the Soviet air forces were firmly committed to developing effective ground attack capabilities. Soviet military doctrine also incorporated the operational concept of rapid concentration of firepower afforded by air power as a basic premise in the conduct of a campaign. These two factors made tactical battlefield support the main thrust of the air force's development, and resulted in the neglect of the strategic potential of air power.

The Bolshevik Revolution, born out of and sustained by a politico-economic ideology, also had a strong impact on the military forces' doctrine and operational ethos. This ideology supported the concept of small, semi-autonomous land force detachments operating independently. While this concept directly contravened the primary tenet of air power employment—centralized control and decentralized execution—it permeated the Red Air Fleet, greatly diminishing its effectiveness. This dichotomy between fundamental air power doctrine and its practical applications was to continue throughout the Soviet era.[16] The rigidity of Communist military "science" could not accommodate the peculiarly flexible doctrinal requirements of air power and, even when tempered by operational experience, tended to envelop strategic doctrine in political ideology. Thus, the Red Air Fleet's organization, command and control, and employment were determined by political imperatives rather than sound and unbiased air power doctrine.

The Red Air Force

The reorganization of the Red Army, initiated by Trotsky, was continued by Mikhail Frunze, who was an advocate of offensive maneuver warfare. This led to an emphasis on offensive operational concepts, increasing the influence of the Red Air Force at the tactical level. However, the guiding principles of Soviet air power were greatly influenced by a nascent philosophical-level doctrine of war that relied heavily on certain unique features of Russian military history and fundamentally conventional military precepts. Political sophistry, combined with revolutionary ideology and dogma, robbed the command and control function within the Soviet military forces of all flexibility, one of the fundamental requirements for effectiveness in conflict. This rigidity precluded the possibility of the air force's benefiting from its lengthy combat experience.

The philosophical and conceptual development of the air force was further thwarted by its system of command and control. The Red Army military district commander had operational and tactical control over all air units in his area, at times further delegating it to subordinate army formations. Although there were air advisers at the military district headquarters, this arrangement diluted the command authority within the air force. As a result, the air force played only a limited and peripheral role in the development of the overall offensive strategy being devised for the newly reorganized Red Army, and was relegated to serving as an adjunct to ground operations.[17]

Although modernization efforts, especially in training and equipment, continued throughout this period, the primary air power doctrine remained fixated on massive air strikes close to the battlefront in support of the land offensive.[18] The Soviet high command did not seriously consider the theories of doctrinal pioneers such as Giulio Douhet, who stressed the decisive strategic importance of long-range air power, and therefore developed no concepts for strategic employment of the Red Air Force. This mind-set was further reinforced by the then-prevalent German influence on the Soviet military leaders. The Luftwaffe of the 1920s did not regard strategic bombing as a war-winning factor, even though its heavy-bomber raids on England in 1917 had demonstrated the effectiveness of strategic air power.[19] This confirmed the Soviet disinclination to investigate the potential of strategic bombing.

However, the Red Air Force was used during this period to spread Soviet influence in Asia. Although initial efforts met with only minimal success, the Russians recognized the potential of air power to serve as an effective tool for spreading political, economic, and psychological influence, which proved to be valuable in later years.

WORLD WAR II

> The German invasion of the Soviet Union was the greatest test in
> its history not only of the capabilities of the Red Air Force and the
> doctrine of combined arms warfare but also of all aspects of Soviet
> State power.[20]

From an air power perspective, the Soviet Air Force differed fundamentally in its operational ethos from other contemporary Western air forces, essentially because it always operated as an auxiliary to the Soviet Army and was never developed as an independent fighting arm.[21] While military doctrine was not completely neglected, the impact of political ideology and economic factors on doctrinal development was far more pronounced in the Soviet Air Force than in its Western counterparts.

When Germany invaded the Soviet Union in June 1941, the Soviet Air Force was in the midst of extensive and fundamental reorganization, retraining, redeployment, and expansion. Organizationally, the most important change was the formation of an independent long-range bomber force (Long-range Aviation) and an airborne force command that was independent of army control. However, their concept of employment centered on deep interdiction and not on operations against strategic centers of gravity, such as production centers. This could have come about because Long-range Aviation was equipped with surplus bombers of limited capabilities, and had inadequate doctrinal direction to undertake a strategic campaign.

Another change being instituted was the creation of independent air armies within the air force, with slightly greater responsibility for the planning and deployment of air assets. Even though this provided limited autonomy, the operational mission priorities of the air units were still determined by the army commander at the front, with the air army chief being subordinate.

Although the beginning of the German invasion interrupted the reorganization of the Soviet Air Force, the Soviets demonstrated great flexibility. The quality of leadership that was initially lacking, especially at the lower echelons of command, improved remarkably with battle experience.[22] According to Adolf Galland, World War II ace and Luftwaffe general, "as an integral part of the Red Army, the SAF [Soviet Air Force] perhaps was better organized and suited to its purpose . . . than the Luftwaffe."[23]

INITIAL COMBAT OPERATIONS

The Soviet Union officially dated World War II from September 1939 to August 1945, dividing it into three distinct phases. The first lasted from June 22, 1941, to November 18, 1942, when the German forces held the strategic initiative. The second began on November 19, 1942, and lasted until the end of 1943, during which time the Soviets contested and then won the strategic initiative. The third comprised the period from early 1944 to the German capitulation in May 1945, which saw the development of the Soviet's offensive concept of operations.[24]

The German attacks that commenced on June 22, 1941, were devastating to the USSR. The Soviet Air Force admitted to the loss of more than twelve

hundred aircraft on the first day of the war.[25] Apart from sporadic and courageous defensive actions, throughout western Russia the first three months saw a
continuous retreat of the Soviet forces.[26] Along with the Luftwaffe's clear technical and operational superiority, surprise enabled the Germans to decimate the
Soviet Air Force. It also helped them that most of the Soviet aircraft were deployed at vulnerable forward bases in accordance with the prevailing doctrine.
Although evidence has become available to indicate that both the United States
and Winston Churchill personally had warned Joseph Stalin of the impending
attack,[27] for some inexplicable reason the field commanders were kept in the
dark. It has been suggested, albeit without any evidence so far, that the Soviet
leadership anticipated that the attack would not take place until the end of July,
which perhaps accounts for the feverish attempts to modernize the units at the
western front.[28]

Despite obtaining air superiority at the outset and achieving large-scale destruction of aircraft, the Luftwaffe was unable to eliminate the Soviet Air Force
entirely. Although their resistance lacked coordination, the Soviets were able to
carry out seventy-three thousand sorties in July 1941 alone.[29]

In early 1941, the Soviet Air Force operated according to outdated and
rigid tactics, its technology was inferior, and its doctrine was not adequately
developed.[30] However, almost immediately after the initial setback, the leadership articulated the need for air superiority, which resulted in an emphasis on the
production of fighter aircraft with better performance. It also led to the study
of German tactics and operational concepts, which were then modified to suit
Soviet operational constraints.

By August–September 1941, when the Germans had overcome the defenses
of Kiev and encircled Leningrad, the Soviet Air Force started to reappear in
strength. The Luftwaffe had not prepared for a long campaign. Further, it was
ill prepared for the harsh Russian winter, which in 1941 was more severe than
normal and had also set in early. By contrast, the Soviet Air Force was better
trained and equipped to fight a winter campaign, purely by virtue of experience.
The Soviet Air Force supported the Red Army's winter counteroffensive with
nearly fifteen hundred ground attack aircraft, mainly Il-2s. While their contribution was not decisive because of the aircraft's limited capability, it indicated the
recovery of Soviet warfighting capacity. This is corroborated by an entry in the
diary of Gen. Franz Halder, chief of the German General Staff, in August 1941:
"The whole situation makes it increasingly plain that we have underestimated
the Russian colossus."[31]

By April 1942 the Luftwaffe could no longer muster sufficient numbers of aircraft to simultaneously maintain control of the air and carry out bombing raids on Moscow and Leningrad.[32] From then onward, Soviet aircraft outnumbered the Luftwaffe three to one.

Two interconnected factors gradually allowed the Soviet air forces to fight back on a more even footing. First was the decision by the Soviet high command to sacrifice territory in order to create distance and buy time to reorganize and reequip both the army and the air force. Second, because of the Luftwaffe's fixation on close air support and battlefield interdiction, it was unable to mount a meaningful strategic campaign, especially at great distances, providing the Soviets with a much-needed respite.

The Soviet Air Force learned four major lessons from this phase of the war. First, the employment of air power was most effective when centrally controlled. Second, a cohesive and flexible reserve force was needed to ensure concentration of forces at the time and place needed in an offensive campaign. Third, uninterrupted logistics support was essential to the effectiveness of an air force, especially when operating from forward bases. Fourth, air-ground coordination was critical to the success of any joint venture.[33]

These lessons led to organizational changes aimed at providing unity of command to the larger air units, which was deemed necessary for future offensive air operations. Intrinsic to the restructuring was the formation of homogeneous air divisions, a paradigm shift from the composite units in existence until then.[34] This transition to integral air divisions simplified operations, logistics, training, maintenance, and command, and provided the operational mobility and flexibility vital to the success of air campaigns.

The creation of these large formations led to the development of a doctrine of the air offensive that epitomized the Soviet concept of air power.[35] The doctrine combined massed action of the Soviet Air Force and the Red Army with continuous air support throughout the campaign—from the preparatory concentration of forces to the culminating phase of pursuit and victory.

By November 1942 the Luftwaffe had a mere four hundred aircraft in theater against more than two thousand Soviet planes.[36] Along with this unassailable numerical superiority, the performance of the Soviet fighters had also vastly improved. During this second winter, Soviet pilots adjusted well to the poor flying conditions and fared much better than the Luftwaffe, which was now suffering from a lack of aircraft as well as trained and experienced aircrew.

TURNING THE TIDE TOWARD VICTORY

During the summer of 1943, the Germans made a last concerted attempt to regain the initiative. The two air forces struggled for air superiority in the Kuban Peninsula, with the pendulum swinging both ways a number of times. The availability of trained personnel proved important: the Soviet Air Force was able to stabilize the number of aircrew available and improve their experience incrementally, while the Luftwaffe suffered from the nonavailability of experienced pilots. Combined with the attrition that it was absorbing, this shortage became a debilitating factor for the Luftwaffe.[37]

The Battle of Kursk in July 1943 saw some of the heaviest air action of the war in Russia. The Luftwaffe was so badly crippled by then that it lacked the numbers needed to provide close air support for all ground operations, let alone to contest control of the air. Kursk proved the decisive battle, both in the air and on the ground, and the turning point in the campaign. The Soviet Air Force gained complete strategic air superiority and never relinquished it. The Soviet summer offensive was supported by more than one hundred air divisions equipped with fighters that matched the performance of the German aircraft. Mobile, flexible, and effective air support was central to the success of these offensive thrusts, which were mounted simultaneously in the north and the south.

By the beginning of 1944, the Soviet forces were only waiting for the Allied invasion of France before opening their main offensive on the central front. By this time, the Soviet Air Force's concept of operations for offensive campaigns had been well refined and was repeatedly and successfully applied throughout the rest of the war. The concept entailed gaining air superiority in a corridor in the direction of the proposed main thrust of the Red Army. The air superiority campaign was conducted in three distinct phases. Phase I commenced as early as three months before the start of the major ground offensive with the buildup of reserve strength and increased reconnaissance activities. Phase II involved increased deep air interdiction in the enemy rear against command and control centers, transportation hubs, and communication nodes. The Soviet Air Force operated as much as 120 miles behind the front line in this phase. Phase III consisted of concerted fighter operations to gain and maintain air superiority, and intensified immediately before the offensive "breakout" of the ground force, which itself was supported by concentrated firepower from the air.

The Soviets also formulated the strategy of "springboard" air support. At the beginning of the offensive, the close air support units were deployed within sixty miles of the front. These units provided initial support for the ground offensive, and as the assault progressed, they moved forward with the army, operating from captured airfields and always remaining in close proximity to the moving battlefront. In places with no preexisting airfields the Soviet Air Force also operated from makeshift airfields, constructed with only rudimentary facilities. While this perhaps represented an adaptation of the Luftwaffe's successful Blitzkrieg concept, it was very effective, and the Soviets refined it to form the basis of their offensive maneuver concept in the postwar period.

In studying the post–World War II Soviet Air Force, the implications of this refinement in strategy are impossible to ignore. Even though the advent of jet fighter aircraft demanded much more sophisticated facilities to operate from very austere airfields, the Soviet jet fighters were designed to be extremely resilient, with minimal requirements for elaborate maintenance. Since they operated so close to the front line, the pilots had to devise tactics and procedures that gave them the necessary lead time to react to the army's needs. From a design point of view, these factors made it necessary to produce aircraft that could be rearmed and refueled in the minimum possible time, that needed very limited maintenance, and that were fast and maneuverable. Further, they did not require extended range or endurance. This design philosophy produced the World War II Yak fighter and continued to influence the design of a whole series of fighter aircraft well into the 1980s. A robust airframe, ease of maintenance, and simplicity of operation became the hallmarks of Soviet fighter design—factors that were unjustly maligned in the West, especially during the Cold War.

Another strategy that the Soviet Air Force developed during its World War II offensive was the rapid redeployment of air units to provide overwhelming and concentrated fire support to the ground offensive. Operations in 1944–45 demonstrated the vastly improved skills of the strategic force commanders and their clear understanding of how a joint campaign developed. The Red Army offensive was characterized by several distinct phases and widely separated drives, which meant that effective air support could only be provided through the swift and well-planned deployment of air units. Throughout this phase of the war, the air commanders were able to ensure the desired concentration of air assets for army support, all the while maintaining air superiority.

By the time Germany surrendered on May 8, 1945, Soviet factories were producing so many aircraft that the air force, which was already at full strength, was holding back newly manufactured aircraft from the front. The Soviet Air

Force and its support infrastructure had reached maturity and self-sufficiency. The recovery of the Soviet Air Force to achieve preeminence in 1944, after it had been almost completely annihilated in 1941, can be counted among the most remarkable achievements of modern military history.

IMPACT ON FUTURE DEVELOPMENT

The Allied victory in World War II was as much a victory of optimized production facilities and logistic supply as of battles and campaigns won on the field. This was most obvious in the air war. The Luftwaffe's inability to destroy the Soviet aviation industry enabled production to continue unhindered, which led to parity in the quality of aircraft and to increasing Soviet numerical superiority. In combination with improved tactics and operational capability, these factors permitted the Soviet Air Force to dominate the air war. Maintaining a domestic capability to produce the hardware needed to fight and win—already embedded in the Soviet Air Force since World War I—now became a cardinal principle.

Soviet military doctrine in the war years emphasized close air support and interdiction. The struggle for national survival was predominantly a ground war requiring the buildup of air power capabilities that could directly affect the outcome of the ground battle. The concept of "air offensives" to support the advancing army was refined through operations in 1944–45, with the Soviet Air Force carrying out concentrated and continuous air operations—direct close air support to the army and battlefield and deep interdiction to influence the course of the ground campaign. Accompanying this preoccupation with support to the ground campaign was a clear appreciation of the need for air superiority, and the struggle for control of the air gradually became an overriding priority. Subsequently, air superiority remained the primary quest of the Soviet (and now Russian) Air Force.

Throughout World War II, the Soviet Air Force lacked the capabilities, doctrine, and concepts for strategic bombing—owing not to underestimation of the impact of strategic bombing, but to unavoidable circumstances and shortsighted decisions at the highest levels. The Soviet forces had to rely heavily on the tactical employment of air power to achieve their primary strategic objective—to stop and then roll back the German advance. Under these extreme circumstances, the Soviet high command was forced to focus on the production of aircraft and the accompanying doctrinal and operational development of tactical aviation as the cutting edge of their military aviation capabilities.

In hindsight, after more than sixty years, it is possible to identify the effect that this necessary focus had on the subsequent development of air power capabilities, concepts, and doctrine in the Soviet Union. On the positive side, the war had demonstrated the inherent flexibility and adaptability of Soviet air power, evidenced by the organizational and conceptual changes, instituted under extremely trying conditions, to adapt the air force to the shifting circumstances in the overall status of the war. At the start of the war, the Soviet Air Force was still in the process of distilling the lessons from the Spanish Civil War and was only beginning to understand the great potential of air power. However, it is testimony to the professionalism of the Soviet Air Force that it was able to rapidly stabilize an extremely precarious situation and then initiate an offensive that led to victory. Luftwaffe Gen. Klaus Uebe best summed up the Soviet Air Force's role in World War II: "As events show, Russian reactions to German Air Force operations, however crude they may have first appeared to their more enlightened Western opponents, proved throughout the course of the war to be highly efficient, effective, and ultimately an important factor in the defeat of Germany."[38]

The Soviet experience in World War II also created an ability within the military forces to think on a vast scale, while taking into account modern warfare's enormous complexity. Soviet military art and science became based on the concept of the strategic offensive, which, in turn, became the foundation for development of doctrine, strategy, concepts, and tactics. The current Russian military system is the product of continuous refinement of the lessons learned in the Great Patriotic War through the analysis of smaller wars that have been fought since then.

At the end of World War II, the Soviet military apparatus had gained greater stature than the Communist Party, inducing Stalin to tighten party control over the military.[39] Stalin ensured the slow degradation of the military's status by appointing prominent marshals to the Supreme Soviet, the ruling body in the USSR, and then removing them in apparent disgrace, while also giving "military" titles to Politburo members. In this surreptitious way, Stalin purged the military of powerful and influential marshals while retaining and promoting more pliable and party-loyal officers. By 1953 the senior military leadership had become an entrenched bureaucracy with limited control over the forces and was almost completely devoid of professional military capabilities. While Soviet foreign policy continued to be bolstered by the visible might of its vast military force, the force itself lacked a cohesive strategic outlook and was mostly controlled by nonmilitary party officials. These circumstances are never conducive to the uninhibited development of doctrine and the concepts of operations necessary to develop a capable strategic air force.

SOVIET AIR FORCES DURING THE COLD WAR

The dominant political and military reality of the second half of the twentieth century was the Cold War, which divided a large part of the world into two blocs centered around the military might of the United States and the USSR. During this long period, almost all information pertaining to the strategic arms race and overt as well as covert operations was classified in the Soviet Union and, to a lesser extent, in the United States. In the absence of primary sources, the analysis of Soviet military policies was always a process of "inferences drawn by long chains of logic."[40]

In 1949 the U.S.-led North Atlantic Treaty Organization (NATO) was formed with seven founder nations and eventually expanded to cover most of Western Europe. Air power in Europe was transformed by domestic development of jet aircraft, supplemented by U.S. sales of the latest jet fighters to some NATO members and by licensed manufacturing. The Soviet Union viewed this extensive rearmament of Western Europe with increasing misgivings. The Soviet leadership made some attempts to ensure a demilitarized West Germany and, ideally, a demilitarized Europe. However, NATO viewed the basic concept of demilitarization as the Soviet Union's one-sided attempt to secure its hegemony in Europe and rejected it outright.[41] Thereafter, the Soviet Union formalized its existing bilateral treaties with its client states in Eastern Europe in the Treaty of Friendship, Cooperation, and Mutual Assistance, which came to be known as the Warsaw Pact. This action clearly demarcated the two sides of the Cold War.

As the Cold War became entrenched, the USSR's military strategy was heavily influenced by the rapid development and deployment of intercontinental ballistic missiles (ICBMs), the development of artificial satellites, and the pursuit of nuclear deterrence. Historically, the Soviet military had often undergone drastic strategic-level changes immediately following internal strife and political power struggles. The Cuban missile crisis led to such a situation, perpetuating the already dominant reliance on nuclear-weapon-carrying ICBMs as the first line of deterrence in a simplistic strategy of mutually assured destruction.

During the Cold War years, a dichotomy existed in the Soviet Union's military strategy. Conventional strategic doctrine emphasized offensive combined-arms campaigns, while political thinking was clearly biased toward defensive nuclear deterrence. There was also a debate within the Soviet military regarding the extent to which nuclear deterrence should influence the overall national strategy as opposed to the role of conventional forces. However, nuclear capabilities had relatively little influence on air power doctrine and strategy.

The limited wars that took place between World War II and the 1991 Gulf War were fought mostly between states that owed their military allegiance to opposite Cold War blocs. An analysis of these nations' training patterns and employment concepts, by extrapolation, provides insight into the doctrinal ethos that prevailed within the two major air forces of that time.

Proxy Wars: The Korean War—Limited Involvement

The invasion of the Republic of Korea (South Korea) by the military forces of the Democratic People's Republic of Korea (North Korea) on June 25, 1950, is widely considered a defining moment of the Cold War. The United States viewed the North Korean attack as a clear case of Soviet aggression.[42] Whether Stalin had expansionist motives in allowing North Korea to initiate the conflict is debatable, but there is no doubt that the Korean War represented a classic case of war by proxy. Thousands of Soviet military personnel were sent to join the North Korean forces as "advisers," although the façade of the independence of the North Korean military was scrupulously maintained.

The Soviet Union had two major reasons for instigating the conflict and providing material support to North Korea. First, the existence of nuclear weapons made the Soviet Union confident that the conflict would not escalate, although the Soviet military doctrine of the time did not consider nuclear weapons a decisive factor in war.[43] Second, by 1950 the Soviet aviation industry had started the large-scale manufacture of the MiG-15 jet fighter, which could match any other contemporary aircraft. This gave the political and military leadership assurance regarding the capabilities of the Soviet forces.

The war provided an opportunity for the Soviet Air Force to evaluate its tactics and improve its operational concepts of air defense, tactics, and equipment.[44] The experience considerably increased the combat effectiveness of the air defense forces and later influenced the development of Soviet military policy.

The initial North Korean advance was halted by relentless air attacks on its stretched supply lines and then gradually pushed back by a ground counteroffensive supported by carrier-based air power. At this juncture, Communist China entered the conflict, changing the course and character of the war. Also at this time, the Soviet-designed MiG-15 jet fighter made its operational debut.[45] However, its effectiveness could not be fully evaluated since political considerations restricted its basing to airfields in Manchuria. This limitation involved a long transit to the battle area and back, which reduced the time available for these aircraft to perform their combat role and directly affected their effectiveness in ground battle intervention.

Political considerations also restrained the uninhibited use of the Soviet Air Force to the detriment of its performance, both in air combat and ground support. In keeping with the cover that the war was being fought exclusively by North Korea, the Soviet Air Force only carried out defensive missions, giving strict instructions to its pilots never to fly over areas where they would fall into enemy hands if they were shot down.[46] This higher level control of operational and tactical aspects of a conflict was to become entrenched in the Soviet air forces within the next decade.

The Korean War ended in a stalemate. However, the United Nations forces were able to dominate the air war, despite being numerically outnumbered by aircraft of similar (or, in some cases, superior) performance, mainly because the adversary was operationally constrained by political embargoes. The experience of the Soviet Air Force in the Korean War thus provides a clear illustration of political ideology directly influencing air power doctrine, curtailing the operational employment of clearly superior equipment to its detriment, and affecting both the immediate and long-term outcomes.

From an air power perspective, the Soviet Air Force learned lessons at the strategic and tactical levels. First, air power was an effective element of national power that could be employed with minimal diplomatic risk and limited possibility of escalating the conflict—in other words, it could avoid mission creep. Second, by intervening directly at critical junctures in the conflict, the Soviet Air Force assisted in aligning China more closely with the Soviet Union. Third, the potential of strategic bombing to create long-lasting effects was clearly demonstrated by the rapid destruction of North Korea's military-industrial complex, although these attacks did not affect the conduct of the war per se.[47] Fourth, at the operational level, it was accepted that air power alone could not win battles, campaigns, and wars, especially if its employment was politically restricted. These lessons were long remembered by the Soviet political leadership.

PROXY WARS: OTHER REGIONAL CONFLICTS

Western domination of the air war in Korea, at least toward the latter part of the conflict, earned Soviet fighter aircraft an undeservedly poor reputation. While the F-86 Sabre was a steadier platform with slightly better avionics and gunsight, the MiG-15 was superior in several aspects of aerial combat—rate of climb, acceleration, ceiling, maximum speed, and turn performance.[48] However, wars fought subsequently in the Middle East and Southeast Asia further added to the comparatively poor reputation of Soviet equipment, exacerbated by the one-sided reporting of their performance.

Superiority of equipment—in quality and/or quantity—definitely contributes to winning wars. However, as has been repeatedly proven throughout history, such intangible and nonquantifiable factors as morale, doctrine, training, leadership, and tactical skills play a more significant role. For example, although the Arab air forces were decimated on the first day of the 1967 Arab-Israeli war because of strategic blunders and a lack of foresight, it is notable that they continued small-scale but effective strikes throughout the war, shooting down a number of Israeli aircraft.[49] The point is that the Arab losses were not incurred because of Soviet equipment, but because of flawed planning, leadership, and operational skills.

By contrast, the effectiveness of Soviet fighter aircraft was amply demonstrated during the period 1969–73, when Soviet Air Force squadrons operating on a "voluntary" basis from Egyptian air bases were able to maintain an overall equal exchange ratio of air combat losses with the Israeli Air Force. While the skirmishes that took place during this so-called War of Attrition had lesser significance in the larger perspective of the conflict in the Middle East, they clearly highlighted two factors: that Soviet aircraft were as effective as the Western aircraft when operated by adequately trained and experienced pilots and that strategic planning and operational innovation were essential to winning battles, campaigns, and wars. Equipment was only one part of a complex warfighting system.

In the final analysis, the development of air warfare concepts and effective equipment within the Soviet Air Force had a much wider global influence. The outcomes of all the limited air wars that took place from the mid-1960s on did not indicate the superiority of a particular aircraft type or design philosophy, but instead demonstrated that intensive and realistic training combined with effective and aggressive leadership and flexible concepts of operations were the combat-winning factors for any air force. In a majority of the wars that took place after World War II, the performance of Soviet combat aircraft and other weapon systems was judged to be inferior only because of extraneous factors, such as inadequate training, a lack of strategic and tactical understanding, and a flawed doctrinal approach to air warfare.[50] Where these issues had been addressed thoughtfully, as in the case of the Indian Air Force, Soviet aircraft have performed as well as their Western equivalents.

For the Soviet air forces, the major lesson that emerged from the protracted wars in the Middle East from 1967 on was the importance of ground-based air defenses in conducting an air campaign.[51] The Soviets had invested in the long-term development of surface-to-air missiles (SAMs) and developed a sophisticated concept of ground-based battlefield air defense. This mobile air defense umbrella proved extremely effective, especially in an open, maneuver-friendly theater such as the Sinai Desert.

The Vietnam War also provided the Soviet air forces with some salient lessons, although their involvement was limited to supplying equipment and advice, and did not include physical participation in the conflict. The war reinforced the effectiveness of SAMs and created the concept of a layered air defense that incorporated weapon systems with varying ranges and operational heights. The proliferation of SAMs and the advent of highly capable radar networks brought about a drastic change in the operational ethos of air strikes. All attacks—fighter interdiction and close air support—had to be delivered from extremely low levels, below the radar and missile envelope, necessitating the development of a different class of tactical fighter.

ICBMs and Nuclear Weapons

Careful analysis of the experiences of other air forces in modern limited wars convinced the Soviet Air Force that conventional wars were more probable than nuclear exchanges. However, nuclear weapons, their delivery systems, and operational concepts still dominated the military's organizational structure and strategy. To ensure that the force was prepared for conventional conflict, the tactical air force developed a dual-track capability that would permit it to start a campaign in the conventional mode and shift to nuclear war when, and only if, the need arose. Although the air force took a long time to institute the structural changes needed to implement this more sophisticated doctrine, it became the underlying principle for the modernization of tactical aviation and subsequently the entire military for the next few decades. The political leadership was slow to reconcile itself to the basic ideology of accepting the possibility of a completely nonnuclear war of uncertain duration. However, once it was accepted, the feasibility of such a conventional option was explored, and doctrinal and technological developments were aligned at a much faster pace.[52]

These factors made it necessary to adapt the Soviet Union's basic military doctrine to encompass the possibility of strategic nuclear strikes in the initial phase of a war.[53] However, it was also believed that, even though the initial phase would be critical, in the long term only a potent ground offensive that seized and held ground would ensure victory. The logical conclusion was that Soviet armed forces should be built around a large, strong, and mobile army supported by tactical air forces and battlefield missiles. The missile force would be based on a balanced foundation of strategic rockets and air forces, and antimissile and air defenses. The navy, heavily reliant on submarines, would be an essential, but subordinate, adjunct.

Even though the Soviet philosophical doctrine emphasized ground offensive at the theater level, the standoff between the USSR and the NATO alliance at the end of World War II saw the emergence of two factors that had a long-lasting influence on the formulation and refinement of Soviet air power doctrine. First, the adversary's main strategic center of gravity had moved beyond the effective reach of the combined air-ground team. This unprecedented situation dictated a reexamination of the strategic doctrine based fundamentally on the primacy of the army in the offensive role and the dedicated support role of the predominantly tactical air forces. Logically, such a review should have recognized the need to formulate a more balanced doctrine, with the air forces undertaking a more proactive and primarily strategic role. However, the strategic utility of air power remained underexploited, mainly because of the overemphasis on ICBMs.

Second, the technology-generated revolution in military affairs gave an unprecedented boost to the development of air power capabilities, incrementally increasing the impact of the air force's employment in the overall context of a military campaign. This forced a radical change in the doctrinal outlook of the Soviet Air Force, although the World War II–era air power doctrine framework was modified only gradually. At the same time, Soviet military doctrine placed greater reliance on missiles, which led to the formation of the strategic missile force. The creation of the missile force occurred at the expense of long-range strategic aviation, which was accorded only secondary importance.[54]

These two air power factors were somewhat sidelined by the army-centric doctrine advocated by the higher echelons of the Soviet military, but they provided sufficient impetus for the Soviet Air Force to continue developing its own doctrine. However, both the organization and the command and control structure of the Soviet air forces tended to dilute this effort. Soviet air assets were distributed among three independent mission-oriented services: naval aviation, which was tactically and administratively controlled by the navy; the strategic air defense force, which operated more than two thousand aircraft and a large number of missiles; and the Soviet Air Force. The air force itself was divided into three large functional commands for which it had administrative authority: Long-range Aviation, Frontal Aviation, and Military Transport Aviation. While the air force also had operational control over both Long-range Aviation and Military Transport Aviation, Frontal Aviation was placed under the operational control of army commanders. It was here that the fundamental weakness of Soviet air doctrine was clearly demonstrated: this arrangement flouted the basic air power tenet of centralized control and decentralized execution, and effectively relegated the air force to a purely supporting role.

A side effect of this arrangement was that the USSR never seriously studied the strategic effectiveness of air power and its doctrinal implications. Since the concept of independent air campaigns had no historical precedent, the Soviet Air Force was viewed as a purely tactical force, even though it possessed the complete spectrum of offensive and defensive capabilities.[55] This is not to suggest that air power professionals were unaware of the doctrinal issues that faced the Soviet Air Force. On the contrary, military theorists not only within the air force, but also in wider military-strategic circles, accepted that air power was a crucial multidimensional element in combat. However, its primary roles were still seen as interdiction and close air support in direct contribution to combined arms theater operations. This perception within the military inhibited the holistic development of the air force's combat capabilities and the support infrastructure necessary for an independent force.

SOVIET AIR POWER DOCTRINE AFTER 1945

During the decades following World War II, three separate factors led the Soviet leadership to remodel the organization and force structure of the air force. First, Soviet Frontal Aviation had proven most effective when its air divisions operated as part of a combined arms force in multifront deep operations. Second, prevailing air doctrine stressed centralized control of air assets to ensure their optimum employment in depth.[56] Third, from a technical perspective, Soviet aircraft design reflected a maturity centered on robustness, dependability, and sustainability in operations.

The Soviet emphasis on the development of a strong and independent military force provided the impetus for the air force to continually strive to improve its capabilities, concepts, and doctrine. However, the acceptance of new concepts and the rate of doctrinal change were slow and sporadic.[57]

The Long-range Aviation arm of the Soviet air forces remained without clear doctrinal guidance. Two factors combined to ensure that strategic bombing never became an element of the Soviet forces' mainstream policy. First, the political leadership supported the ballistic missile force as the preferred option for deep strikes against all adversaries. Second, the Soviet General Staff, dominated by army officers, did not collectively understand the potential of strategic bombing; therefore, as noted above, this option was ignored in the development of military doctrine. This approach stood in contrast to that of the United States, which continued to invest heavily in manned strategic bombers.

Three additional factors significantly affected the development of Soviet military doctrine, including air power doctrine, during the Cold War years. First, the preoccupation with the development and deployment of nuclear weapons and their long-range strategic delivery systems produced a somewhat sketchy strategic doctrine that affected the operational doctrine of the conventional forces. Second, the technological innovations that rapidly improved the design of conventional weapon systems, including aircraft, affected not only the operational warfighting capability of the force, but also the development and application of strategic doctrine. Third, the constantly changing international politico-economic environment made it imperative to retain versatility in the basic doctrine of the force.

Under Stalin, the Soviet Union continued to believe that it had a fundamental responsibility to spread Communism throughout the world and, therefore, that war was a distinct possibility. Accordingly, the basic doctrine remained one centered on offensive campaigns built on a strategy of rapid mobility and overwhelming force projection on the ground and in the air. Lenin wrote, "One who knows how to advance and has not learned how to withdraw will lose the war. Wars that have begun and ended with a continuous victorious offensive are not evident in history or, if they occurred, are exceptions."[58] Drawing on Lenin's military theories and the lessons of World War II, Soviet doctrine therefore also included the concept of a premeditated withdrawal or retreat, even at the theater level of operations. However, even the defensive part of Soviet doctrine—not applied in practice after World War II—emphasized firmly holding territory and perimeters while ensuring maneuver options. Air power's versatility, flexibility, and ubiquity were expected to provide mobility options to the entire force.

In January 1960 the USSR outlined a radically different doctrine, which emphasized that the nation did not believe that war was inevitable. Further, the prevailing belief was that a conflict would commence with nuclear rocket strikes deep in the interior of the adversary as opposed to an invasion across geographic borders. The leadership did not discount the possibility of a surprise attack on the Soviet Union, but considering the strategic-geographic depth of the nation, it felt that such an attack would not be decisive, providing the Soviet military with the opportunity to retaliate massively. While the United States developed the nuclear doctrine of mutually assured destruction, the Soviet Union never fully subscribed to it.

The political leadership of the Soviet Union believed that ICBMs provided an offensive edge, which gave rise to the idea that conventional air forces could be substantially cut without diminishing overall military capability and potential.[59] The reliance on ICBMs made the concept of obtaining control of the air almost irrelevant. At the strategic level, the air force's primary role was to provide a second-strike option to destroy the enemy's nuclear capabilities, which meant that the air force had to develop at least limited strategic attack capabilities. The government accepted an altered doctrine based on the deterrent power of nuclear weapons and the use of strategic missiles, which assumed the strategic strike role of the air force in the European theater. The air force was to concentrate on support to the army through interdiction and close air support, which had profound doctrinal and organizational implications from an air power perspective.

In fact, the Soviet Union was prepared to fight a war, shifting toward a nuclear-retaliation–based deterrent doctrine, regardless of whether the attack was conventional or nuclear. The doctrine tacitly accepted the primacy of deadly aerial firepower over conventional fielded forces, leading to almost a decade of stagnation in doctrinal thinking within the Soviet Air Force. Fortunately, this belief in the primacy of nuclear weapons within the nation's security doctrine eroded with time and with the realization that conflicts were much more likely to take the form of conventional war than an assured nuclear exchange.

Until 1964 the Soviet political leadership aggressively pursued an ideology of nuclear standoff and deterrence with the United States that forced the military to become fixated on a single nuclear warfighting posture. This represented a classic case of political ideology overriding the strategic and philosophical doctrine of the military forces and discounting theories that were being developed.[60] It is now generally accepted that the one-sided emphasis on ballistic nuclear weapons, which caused the leadership to discount both the relevance of limited conventional warfare theories and the lessons learned in World War II, was detrimental to the doctrinal and operational development of the Soviet air forces' conventional capabilities.[61]

Post-1964 Doctrinal Shift

Even though nuclear doctrine had taken center stage, the Soviet military had also maintained a traditionalist school of thought, supported by some members of the political leadership, that continued to advocate the conventional view of military preparedness. This group argued that an effective military doctrine could not rely solely on one weapon system and emphasized the need for an approach wherein all elements of the military would combine to provide a balanced doctrine. With the passage of time, the belief in the inevitability of nuclear war was slowly tempered by the recognition that a war might start with the use of conventional weapons. Therefore, when the Soviet military recognized that the U.S. policy of flexible response ensured that any conflict with the West would result in conventional warfare for an indefinite period, some doctrinal groundwork was already in place to facilitate changes in military thought.[62]

The buildup of a nuclear arsenal was, therefore, paralleled by an equal investment in conventional forces. Following Nikita Khrushchev's fall in late 1964, the USSR started an unprecedented military expansion. The need to achieve parity in conventional capabilities became paramount. So vast was the buildup that it played a leading role in destroying the economy, which, in turn, was a major contributory factor in the collapse of the Soviet state.

The Soviet military fully understood the synergy between tactical aviation and ground operations, and the mutually supporting roles that tactical aviation and strategic rocket forces could play in the pursuit of ultimate victory. The flexible maneuverability of tactical forces, when combined with the ability of the rocket forces to strike rapidly over long distances, could produce devastating effects.

By 1968 the concept of command of the air that had been relegated to the background in the 1950s returned to the forefront of Soviet military doctrinal writing and thinking. Colonel N. Semenov clearly indicated this shift when he wrote:

> it is becoming quite obvious from the above [a discussion of the increased capabilities of modern aircraft] that the necessity of gaining air supremacy in conducting military operations without the use of nuclear weapons in modern conditions is becoming even more acute than in the past. However, it is clear that it will be considerably more difficult to resolve this problem. It will require a re-evaluation of many factors and a different approach to the use of forces and means.[63]

The tactical air force was given primary responsibility for obtaining and maintaining the required level of air superiority on a case-by-case basis. The second mission for tactical air forces was delineated as operational cooperation with the ground forces and dedicated support for their offensive. The two missions in combination formed the classic concept of an air offensive with optimized employment of all air assets across the entire theater.

Throughout the 1970s and early '80s, the Soviet Union pursued an ideology-dominated agenda that used the military buildup to project itself as a capable superpower and a counterbalance to the United States in international politics. This strategy hinged almost completely on the establishment of a large and powerful conventional force, which not only gave the nation a "sense of power," but also brought about a gradual but perceptible shift in the balance of power. The Soviet intention was to achieve "equal security" with the United States; equal security being defined not merely as parity in numbers, but also assured equality in the crucial factor of weapons technology.[64] The leadership measured national security in terms of political and ideological perceptions and viewed the military as a necessary tool to ensure the appropriate international image.

THE AFGHANISTAN EXPERIENCE

The provision of military advisers and arms to the nations of the developing world played a growing role in the overall Soviet strategy during the 1970s and 1980s. Although the performance of the equipment can be rated as mixed at best, these cooperative arrangements generated markets for Soviet weapons sales.

The Soviet Union's activism in the developing world finally embroiled the nation in the disastrous civil war in Afghanistan.[65] During the nine-year conflict, the Soviet Union supported the Marxist government of the Democratic Republic of Afghanistan against the Islamic mujahideen resistance. Deployment of Soviet troops began on December 24, 1979, under Leonid Brezhnev. The final troop withdrawal occurred gradually, commencing on May 15, 1988, and ending on February 15, 1989, under the last Soviet leader, Mikhail Gorbachev.[66]

The Soviet intervention was spearheaded by airborne divisions, followed by the ground forces. These forces relied on airborne supplies in the initial phase; in the second week alone, the Soviet Air Force flew around four thousand supply missions. Soviet reports indicate that the armed forces had difficulty fighting in the mountainous terrain, and the air force played an important role throughout the operations in sustaining the deployed forces through airborne supplies and the airlift of personnel. As the insurgency developed, the air force became a critical component in the combined arms operational concept by providing fire support. Initially, this close air support was supplied by helicopter gunships, but as the war progressed, fighter aircraft were also used extensively. Unofficial reports indicate the loss of nearly 300 helicopters and around 140 fixed-wing aircraft.

Soviet air operations in Afghanistan were conducted almost completely in support of ground operations. In essence, the Soviet Air Force demonstrated its flexibility in adapting to changed operational conditions, much as it had done during World War II. However, the adaptations were tactical in nature and had no long-lasting effect on the strategic outlook of the force.

By the time of the withdrawal from Afghanistan, the Soviet Union's imminent collapse meant that the military was also in turmoil. The breakup of the Soviet Union immediately following the withdrawal gave the military no opportunity to evaluate the performance of the different arms and services, carry out a comprehensive study of the lessons that could be distilled from the Afghan experience, and institute strategic remedial actions. Even so, some analysis was performed. The major lessons learned were at the operational and tactical levels, especially since a strategic reorganization of the military was being envisaged. First, the conflict drove home the criticality of airlift in sustaining expeditionary operations, especially when fighting an insurgency in inhospitable terrain. Second, the leadership came to recognize the vulnerability of helicopter gunships to heat-seeking man-portable antiaircraft missiles, and Soviet designers and manufacturers were able to rapidly modify the heat shields around the engines to reduce the loss rate. Third, the tactics employed by the helicopters were modified to reduce losses further. Fixed-wing aircraft were also used in interdiction and pre-assault "softening-up" missions before a ground offensive.[67]

THE LESSONS OF THE COLD WAR

The international geostrategic circumstances of the 1970s and 1980s, which included comparative stability in the security environment, permitted the USSR to gradually shift its ideological stance regarding its perception of future wars. Furthermore, the study of localized wars in Vietnam, on the Indian subcontinent, and in the Middle East raised four issues critical to optimizing the warfighting capabilities of the Soviet Air Force. Although created by the ideological pressures of the Cold War, these issues demanded doctrinal solutions to ensure the appropriate direction of future air power development.

First, the fighter-bomber aircraft of the 1950s were not optimized for any one dedicated role, having been designed to fulfill multiple roles.[68] With the doctrinal shift toward conventional warfare and the acceptance of the need to control the air, the military leadership recognized that the design philosophy would also have to shift toward the development of different kinds of aircraft optimized for different roles, such as air superiority, interdiction, and close air support.

Second, even the monolithic and state-controlled defense aviation industry recognized that the cost-intensiveness of manufacturing modern military aircraft precluded the capability to produce unlimited numbers of planes to act as backups for combat losses. The result was a focus on the development of standoff precision-guided munitions, which increased the survivability of expensive fourth-generation aircraft by keeping them outside the lethal envelope of enhanced battlefield air defense systems.[69]

Third, it became necessary to recast air defense concepts to create an integrated air defense network. The layered combination of antiaircraft guns, missiles, and interceptors with sufficient mobility provided the surface forces with enhanced maneuver capabilities. The outcome of this "radical" thinking was the move toward the design of fighters suited to contesting air superiority within the concept of a forward air defense. The thread that connects ideological change emanating from a stable geostrategic environment in the post-1973 period, through doctrinal transformation, to the operational employment of weapon systems is perhaps most clearly evident in this process, which altered the Soviet Air Force.[70]

Fourth, the Soviets realized that the air forces they had trained and equipped tended to come out second-best in most of the wars they fought. The Indian Air Force was a notable exception, but its better performance was attributed to domestically developed training and doctrine.[71] Further, the operational introduction of beyond-visual-range air-to-air missiles and sophisticated electronic warfare capabilities made it mandatory to review the training regime, especially in the realm of air combat.

This move was a doctrinal turning point. It began the process by which the strategic and operational ethos of the air force gradually changed. Even though the emphasis remained on offensive operations, command of the air over the theater was recognized as a prerequisite for military success. According to the concepts underlying Soviet doctrine, command of the air was achieved through a three-pronged offensive strategy: destroying the enemy's aviation assets, including support facilities and bases; defeating the opposing air defense systems within and beyond the theater; and neutralizing the command and control infrastructure. The core element of offense was retained, ensuring continuity in doctrinal development while implementing ideological and technological changes.

Doctrine, Training, and Aircraft Design

Air power is a cost-intensive capability. Even the richest nation cannot afford to build air power weapons systems without aligning its capabilities with the doctrinal requirements that guide its air forces. The Soviet Union was no exception. Throughout the Cold War, the Soviet aircraft design bureaus consistently produced aircraft that were aligned with the air force's concept of operations and comparable in performance to contemporary Western designs.

Every air force trains and fights according to accepted contemporary doctrine. Only by following broad doctrinal guidelines can an air force accomplish individual missions that collectively produce a successful air campaign. This was as true for the Soviet air forces as it was for the Western or NATO forces of the time. Therefore, even if one acknowledges that the Soviet air forces' performance was substandard—for which there is only convoluted and biased evidence—the fault would lie in the doctrine developed by the force and not in the training regime that faithfully ensured its alignment with the approved doctrine. It is more than likely that the fallacy of the "inferiority" of Soviet Air Force training arose from wishful thinking and subsequently became perpetuated as a propaganda initiative.

Western analysts have perceived two major shortcomings in the Soviet air forces, particularly in the Frontal Aviation forces.[72] However, when explained and understood through the prism of doctrinal interaction with training requirements and aircraft design, these do not feature as vulnerabilities of the force.

The first shortfall was the perceived qualitative inferiority of several aspects of the Soviet pilot training program. The Soviet training regime was very rigid and did not encourage individual initiative. The pilots' average annual flying time was some 40 percent less than that of their Western counterparts. More importantly, all missions, including air superiority ones, were controlled by ground radar, which deprived even formation leaders of the opportunity to exploit the inherent flexibility of air power so critical to mission success. Analysis from the perspective of Western air power doctrine supports these allegations. However, Soviet pilot training and operational employment were based on the doctrine of the Frontal Aviation command, which itself was only a part, albeit acknowledged as a critical one, of the air-ground operational team. Soviet doctrine required each pilot to carry out a given mission without fail, and tasking orders were given in such a way as to avoid the need for the pilot to make any individual decisions. The rigid training regime supported this fundamental doctrinal need for blind adherence to orders.

The comparison of flying hours is also lopsided; as previously mentioned Soviet Air Force missions were far shorter than Western missions. Soviet doctrine envisaged a relentless ground offensive that would also include the rapid capture of enemy airfields. The air force concept of operations was one of "leapfrogging" over these forward bases in direct support of the army. This did away with the need for lengthy transit to the target area and back, thereby reducing the total flight time of a mission while not substantially reducing the time over the target. It also had the advantage of precluding the need for the air-to-air refueling of fighter aircraft to ensure sufficient staying power. A more realistic basis of comparison would, therefore, be the number of sorties that each pilot flew, which would perhaps favor the Soviets considering the long duration of the missions of Western air forces.

The second shortcoming often mentioned is that Soviet aircraft were technologically inferior to the Western models, especially in avionics and weapon effectiveness. Even this argument is one-sided. The foundation of Soviet military strategy is the army concept of overwhelming the opposition by using comparable capabilities, but with total numerical superiority. The air power doctrine that developed from this broad doctrinal base emphasized the provision of adequate air support to the ground offensive, if necessary making even the quest for air superiority a lesser priority. This is the direct opposite of Western air power doctrine, which centers on control of the air. In the Soviet context, air superiority was contextual—in time and space—obtained through the employment of a concept of ground-controlled "spot-interceptions" carried out by attached fighter assets.[73] This tenet served to align the Soviet Air Force doctrine with the military concept of operations.

Thus, the design of fighter aircraft and their weapon systems in the Soviet Union clearly demonstrated the impact of strategic doctrine. In essence, Soviet fighter aircraft designs were driven by the air force's strategic and operational requirements, which, in turn, were fundamentally aligned with the prevailing doctrine. Soviet air superiority doctrine depended heavily on numbers to succeed. This was made possible by locating fighter units close to the ground battle and by ensuring that the design of both the aircraft and the weapon system emphasized ease of maintenance. Standardization and simple design of fighter aircraft supported the doctrine of operating at high intensity close to the battlefront. This requirement also led to the development of centralized but mobile maintenance units collocated with operational units, which gave the units much greater flexibility and deployment mobility. Almost all Soviet fighter aircraft are capable of operating from semiprepared airstrips and can sustain operations under very primitive maintenance conditions. For example, a MiG-21 squadron of twenty aircraft was expected to maintain a normal daily serviceability state of 75 percent and to be able to produce a minimum of ninety sorties in a day without having to surge. These expectations very clearly demonstrate a design philosophy that supports a doctrinal concept of numerical superiority and the saturation of enemy air defenses.

PERESTROIKA: COPING WITH CHANGE

By the mid-1980s, NATO doctrine had evolved to encompass the offensive concepts of Follow-on Forces Attack and AirLand Battle supported by greatly enhanced conventional capabilities. A pragmatic assessment of these changes compelled the Soviet Union to revise its military doctrine to incorporate defensive postures within the overall offensive stance. This viewpoint was anathema to the Soviet military, which was totally reliant on an offensive doctrine. However, the reality of the nation's economic weakness had become apparent to the highest levels of leadership and the doctrinal change was instituted, although it percolated down to the defense budget only gradually.[74]

Historically, the development of Soviet military doctrine had always been an evolutionary process, essentially retaining basic continuity despite changes resulting from the direct and indirect influence of political, economic, geostrategic, and technological developments. The changes instituted in 1985, however, were revolutionary. Therefore, this year is considered a turning point in the development of Soviet military strategy. The primacy of the military in all matters concerned with the nation's security was altered irrevocably.

As previously noted, military doctrine in the Soviet Union had always been oriented toward moral-political indoctrination of the general public, which included mentally preparing citizens for the prospect of future war. The Marxist-Leninist philosophy of war clearly identified the direct impact of economic, scientific-technological, moral-political, and military potential on the war-waging capacity of a nation.[75] The outcome of any war was considered to be the result of applying the nation's combined capabilities.[76] Marxism-Leninism therefore required that the Soviet Union always retain superior war-winning capabilities, and this aim was seemingly what *perestroika* (restructuring) set out to accomplish.[77]

Mikhail Gorbachev, who came to power in March 1985, was pragmatic enough to realize that the national security concept of relying exclusively on military forces, however capable they might be, would not be viable as long as certain other elements of national power, such as the domestic economy, remained inherently weak. He was committed to gradually reducing the financial outlay on the military and transforming Soviet forces so that they could adhere to the defensive doctrine that he ardently advocated.[78] In a historic speech to the United Nations General Assembly on December 7, 1988, Gorbachev stated that "their structure [the Soviet military] will be different from what it is now" and "after a major cut back [*sic*] of tanks their purpose will become clearly defensive."[79]

The changes in Soviet policy also altered the country's basic evaluation of the international political, social, and economic developments taking place. From a military perspective, the concept of nuclear deterrence started to lose its hold. The economic realities that made military restructuring a necessity also made it essential to accept that the military could play only a reduced role in providing security or supporting other national objectives. Soviet leadership expected the inevitable loss of status as a military superpower to be compensated through the success of the domestic social, economic, and political transformation that was to take place.[80] That the attempt to institute perestroika failed to stem the downward economic slide, leading to cataclysmic changes, is a separate issue.

The economic turmoil within the nation made it imperative for the government to embark on a modernization program under *glasnost* (openness or transparency) and then under perestroika to rebuild a stagnant economy. As perestroika progressed, the inadequacies of the nation's science and technology infrastructure were exposed and identified as a major weakness of the military-industrial complex, directly threatening the Soviet Union's superpower status. The acknowledged deficiency in the science-technology area led military leaders to believe that the military superpower status of the Soviet Union was endangered. From a military perspective, perestroika came to be viewed as a solution to this grave problem.

Gorbachev's foreign policy initiative, "new thinking," manifested itself in the imposition of economic constraints that proved an important factor in effecting doctrinal changes in the military. Further, the new foreign policy was designated as a principle on which the idea of "reasonable sufficiency" in military capabilities—which implied the unilateral reduction of military forces to a level sufficient for effective defense—was subsequently developed.[81]

By 1988–89, the military had undergone substantive restructuring, which led to the redefinition of some basic doctrinal foundations. The concept of "strategic stability," wherein military doctrine incorporated not only the concepts of waging and winning wars, but also of preventing war, became the watchword, although the term itself had entered the military lexicon only in 1986.

THE TRANSFORMATION OF THE RUSSIAN AIR FORCE

War prevention was essentially a defensive doctrine, although the rank and file of the military, brought up for generations on the premise of an offensive strategy, found it difficult to understand and, therefore, effectively implement. The new doctrine emphasized using political measures to build confidence and security, pursuing arms control and arms reduction, and, most importantly, seeking political solutions to hostilities. As a result, Soviet military doctrine was transformed from one of seeking dominance through escalation, as was practiced until then, to one of avoiding escalation in defensive battles and campaigns.[82] These changes, although subtle in terms of the nuances of the concepts, required that the air force reorganize both its strength and structure to adjust to the broader doctrinal shift. Avoiding escalation meant that the air force would have to fight a holding battle and retain the capability to cease fighting at the earliest possible moment. This new doctrine required a force structure different from the one that was optimized for predominantly offensive operations.

By the early 1990s, the economic and social uncertainties being felt in broader Russian society were reflected in the noticeably growing confusion in military thinking. The USSR began to revise military science—a "system of knowledge" regarding the characteristics of war that, while subordinate to doctrine, was also a separate entity and an essential tool for its correct interpretation—to accommodate the reduced availability of resources and the diminished status of the military establishment.[83]

Shifting from offensive to defensive doctrinal thinking is never easy and requires simultaneous implementation of several radical changes. The Soviet armed forces were slow to put the necessary changes into effect, prompting external observers to predict the complete collapse of the Soviet military.[84] In fact, the lethargy was that of a monolithic institution coming to terms with altered circumstances and calling on its inner strength to address the changes. That strength resided in the traditional ideology of the Marxist-Leninist thought process, the inherent flexibility of which had been constrained over a period of time. Therefore, the "changed," emerging doctrine and strategy of the new Russian armed forces are as clear and resilient as they were in the forces of the erstwhile Soviet Union.[85]

RUSSIAN AIR POWER AND THE NEW AIR WAR

In Soviet-Russian military thought, doctrine had to accommodate future technical capabilities, and a war-winning force had to be structured in accordance with the characteristics of future war. This is a sophisticated philosophy. Even while substantial doctrinal and organizational changes were being instituted within the military, strategic thinking centered on emerging technologies perceived to be at the heart of future military capabilities.[86] The military conceived a new concept of operations to support this doctrinal foundation, which was based on highly effective air- and space-based intelligence, surveillance, reconnaissance (ISR), and target-acquisition systems; overwhelming and precise firepower with great range and destructive capability; and automated command and control capabilities that could ensure a real-time strike. Incorporating these capabilities into a strategic philosophy provided the basis for a gradual long-term move toward waging an essentially new type of war—"aerospace war."[87] The concept of such a war was further refined by the creation of what came to be called "reconnaissance-strike complexes" and their successful integration into unified systems at the large formation level. This new concept brought about a fundamental change in the Soviet perception of the character and conduct of war and its expected outcome.

The doctrinal shift was clearly apparent in the changed modus operandi of the military forces. Under the new guidance, politico-military objectives would be achieved not by seizing territory and enemy centers of gravity, but by destroying the adversary's war-making capabilities and potential by making use of technological innovations.[88] From a basically two-dimensional view of war, with air power as a mere adjunct, the Russian perception changed radically to one of three-dimensional warfare based on air- and space-based systems. Russia interpreted the 1991 Gulf War as the first demonstration of the transition between the old and the new. The central lessons that the Russian military took away from that war were the importance of surprise (even though the military buildup had taken over three months) and of the coalition's command of the air, established from the very outset of the campaign. The Gulf War was characterized as an air war conducted according to a contemporary version of Douhet's strategy, made possible by greatly enhanced air power capabilities. The Russians surmised that the availability of improved technology had facilitated the conversion of this traditional concept into a war-winning air doctrine.

The reform process and changes in doctrinal development within the Russian Air Force started almost immediately after the Gulf War ended. The first step undertaken was a brutally honest appraisal of the performance of Russian equipment and a careful analysis of the West's operational concepts. The air force accepted a number of "lessons," however reluctantly, that indicated a changed strategic thinking among the air force's top leadership. ISR and precision weapons were identified as key operational factors underlying the success of the air campaign. It was tacitly acknowledged that the Allied air forces displayed great professionalism across all levels—from the strategic level of planning and control, through decentralized operational execution, to the tactical level of mission completion.

Although Russia's military hierarchy was slow to acknowledge the dominant role air power had played in winning the Gulf War, and even more reluctant to accept the relegation of the ground campaign to a mopping-up operation, the more progressive generals recognized the need to align military doctrine and structure accordingly. This acknowledgment was sufficient to break the tight control that the army had so far exercised over all aspects of the air force's development. For the first time in its history, the Russian Air Force cast off its shackles and was allowed to develop as an independent force. After seventy years of being undervalued and externally controlled, the air force now proceeded to eliminate political and ideological influences in the development of its doctrine and force structure. The Russian Air Force initiated and internalized a paradigm shift in air power doctrine and strategy that facilitated its emergence as a primary force operating at the high end of technology.

At the doctrinal level, the main fallout of the Gulf War was that, by the late 1990s, the Russian military had almost completely invalidated the 1985 shift to a defensive doctrine in both the military-technological and sociopolitical spheres. Russian military doctrine was refocused by the insistence that an inherently defensive doctrine meant neither an exclusively defensive strategy nor a rejection of offensive operations.[89] In typical Russian fashion, this also represented an evolutionary change that maintained an underlying continuity in the philosophical doctrine at all times.

DEVELOPMENT OF A NEW DOCTRINE

Political and economic imperatives drove many of the hard decisions taken by the Russian Air Force leadership after the Soviet Union's collapse. Attempting to come to terms with numerous challenges in both domestic and international arenas, Russia issued its new national security concept and military doctrine in 1993 and revised them in 1997 and 1999.[90]

Unlike the Soviet Air Force, the Russian Air Force kept ideology in the background. The influence of ideology on the formulation of core doctrine was replaced by national security imperatives and strategic initiatives. The air force made pragmatic decisions to ensure that the military aviation industry, one of the most developed sectors in the erstwhile Soviet military-industrial complex, would survive the economic downturn. The Russian Air Force reorganized itself through large-scale changes in personnel policies, equipment numbers, and capability requirements, and through doctrinal reorientation.

Russia's current security perceptions are influenced by a complex military and security environment resulting from the diversity of new states, their proximity to vastly different ethnic and religious regions of Europe and the Middle East, and Russia's own ability to create viable military forces.[91] Russia is now grappling with issues that include the serious erosion of its international power, status, and influence; the loss of diplomatic and economic leverage; the increasing disparity in military power with not only the United States, but also China; the rise of a secessionist movement in Chechnya; terrorism and the growth of fundamentalism in the border states; and the increasing alienation of the newly independent republics, with their tendency to embrace the Western model for economic prosperity. The most recent national security concept reflects five major factors with a strong impact on Russian security thinking: NATO expansion to encompass what is viewed as Russia's area of influence, the deployment of U.S. ballistic missile defenses in Poland (which has since been canceled), separatist movements and the rise of fundamentalist Islam in Chechnya and other areas on the Russian periphery, the growth of terrorism as a cross-border security threat, and the unease resulting from the actions of the United States and its "coalition of the willing" in initiating and pursuing unilateral action, bypassing the United Nations.

THE RUSSIAN AIR FORCE TODAY

The Russian Air Force has successfully undergone a transition from a quantitatively massive superpower organization to a much smaller force. It is also presumed to be definitely "meaner" and more competent, although there is no way to ascertain the altered capability, since it has not been tested. However, a minimal dichotomy remains in evidence. Russia continues to maintain some of the Soviet-era infrastructure required for global power projection, particularly ICBMs, although other capabilities that the Soviet Union considered necessary have been allowed to deteriorate. This may indicate that the long-term objective of the Russian government is to reestablish the nation as a global power. For the time being, even though sufficient nuclear deterrence is steadfastly maintained, the Russian Air Force doctrine is clearly oriented toward managing and containing regional conflicts in the nation's "near-abroad."

The Russian Air Force was used most recently in combat operations in the short conflict with Georgia in August 2008. Different perspectives have been reported regarding its performance in the conflict. The indisputable fact is that the air force was able to command the air effectively, albeit against very limited opposition, and deliver the effects needed to achieve its aims. It is also certain that the air force is remedying a number of deficiencies that were recognized and formulating improved operational procedures.

CONCLUSION

The policy that governs the employment of military forces in a nation-state is a calculated function of many factors that are debated and decided at the highest level of government.[92] Strategic doctrine, operational concepts, and warfighting capabilities of military forces derive from decisions that governments make to ensure national security and are defined contextually. The Soviet/Russian Air Force is no exception.

That ideology plays an important role in the development of military capabilities is not by itself surprising. It is the overriding impact of ideology on the concepts underlying employment of air power at both the strategic and tactical levels that makes the Soviet Union a case in point. For more than sixty years, ideology was in the driver's seat, guiding the development of the doctrine, strategy, and even operational tactics of the Soviet Air Force. The result was relatively low attention paid not only to development of air power capabilities, but also to the design and development of aircraft and air weapons.

Although the Soviet leadership understood the importance and potential of air power, unrelated circumstances and factors made the development of the Soviet Air Force uneven from its very beginnings. The air force was considered a subordinate capability, and its strategic doctrine was heavily influenced by the army, becoming stilted and biased in the process. Even though the Soviet Air Force at first performed abysmally during the German invasion of the Soviet Union, it emerged from World War II as competent as any other air force. However, the doctrine that relegated it to the position of a necessary support to the ground campaign was extremely detrimental to the balanced long-term development of air power capabilities. This doctrinal flaw plagued the Soviet air forces for the next four decades.

Even a cursory study of Soviet military forces reveals that the strategic military leadership recognized the potential of air power as a potent capability for force projection and deterrence. However, air power doctrine was greatly manipulated to suit the ideological requirements that echoed the greater Soviet ideal. Military inputs into the grand strategic doctrine were always army-centric, and the Soviet Union therefore never articulated a clear air power doctrine at the strategic level.

The spectacular success of the Western coalition in the Gulf War in 1991 proved a turning point for the Soviet Air Force. For the first time, the military hierarchy accepted that air- and space-based systems provided a third dimension to war and recognized the need for deep-strike standoff weapon systems and information warfare. This also represented the first acknowledgment of the concept of an independent air force equipped with an articulated doctrine of its own.

Currently, Russian Air Force doctrine is almost completely aligned with the global air power concepts accepted by other major air forces. The transition has been achieved through carefully considered inputs into the development of doctrine, rather than the purely ideological inputs that drove doctrine even a decade ago. This shift exemplifies, as nothing else can, the resilience and cohesiveness of the Russian Air Force, and demonstrates the leadership's exceptional understanding of air power issues. The Russian Air Force in its new incarnation is a force to be reckoned with at all levels.

5

INDIAN AIR POWER

Jasjit Singh

Like that of most countries, India's air power is not confined within the air force, although the Indian Air Force naturally constitutes the major element of the military air capabilities of the Indian armed forces' three main components. The Indian Army has acquired nearly two hundred helicopters since 1986, but does not possess combat aircraft. The Indian Navy has a mixture of air power capabilities, ranging from an aircraft carrier with combat aircraft to light surveillance aircraft and helicopters, most of which are now embarked on its warships. This chapter, however, will focus mainly on the Indian Air Force, while addressing relevant issues concerning the other components of the military. For reasons of space, this chapter does not cover civil aviation, though it, too, represents part of a nation's air power.

THE BIRTH OF THE INDIAN AIR FORCE

Formally established by a legislative act in 1932, when Britain was still ruling India, the Indian Air Force is one of the oldest air forces in the world today. Contrary to conventional wisdom, it was an independent military force from its very inception. It was patterned on the Royal Air Force (RAF), which had emerged from the erstwhile Royal Flying Corps of the British Army to become an independent component of military power in 1918. The Indian Air Force was always in a state of readiness; its people were in the air force from the start and not transfers from another service. (The other independent air forces of consequence in the 1930s were the RAF, the Royal Canadian Air Force, the Royal Australian Air Force, and the South African Air Force.) The United States Army

Air Forces (USAAF) was in the making, albeit as a component of the U.S. Army, as the name implied. The German Air Force and some of the other significant air forces had been disbanded after the First World War, although most of them were later reestablished.

The influence of the British Empire was obvious, as was the model of the RAF in the formation and formative years of the Indian Air Force. This is hardly surprising, since the RAF was deployed in India when the country was under British rule. After the First World War, Britain was concerned about Soviet expansion southward into Afghanistan; while British rule of India was secure, the northwest frontiers of India were unstable and needed protection. The grand strategy of Lord Curzon, India's viceroy at the beginning of the twentieth century, had already pushed India's frontiers as far as possible into inhospitable regions by establishing buffer zones and states to provide additional security to India. But an adequate frontier policy had to be put in place to ensure the viability of both new and old frontiers, and the air force was thought to be a cost-effective solution.

The 1920s were a period when the British Empire, exhausted by the Great War, developed what came to be known as "air control" as an efficient and economical strategy for maintaining a hold over the empire's thinly populated, but turbulent and violent, tribal regions. The RAF had only recently emerged from the army's control and was keen to find a justifiable peacetime role that would help maintain it and its independence in the coming years. Air control offered such an opportunity and role, especially at India's Northwest Frontier and in Iraq. Winston Churchill, as secretary of state for the colonies, became an ardent advocate of the air control strategy, and Marshal of the RAF Hugh Trenchard embraced it enthusiastically.[1] In 1921 Trenchard and Churchill secured formal approval to police the Middle East (particularly Iraq) with a force of eight squadrons—four bomber, one fighter, one fighter-reconnaissance, and two transport squadrons. With that, a new role for air power was born, which continues to the present day in the form of applying air power in asymmetric warfare and imposing "no-fly zones." In India, a new mission for air-ground cooperation also evolved during the interwar years in the form of close air support.

The Indian National Congress Party spearheaded the struggle for independence from Britain. While the party espoused a nonviolent ideology, it also clearly recognized that Indians would have to control the country's military power if India were to obtain independence and maintain its sovereignty.[2] The Indianization demands focused on commissioning Indians as officers in the military services so that leadership of the armed forces would sooner rather than later pass into the hands of the Indians themselves. The British imperial policy from 1918 onward of making Indians eligible for the king's commission only in the infantry and the cavalry fell far short of meeting these demands. On March 28, 1921, the Indian Legislative Assembly passed a resolution that included the demand that Indians have a greater share of the commissioned ranks in all arms of the British fighting forces, including the RAF. The assembly reiterated this demand later in 1921 and in 1925.

The British Government of India, responding to growing public and political demands, appointed what was called the Indian Sandhurst Committee (formally known as the Skeen Committee after its chairman, Lt. Gen. Sir Andrew Skeen, KCB, KCIE, CMG, chief of the general staff) in June 1925 to "examine the means of attracting the best qualified youths to a military career and of giving them a suitable military education."[3] The committee knew that India was not new to aviation; in fact, aviation had come to India within a few years of the first manned heavier-than-air flight by the Wright brothers in the United States. A number of Indians had served with great distinction with the British Royal Flying Corps during the First World War; for example, Lt. Indra Lal Roy won the Distinguished Flying Cross (DFC) for his exemplary performance in the war in Europe. Faced with such robust evidence of the Indians' performance in the Great War and in civil flying, the Skeen Committee, not surprisingly, came to the firm conclusion that refusing commissions to Indians in the air force was wholly indefensible.[4]

On October 8, 1932, the legislative assembly passed the Indian Air Force Act (XIV of 1932), and the governor-general ordered the establishment of the Indian Air Force effective as of that date. Meanwhile, pending the formation of the air force, twenty-nine of the *Hawai Sepoys*, or "Air Warriors," had been enrolled as apprentices and incorporated into the Indian Air Force wing of the Indian Technical and Followers Corps in September 1931, since their training period would be longer than that of pilots whose training commenced the following year. Previously, the Federal Public Service Commission had selected six Indians for a two-year flying training course at RAF College Cranwell, to be followed by training at RAF establishments; they had proceeded to England in

September 1930. On their return, the first Indian Air Force squadron, No. 1 (AC) Squadron, was raised on April 1, 1933, with the formation of its "A" Flight (four Westland Wapiti aircraft) at Drigh Road, Karachi.[5] The squadron, when fully staffed, was to be located at Ambala, alongside an RAF squadron. Flight Lt. Cecil A. Bouchier, DFC, seconded from the RAF, served as the first squadron commander.

In 1936 the then–wing commander John Slessor wrote what was perhaps the first manual on air-land joint operations, titled *Close Support Tactics—Provisional*, as a guide for an exercise.[6] He put his concept into practice—intertwined as it was with the RAF's air control strategy—during the 1936–37 campaign against the tribal fighters in Waziristan. Slessor was convinced that in low-intensity operations of the sort that characterized the conflict at the Northwest Frontier, "air and army forces had to work closely together."[7] Slessor's concept and the RAF and Indian Air Force operations during the Waziristan campaign altered the land-air warfare concept as practiced in Europe and taught in their staff colleges. A two-seater aircraft, such as the Westland Wapiti, would dive on enemy troops, strafing with its forward-firing Vickers gun, and release its twenty-pound bombs. During the vulnerable pullout, the rear gunner would spray the enemy with fire from his Lewis gun. Operationally, the solitary Indian squadron, with its Wapiti aircraft, was blooded in these actions within a short time after its establishment, before it had reached full strength. Along with No. 28 Squadron (RAF), it played a key role in establishing the tactics and procedures for close air support of the land forces.

On March 16, 1939, Squadron Leader Subroto Mukerjee (later air marshal, commander in chief of the Indian Air Force, and chief of the air staff) took over command of No. 1 Squadron—the first Indian to be appointed to this post. With the war clouds gathering over Europe, the Government of India set up the Expert Committee on the Defence of India (1938–39), otherwise known as the Chatfield Committee.[8] This committee recommended that, in addition to the regular air forces proposed for the local defense of India, five flights should be established on a volunteer basis for coastal defense duties. This proposal laid the foundations of the Indian Air Force Volunteer Reserve with coastal defense flights at Karachi, Bombay (now Mumbai), Cochin (now Kochi), Madras (now Chennai), and Calcutta (now Kolkata). Thus, the naval aviation component of the country's military air power was born as part of the Indian Air Force, albeit with various land-based aircraft, two decades before the Indian Navy established its naval aviation wing.

Meanwhile, the Indian Air Force had begun a modest expansion, officially only for support roles. With the outbreak of World War II in Europe, the Indian government calculated the requirements of its air forces to be twenty-one squadrons and the five coastal defense flights, with a total of 282 aircraft. Of these, the RAF was to maintain eleven squadrons, while the Indian Air Force would initially have ten squadrons.[9] Three new squadrons were to be formed by January 1942, with the five coastal defense flights progressively expanded to squadron strength. In fact, the Indian Air Force only achieved its planned force level by 1946, and its strength declined with the partition of the country in 1947, with the force level dropping down to six and a half squadrons, each having less than half the authorized number of aircraft. So began the cycle of the Indian Air Force's attempt to consolidate, but it was disrupted by factors beyond its control, until the next phase of consolidation was followed sooner rather than later by another disruption, and so on. Among these challenges, we must include the imperial legacies that burdened the Indian military long after the country became independent.

WORLD WAR II COMES TO INDIA

The Indian Air Force confronted enormous challenges during the eight turbulent years from the formation of its first squadron to its immediate involvement in the Second World War, when it faced the then-first-rate military power of Japan, which was undefeated in Southeast Asia. By late 1943, the Japanese had consolidated their position in Southeast Asia, even occupying the Andaman and Nicobar Islands in the Bay of Bengal.

During the Second World War, two dozen Indian pilots first saw action during the Battle of Britain and in bombing missions over Germany and German-held territories. Of the original twenty-four pilots sent to England in September 1940, eight were killed in operations; the survivors returned to India after the war in Europe ended but continued in the East.

The dichotomies of imperial policies can best be seen in the role that the Indian Air Force played in Burma. The Japanese attack on Pearl Harbor coincided with the massive invasion of Southeast Asia. Britain's belief in the impregnability of the region's defenses, thanks to the presence of the Royal Navy, rapidly crumbled with the sinking of the battleship HMS *Prince of Wales* and the battle cruiser HMS *Repulse,* the two capital ships that formed the core of the Royal

Navy in the East. The British Burma Army fought well but had to keep retreating from the Japanese juggernaut until the latter had halted along the river Chindwin in Burma.

It was during this "great retreat" that the Indian Air Force, still in its very early formative stages and possessing only the most obsolete aircraft in the British inventory, entered the Second World War in the Eastern theater. The solitary squadron, No. 1(AC) Squadron (Indian Air Force)—equipped with slow-flying, lumbering Lysander reconnaissance aircraft and commanded by young Squadron Leader KK "Jumbo" Majumdar—was quickly moved from Peshawar in Northwest India to Toungoo in eastern Burma. The squadron carried out the move in two days, traversing a distance of more than 3,500 kilometers and arriving at Toungoo on February 1, 1942, with all its 12 aircraft. The Japanese, already entrenched in Thailand, promptly bombed the airfield, but failed to damage any of the squadron's planes, which had been camouflaged immediately after landing. Thus began the Indian Air Force's operational history in a world war.

Majumdar decided to retaliate against the Japanese bombing. At Majumdar's command, by the following day Indian technicians had modified the Lysander reconnaissance aircraft to carry two 250-pound bombs under their wings. On February 3, Jumbo led his squadron to bomb Mae-Haung Saun, the air base from which the Japanese bombers had taken off to bomb Toungoo on February 1. The Indian Air Force achieved significant results. Thus, ironically, a squadron formed and trained in army cooperation carried out its first operational mission in a regular war by attacking a foreign air base. Majumdar flew fifty-six operational hours and was decorated with a DFC by the British.[10] More importantly, the bombing missions set the tone for an expanding role for the Indian Air Force despite its being equipped only with obsolete, slow-flying reconnaissance aircraft.

As the Japanese advance continued, No. 1 Squadron was withdrawn to rear bases and finally to Rangoon; it was the last unit to evacuate the city after the Japanese forces had entered the capital. Meanwhile, six coastal defense flights had been established using whatever type of aircraft (from Wapitis to Blenheims) the authorities could lay their hands on. The Japanese fleet had entered the Indian Ocean and, curiously, sailed up the Bay of Bengal, along the Indian coast, bombing the cities of Madras, which the British governor had ordered evacuated; Vishakapatnam; Kakinada; and continuing all the way to Calcutta. But the

Japanese made no effort to occupy any territory in mainland India, which was in no condition to defend itself against the onslaught. An occupation of India and/or Ceylon at that stage would have made it impossible for the Allies to recover the subcontinent later. Winston Churchill believed this to be the turning point of the Second World War.

Two years later almost to the day, the No. 1 Squadron, now equipped with Hurricanes, was back on the Burma front with the Fourteenth Army at Imphal, arriving just in time to take part in the famous Battle of Imphal-Kohima.[11] By that time, India was raising more squadrons by converting the coastal defense flights, which had grown to eight. The air force was rapidly expanded to nine squadrons, all of them deployed at one time in the Burma theater in 1944–45. While the air force made a seminal contribution to rolling back the Japanese invasion of India, its No. 1 Squadron in particular played a crucial role in the Battle of Imphal-Kohima, while itself operating under the Japanese siege at Imphal airfield.[12] Lord Louis Mountbatten, Supreme Allied Commander, South East Asia Command, specially flew into Imphal to award the DFC to the twenty-four-year-old squadron commander, then–squadron leader Arjan Singh (now the sole marshal of the Indian Air Force). The ceremony took place on the air base tarmac.

During 1944–45, the Indian Air Force's role and operations had inevitably expanded to include bombing (with Vulti Vengeance light bombers) and Spitfire and Hurricane fighter aircraft that escorted RAF and USAAF transport aircraft on the Burma front against the Japanese. The half-dozen operational squadrons flew a total of sixteen thousand hours in operations in the short time they were deployed on the Burma front to support the Fourteenth Army. Indeed, all of the air force's nine fighter squadrons played a stellar role in Burma, and this was recognized by assigning No. 8 Squadron of what at that time was known as the Royal Indian Air Force to Japan as part of the Allied occupying forces.

INDEPENDENCE AND AFTER

Here, it may be useful to touch briefly on India as a nation and a society. India emerged from colonial rule in 1947 as an agrarian economy unable to feed itself. What little industry it had was mostly small scale as a consequence of deindustrialization over the previous three centuries.[13] India set about firmly establishing itself on the path to what has become history's most ambitious experiment:

transforming a stratified, discrimination-based society into an egalitarian society, and an agrarian economy into a modern, balanced economy characterized by social justice, and doing so through consultative politics while maintaining secularism and tolerance. The democratic principle required that the leadership make the development of India's large population (now one billion) its highest strategic priority. Moreover, defense itself would require a strong economic production base.

The government adopted central guiding principles that rested firmly on nonalignment and self-reliance. Nonalignment should not be confused with the Nonaligned Movement. Contrary to conventional wisdom, the Cold War was neither the sole nor the most critical reference point of nonalignment, which emanated from and grew as part of a wider complex of national and international factors in the post–World War II period. In India, the roots of the nonalignment policy can be traced back to at least 1939, when the leading Indian National Congress, at its session in Haripur, passed a resolution: "India was resolved to maintain friendly and co-operative relations with all nations and avoid entanglements in military and similar alliances which tend to divide up the world into rival groups and thus endanger world peace."[14] Self-reliance became a natural partner to the policy and strategy of nonalignment, and was to be achieved through deep engagement with any and every country. Thus, the country's military policy was founded on the premise of a defensive orientation, a minimalist approach to investments in defense, and a firm belief that the use of force *must* remain an instrument of last resort.

STRATEGIC CULTURE AND THE ROLE OF FORCE

In the final analysis, the behavior of countries is strongly influenced by cultural factors, including historical experience, perceived national interests, and capabilities that control their values and belief system. The roots of Indian strategic thinking go back to ancient times, and some aspects are relevant to conflict prevention and management, especially regarding the role of force. While the concept of force occupied an important position in the political theory of ancient India, it had to be tempered by the concept of *dharma* (duty), which itself was regulated by many checks and balances.

The theory emphasized two aspects of the actual and threatened use of force. First, several alternatives to force were accorded a higher priority than the actual use of force.[15] However, the theory accepted force as one of the state's principal branches and therefore acknowledged the importance of powerful armed forces that should be supported to keep peace within and outside the state.[16] Manu (ancient Indian political-social thinker and adviser to kings around 3,000 BC) described the four political tools at the disposal of the state as conciliation, diplomacy, or *sama*; concession, gifts, grant in aid, or *dana*; sowing dissension, or *bheda*; and war or use of force, or *danda*, in that order of merit. The use of force was the last option. Even within the use of force, Manu held that there were six expedients, which included threats to resort to force such as *yana* (mobilization) or *asana* (readiness to attack), that could be used before resorting to the last alternative, which involved actual physical force.[17] These concepts—the primacy of diplomacy, conciliation, and the search for alternatives to the use of force—appear to have resurfaced, at least at the subconscious level, with the emergence of India as an independent nation-state.

The second aspect concerns the strong linkage between the use of force and the principles of duty and justice. This, as Dr. Nagendra Singh has pointed out, has historically put heavy restraints on the use of force:

> A study of the constitutional history of India reveals that, despite the dictatorial nature of the authority arising out of control over the armed forces, political theorists have tried to impose certain checks, based on what was universally accepted [as] reasonable and what could be justified by considerations of *law*, whether founded on the concept of *dharma*, or the authority of the *shariat*, or, in later times, with the advent of the British, by their strong sense of legalism. These fetters placed on the authority that wielded the armed might of the State may be said to have generated the forces which ultimately crowned *lex* as *rex* and thus subordinated the armed might of the State to *dharma* (duty) or *sunna* or law.[18]

Seen against this background, India's politico-military doctrine faces an apparent contradiction: the creation of a credible military force and the negation of force as an element of interstate relations. The latter principle, however, is based on long-term idealism, while the former recognizes the imperatives of short-term pragmatism. The issue is not so much one of maintaining military power, but of determining how it is to be used.

During the first two decades of independence, India relied primarily on the alternatives to force and thus left itself underprepared to meet aggression from countries that did not share the same belief system. In more recent years, there has been a growing realization that an adequate and credible defense capability is an essential deterrent against aggression and adventurism.[19] However, the governing principles and restraints on the use of defense forces remain unchanged, especially with regard to using those forces in roles other than territorial defense.

India has also been forced to pay increasing attention to the expansion of its interests as a consequence of its robust economic growth, which, in the last decade, has hovered near 8 percent. These interests require protection, potentially by military power, as does the expansion of military tasks, where the Indian Air Force is most likely to be the major instrument of choice.

THE PEACE IMPERATIVE

India has made tremendous progress since it became independent more than six decades ago. Before it gained independence, India had been deindustrialized for more than two centuries. But, at the time of independence, it also lacked food for its people, and the shortage persisted for two decades, so long that India was living what was termed a "ship-to-mouth existence." Since then, tremendous social and economic changes have taken place. The proportion of people below the poverty line has been reduced from 76 percent to 18 percent, but hundreds of millions of people still remain poor. Since the political system depends on the will of the people, exercised through the electoral democratic process, the national policies naturally reflect the people's needs and aspirations. Thus, as noted, the central priority for India has been, and for the foreseeable future is likely to remain, the comprehensive socioeconomic development of its large population. The experiences of the past decades clearly point to peace as a prerequisite for such development. Hence, deterrence will remain the central doctrine of its military power.

THE INDIAN AIR FORCE IN WARS SINCE INDEPENDENCE

The Cold War provided the context in which India had to plan for its security and defense during its first forty-five years of independence. The policy and strategy of nonalignment served as powerful instruments for enhancing national security during this period. At the same time, as outlined above, India's approach to conflict and to the use of force meant that defense doctrine and policy were heavily weighted in the past toward a defensive philosophy.

Unfortunately, this posture often encouraged aggression. For example, India unilaterally restricted military operations to the territories of the state of Jammu and Kashmir during 1947–48, even after obtaining clear evidence that Pakistani military forces were directly engaged in fighting against Indian forces.[20] This apparently led the Pakistani leadership to conclude that India would not cross the international border to apply military pressure on Pakistan, even in cases where the risk of losing Kashmir was high, and would instead prefer to seek political-diplomatic intervention by external powers. This perception was reinforced by the Rann of Kutch conflict in April 1965, which appears to have been a rehearsal for the war that was to follow a few months later.

PAKISTAN'S FIRST KASHMIR WAR, 1947–48

Before the Indian Air Force could recover from postwar demobilization and stabilize its ten-squadron force, India's partition led to a similar division of the air force, leaving India with six of the ten squadrons, while the RAF pulled out its eleven squadrons. Within weeks after the state of Jammu and Kashmir had formally acceded to India in accordance with the provisions of the transfer of power arrangements laid down by the British, Pakistan launched a covert war in Kashmir,[21] which was rapidly followed by its overt military involvement. Pakistan continued to occupy nearly one-third of the state, violating United Nations (UN) resolutions as the result of its aggression and doing so with the support of the Western powers (essentially the United States, following the advice of the British foreign minister), which saw the conflict through the tinted prism of the Cold War.

In this conflict the Indian Air Force played a crucial strategic role by airlifting troops to the only grass strip at Srinagar, while the raiders were already at the gates of the capital city. The solitary transport squadron, equipped with DC-3 Dakota aircraft (designed in the 1930s), became legendary thanks to its regular supply runs and landings at small fields at Punch (to deliver guns to the army during the dark of night) and at Leh (at an altitude of over ten thousand feet). The squadron accomplished this despite having no navigation facilities and no pressurized aircraft, and being deprived of oxygen after flying over peaks fourteen thousand feet high. Meanwhile, the small force of three Tempest squadrons provided sustained support to the Indian Army.[22]

The Pakistan Air Force (PAF) did not attack the Indian Air Force's operations or support its own army during the fourteen-month war. An Indian Air

Force Tempest did shoot down one PAF transport aircraft on a supply mission inside Kashmir.

THE SINO-INDIAN WAR IN THE HIMALAYAS, 1962

Triggered by China in 1962, the Sino-Indian War caught India totally unprepared on land and in the air.[23] The Chinese launched the war to "teach India a lesson" because, in the words of Zhou Enlai, India "was moving too close to the United States."[24] Mao Zedong had earlier stated (in 1959) that "[t]he border conflict with India—this is only a marginal border issue, not a clash between the two countries" but "we should crush him [Prime Minister Jawaharlal Nehru]."[25] But in retrospect a major factor in China's decision was clearly the issue of Tibet and the revolt in 1959, which also led to the Dalai Lama's seeking refuge in India (although some accounts claim that this move was facilitated, if not actually organized, by the Chinese authorities to remove a rallying point from the province).

Recently declassified documents from the Cold War period reveal that Mao Zedong and other Chinese leaders stated on October 2, 1959, that "nobody knew precisely what actually occurred on the China-Indian border" earlier that year that led to Chinese guards killing twelve Indian policemen. This incident triggered complications in the relationship between the two countries. Zhou Enlai admitted that the government in Beijing had been unaware of the action, saying that the local authorities had undertaken all the measures there (i.e., the killing of Indian policemen) "without authorisation from the centre."[26] But by that time Mao and his senior advisers, in their drive to export the socialist revolution and to secure a global leadership role for China and Mao himself, were clearly looking for an excuse to humiliate India—and Nehru—using military power. They were no doubt fully aware that India was not militarily prepared to defend the Himalayan frontier, largely because there were no roads in the High Himalayas capable of carrying military vehicles or keeping ground troops supplied for a war with a major country. Roads began to be constructed in 1959, but it would take more than a decade before they became fully usable. In fact, the responsibility for border defense, previously under the jurisdiction of the police and the intelligence bureau, had only been transferred to the Indian Army in 1961.

During the conflict, India used only the Indian Air Force's airlift component, which was employed intensively to provide logistics to the army. While the air force had started its expansion program from twenty-five to forty-five squadrons, the twenty-odd combat squadrons available at the time were not called upon. India had several reasons for not employing this potent combat force. First, there were concerns that the Chinese would bomb Indian cities, especially Calcutta—a significant danger, given the major weakness of Indian radars and interceptors to respond to any potential bombing missions. Second, the army leadership feared that the use of combat force would lead to retaliation and that the Chinese would target the aircraft supplying the army that India was rapidly introducing into the Himalayan battleground. Given the lack of roads, this would significantly reduce the army's ability to fight. Third, the air headquarters supported the nonuse because it believed that the Indian Air Force should focus on only one mission, close air support, and this would not be effective in the Himalayan jungle terrain at very high altitudes. Fourth, India's decision makers shared an overall belief that using the air force would "escalate" the war, which, in turn, would harm Indian political and national goals. The U.S. ambassador at the time, John Kenneth Galbraith, reinforced this viewpoint.[27]

Fearing the possibility of Western intervention and of a volte-face by the Soviet Union, which might decide to support India, China declared a unilateral cease-fire and withdrew from the territories it had occupied during the war but that it still claims. In a sense, China denied India the war for which it had prepared, and the territorial dispute continues despite decades of dialogue and negotiations.

Pakistan's Second Kashmir War, 1965

In 1965 Pakistan launched a three-phased, integrated, covert-overt military aggression. The crucial thrust came during the third phase, on September 1, 1965, in the form of an armored blitzkrieg, supported by corps-level artillery and American-supplied F-104 and F-86 fighters; its objective was to take Akhnur in the state of Jammu and Kashmir and cut India's road link into the state. India, in the words of its western army commander, had only a "truncated brigade" in the area. By late evening, the Pakistan Army had made major gains. At this point the air force went into action. This was the first full-fledged war that the Indian Air Force had been involved in since India's independence, and it took place when the air force was in the middle of a massive expansion and reorganization. Twenty-six fighter sorties destroyed more than one-third of the Pakistani Patton tanks and caused extensive damage to artillery and vehicles before nightfall. This blow to Pakistani forces blunted the blitzkrieg, and it took them thirty-six hours to reorganize and resume the advance. By this time the Indian Army had moved in more troops and artillery for defense. The Pakistan Army was unable to reach Akhnur, and the PAF lost two F-86s to Indian Air Force Gnats and abandoned its aggressive tactics, switching instead to defensive ones.[28]

India's counterattack came on September 6. It was circumscribed by significant political restrictions that limited the war to the aims approved by the Indian prime minister on September 3, 1965:

1.　To defend against Pakistan's attempts to annex Kashmir by force and to make it abundantly clear that Pakistan would never be allowed to wrest Kashmir from India.
2.　To destroy the offensive power of Pakistan's armed forces.
3.　To occupy only the minimum Pakistan territory necessary to achieve these purposes which would be vacated after the satisfactory conclusion of the war.[29]

The war ended on September 23 after a UN-brokered cease-fire, with the Indian Army occupying large tracts of territory across the border. The Indian Air Force had won a resounding victory against the PAF, which fought well using its high-technology aircraft and radars. The air-to-air combat exchange ratio was three PAF fighters lost to one Indian. At the same time, the Indian Air Force provided 90 percent of the air support required by the army.[30]

According to its official history of the 1965 war, we find that the PAF flew a total of 2,364 sorties.[31] Indian official estimates place the PAF's total aircraft losses at forty-three aircraft, which equates to the PAF having lost 1.8189 aircraft per hundred sorties flown, that is, an attrition rate of 1.82 percent. Conversely, the Indian Air Force, with ten of its squadrons assigned to the eastern sector but not permitted to take offensive action, flew 3,937 sorties in the western sector alone, and lost a total of fifty-nine aircraft in the air and on the ground, in the western and eastern sectors combined.[32] This works out to 1.4986 aircraft lost for every one hundred sorties flown. In other words, the Indian Air Force suffered an attrition rate of 1.50 percent compared to Pakistan's 1.82 percent. The difference is not so marked that one can draw an unambiguous conclusion about the winner and the loser, but a more than 21 percent differential in the attrition rates of aircraft would matter considerably over time.

Further, these numbers do not take into account the political restrictions on the Indian Air Force, which prevented it from taking offensive action in East Pakistan. This would have reduced losses in the east once the air force was permitted to go on the offensive (with ten squadrons deployed in that sector against the single squadron of PAF), leading PAF attrition to balance the Indian Air-Force losses. The attrition rate would obviously have been greatly in India's favor.

Even so, during the war, the Indian Air Force lost an unacceptably large number of aircraft on the ground to enemy action (thirty-five out of a total of fifty-nine). Many of the planes were lost to two PAF strikes: one on Pathankot on September 6, and the second on Kalaikunda (in the eastern sector) on the morning of September 7. The reasons are not hard to find: they include a lack of early warning radars, inadequate dispersal and deception, and carelessness and complacency at the air base. Pathankot had not launched its fighters despite being alerted to an incoming raid, and its antiaircraft guns did not open fire even though the F-86s carried out multiple passes. Kalaikunda had not taken even the most basic precaution of air defense; its antiaircraft guns arrived only on September 6, and, except for three guns, they had yet to be sited. Hence, the terminal defenses failed at both places.

While the PAF's higher technology was a source of concern, it did not prevent the Indian Air Force pilots from engaging PAF F-104s and F-86s. By the time the war had reached the halfway point, these high-performance, missile-firing aircraft had lost their mystique, thanks to the better skills of the Indian Air Force pilots. In fact, an F-104 that had bounced a Mystère formation after its daylight strike on Sargodha on September 7 was finally shot down by Squadron Leader A. B. Devayya in a classic low-level dogfight. The Mystère was essentially a ground attack aircraft, and to fulfill this role it was the most heavily armed aircraft in the Indian Air Force, which made this "kill" even more significant.

Setting aside the aircraft lost on the ground due to enemy action, we find that India lost a total of twenty-four aircraft in air-to-air action and to ground fire. Measured against the 3,937 sorties flown, this amounts to a loss rate of 0.6096 aircraft per 100 sorties flown. In comparison, the PAF lost forty-three aircraft. It admitted to losing only one F-86 on the ground to Indian Air Force action (at Sargodha). Taking this at face value, against a total of 2,364 sorties flown in the war, the PAF's loss rate in air warfare would amount to 1.7766 aircraft for every 100 sorties flown. In other words, the PAF was losing aircraft at nearly three times the rate of the Indian Air Force in air-to-air engagements (1.78 percent vs. 0.61 percent). After President Field Marshal Ayub Khan made a secret visit to Beijing, Air Marshal Nur Khan, commander in chief of the PAF, advised him to ask China to exert stronger pressure on India. Otherwise, he cautioned, the PAF would cease to act as an operational force if the war were to continue for another two weeks or so.

Looking closely at the counterair operations, we find that India's defense minister only authorized strikes on Peshawar on September 12 and that the bulk of the Canberra light bomber/interdiction fleet was being used for close air support. Pakistan had been withdrawing its aircraft to Peshawar and other airfields deeper inside its territory (even to Zahedan in Iran) since the first day of the war. They were clearly beyond the range of Indian Air Force fighter aircraft and could be attacked only by the Canberra force. Fortunately, the chief of the air staff was finally able to persuade the government to allow the strikes. On the same night, Indian Air Force Canberras struck Peshawar in what the pro-Pakistani author John Fricker called "the most effective Canberra attack of the war."[33] Altaf Gauhar, information secretary to President Ayub Khan, reported that in a cabinet meeting soon after the 1965 war the president stated emphatically: "I want it understood that never again will we risk 100 million Pakistanis for 5 million Kashmiris—*never again*."[34]

THE INDIA-PAKISTAN WAR, 1971

The war in 1971 resulted from severe political instability in Pakistan owing to the continued discrimination against non-Punjabis, especially the majority Bengali population of the country. The military-political complex in West Pakistan refused to implement the results of the country's first free and fair elections based on universal adult franchise, which would have brought the Bengalis to power. Instead, the leadership of West Pakistan authorized what independent observers have termed the genocide of the Bengalis by the Pakistani Army in the spring of 1971. Indian diplomatic efforts to broker a negotiated political solution received no support from the international community. Over ten million refugees poured into India from East Pakistan, trying to escape the holocaust that the Pakistani army had perpetrated to suppress political dissent. While the international community, especially the liberal democracies, did nothing during the eight months to persuade Pakistan to implement the election results, the internal standoff in Pakistan continued. The subcontinent was clearly facing a progressive deterioration in the situation and, as a result, the possibility of sliding into war.[35] Given strong leadership and the time available for preparation, the high-level defense management in India worked well in spite of occasional interservice hiccups.[36]

As it became clear that Pakistan was not willing to implement the results of the election, East Pakistani leaders, with the help of Indian agencies such as the border security forces, trained a militia called *Mukti Bahini* (Freedom Fighting Force). By October 1971 this force started to retaliate against the Pakistani Army in East Pakistan. Meanwhile, after Gen. Yahya Khan refused to call the parliament into session, the Awami National Party (which had won all but two of the seats in East Pakistan and held the majority in the Pakistan parliament) declared the region's independence. By November 19, cross-border skirmishes had started in the east. On November 22, the Indian Air Force shot down three out of four PAF aircraft formations (F-86 Sabres) that had violated Indian airspace, possibly with the intention of launching an attack on the *Mukti Bahini* camps.

As regards the Indian Air Force strategy, it is useful to quote Air Chief Marshal PC Lal, then chief of the air staff, who later wrote:

Early in 1969, at the Commanders' Conference, we decided that the priorities for air operations had to change. Air defence of our homeland and the air bases remained priority one. The next most important job was support of the Army and Navy, the Army taking precedence over the Navy. Bombing, especially as a weapon to neutralise or counter enemy air, came third in our list of priorities. The other operations like paratrooping, transport and so on came thereafter. . . . As we saw it, the Air Force was primarily, and essentially for air defence.[37]

It must be noted that there was no written and commonly understood air power doctrine in 1971, although the Indian Air Force had been in existence for four decades.[38] The dichotomy becomes obvious through PC Lal's views on air war priorities, when he wrote in the same volume that "[t]he basic requirement of air warfare is air superiority: in other words the freedom to operate where one pleases without undue interference by the enemy." The second priority he listed was interdiction, and the Indian Air Force's success in this mission resulted in an acute fuel shortage in Pakistan, forcing it to import fuel "in tankers by road from Iran."[39]

Ultimately, Pakistan launched a surprise air strike on Indian Air Force airfields in the western sector on December 3, 1971, which led to a full-fledged war. Pakistan's military grand strategy was based on the concept of initiating an "all-out" coordinated two-corps offensive deep into India (in the Punjab-

Rajasthan sector) under the command of one of Pakistan's outstanding military leaders, Lt. Gen. Tikka Khan.[40] The reserves located south of the Ravi River, consisting of one armored division and two infantry divisions, would form the spearhead and the Army Reserve North would strengthen the offensive. The PAF air headquarters had agreed to cooperate closely in air defense for the offensive, attack Indian Air Force forward-deployed airfields, and provide close air support to the advance where necessary.

According to official accounts, "The overriding priority of PAF (in 1971) was to give maximum support to General Tikka Khan's proposed offensive into India; every other air force objective was to be subordinated to this requirement."[41] The planning staff at Pakistan's air headquarters estimated that the cost "worked out to 100–120 combat aircraft and pilots over the projected 7–10 day period." In the end, the offensive never began, since the Indian Air Force had put a huge dent in the armor of the Army Reserve South as it moved up to its launch position near Suleimanki. Farther south, Pakistan's 18 Division, with two armored regiments, launched an offensive to capture the Jaisalmer air base; however, a handful of Indian Air Force Hunters, in a classic and unique aircraft vs. tank battle, destroyed the force at Longewala in the Rajasthan desert.

On December 12, acting on the basis of intercepts by its electronic intelligence units, the Indian Air Force launched a four-aircraft mission (MiG 21FLs) to strike at the East Pakistan governor's residence, where the top leadership was to meet. During the meeting, the aircraft hit the target with rockets, after which the East Pakistan government decided to surrender. As a result, ninety-four thousand prisoners of war surrendered to the Indian Army. The war ended in the emergence of an independent Bangladesh.

The air force had played a major role in India's victory. The total attrition rate in the two air forces was 0.48 percent for the Indian Air Force and 3.2 percent for the PAF. In the east the Indian Air Force acquired air supremacy within twenty-four hours and after that concentrated on supporting the land forces. The air-to-air combat exchange ratio was similar to that of 1965, with the PAF losing 2.14 combat aircraft for every Indian Air Force fighter downed.[42] Pakistani major general Shaukat Riza was to write later that "by December 14, [the] Indian Air Force had achieved alarming success."[43]

The Indian Navy had obtained its organic air power (Alize and Sea Hawk aircraft, and a few light helicopters) in the late 1950s when it acquired an aircraft carrier, renamed INS *Vikrant*, from the United Kingdom. The aircraft carrier was under maintenance when the 1965 war took place and had no part in that conflict. In the 1971 war with Pakistan, INS *Vikrant* and its aircraft played a seminal role, with its aircraft attacking targets in East Pakistan while operating from the Bay of Bengal. Indian Navy pilots on exchange duties with the Indian Air Force flew Hunter aircraft during the war. One of them, young Lt. Arun Prakash (later admiral and chief of naval staff) was awarded the Vir Chakra for gallantry in the face of the enemy during the war. Since then, the navy has modified all its warships to carry helicopters and the indigenous destroyers of the INS *Delhi* class carry two helicopters each.[44]

THE KARGIL WAR, 1999

In the summer of 1999, Pakistan, which had been backing up its covert war through terrorism in the Punjab and the state of Jammu and Kashmir since 1983, launched a major military offensive across the line of control in the high-altitude and largely unpopulated areas of the Kargil sector. Pakistani troops occupied a substantial bridgehead nearly 120 kilometers (km) by 6–9 km, at altitudes of eleven thousand to eighteen thousand feet. From there they could dominate the only road link between Srinagar and Leh, through which all civil and military supplies were transported into Ladakh and the Siachen Glacier region.[45] Fighting uphill and facing sophisticated firepower from the bunkers built by the Pakistani troops, the Indian Army repelled the aggressors. Gen. VP Malik, then chief of army staff, later wrote:

> As the magnitude of the intrusion became clear, it became necessary to employ air power for various purposes: to support ground operations; to carry out reconnaissance; to interdict enemy supply routes and logistics bases; to destroy enemy footholds; *and, most importantly, to establish strategic and tactical superiority over the enemy.*[46]

The first-ever use of the air force at Himalayan altitudes was a key factor in defeating the Pakistan Army in the Kargil sector. The Indian Air Force provided invaluable support in a very restricted area—a valley between steep mountains—and under tight political constraints imposed by the government, including orders not to cross the line of control.

The Indian Air Force flew over twelve hundred fighter sorties during the war and lost one fighter on the first day to surface-to-air missiles. However, the PAF did not contest air superiority in the Kashmir region, even though this might have made a significant difference, since India's artillery deployments presented lucrative targets, being sited in the few open places among otherwise steep mountains. Nor did the PAF try to provide air support to its army, which had come under severe pressure from the Indian infantry, artillery, and air force.

It seems strange that Pakistan did not employ the PAF's combat component in support of its army, which has always enjoyed a preeminent position in the country's political system and had lost 720 men in the short war. The PAF's F-16s flew over Pakistan-occupied Kashmir, and its helicopters provided logistics support across the line of control. One possible conclusion is that Pakistan, overconfident because it had possessed nuclear weapons since the previous year, had not taken into account the likely use of the Indian Air Force. Even if it did consider this possibility, the leadership may have concluded that the Indian Air Force would be ineffective at such altitudes, since there was no historical precedent for these operations. It is also possible that Pakistan was deterred by diplomatic intervention from other countries.

THE INDIAN AIR FORCE IN OPERATIONS OTHER THAN WAR

India has used its air force extensively in two roles other than war: peacekeeping/support operations, and disaster relief and mitigation missions. The country's earliest major commitment to UN peacekeeping operations was undertaken by a Canberra squadron in the Congo in 1961–62, a good eight thousand km from its home base. Here the Canberra B-58(I) Interdictor bomber aircraft of No. 5 Squadron filled the air superiority role, escorted UN Globemasters, and carried out strike operations against the rebel province of Katanga. Since then, India has taken a large part in UN peacekeeping operations. In recent years, Indian Air Force attack and utility helicopters have been integral to the Indian contingents.

The Indian Air Force has also provided support to friendly governments since 1950. The most dramatic example was the rescue mission to the Maldives, with Il-76 aircraft airlifting a battalion of paratroopers two thousand km to restore President Maumoon Abdul Gayoom's control over his country within twelve hours of his telephone call to Prime Minister Rajiv Gandhi. The other notable instance was the air force's role as part of the Indian peacekeeping force in

Sri Lanka. Here the air force flew over 76,000 sorties (including 16,000 helicopter sorties) during the peacekeeping and peace-enforcement operations between July 1987 and January 1990.

For decades the Indian Air Force has also been engaged almost routinely in the extensive air maintenance of army and civil organizations in the High Himalayas. Similarly, it flies hundreds of sorties every year for casualty and medical evacuation from inaccessible areas. The air force has always played a key role in disaster relief and mitigation, for example, after floods, cyclones, and earthquakes. The most dramatic example was the air force's role during the Indian Ocean tsunami of December 2004, when it provided emergency assistance and relief. The 1990 airlift of 117,000 Indian refugees from Iraq and Kuwait, who were evacuated from Jordan after the Iraqi invasion, stands out as the second-largest strategic humanitarian airlift in history after the Berlin airlift. During the second Lebanon war in 2006, the Indian Navy transported two thousand Indians from Lebanon to Cyprus, and the Indian Air Force then flew them to India.

DOCTRINE AND STRATEGY

Until 1995 the Indian Air Force never had a written document spelling out air power doctrine. But the brief summary earlier clearly indicates a central theme in the beliefs that have directed India's employment of air power. To understand the evolution and development of these beliefs, as well as the changing doctrines and strategy that result from it, one must examine the past.

IMPERIAL LEGACIES

The Indian military was part of the British forces during the colonial period. It was therefore inevitable that India's defense doctrine and strategy would be deeply influenced by imperial policies and interests. This process created many paradoxes that persisted for a long time, even after the British had left India and the country had become independent.

British military policies since the early eighteenth century, when the first Indians were recruited to guard the factories of the East India Company, were based on the strategy that the weapons and equipment of the native forces would be superior to those of likely enemies, but inferior to those of the company's European troops. It is not surprising, therefore, that the Indian Air Force's first operational formation was designated an "army cooperation" squadron. This has had a profound impact on the evolution of the Indian Air Force and its role during the subsequent decades, and the effects have not yet worn off completely.

As part of this basic policy, the newly created Indian Air Force was restricted to support roles, while the RAF in India was assigned independent and strategic roles, namely, fighter defense and independent bombing. The Indian Air Force's eight operational units, formed after No.1 Squadron, were initially all coastal defense flights. The outbreak of World War II had an impact on the air force's force structure and its role in the defense of India, but the British still patterned the expansion of the air force on the broad strategic policy guidelines of the earlier centuries. Under the 1941 air power plan for the defense of India, all ten Indian Air Force squadrons were to be used in tactical army support, which in those years essentially meant artillery spotting and visual/photo reconnaissance, while the RAF would consist of fighter and bomber squadrons.[47]

The expansion of the Second World War in the Asia Pacific theater, especially after Pearl Harbor and the fall of Singapore, changed many assumptions underlying the earlier plans. The European war effort left little margin for the defense of India, and the British were forced to rely increasingly on Indian resources. The growth of the Indian Air Force was inevitable under the circumstances. But the guidelines of the earlier periods still appeared to exert a great influence. Even under the pressure of a direct threat to the defense of India, which constituted a key source of strength for imperial defense—the proverbial "jewel in the crown"—the government's policy of retaining all strategic elements of military power within the British forces continued. In particular, very few Indian officers were appointed to planning, intelligence, and air strategy positions at air headquarters. Thus, while the strength of the Indian Air Force was increased to nine squadrons, all of them were assigned to basic army cooperation roles. That is not to say the Indian officers accepted this situation; in fact, they constantly struggled to expand the Indian Air Force's role and capabilities.[48] But the dichotomy in approach and philosophy remained.

The experience of the Second World War also had an impact on the doctrine and employment of air power in independent India. First, the massive amounts of resources invested in air power during the war led to doubts whether similar employment was even feasible for countries such as India, which had limited resources and heavily competing priorities. Air power, by definition, generally operates at levels of technology much higher than those found in a country as a whole. This also makes air power costly. Air power capabilities, therefore, remained suboptimal, constrained by resource limitations and the army's taking the lion's share of military budgets.

Second, postwar Europe, especially the United Kingdom, adopted the view that in the nuclear age there would be neither the time nor the resources for conducting the prolonged battles that had characterized the employment of air power in the Second World War, whether struggles for air superiority or strategic bombing campaigns. Air power could now, so the argument went, perform its task much faster and with much smaller, nuclear-armed forces. Thus, conventional air power in the nuclear age was seen primarily in terms of a support role for land forces. The Indian Air Force had grown up with British traditions, and almost all of its military strategic literature was derived from British sources. As in independent India, most strategic thought in the world favored an essentially tactical role for conventional air power.

Third, in a country where the thinking was dominated by continental/territorial defense the ground forces naturally assumed a higher priority than air forces, except where they provided direct support to the land forces.

Fourth, since the RAF had retained a strategic and independent role in World War II (although it shared this role with the USAAF in India), India's equipment and infrastructure were designed to meet only tactical objectives, and then mostly in the Northwest Frontier Province. Under the circumstances, it appears almost a miracle that a large section of the Indian Air Force leadership, no doubt seeing the RAF and the USAAF play a strategic role, believed in an independent and strategic role for air power.

Concurrently, other developments were to shape thinking about the role and employment of air power. India's political leadership had current knowledge of the Second World War. Radio and newspaper reports had reminded them almost daily about the destruction wrought by indiscriminate bombing of civilian targets and economic infrastructure under the rubric of "strategic bombing"—a term that now implied nuclear bombing. It is not surprising, therefore, that abhorrence of bombing and deep concern for population centers characterized the attitudes of the political and military leaders of independent India.[49] The existence of nuclear weapons and the strategies of area bombing, almost invariably delivered by air power, only strengthened India's distaste for bombing, which technological limitations almost invariably rendered indiscriminate.

DEFENSE GUIDELINES

Defense policy requires a clearly articulated strategic doctrine around which a national consensus can be built and sustained. Such a policy must serve core national interests. In 1995, following precepts laid out by Prime Minister Narasimha Rao in his speech in the parliament, the ministry of defense stated to the Standing Committee of the Parliament the official position on India's national security interests and guidelines for defense policy:

1. [To provide] defence of national territory over land, sea and air, encompassing among others the inviolability of our land borders, island territories, offshore assets and our maritime trade routes.
2. To ensure an internal environment whereby our nation state is insured against any threat to its unity or progress on the basis of religion, language, ethnicity or socio-economic dissonance.
3. To enable our country to exercise a degree of influence over the nations in our immediate neighbourhood and to promote harmonious relationships in tune with our national interests.
4. To be able to effectively contribute toward regional and international stability.
5. To possess an effective out-of-the-country contingency capability to prevent destabilisation of the small nations in our immediate neighbourhood that could have adverse security implications for us.[50]

As can be seen, these goals are too broad to provide specific and effective direction to a country's defense doctrine and strategy. It must be emphasized that India's primary *strategic priority* can only be the rapid *comprehensive national development* of its people. Hence, this fundamental purpose must guide Indian foreign and defense policy. In the author's opinion, some of the key factors that deserve mention in this regard are:

1. Seeking a cooperative peace with all countries, especially major players in the transitional international order. A closer look at the imperatives of Indian interests would indicate that the nature of relations among the United States, China, and India, as well as their individual relations with all other countries, would have a profound impact on future events.
2. Achieving defense through credible minimum deterrence (at both the conventional and the nuclear levels) as the core of defense policy. The credibility

of that deterrence would actually depend on the capabilities, doctrine, strategy, and the resolve to *use (appropriate) force if deterrence fails.* Hence, the core defense doctrine had to shift from the purely defensive posture articulated by the defense minister in 1989 to one of a credible counterstrike.

3. Completely understanding that the use of military power in essence must serve political goals and that, in a democracy such as India, military commanders will have to function under political direction and even constraints. This also necessitates eliminating disjunctions at the higher echelons of defense management and providing for a direct interface and interactions between the political and military leadership, as visualized in the original structure for higher defense management.

The Indian Air Force's role, and the roles of the air power components of the army and navy, must be defined within this paradigm. Given the circumstances and the policies that evolved during the creation and the early growth of its air force, India would inevitably experience a certain level of dichotomy, and even an increasing degree of tension, in the concepts and doctrines governing the appropriate use of air power. Conceptual divergence and debates have continued within the Indian Air Force and have been the topics of incessant (and often inconclusive) debates and discussions in crew rooms and briefing sessions. The dichotomy has in essence revolved around two interconnected issues. One concerns whether the Indian Air Force should function as a strategic component of national air power or should remain what the British had designed it to be: a tactical air force. The other concerns the appropriate role of Indian air power as strategically defensive or offensive.

It is evident that a large section of the leaders of the small air force firmly believed that the Indian Air Force must be equipped, trained, and used in a strategic role, and, hence, advocated a "balanced force." But the air force faced two problems in adopting a fully strategic doctrine. One arose from inherited concepts and internal divisions on the issue. The other was the lack of suitable equipment after 1962, when a new theater of war had opened up in the north after China had initiated hostilities, and plans called for the Indian Air Force to be expanded from twenty-five to forty-five squadrons as an interim step, with a final goal of sixty-four squadrons.

This situation also divided the air force leadership into those who favored acquisitions from the West and those who saw little option but to seek Soviet aircraft and equipment. India's efforts to obtain U.S. weapons and equipment for the expansion of the Indian Air Force and the other armed forces after the 1962 Sino-Indian War resulted in promises that were never kept. Less than three years later, when Pakistan attacked India, the United States imposed an arms embargo on both nations.

Thus began the military-technical cooperation between the Soviet Union and India. Using the Soviet Union as a supplier met a number of India's requirements. Moscow was ready to supply arms on rupee-ruble terms, with India offsetting the cost of imports by commodity exports to the Soviet Union, and was willing to license manufacture of Soviet weapons in India. Moreover, Soviet arms were of a fairly high technological standard and could be improved to meet Indian needs. Thus, even though India was a poor, developing country with enormous defense commitments along its vast (and partly disputed) land borders, its coastline, and the Andaman and Nicobar group of islands, Soviet equipment made it possible to maintain an affordable and credible defense.[51]

The high level of dependence on Soviet aircraft brought its own problems. Contrary to Western perceptions, Soviet military systems were designed mainly in conformance with a defensive philosophy; the Soviet Union compensated for any technical shortcomings with large numbers. Until the late 1980s, Soviet fighters had a short range and carried a small payload. Therefore, the Indian Air Force had no option but to deploy its combat aircraft well forward and acquire surface-to-air missile systems to compensate for the short reaction time for air defense. As a result, there were thirty squadrons of Pechora (Soviet SA-3 surface-to-air missiles modified in India) systems in service by the end of the 1980s. India could not afford large numbers of aircraft; hence, the Indian Air Force was never increased to the sixty-four-squadron level authorized by the cabinet in 1963.

Unlike almost all other countries that acquired Soviet weapons, India did not simply adopt Soviet doctrines and tactics for the employment of air power systems, and never even had a single Soviet military "adviser."[52] Instead, the Indian Air Force innovated and evolved its own doctrine to suit the Indian way of fighting.[53]

The volume of India's purchases increased Soviet dependence on India. Not surprisingly, therefore, a stronger political relationship also developed between the two nations. The Soviets also benefited from having India as its customer, since this allowed them to produce further models and upgrades to existing systems. The MiG-21 is a case in point: Indian acquisitions resulted in some twenty-three variants, from the original design for a high-altitude supersonic interceptor against nuclear bombers to a low-level multirole combat aircraft, albeit with a limited range and payload.[54] The same process had begun to influence the MiG-29 when it was interrupted by the collapse of the Soviet Union.

NUCLEAR DOCTRINE

After the nuclear tests of 1998, the government of India set up a National Security Council with a National Security Advisory Board, which produced a doctrine that was released to the public as a draft document. Its main elements were formally reported in a government press release on January 4, 2003.

The dynamics of deterrence merit brief examination. India has adopted a nuclear strategy of counterstrike: that is, no first use of nuclear weapons, but assured retaliation "leading to unacceptable damage" to any nation that uses nuclear weapons against India and its forces.[55] The credibility of deterrence rests on India's capability and resolve to achieve its objectives even if deterrence fails, so that the aggressor does not attempt a military adventure. To that extent, India has a credible nuclear deterrent with respect to Pakistan, and a limited but evolving deterrent (in terms of the range of operationally available ballistic missiles) against China.

For the past two decades, the Indian Air Force has served as India's most important means of deterring the use of nuclear weapons. The army (and later the navy) will increasingly share this role as part of the triad, as land-based, long-range (5,000-km) ballistic missiles and ballistic missile submarines enter operational service.

The critical element in this doctrine is survivability. One can expect that an adversary would view Indian Air Force air bases as prime targets to neutralize and thereby degrade the air force's nuclear and conventional delivery capabilities. In my view, therefore, India's future strategy would have to rely on two imperatives:

1. Exploiting strategic depth to disperse and relocate air bases to central regions of the country, rather than maintaining nearly 80 percent of the air force's operational bases within three hundred km (many of them a mere seventy km) of the borders. This positioning, as noted, was necessitated in the past by the short range of India's combat aircraft.

2. Developing ballistic missile defense systems (including missiles for deterrence) to protect the air bases against missile attacks, and thereby ensure a guaranteed launch of nuclear-delivery aircraft for a counterstrike.

At the conventional level, it is obvious that any ground war between nuclear powers can at best result in a stalemate if the risk of escalation to nuclear levels is to be avoided. In this scenario, deterrence would rely on rapid reaction through the Indian Army's "Cold Start" doctrine, which a former chief believes would require greater close air support by the Indian Air Force from the very beginning,[56] combined with limited success, such as shallow penetration and the selective destruction of the aggressor's military power. This is unlikely to produce the punitive effect of changing the adversary's policies, but a decisive military victory that could produce such effects would risk an escalation to nuclear levels.

Covert subconventional wars have posed a different type of challenge during the past quarter century, particularly since 1987, when Pakistan acquired a nuclear deterrent. India has responded essentially with a defensive counterterrorism strategy that is confined within India's own territory, even though the sanctuary for the terrorists, as well as the sponsoring organizations and infrastructure, are across the border in Pakistan.

EMERGING CHALLENGES FOR THE INDIAN AIR FORCE

As it plans its future defense, India must contend with the potential for tipping the operational balance. The enhanced role of aerospace power in future armed conflicts must be factored into any military capacity and capability building. Three issues merit particular consideration: the modernization of the Chinese Air Force (formally known as the People's Liberation Army Air Force [PLAAF]); the rapid modernization of the PAF; and the possibility that China and Pakistan might take a collaborative military posture/action against India.

CHINA: CHANGING PARAMETERS

Unquestionably, India must continue to improve its relations with China, and thereby reduce the potential for disagreements and possible conflicts. It would not be in India's interests to think of China in adversarial terms. At the same time, it would be less than prudent to ignore the changing realities of military power, which could provide the capabilities leading to altered intentions. Factors beyond India's control could propel the two countries into a possible conflict situation.

The war in 1962 and its outcome were aberrations that will not recur unless India neglects its defensive capabilities significantly and/or China arrives at flawed conclusions as it did in 1962. Unfortunately there have been signs of both trends in recent years. China has grown in power and self-confidence during the past two decades. Its grand strategy remains the quest to become a global superpower. In concrete terms, it is much closer to reaching that goal than ever, but still needs more time. Meanwhile India is reemerging as a global player and is being wooed by the sole current superpower. Conceivably, China could make the same mistake it did earlier and try to use force—ostensibly in the context of resolving outstanding territorial disputes along the frontiers—to deter India's rise and/or its close relations with the United States.

China's greater assertiveness in its attitude toward India, especially since 2005, gives some evidence of this propensity. For the first time, an authoritative report by the Washington-based National Intelligence Council concluded that global power was shifting from the West to the East, due primarily to the rise of both China and India as future global players.[57] This report was soon followed by a significantly improved relationship between the United States and India. These closer ties were particularly clearly encapsulated in the new defense framework of June 2005 and the joint statement of July 18, 2005, which set out cooperation on a range of issues, especially nuclear power.[58]

China's sense of vulnerability was palpable in the early 1990s, when the Soviet collapse followed soon after the Tiananmen tragedy. China sought agreements with India in 1993 and 1996 to establish "peace and tranquillity," including the demarcation of the line of actual control to avoid any misunderstanding and potential for clashes and conflicts. There now seems to be a domestic consensus in China not to implement these agreements.

China's newfound confidence in its power contains the risk of overreach. China tried to block actions such as the Nuclear Suppliers Group's 2006 waiver, which made India eligible for nuclear-related exports and basically treated India as a de facto nuclear weapon state. China remains deeply concerned about the obvious increase in Indian access not only to U.S. weapons and military technology, but also to broad-based cooperation in the military and civil sectors.

At another level, India must take into account the strategic nexus between China and Pakistan, although it is aimed more at the United States than India. Pakistan's motivations for acquiring nuclear weapons are the perceived threat posed by India and the desire to acquire strategic autonomy from U.S. pressure.

As will be discussed later, China has actively assisted Pakistan in all aspects of its nuclear program.

In 2002 the Pentagon's annual report to Congress on China's military power stated that "[t]he principal area where China appears to be making advances in coercive military capabilities involves airpower, to include missiles and information operations."[59] The Pentagon's annual report for 2009 goes well beyond these understatements to express concerns about changing regional military balances that have implications beyond the Asia Pacific region. From India's perspective, the central implications that merit attention are:

1. During the past fifteen years, qualitative changes have been taking place in the People's Liberation Army (PLA) as a result of access to Russian military technology, which provides the PLA with potent power projection and long-range strike capabilities.

2. China leads all countries in Asia in military expenditure, even when the official data are accepted.

3. As a consequence, the PLA has become the largest military force in Asia and outguns any of its immediate neighbors by a factor of five to five hundred.

4. Over 96 percent of China's nuclear arsenal has relevance only for its neighbors. The Intermediate-Range Nuclear Forces Treaty left China as the undisputed dominant missile power in Asia. China has been working on a Chinese version of the "theatre missile defence system." Such a missile was successfully tested on the Tibetan Plateau in 1998.[60] Chinese ballistic missile defense would have a profound impact on Indian nuclear strategy, forcing India to expand its arsenal.

5. The prominence that China accords to aerospace power in doctrine and policy indicates its role as the instrument of choice for winning a local border war and for employing coercive force for political reasons (such as "teaching lessons").

By contrast, Indian military modernization has suffered greatly during the past two decades for various reasons. The resultant military-power equation is increasingly adverse for Indian security.

The PLAAF is undergoing phenomenal changes. The commander of the PLAAF announced in 1999 that "[a]t the turn of the century and in the early part of the new century, the Air Force will have a batch of new types of early warning aircraft, electronic-equipped fighter planes, and ground-to-air missiles" and that the air force "must give more prominence to air offensive, gradually integrate offensive and defensive, and build up a crack, *first-rate air strike force.*"[61] Current assessments indicate that China has exceeded this goal within the first decade of the twenty-first century.

By 2010–15, China will be capable of deploying nearly 300–500 modern, fourth-generation, high-performance, high-technology, multirole combat aircraft of the Su-27/30 class with long-range precision-strike and air dominance capabilities. These will be supplemented by combat support systems, such as the airborne warning and control system (of which four are being supplied to Pakistan), aerial refueling (which liberates the PLAAF from having to use high-altitude airfields in Tibet, although these facilities are being expanded), and "informationalization" (including network-centric capabilities). In the official defense policy statement in the White Paper of December 24, 2004, the Chinese government, while announcing a reduction of two hundred thousand army troops, unequivocally pointed out:

> While continuing to attach importance to the building of the Army, the PLA gives priority to the building of the Navy, Air Force and Second Artillery Force to seek balanced development of the combat structure, in order to *strengthen the capabilities for winning both command of the sea and command of the air, and conducting strategic counter-strike.*[62]

The doctrine's use of the term "command of the air"—a concept that has not been in use for nearly nine decades—is significant. This goal seems so ambitious that many people may be tempted to dismiss it as unrealistic. But no one can say that Beijing has not put its neighbors, if not the world, on notice regarding how it plans to use its aerospace power and to what purpose. China's ground-based missile strike that destroyed a low-earth-orbit satellite in January 2007 was in keeping with this doctrine. Given trends in technology and doctrine, China would clearly employ its aerospace capabilities for offensive operations if and when it employs military power. This could prove decisive in achieving the desired outcome in a local border war where the ground forces would be under severe constraints.

China has the third-largest nuclear missile arsenal in the world, and it has been developing more accurate mobile ballistic missiles. Over eleven hundred ballistic missiles are deployed on the mainland opposite Taiwan, an increase from

eight hundred only three years ago. Mobile missiles can also be quickly redeployed to China's other borders. Such a large deployment, and the possible
employment of ballistic missiles armed with conventional warheads by a nuclear
weapon state, would increase operational ambiguity and erode strategic stability. At the same time, the potential for resorting to preemption to resolve that
increased ambiguity would increase instability in a crisis.

There are serious questions and disagreements about the level of China's
investments in its military power and its modernization. But even the official
data indicate that, for the past two decades, China's military spending has been
increasing at an annual average rate of almost 14 percent in real terms.[63] This
rate of sustained annual growth far exceeds that of any other country. Even in
absolute terms, the *official* military expenditure data show China to be the largest military spender in Asia. By contrast, India's defense expenditures in 2008
amounted to 1.9 percent of gross domestic product (GDP), which is much
smaller than China's.

These percentages do not necessarily imply that China is about to embark
on aggressive military wars. In fact, should the use of military power become
necessary, China can be expected to employ it with great care and only after
pragmatic deliberation. The central strategy is likely to revolve around three key
principles: (1) using military power where the chances of success are high, (2)
ensuring that the use of overwhelming force would reduce the risk of failure,
and (3) aligning the use of military power with political objectives. The last
factor deserves special attention since, as a RAND study argues, China may use
force for reasons that have little to do with its immediate tangible interests—for
instance, in territorial disputes.[64] The Sino-Indian War of 1962 offers a clear
example in this context.

THE PAKISTAN AIR FORCE

Since 1947, Pakistan, India's western neighbor, has consistently pursued a strategy of covert war followed by overt war.[65] Pakistan has waged four wars of aggression against India and now occupies one-third of Jammu and Kashmir in
violation of UN Security Council resolutions. The most recent overt act of aggression—in the Kargil sector in the summer of 1999—was based on a plan
devised more than a decade earlier.[66] More recently, Pakistan has made no secret
of the rationale behind its nuclearization, which is specifically targeted at India.

Since the Kargil conflict, Pakistan has undertaken a massive effort to modernize its air force and naval air strike force, and to acquire force multipliers such

as airborne early warning and control, aerial refueling, and long-range strike capabilities. This modernization has received added impetus from arms supplied by the United States after 9/11, largely at subsidized costs and even cost free. In addition, China continues to offer Pakistan an alternative choice as an "all-weather friend," while most Pakistanis consider the United States an unreliable partner.[67] As a result, in another five or so years, Pakistan will have a modern, high-technology air force backed by modern force multipliers.[68]

The implications of the current trend require careful attention, especially in view of the Indian Air Force's reduced combat force level. If the PAF were the only challenge, then even a reduced force level could contain an aggressive PAF. But, as highlighted previously, India must take into account the China-Pakistan strategic nexus. Since 1965, China has provided Pakistan not only with conventional weapons, but also with nuclear weapons technology to increase its autonomy in dealing with the United States. It may be recalled that China applied strong political and military pressure on India (including resorting to infantry and mortar firing during the 1965 war initiated by Pakistan). In the 1980s, China posed a challenge to India by providing "proven nuclear weapon design and enough enriched uranium for two devices," and continued to provide assistance to Pakistan's nuclear weapons program during the 1990s.[69] Numerous reports have indicated that the Pakistani device was tested at Lop Nor in China in 1983; the Pakistani nuclear scientist Dr. Samar Mubarakmand, who was in charge of the nuclear tests in May 1998, has stated that this was the case.[70] China has also been supplying ballistic missiles to Pakistan since 1987, when Pakistan was believed to have acquired its nuclear weapons capability,[71] and has increased these sales since 1991.[72]

Pakistan has also prosecuted a vicious proxy war against India through jihadi terrorism, on the assumption that India would have no scope to apply coercive force (as it did in previous wars). Pakistan sponsored and assisted ideologically and politically motivated terrorist operations in the Punjab between 1983 and 1993, in Jammu and Kashmir since 1988, and more recently in cities throughout India. Pakistan has justified the terrorists' actions by calling them "freedom fighters," although there was incontrovertible evidence that at least 60 percent of the perpetrators have been "foreign" (Afghans, Arabs, etc.), and that a large proportion came from Pakistan-occupied Kashmir. Further, state elections in 2002 and 2006, as well as those to the Indian parliament in 2008, recorded that 45–60 percent of eligible voters cast their votes to elect the government of the state. Even so, the radical Islamists have been encouraged and trained to carry out acts of violent terrorism that have killed more innocent people than the atomic bombs dropped on Hiroshima and Nagasaki. The latest major attack, which took place in Mumbai on November 26, 2008, killed nearly three hundred people in several five-star hotels and at the Jewish center.[73]

In these operations, Pakistan has sheltered under its nuclear "umbrella," knowing that this situation could carry the risk of a nuclear escalation that India seeks to avoid. Thus, the Pakistani elite has perceived the terrorism war as a low-cost option to "bleed India through a thousand cuts."

As noted previously, it is generally accepted that a ground war between two nuclear neighbors can, at best, produce a stalemate, unless one side is willing to run the risk of a nuclear escalation and a holocaust by pushing for a "decisive military victory." But the Indian formulation of a conventional riposte *below* the nuclear threshold, termed "limited war," has challenged Pakistan's traditional assumptions. Air forces would play a crucial role in the strategic space available between the nuclear and terrorism levels. The highly effective use of the Indian Air Force in Kargil seems to have prompted Islamabad to make every effort to enhance, modernize, and even expand its air force with modern U.S. and Chinese weapons to ensure an adequate capability to deal with the Indian Air Force. The salient issue is not so much who will win an air war as the possibility that Pakistan might again assume it now had a sufficient capability to bring about a Pakistani victory. This might create a continuing temptation to persist in, and even escalate, the covert terror war, and perhaps even to increase the limited conventional use of military power. The Indian Air Force must therefore ensure that India factors adequate superior aerospace capability into its plans.

LOOKING INTO THE FUTURE

Indian air power not only faces the challenge of an increasingly adverse balance of power, as previously summarized, but must also be able to take on the additional tasks and responsibilities that it will inevitably face as a result of changing geostrategic factors.

CHANGING POWER EQUATIONS

One of the key factors that will affect all of Indian air power (army, navy, and air force) is the altered geopolitical situation. For the past two decades, my own assessment has been that global affairs will ultimately be determined by a strategic triangle of three powers: the United States, China, and India. Meanwhile, other middle—and even erstwhile major—powers will have their own regional roles and influence on the three pivotal powers. Two other issues are relevant to this current study.

First, India's economic growth, averaging 6 percent annually in terms of per capita income and expected to increase to around 8–9 percent over the next two decades, will carry with it greater responsibilities and interests beyond the immediate "South Asia" region. India will need to protect these interests, perhaps even with military power. For example, 4.5 million Indian citizens work in the Gulf Cooperation Council countries and are deeply embedded in their economies. Any instability in the region could create serious challenges that would require appropriate Indian responses, optimally in coalition with other countries and desirably with a UN mandate. That India faced such a situation after Iraq invaded and annexed Kuwait two decades ago, which resulted in 350,000 Indians becoming Saddam Hussein's hostages, is symptomatic. Ultimately, nearly 117,000 Indians, transported by road to Jordan, were evacuated by air. Hence, the Indian Air Force will have an increasing need to develop the requisite strategic reach and capability.

Second, nearly 75–80 percent of the Indian armed forces' weapons and equipment are of Soviet-Russian origin. Although many of these are manufactured in India, high dependence on Russia for the next four or so decades would obviously not be in India's interests when international power equations are in transition. In the past, political and financial pressures limited India's sources of supply to the Soviet Union and a few European countries, such as France and the UK. However, new opportunities are opening up for India to obtain weapons and equipment from the West in general and the United States in particular. In my considered view, India faces a strategic need to ensure that any future weapons and equipment acquisitions significantly reduce import dependence on Russia well below 80 percent.

ENSURING CREDIBLE, AFFORDABLE DETERRENCE

Deterrence will remain especially critical in the conventional area of any war-prevention strategy, although deterrence in the nuclear arena will inevitably have more apocalyptic dimensions. Deterrence in a nuclear environment, where India seeks to minimize the role of nuclear weapons, will require a conventional military capability that is qualitatively and quantitatively superior to those of India's likely adversaries.

During the past nearly thirty years, India's economic growth and human de-velopment indices have improved steadily, even if not dramatically. But, in the context of affordability, it is necessary to note that India maintained a level of defense spending at an average of 3 percent of GDP over that period. This level of spending has provided a capability that enabled decisive military victories in 1965 and almost equally dramatic ones in 1971 and 1999. But, above all, this capabil-ity has also provided deterrence against foreign military adventures. The clear evidence is that, in contrast to the earlier era, no country has waged war against India since 1971.[74]

However, most of the factors that ensured such military capabilities have changed in recent years. India's defense spending has averaged a mere 2.24 per-cent of GDP during the past fifteen years even though the value of the Indian rupee against foreign currencies dropped and arms therefore became more expen-sive. This, in turn, has had a deleterious effect on the force modernization of all three services. We must consider whether this level will be more or less adequate to maintain a credible capability in relation to the environment against which India must protect itself. The major factors that had kept defense costs low were low manpower costs; the modest cost of weapons and equipment acquired from the former Soviet Union and mostly manufactured in India itself, the poor cost consciousness of India's defense-industry complex notwithstanding; the relative absence of strain on scarce hard currency for foreign exchange, since the bulk of the acquisitions were in the rupee account, payable at low interest rates and with long-term credits; and the decline in China's military capability since 1962, which reduced the level of deterrence capability required. Pakistan's military capabilities had also suffered occasional declines due to erratic U.S. supplies and reliance on low-technology weapons from China.

The drastic shifts in recent years have had a negative impact on India's force modernization. India's defense posture must now take into account the implica-tions of Pakistan's having acquired nuclear weapons and the resulting impera-tives of limited war in the conventional sphere. China's defense modernization program is creating new options in the conventional sphere that had not been available to China since the 1950s. In addition, the economic recession and re-trenchment in the global defense industry are raising the costs of weapons and equipment.

Moreover, while an average of 2.24 percent of GDP for defense would be affordable in economic terms, it may be difficult to achieve politically. In 1995–

1996 the Indian parliament's Standing Committee of Defence recommended increasing the defense budget to 4 percent of GDP,[75] and the most recent committee has recommended defense spending at 3 percent of GDP. However, defense spending in the coming decades will, in all probability, remain at around 2.5 percent of GDP.

REFORMING INDIA'S HIGHER DEFENSE STRUCTURE

In the past the Indian Air Force faced serious difficulties because of a lack of joint planning and had to compensate for this disadvantage during actual operations. While a cabinet decision established the higher defense organization of the armed forces just over a month after India became independent, the Defence Committee of the Cabinet (DCC) functioned effectively only in the first Kashmir war, when the Indian Air Force responded within hours to land troops in Srinagar. Any delay would have resulted in the map of the Indian Subcontinent being different from the present one. But the DCC hardly functioned after the mid-1950s, when Krishna Menon took over as the defense minister.

The original cabinet decision to set up the higher defense organization had mandated that the Chiefs of Staff Committee (COSC), comprising the commanders of the three services, give military advice to the DCC (normally through the defense minister), and that the individual members also command their respective forces and oversee the future force planning for their respective services. For this purpose, two important committees were created under the COSC: the Joint Planning Committee (JPC) to provide a permanent staff, and the Joint Intelligence Committee (JIC) to provide intelligence assessments of military significance for operational planning.

Unfortunately, the JPC staff was not established until the Defence Planning Staff was created four decades later and was soon superseded by the Integrated Defence Staff of the COSC in 2001. The JIC was removed from the authority of the COSC after the 1962 debacle. Worse, the army, as the senior service, sought to place the air force under its authority and still viewed its role merely as that of a supporting service to ground operations. For example, in 1961 the army chief instructed his operational staff not to divulge the defense plans for the operations in Goa to the Indian Air Force and the Indian Navy. In 1962 the army also failed to appreciate the role that the air force could play through interdiction. When Pakistan launched its armored attack on September 1, 1965, the Indian army initiated its counterattack five days later without a joint plan. In fact, the Indian government, suspecting that Pakistan would follow the aggression that began on August 6 with a larger military offensive, had authorized the armed forces to choose their own time and place for a response a week before Pakistan's attempted blitzkrieg. However, the air chief was not informed and only found out about the authorization through casual conversation ten days into the war.

Efforts to instill jointness into the three services have made significant progress compared to these earlier days. Currently, however, they seem to be mired in the army's desire to create a chief of defense staff and unified theater commands for the territorial defense of the country. Even so, there have been few, if any, instances of a lack of tactical and operational jointness despite the absence of formal structures for joint force planning and operations. Joint planning still suffers from the disjunction between the civilian Ministry of Defence and the military-staffed service headquarters. India must find ways and means to eliminate this barrier at an early date if joint planning and operations are to reach their desired levels.

Notwithstanding these handicaps, the Indian Air Force remains committed to the major roles of air dominance (air defense, air superiority, etc.). In essence, the air force is currently heading toward a transformation from a predominantly tactical strike force that has tried its best to achieve strategic effects to a truly strategic force with a strategic reach and effect. This transformation would actually enhance its ability to undertake tactical air campaigns, especially in support of surface forces. Only this approach to strategy can constitute the rational basis of any credible deterrence capability against China's stated strategy and known acquisition processes. Doctrinally, this implies that the air force's fundamental purpose, previously the air defense of India, would now become the defense of the nation. Stated differently, in my opinion, the primary role of the Indian Air Force in the future would be to win the nation's wars, jointly with other services where required and by itself where possible.

Political and military leaders have not yet grasped this role fully, in spite of the almost-dramatically demonstrated preeminence of air power in all the wars and armed conflicts of the past quarter century. The sole superpower, with its immense military-technological advantage over the rest of the world, is not the only country whose air power can achieve such an impact. Israel's employment of air power in the Bekaa Valley in 1982, and in its earlier wars, serves as an illustration. Only the air force leaders have been increasingly in tune with this line of thinking, since they have studied the growing impact of air power more closely. The Indian Air Force's future force planning policies reflect this viewpoint.

Accepting this assessment of air power would, for example, require the air force to focus on the adversary's army reserve formations to neutralize, immobilize, or degrade them before they could reach the battle zone. In other words, the Indian Air Force would seek to ensure that hostile army reserves would be unable to support a ground offensive and contribute to joint operations. This objective would, no doubt, require air intelligence of a very high order in terms of both accuracy and timeliness, but technology, especially space capabilities, has now put this within reach. Long-range precision weapons would also add to the overall impact of such an employment of air power.

Such a doctrine and strategy would have to be based on the doctrine of air dominance, a term that here is meant to encompass both air-to-air superiority and air-to-ground dominance. The air force, and to a great extent the Indian Navy, have largely internalized this concept. The Indian Army remains hesitant, since this also implies a larger role for air power. Air dominance in both dimensions of the concept also implies that India would have the ability to achieve coercive capabilities that could be calibrated in time and space to exploit the strategic space below the nuclear level and above the level of subconventional war. It is important to note that Pakistan has continued (and China could be tempted) to wage such a war through terrorism in the belief that India can no longer undertake any punitive action, as it did in 1947 and 1965, without risking escalation to the nuclear level. However, long-range punitive strikes designed to achieve not military effects but direct economic and political effects could provide the necessary scope for punitive retaliation against the terrorism that Pakistan has sponsored for a quarter century.

The conclusion is inescapable: if India is to maintain a credible and affordable defense capability, force modernization must receive substantially more resources, and military decisions must be made faster than in the past, both of which require important changes in defense management. Although events of the past quarter century clearly established the dominant role of air power in military conflicts, high-powered task forces and the group of ministers have almost totally ignored this crucial factor in defining the management of India's defense nearly a decade ago. The longer it takes India to come to grips with the issues involved, the more difficult and costly the decisions will be.

However, developments both inside and outside India also offer opportunities to make some fundamental changes that will give India a more affordable and credible defense capability to meet future challenges. Current trends, and the adverse air power balance, pose serious challenges to India's air power, but may prompt the country to alter some of its past approaches. In essence, India's air power in general and the Indian Air Force in particular must undergo a fundamental transformation in every aspect.

The innovative spirit shown by "Jumbo" Majumdar in 1942 still characterizes the Indian Air Force today, and innovation is intrinsic to achieving more with less. It serves as the foundation on which the air force bases its current transformation, as has been demonstrated in air exercises with other nations during the past six years. Hence, the air force should have little difficulty in transforming its air power in the future. What it must remember is that transformation of mind-sets is the most important aspect of force transformation. This would require military, and specifically air power, leadership with a clear vision of the future in both the geopolitical and technological-operational spheres, and professional education that equips Indian Air Force officers to deal with the inevitable uncertainties of the future.

6

CHINESE AIR POWER

Xiaoming Zhang

On November 21, 1951, Gen. Hoyt S. Vandenberg, chief of staff of the U.S. Air Force (USAF), alarmingly announced that Communist China had "become one of the major air powers of the world" almost overnight, and was seriously challenging United Nations (UN) air superiority in Korea.[1] In reality, since this announcement was made, China has struggled for more than half a century to try to build a modern air force. It was not until recently that the emphasis placed on the development of air power has achieved some tangible results and the People's Liberation Army Air Force (PLAAF) is evolving into an offensive air and space power. During the last sixty years, the young Chinese Air Force was first baptized in the Korean War (1950–53), and then involved in numerous air combat engagements with the Chinese Nationalists over the Taiwan Strait in the 1950s, and with intruding U.S. warplanes and drones during the Vietnam War (1964–69). Despite its expansion throughout these years, the PLAAF remained a defensive force rather than an offensive one, and China steadily fell behind in the air power race due to a backward military philosophy and a lack of access to modern industry and technology. Yet Chinese efforts to build a modern air force have never ceased.

The rise of China as a global economic and political power in recent years raises concerns for many policy makers, strategists, and scholars about Chinese military modernization—concerns that might provide a new perspective on global security for years to come. At the center of this concern is that the PLAAF has gained offensive capability by equipping itself with an increasing number of third- and fourth-generation aircraft and long-range antiaircraft missile systems.

But what matters most is not so much the growth of Chinese air power capability per se; rather, it is how China might use its new military strength, especially its air and naval power.

The following overview of the development of Chinese air power will provide a historical perspective on how the leaders and airmen of the Chinese Air Force molded the PLAAF's organization, equipment, doctrine, and tactics to implement their concepts of air war. Certain factors, such as the People's Liberation Army's (PLA's) traditional military thinking, its combat experience, Soviet influence, and the interference of the Chinese Communist Party (CCP) leadership, appeared to play a significant role in influencing China's quest for the development of air power. This chapter will carefully examine these factors and attempt to address several questions: How did the air force become a major service of the PLA, but still play a very limited role in Chinese modern warfare? How did traditional military thinking, including China's strategic culture, influence the quantity and quality of the Chinese Air Force? And finally, what problems did the Chinese Air Force confront in the past, and what problems will it continue to face today and in the future? In seeking to answer these questions, the discussion suggests that, although the Chinese Air Force has recently adopted an offensive concept of air power, it will take far longer for the PLAAF to nurture an institutional culture that can embrace both offensive and defensive capability and "bear the brunt of, and play a sustained, independent role" in modern warfare.[2]

INITIAL QUEST FOR AIR POWER

PLAAF literature records fascinating folk legends that capture China's earliest attempts at flight. The legend of Chang'e flying to the moon is one of these ancient and beautiful Chinese fables; China's current lunar exploration program has adopted "Chang'e" as the name for its moon orbiters, symbolizing the effort to actually realize the millennium dream of flying beyond earth's orbit.[3] Apparently, the legends recorded in PLAAF literature represent an emotional and cultural bond that helps the Chinese link their past civilization to modern aviation and inspires them to take further strides toward the pursuit of aerospace power.[4]

Even though history witnessed China's stagnation and decline in the early twentieth century, Chinese publications continued to credit the success of Feng Ru, a Chinese American, in building an aircraft and flying it for 2,640 feet in a suburb of Auckland on September 21, 1909, as a Chinese achievement in mankind's quest for aviation. The first Chinese to really appreciate the potential of air power was Sun Yat-sen, the founding father of the Republic of China, who advocated using "aviation as the nation's salvation."[5] Sun had studied medicine and had no military credentials whatsoever. Still, the overseas experience that had exposed him to Western influence made him the first Chinese to espouse a view later advocated by the Italian air power champion Giulio Douhet and the American aviation crusader Brig. Gen. William "Billy" Mitchell: that aircraft would play an increasingly decisive role in warfare.[6] Since China had no industry, Sun suggested that China first buy naval, army, and air weaponry from the Western powers to strengthen the armed forces, and then try to copy and manufacture the weapons to meet China's own needs. This became the essence of Sun's theory of saving the nation through aviation, and it appeared applicable to the Chinese situation for many years to come. Sun also believed that the concept of territorial sovereignty would expand from land and water to include airspace, and that air power would dramatically change the manner and nature of combat.[7]

The influence of Sun's view on the development of air power in China is difficult to gauge accurately. PLAAF literature regards Sun as one of the few Chinese able to articulate an understanding of the potential role of air power, even though his view on the subject was indeed sporadic.[8] Sun died in early 1925, before he could put his theories into practice.

From the late 1920s on, continuous civil strife prevented the newly established Nationalist government in Nanjing from claiming effective control over the whole country. Local warlords remained independent of the central government's control, while the Communists instigated armed revolution across the country. The Nationalist leadership came to realize that no military arm offered more potential for compelling officials in isolated provinces to obey the government and unifying China than a well-organized and effective air force. The Chinese demand for an air force was further enhanced by the Japanese occupation of China's northeast in September 1931 and military attacks on the city of Shanghai in early 1932.

Nationalist strategists knew it would take years, and enormous funds, to build an effective navy, so the creation of a modern air service became the focus of China's national defense efforts in the 1930s.[9] However, even after the Nationalist government put tremendous effort and investment into air power, the air force used to subdue domestic opposition was destroyed by air operations within weeks of the outbreak of hostilities between China and Japan on July 7, 1937.[10] Thanks to Soviet military assistance in late 1937, followed by U.S. military assistance in 1942, China was able to keep an air force in service against the invading Japanese.[11] Ironically, in the ensuing civil war with the Chinese Communists, air power could not prevent the Nationalists from losing control of the mainland in 1949, and the bulk of the Nationalist air force fled to Taiwan. The Communists used the planes, airfields, repair shops, and other aviation facilities abandoned by the Nationalists to help build their own air force.

The CCP's initial quest for air power focused on sending its members to study aviation science and technology in Moscow and managing, with Soviet assistance, an aviation training program at a small military academy in Xinjiang. The CCP then established an aviation school in northeast China, employing Japanese aviators to help. The personnel who participated in these programs became the first group of Red pilots and the initial core members of a Red air force.[12]

Nevertheless, the experience of the Communist leaders during their revolutionary careers had never included air power, except as victims of the Nationalist and Japanese air bombardment. Unlike Sun Yat-sen, who stressed the importance of an air force even before he actually possessed a military force, they demonstrated their pragmatism during their military careers by adapting strategy and ideas to suit reality. For example, Mao Zedong, the founder of the PLA, did not give the establishment of Communist air power serious consideration until 1947, when the CCP's victory was on the horizon.

Mao's military ideology was dichotomous. His guerrilla warfare experience convinced him that a weak army could prevail in a war against a strong enemy. To Mao, the Communist way to deal with enemy air power was to "not fear death, but be brave, and dare to sacrifice lives."[13] However, in his famous essay on military affairs, "On Protracted War," he conceded that military modernization would be a prerequisite for defeating the Japanese aggressor. According to Mao, Communist forces, armed with new weapons, would be transformed into "a world class" military power.[14] As a pragmatist, Mao often expressed his concern for Chinese Communist aviation in a less obvious manner. He supported the idea of sending Red army soldiers for flight training in Xinjiang, and ordered that aviators receive special treatment despite the poor living conditions in Communist-controlled areas. By showing parental concern for them, Mao demonstrated his support for a CCP air force at a time when the reality lay far in the future.

It was not until the long contest with the Nationalists reached a turning point in late 1948 that Mao began to predict a Communist victory and ponder how to take over cities and consolidate control. He believed the most difficult tasks remaining were the crossing of the Yangzi River and the seizure of large cities such as Nanjing and Shanghai. He also felt that taking Tibet and Taiwan would pose a great challenge for the PLA.[15] Perhaps more important was his recognition that the new Communist regime would need to be defended. As Sun Yat-sen and the Nationalist leaders had realized, aviation appeared to be the single most effective tool that could annihilate distance and bind the fledgling Communist society together.

As they considered a new China continuously surrounded by enemies, both foreign and domestic, Chinese leaders could not defer a long-term national defense program to modernize the armed forces. A powerful air force, acting as a deterrent and presenting a threat of massive retaliation, symbolized modernization for the PLA. Drafting a document for the CCP's task in 1949, Mao remarked, "We must be devoted to the construction of an air force so that [we will be able to] use it [in our military struggle] in 1949 and 1950."[16]

At the time, China had no industry, and its economy was bedeviled by political squabbles and military conflicts. Soviet assistance became the only hope for the Chinese Communists to achieve their ambition to build an air force. Thus, the PLAAF would inevitably be influenced by Soviet air doctrine.

THE CREATION OF THE CHINESE AIR FORCE

Despite any early discussion about possible Soviet assistance during the difficult years of the Chinese Revolution, it was not until the CCP leadership began to consider a plan for the invasion of Taiwan that they sought Moscow's assistance in air power. On July 10, 1949, Mao issued an instruction that stressed that "it is impossible for our air force to overwhelm the enemy in the short run (such as a year); but we may need to consider sending three to four hundred of our people to study in the Soviet Union for six to eight months, and at the same time to buy about one hundred planes . . . they will form an offensive unit to support the cross-strait campaign and prepare to seize Taiwan next summer."[17] This was the earliest evidence of the Chinese leader's interest in building an air force and obtaining Soviet assistance. Moscow responded to the request positively, believing that the CCP should have made such an appeal a year earlier, but suggested training the Chinese pilots inside China rather than in the Soviet Union.[18]

After the Soviet Union agreed to assist, China officially established the PLAAF on November 11, 1949. The first decision was to set up seven aviation schools (four for fighters, two for bombers, and one for transport) to train Chinese pilots on Soviet aircraft and under Soviet instructions. The trainees from the first and second classes were expected to graduate in May and October 1950, respectively.[19]

As early as 1947, the basis for the future training and doctrine of the air force was established after a serious debate about what air training systems the Communist Chinese should use at the Northeast Aviation School. The school opened with some Japanese instructors who had surrendered to Communist troops at the end of World War II, defectors from the Nationalist air force, and others who had received flight training in the United States, along with personnel who had received training in the Soviet Union. Great confusion existed as to whose training system was best.

Turning to the Soviet Union for help with aviation meant that the development of the Chinese Air Force would fall under the influence of Soviet doctrine and technology. The Soviet-trained instructors argued that any Communist air force should follow the Soviet model and that the training program should use Soviet techniques. Obviously, they sought to remain politically correct. The CCP and military authorities in the northeast supported this position. Those who had been trained by the Soviets soon became the dominant force at the school, which adopted Soviet teaching materials and techniques. By the time the CCP and the Soviet Union forged a new relationship in late 1949, the Chinese were ready for Soviet aviation assistance and receptive to Soviet doctrine.[20]

Three factors were instrumental in the development of the PLAAF in its fledging years as it was shaped by the Soviet influence. First, and most decisive, was the personality of the air force leaders—especially Liu Yalou, commander of the air force—who had long since outgrown the guerrilla military personae to which they had become accustomed as senior Communist army commanders. They tended to view policies strictly in their political dimensions, which rendered them incapable of logical military thinking. When assessing U.S. and Soviet air doctrine, they attributed the differences simply to the different social systems, criticizing the United States for being an imperialist country and, therefore, needing more bombers than fighters so that it could invade other nations.[21] The second factor was the newness of the organization and its conspicuous lack of personnel who were familiar with the roles that an air force should play and knowledgeable about the resultant requirements. Copying the Soviet system blindly or mechanically was therefore inevitable, and entailed no political risk at a time when the Soviet socialist system was regarded as the model for the future of China.

The third factor was the influence exerted by the Soviet Air Force, whose doctrine stressed air superiority, air defense, and ground support missions, and never embraced the concepts of independent air operations or of strategic bombing.[22] From the PLAAF's earliest years, Chinese military leaders perceived the air force as a support unit of the PLA, and considered air power essential not in a strategic sense, but only in respect to the tactical support it could provide to the ground forces during operations. No consideration was given to making the air force a service independent of the army. It was under the influence of these factors that the Chinese Communists began to plan an air force and its organization, type of aircraft, training, and concepts of operations.

Little in existing Chinese guerrilla military doctrine addressed the use of air power, but the doctrine's defensive nature did exert an influence on Chinese thinking about the air force. The PLAAF leadership preferred to build an air force similar to that of the Soviet Union: one that possessed more fighters than bombers. Its theory was that the role of fighters dovetailed well with the defensive cast of Chinese thinking. Bombers attacked enemy countries and territories—an aggressive act—but fighters were defensive in nature and their success in fending off attacks would lead to air superiority.[23] The PLAAF's immediate mission, therefore, was to attain air superiority over the Nationalists, provide support to the planned amphibious assault on Taiwan, and then develop itself into a force capable of defending China's airspace and waters.

This mind-set gave priority to the development of fighter units.[24] In June 1950 the first PLAAF aviation unit was established in Nanjing. It consisted of two fighter regiments, one bomber regiment, and one air-to-ground attack regiment, and was equipped with Soviet-made aircraft (La-11s, Tu-2s, and Il-10s). The unit became operational a few months later.[25]

The Korean War provided the impetus for the rapid expansion of the air force in terms of both aviation personnel and equipment. Shortly after the Chinese intervened in the conflict, a huge influx of Soviet assistance enabled a greater enhancement of the PLAAF's capabilities in both quantitative and qualitative terms. Between November and December 1950, thirteen Soviet air divisions arrived in China, including eight MiG-9 divisions, two MiG-15 divisions, an La-9 division, an Il-10 attack division, and a Tu-2 bomber division.[26] Their main responsibility was to help China with air defense, but they were also charged with training Chinese air and ground crews.

The urgent situation in Korea left the Soviets with two unpalatable options: either develop Chinese air capabilities or prepare to intervene themselves. They chose the first alternative. Their primary mission was to train Chinese crews quickly and then hand their aircraft and equipment over to the Chinese. The large-scale transfer of Soviet air power to the PLA was undertaken in late 1950 and early 1951. All Soviet units turned their aircraft and equipment over to the PLAAF at the end of their mission. In the short span of six months, the PLAAF would expand dramatically to sixteen air divisions, and appear almost ready for action in Korea.[27]

However, some problems concerning the nature of Soviet aid to China at the time merit further analysis. That the Soviet Union had supplied China with such a large number of MiG-9s raised Chinese suspicions about Soviet sincerity. The MiG-9, powered by two engines copied from the Germans, was the first jet fighter to enter service in the Soviet Air Force in 1946, and had a very short range and poor maneuverability. While upgrading its own air force with the advanced MiG-15bis, the Soviet Union had apparently seized the opportunity to dump its retired MiG-9s on China.[28] When Chinese leaders later began planning to send their air force to Korea, they were disturbed to discover that the PLAAF had more MiG-9s than MiG-15s. They subsequently complained to the Soviets that these obsolete jets were not frontline aircraft. Russian advisers appeared to resent this negative view of the quality of Soviet planes. The top Soviet leadership intervened and ordered the delivery of additional MiG-15s to the Chinese without charge. However, this delay prevented the early entry of the Chinese Air Force into the war.[29]

THE PLAAF IN THE KOREAN WAR

The Korean War saw the PLAAF grow into one of the largest air forces in the world, and gave China's leadership a clear idea of how to develop an air force that would conform to Chinese thinking about the role of air power in modern warfare. The official PLAAF history conventionally divided the air operations of the Korean War into three phases: preparation, learning period, and engagement. It was not until the last year of the conflict that the Chinese Air Force was able to conduct operations more independently of the Soviets and to emerge as an experienced fighting force. Still, its involvement remained limited.

During the months from late October 1950 through July 1951, the PLAAF prepared for intervention under an air plan stating that its primary mission in Korea would be to provide support for ground troops. As noted previously, the air force had just been established and consisted of a few operational units, totaling about two hundred combat aircraft.[30] Chinese pilots had less than one hundred hours of flying, and most of the MiG pilots had just started to fly solo. More to the point for the PLAAF, no Chinese pilots or commanders had any air combat experience.[31] Evidently, any rash decision to throw the young air force into action would court defeat, given the overwhelming disparity with the strength and experience of the U.S. Air Force (USAF). The debate over the correct air war strategy in late 1950 thus became critical to the PLAAF's survival as well as to its future development and mission.

On October 30, 1950, the PLAAF convened a meeting focusing on when and how the air force should enter the war. Participants concurred that the Chinese ground forces were at a clear disadvantage in Korea because the United States was effectively in control of the air. Chinese pilots needed additional training, but it was argued that the PLA's history had shown that Chinese airmen could learn warfare through hands-on training under actual combat conditions. The PLAAF leaders understood that they could not compare with the USAF in terms of the numbers of aircraft available or technology. Based on their confidence in the CCP leadership, their countrymen's support, and their own ground war experience, they were, nevertheless, certain that the PLAAF could hold its own against the USAF, and that ultimate victory would depend upon ground operations. The Chinese Air Force, the participants concluded, should play only a supporting role in meeting the needs of the ground troops.[32]

Such an analysis reflected the traditional dominance of the army within the Chinese military. This is not surprising, since the majority of air force personnel were transfers from the ground forces. The army bias also accounts for the PLAAF leadership's inclination at the time toward an air doctrine stressing direct support of ground forces as its major mission. Still, the PLAAF leadership faced two diverging opinions about how the air force should be used in the Korean conflict. One faction suggested that fighters, bombers, and attack aircraft should move into Korea to provide direct support for Chinese ground operations. The other recommended that air force units be based only inside China to engage

American aircraft over Korea. The former approach would give ground troops more direct support than the latter, but it would also entail a greater risk of losing the entire air force. Such a lesson could be easily drawn from the experience of the North Korean Air Force, which had been destroyed by the Americans during the first month of the war.[33] The leaders of the Chinese People's Volunteers, particularly Peng Dehuai, however, demanded direct air force support for ground forces.

In early December 1950, the PLAAF drew up a plan for air warfare that opposed a piecemeal approach to employing the air force, and favored a strategy that emphasized the accumulation of strength and its concentrated employment in timely attacks. Mao was highly complimentary about this prudent strategy, which he endorsed.[34] PLAAF scholars later regarded the concentrated and safe employment of the air force (*jizhong wentuo shiyong*) as one of the key elements of Mao's military thinking on air power that shaped the direction of the PLAAF at the time.[35]

Guided by this principle, the PLAAF began to prepare for an intervention with two stages. First, it planned to commit four air regiments to provide protection for transportation lines between the Chinese border and Anju in North Korea by early 1951. During this period, PLAAF planners hoped that Soviet air units assigned to defend Chinese supply lines over the Yalu River would coordinate the operation, letting the newly trained Chinese pilots fly with them in combat to gain experience. Second, the PLAAF needed to construct airfields inside Korea to allow use of the air force in direct ground support. When these forward airfields became operational, the Chinese planned to deploy a significant number of fighters, bombers, and attack aircraft to commence full-scale air war against U.S. and UN forces in the spring of 1951.[36]

Chinese leaders soon found out that U.S. air superiority made attempts to provide direct ground support impractical. The USAF carried out sustained bombardments of airfields in North Korea, which prevented the Chinese Air Force from carrying out its planned deployments. As a result, the PLAAF had to abandon its plan for air-to-ground support and was limited to airfields inside China from which to engage the Americans over Korea.[37]

Beginning in September 1951, more and more Chinese air units were sent to air operations in Korea. The need for a joint command to coordinate the Communist air forces' effort in Korea was obvious, but the Soviets persisted in acting alone. However, a joint command was set up for the Chinese–North Korean air army. The Chinese apparently had no authority over North Korean air units. Plans for using them had to be negotiated between Beijing and Pyongyang.

During the war, two additional factors determined the PLAAF's limited air defense role in Korea: the Soviet involvement was limited to air defense in the rear, making little attempt to coordinate Soviet efforts with Chinese ground operations; and the Soviet-made MiG-15 had a relatively short range that also confined Chinese air operations to an area not far south of the Yalu River. Throughout the war, the PLAAF basically acted independently, with its air operations concentrated on protecting key transportation lines and military and industrial targets in Korea, despite the ground forces' desperate need for air support. PLAAF literature argues that air combat against U.S./UN forces to gain air control over northwest Korea constituted indirect support for ground units, but also accepts that any vigorous competition for air superiority with the Americans would have cost the PLAAF many more aircraft and lives.[38]

The PLAAF claimed to have shot down 330 UN aircraft with 231 of their own losses. What China furthermore gained from the three-year Korean War was a rapid expansion of the air force from four regiments at the beginning to sixty-six regiments—a total of more than three thousand aircraft—at the end. Equally important was an opportunity for some eight hundred pilots and fifty-eight thousand ground personnel to take part in air operations in Korea. This produced mature Chinese combat pilots and prepared the Chinese Air Force for future conflicts.[39]

The Korean War experience also helped the PLAAF develop its own combat tactics, known as "multi-layers, and four aircraft formation in one zone." The Chinese believed that this tactical principle of dividing Chinese formations into attack and cover pairs significantly improved flexibility and safety in combat against American aircraft in comparison with their initial usage of Soviet World War II tactics, which involved flying large formations that had appeared too rigid and extremely vulnerable to enemy attacks.[40] Nevertheless, the memories of the U.S. and UN air attacks in Korea and the threat of American nuclear-armed bombers continued to drive the PLAAF to emphasize air defense, and fighters constituted the largest and most important element of the Chinese Air Force.

The Chinese leaders gleaned a mixed understanding of air power from the Korean conflict. While recognizing America's air superiority, they discounted the role air power had played. Mao found it particularly interesting that air bombardment inflicted fewer casualties upon Communist forces than ground fire. Even though one of his own sons was killed during an air raid in Korea, Mao appeared extremely gratified to learn that enemy aircraft had less effect on ground operations than he had previously imagined.[41] Given his confidence in the human factor—that men could overcome weapons—and his own guerrilla war experience, Mao remained convinced that Chinese ground forces could overwhelm stronger opponents and win the war.[42]

The UN's air superiority and the PLAAF's failure to provide air support for ground operations during the three-year Korean conflict appeared to have little impact on the Communists' claim of victory. It is thus not surprising that Chinese political leaders and generals maintained their view that future wars would be conducted in the context of ground operations, with air power used to supplement the power of the army. As we shall see, the Communist air defense experience thus resulted in the PLAAF's continuing to emphasize an air defense strategy and the development of fighter planes, radar, and ground antiaircraft systems, while devoting a small portion of the overall force structure to delivering limited air-to-surface ordnance.

Nonetheless, Chinese leaders and the PLAAF could not delude themselves about the capabilities and performance of their pilots and aircraft during the war. American air superiority had not only prevented the Chinese Air Force from providing ground support, but also inflicted heavy casualties and material losses on the ground. Consequently, the Chinese did not achieve the total victory their leaders had so eagerly sought in Korea. The PLAAF might have become one of the largest air forces in the world by the end of the war, but it was far from being the best. Chinese leaders continued to express an interest in building a strong air arm in the 1953 "First Five-Year Plan," which called for 150 aviation regiments with more than 6,000 aircraft, and the construction of 153 new airfields. Although this plan was revised to a more realistic schedule after the war, the air force continued to receive priority funding in the 1950s and 1960s.[43]

The Korean War experience further increased Chinese acceptance of Soviet doctrine, which emphasized air defense, followed by air superiority, and then offensive air support. The Chinese military became obsessed with air defense—a mind-set stemming from the defensive nature of the "people's war" doctrine and the strong air threat from both the United States and Taiwan since the founding of the People's Republic. From day one, the Chinese Air Force was not designed to function as a separate strike force, but rather to act as a subsidiary *junzhong* (armed service) of the PLA. The existing ground force structure was simply grafted onto the air force, and army officers were chosen to command the air force. When the role of the PLAAF in war was discussed, its leadership made it clear that the air force had to focus on how it would relate to ground operations, or how it could best aid ground forces.[44]

The Korean War experience also taught Chinese leaders that a greater reliance on air power could have escalated the conflict. The Korean War was the first limited war of the nuclear age, fought within a limited time and space, and with limited means and objectives. Both the Communists and the UN allies restrained themselves, keeping the conflict confined to the Korean Peninsula. The USAF was prohibited from attacking Communist air bases in northeast China. Restraints also prevented Chinese Communist leaders from allowing their air force to conduct any offensive action against the U.S. safe haven south of the thirty-eighth parallel. Apart from making an initial request for Soviet air support for Chinese ground operations, Chinese leaders apparently shared Moscow's concern that air offensives might escalate the conflict, and, therefore, made no further appeals.

It is interesting to note that, during China's thirty-day invasion of Vietnam in 1979, Chinese leaders still believed the use of the air force was likely to escalate conflicts. Chinese strategic thinking evidently still regarded air power as a defensive, rather than an offensive, force or as a deterrent against threats to China's national security interests during the Cold War.[45]

THE PLAAF'S IDENTITY STRUGGLE

Throughout the rest of the 1950s, following the cease-fire on the Korean Peninsula, the PLAAF continued to define its identity and characteristics. Two developments apparently further influenced Chinese air power thinking that led to the development of the PLAAF into a defensive force. One was the ongoing debate about whether the Chinese military should definitely follow the Soviet model of professionalism and modernization. To many Chinese, Soviet experience and technology were the best basis for PLA reform. The question was what examples the PLA should copy from the Soviet model. The other factor was the continuing struggle against the Nationalist air force for the control of the air over the coastal region in southeast China and the PLAAF's later efforts against U.S. air intrusions. Again viewing itself as a defender, the PLAAF would further stress its defensive role in China's national security.

The experience the PLAAF gained in the 1950s also led the air force to become the vanguard of the movement to politicize the military with Mao's radical ideology throughout the following decade. From this same experience the PLAAF built its own tradition, which continued to influence Chinese thinking and efforts in the nation's current pursuit of a modern air force.

The debate about whether the Chinese military should follow the Soviet model and how to learn from the Soviet experience took place at a time when China had become increasingly reliant on the Soviet Union for aid and military hardware. By the end of 1955, 186 army and air divisions of the PLA were equipped with Soviet weapons and aircraft.[46] As more and more Soviet technology was introduced into the PLA, along with parallel developments in the methods of military command, control, and communications within the PLA, standardization and professionalism became necessities in fighting a modern conventional battle against a modern conventional enemy. Mao particularly believed that the study of Soviet experience and advanced technology would "change our army's backward conditions" and would "reconstruct our army as the second finest modern army in the world."[47] From the early 1950s on, therefore, Chinese leaders repeatedly urged the Chinese military to study the Soviet experience with modern warfare and armed forces. The effort to learn from the Soviet military came to a climax in 1955. From 1954 to 1958, the PLA carried out a series of reform programs that used Soviet military organization and doctrine as the models for Chinese military modernization; the programs included major institutional changes, a centralized command system, technological improvements, advanced training and educational programs, and the reorganization of the defense industries.[48]

The PLAAF was no exception, copying indiscriminately the model of the Soviet Air Force, adopting its military structure, and translating all of its regulations, manuals, curricula, and handbooks for command and control as well as training and operations.[49] Under the influence of Soviet air power thinking, the Chinese firmly rejected all arguments favoring the dominance of air power in warfare, believing that the only role of the air force was to assist the army and navy to accomplish the mission of national war. The 1958 *Kongjun zhanyigaize* (General Rules for Air Force Campaign), compiled by the PLA Military College, strongly criticized those who championed air dominance as the key to victory, contending that seeking to use machines as a substitute for the role played by humans in the war would delude a country into waging wars of aggression.[50]

While accepting that the Soviet Red Army was the best military in the world, many senior PLA officers questioned the effort to introduce the Soviet system into the Chinese military, and cautioned that the PLA should not lose sight of the people's war doctrine. This resulted in debates about whether Soviet doctrine and experience were suitable for the Chinese military or whether the PLA should maintain its own traditions and practices that had proved successful. Criticism centered on Soviet military organization, regulations, and the one-man command system claimed that the Soviet system was not appropriate to the Chinese military. In September 1956, at the Eighth Party Congress, the Chinese military leadership reached a consensus, reiterating that the PLA must keep the command and control system under the party committee's collective leadership, which divided responsibilities between military commanders and political commissars. New principles for the PLA's modernization over the next decades made it clear that the PLA would adopt only that part of the Soviet military experience that was applicable to China's situation.[51]

In the 1950s questioning the Soviet experience also created confusion within the PLA, particularly in the PLAAF. The lack of standardization and professionalism in modern warfare and training had proved deadly. Soviet regulations and rules often appeared to contradict the PLA's customs and tradition. The introduction of Russian military terminology also proved to be disastrous. On July 23, 1954, a ground control intercept commander who did not speak Russian failed to use standard military terminology. As a result, two La-11 pilots mistakenly shot down a Cathay Pacific Airways aircraft over the South China Sea, killing ten people on board.[52] In the midst of the 1958 antidogmatic campaign throughout the PLA effort to free itself from the Soviet model, demands were made for the regulations and rules to be compiled by the Chinese themselves. Beginning in 1959 the PLA endeavored to write its own operational regulations while revising the existing intra-service regulations, disciplinary regulations, drill regulations, and political work regulations.[53] Similarly, the PLAAF began to produce its own operational and flying regulations, which noticeably defined Chinese air power thinking as a defense force.

The immediate impact on the PLAAF was the merging of the PLA Air Defense troops—a service equivalent to the air force and navy under the Soviet system—with the air force. The Air Defense troops did not possess any aviation assets as the Soviet forces did, while these two services of the PLA shared many similar responsibilities for air defense. This reorganization eliminated redundant elements in the administrative and operational structure. The PLAAF became a mixed force, consisting of twenty-eight air divisions, six independent air regiments, eleven antiaircraft artillery divisions, twenty-five radar regiments, six searchlight regiments, and one airborne division,[54] with more than five thousand combat aircraft in its inventory.[55]

Simply bringing aviation and air defense assets together did not produce an effective air defense system. During the air battles over the Taiwan Strait in 1958, a lack of coordination between friendly aircraft and antiaircraft artillery units resulted in at least five incidents that involved Chinese fighters coming under friendly fire from antiaircraft artillery.[56] Perhaps even more influential was that the merger reinforced the PLAAF's air defense characteristics and made it difficult for the PLAAF to develop its air offensive doctrine and capabilities for more than four decades.

The political work system was a strong tradition in the PLA. The centralized Soviet command system, which placed political work under the control of the military commander, was criticized as counter to the PLA's rich experience of performing political tasks, which was perceived as essential to Chinese Communist success. Since the inception of the PLAAF, the leadership had recognized the importance of political work, emphasizing that the political cadres must regard security and safety issues as their major job during flight training. Regulations empowered political officers to exercise control over every aspect of military affairs within the air force. The PLAAF focused particularly on ensuring the political reliability of aviators. Emulating the Soviet model, the PLAAF developed a series of measures that included selecting pilots from peasant and working-class families and offering pilots a higher standard of living and other preferential benefits. One unique institutional characteristic required all pilots not only to work and fly together, but also to live together in on-base housing, regardless of their ages, to ensure political cohesion and loyalty. Political commissars at the regimental and division levels had the authority to determine whether a pilot was suited to fly training or combat missions. Criticism of the Soviet one-man command system further affirmed the political control measures adopted by the PLAAF throughout the air force.[57]

While continuing to search for its identity, the PLAAF constantly engaged in air combat against the Nationalist air force for the control of airspace over the coastal areas of Zhejiang and Fujian provinces.[58] As the air battles over the Taiwan Strait intensified in the summer of 1958, Chinese leaders increasingly feared that military operations against the Nationalists might escalate into an international crisis. Beijing authorities established strict rules of engagement for the air force: aircraft should not conduct any operations over the high seas; the air force should not bomb the Jinmen and Mazu islands if the Nationalists did not bomb the mainland; and air force pilots should not initiate attacks on U.S. aircraft unless they had violated Chinese airspace.[59] These self-imposed rules reflected Beijing's political concern about Washington's possible involvement; however, Chinese archival sources that have recently become available also suggest that the Chinese leadership was very uncertain about the air force's ability to counter Nationalist air bombardment of the mainland.[60]

In response to this situation, the PLAAF defined a set of operational principles along the lines laid down by the Chinese leadership, which stressed the use of overwhelming force to achieve the protection of friendly forces and the destruction of enemy forces; the subordination of military objectives to political ones through strict adherence to the central authority's operational policy; and the study and application of PLAAF experience and tactics drawn from the Korean War.[61] Preparing for Nationalist bombing attacks, the air force developed some countermeasures, including mental preparation, camouflage, quick repair, and the deployment of Chinese bombers to the region at risk to strike back. The PLAAF leadership appeared confident that the enemy bombers would become easy targets for PLAAF pilots and that, after some bomber losses, the Nationalists would not dare to continue bombing the mainland.[62]

The PLAAF deployed twenty-three fighter regiments to the southeast coastal area, with six regiments in the forward airfields across the strait and the rest as the second echelon in the surrounding provinces. From late July to mid-October 1958, the PLAAF managed to fly 3,778 sorties and engaged in thirteen air battles, claiming to have shot down fourteen and damaged nine Nationalist aircraft, while losing only five aircraft with five damaged.[63] From that time on, the PLAAF firmly controlled the airspace over southeast China.

Nevertheless, PLAAF operations during the Taiwan Strait crises revealed inherent problems. PLAAF air power thinking remained defensive, as the PLAAF deployed large numbers of fighters to the region but could not capitalize on its numerical superiority.[64] Fearing a Nationalist attack on its airfields, the PLAAF leadership used only half of its strength to fly fighter patrols, while the other half was held back to protect home bases. As during the Korean War, the PLAAF took a passive stance and waited to respond to intrusions by the Nationalist air force, which was much smaller, but was free to choose the time and method of aerial combat. PLAAF, by contrast, had to depend on the ground control intercept to scramble its fighters. Furthermore, the capability of the air force was restricted by political considerations and the limited range of the MiG-17.

Operations against the Nationalists over the southeast coastal areas in the 1950s gave the Chinese more experience in employing air power in air defense. It was in such a context that the PLAAF began to write its own doctrine and manuals.

CHINA'S AIR DOCTRINE

Western studies have concluded that the PLAAF did not develop "a formulized, written doctrine" and in particular did not have a strategic air defense policy until the 1990s.[65] This belief can perhaps be largely attributed to the restricted access to the PLAAF's official documents, and to unfamiliarity with Chinese terminology.[66] The Chinese often adopted the alternative term—*military thought*—to represent their thinking on the air force and its employment, yet that thinking was also imbued with the PLA's political jargon as well as the PLAAF's own experience, which may not have been honestly defined.

In fact, the PLAAF has never included the term "doctrine" in its lexicon. Instead, it has used the word *tiaoling* (field manual) to represent official beliefs, warfighting principles, and terminology for use by the air force in military operations. In May 1959 the PLAAF organized a team of more than one hundred officers to compile *kongjun zhanduo tiaoling* (The Field Manual of Air Force Operations). It was not until 1963 that the air force officially accepted this field manual, along with several other regulations and handbooks, as its official guidance.[67] Even now, documentation on PLAAF manuals and regulations remains fragmentary. The recent publication of books, academic treatises, and individual recollections provides illuminating insight into the formation and development of the PLAAF's strategy and basic models and principles for air force operations.[68]

The 1963 field manual was the nearest equivalent the Chinese had to a Western doctrine manual, presenting the fundamental concepts that govern PLAAF operations at the various levels of war, the application of operational theories, and the specifics of how to apply military force at each level of warfare. The manual emphasized that the air force's primary mission was to carry out protracted defensive operations to wear down the adversary while the enemy was strategically on the offensive and attacking. This defense strategy also specified operational principles with the following main points: the air force should try to gain the initiative with a counteroffensive to seize air superiority; the entire air force should be used in a concentrated fashion instead of dividing the forces; the air force should operate in a surprising and concealed manner, and with a flexible strategy and tactics; careful planning and total preparation were necessary before each operation; the air force should implement positive measures to enhance its protection in order to sustain combat capability; and the air force should maintain a brave and tenacious combat style, and attain perfect mastery of combat skills to defeat the enemy.[69]

The political culture of the PLAAF viewed the central leadership's instructions and directives concerning the construction and use of the air force as codification of Mao's military thought. While Mao wrote extensively on war and warfare, his instructions for the air force never appeared systematic or theoretical. Nonetheless, the PLAAF leadership adopted Mao's 1950 epigraph in the inaugural issue of the air force newspaper—"eliminating the enemy remnants, strengthening national defense"—as the strategic mission of the air force.[70] Based on Mao's comments that the air force should take a prudent approach, the PLAAF established its strategy for Korea: "to accumulate its strength and then to employ it in a timely, concentrated, and safe manner."[71] Mao's 1958 instruction on "going out with all our strength to destroy the invading enemy" (*quanli yifu, wujian ruqin zhidi*), prompted the air force to pay further attention to territorial defense operations.

In reviewing the air force's own experience, PLAAF literature argues that the air force's military practices serve as sources for summing up Chinese military thought on air force, and were a driving force in the development of their military thought on the air force. Perhaps even more important, the experience of the air force became the only standard for examining the validity of military views on the air force.[72] As it compiled the first air force field manual, the PLAAF utilized Mao Zedong's military thought and instructions as theoretical guidance, while drawing only on the PLA's fine traditions and the PLAAF's decade-long experience as references.[73]

The PLAAF, like most of the world's air power nations, regarded command of the air as the air force's primary mission. The field manual stated that, during the struggle for the command of the air, the air force should keep a lookout over the major airspace, and seize and maintain command of the air at critical moments over certain areas. The foremost difference from equivalent guidance in other countries was that China considered command of the air as restricted to its own airspace. The air force's critical role was to protect vital domestic objectives and ensure that the main military forces were not subjected to the enemy's systematic air attacks.[74] Believing that China still faced the danger of a large-scale invasion by powerful enemies, the PLA strategy required the air force first, to participate in resisting the enemy's strategic surprise attacks to shield the nation during its transition from peacetime to wartime; next, to provide air cover for the strategic deployment of massed troops; and finally, to enhance domestic air defense, with special focus on defense of strategic points, while providing support in land and naval battles.[75]

Based on these guidelines, the PLAAF, for the first time, determined the principles underlying the specific utilization and development of the air force. Even so, no matter how the PLAAF attempted to define its strategy and doctrine, it never viewed itself as an independent arm, but instead as a subsidiary service for both the army and navy.[76] The PLAAF's two primary roles would be to support the ground forces, claiming their victories as its own victories, and to maintain local air dominance through air defense. Given that the PLAAF remained a relatively weak force, the preservation of its strength became another important principle, emphasizing commitment to protracted operations and patience in awaiting opportunities to destroy the enemy. As a result, the PLAAF concluded that the air force had to continue attaching importance to building up air defense troops on a fairly large scale.[77]

While the PLAAF officially adopted this strategy and doctrine in 1963, it had never found an opportunity to implement them. China continued to perceive a U.S. threat coming from South Korea, Taiwan, and Indochina, but no large-scale invasion of China took place during the 1960s. Despite sending a significant number of air defense troops to help North Vietnam resist U.S. air bombardment, the PLAAF found itself engaged in no meaningful air combat, other than constantly scrambling its fighters to intercept intruding aircraft, many of them spy planes.[78]

As a result, the 1963 field manual, slowly but surely, became irrelevant. One major focus of the PLAAF's day-to-day activity was on finding ways to bring down the manned reconnaissance aircraft (initially, the PV2 and then the U-2) and unmanned drones that routinely flew over Chinese airspace. Failure to intercept and shoot them down would demonstrate the PLAAF's inability to defend China. The Chinese leadership would have found this humiliating, because any such incident would create "a bad impression" of China's air defense capability.[79] Engaging intruding reconnaissance aircraft was considered a rare combat opportunity for Chinese pilots to sharpen their combat skills. The PLAAF organized special air combat units (including surface-to-air missile units) and deployed them to several key locations around the country. Despite frustrations over many unsuccessful attempts to shoot down the spy planes, the air force claimed that it had brought down more than two dozen manned and unmanned spy planes throughout the 1960s.[80]

The incidents of intruding overflights took place in the midst of an upsurge in political radicalism that emphasized political factors in the command and promotion of Mao's cult of personality. The downing of every intruder was described more like a political victory than a military one. Celebrations were held and awards were given to those involved in combat actions. Senior party and state leaders, including Chairman Mao and Premier Zhou Enlai, always received the men responsible for the shoot-downs, making headline news across the country. Senior military leaders also used these events to promote the air force, proclaiming, "all military services must learn from the air force."[81] Study of Mao Zedong's thought and his people's war theory dominated the PLA's daily training and teaching. Field manuals and other air force teaching materials compiled in the early 1960s were refuted; many copies were even burned.[82] Although flight training required each fighter pilot to fly an average of 122.25 hours per year, by 1970 Chinese pilots flew on average only 55 hours a year because political teaching focused on the study of Mao's works had become more important than flight training.

The zealousness of Mao's people's war theory further fostered unrealistic thinking about the development of the air force in order to satisfy radical political sentiments. In the late 1960s, the PLAAF came up with a new vision that urged the air force to "fit in with the requirements of the people's war and carry out guerrilla warfare in the air."[83] It then requested the aviation industry to develop and then build a supersonic fighter that was small, light, and agile, and had short-takeoff-and-landing capability.[84] The first prototype made its inaugural flight at the end of 1970. In a later demonstration flight, senior PLA military leaders named the plane "Li Xiangyang in the Air" after a famous guerrilla fighter character in a popular anti-Japanese war movie at the time. The aircraft, however, never entered service because of its inadequate thrust and firepower.[85] This failed project unmistakably suggested that it was impossible to develop any healthy and objective air power theories in a political system dominated by absolute authority and arbitrary decisions by key individuals.

Although the PLAAF failed to fully implement its strategic vision for the air force, by the early 1970s it had grown into a force with fifty air divisions, of which two-thirds flew short-range fighters.[86] Worship of the leader during the ten years of political upheaval brought about by the Cultural Revolution (1966–76) not only prevented the PLAAF from further defining and perfecting its strategy and doctrine, but also bequeathed a legacy that continued to convince PLAAF scholars to regard the instructions of the senior leaders as an important component of Chinese military thought about the air force. For this reason, Mao's thinking on war and warfare, especially his people's war strategy, remains relevant to the PLAAF even today as the air force seeks to develop its own strategy and doctrine.

THE 1979 WAR WITH VIETNAM AND ITS RAMIFICATIONS

The PLAAF could be seen in combat action when China invaded Vietnam in 1979. During a month of ground operations, from February 17 to March 16, Chinese Air Force fighter units flew 8,500 sorties in air patrol missions, while transport and helicopter units made 228 sorties in airlift missions. Political considerations, however, prevented the PLAAF from flying any ground support missions and engaging in air combat. Deng Xiaoping, the post-Mao Chinese leader, did not want to escalate the conflict to an unmanageable level by using air power. He instructed the air force to confine its activities within China's territory, while preparing to support ground operations "if necessary," though without giving a clear definition of "necessary." The air force responded with a strategy that deployed as many as seven hundred warplanes to the border region, requiring them to be combat-ready for air defense and ground support. When the ground assault started, the air force was ordered to fly as many sorties as possible over the border airspace to deter the Vietnamese Air Force from taking action against China.[87] On the occasions when ground troops badly needed air support, Beijing authorities refused to grant such permission, instead ordering the army to rely exclusively on artillery for fire support.[88]

Western analysts have consistently attributed the air force's negligible role in the conflict to China's perception that its air force could not win.[89] The real problem was a failure on the part of Chinese leadership to appreciate the critical role of air power in modern warfare. China maintained that its numerical superiority had demonstrated the might of the PLAAF and, therefore, deterred the Vietnamese Air Force from daring to engage the Chinese aircraft.[90] Marshall Ye Jianying even ridiculously commented that China's phony air operations in the war against Vietnam were an "ingenious way of employing the air force."[91] Such a fallacious view again originated from the people's war doctrine, which did not envisage the need for offensive air power.

The PLA's modern war experience had suggested that air power had little impact on the victory claimed by China (i.e., the Korean War) in the past. It is thus not surprising that Chinese political and military leaders maintained the view that war would continue to be conducted in the context of ground operations, with air power used only to supplement the power of the army. Furthermore, given their confidence in the human factor—that men could overcome weapons—and their own guerrilla war experience, Chinese leaders were convinced that their ground forces could overwhelm any opponent and win any war. Consequently, the PLAAF had long argued that ground operations would determine the air force's victory.

The 1979 war experience apparently contributed to the Chinese view that air power served to deter an enemy from using air attacks against China. The development of such thinking was supported by the objective reality confronting the PLAAF. While the PLAAF was one of the world's largest air forces, its equipment was outdated and limited in capability, and not even equal to that of some countries surrounding China.[92] To address technological deficiencies and maintain the air force's overall combat capabilities, China favored an air force based on quantity instead of quality.[93] In the early 1980s, the Chinese aviation industry continued to manufacture the obsolete Jian-6 (Chinese MiG-19). A total of more than four thousand of these planes had entered the PLAAF's inventory by 1986.[94] Numerical superiority served as the indicator that convinced the Chinese at the time that China had built a credible air defense force capable of deterring and, if necessary, resisting any attack on China.

Despite this, the PLAAF also realized that China faced, and would continue to face, the major challenge of preparing an inferior force to use whatever resources it had to fight against a superior enemy.[95] Having seen how rapidly the Soviet Union invaded Afghanistan in December 1979, the central military leadership adopted, in 1980, an active defense strategy against a possible similar invasion from the north.[96]

In response, the PLAAF continued to follow its strategy and doctrine, implemented in the 1960s, as reflected in the North China exercise in 1981. Its mission in this exercise still focused on ways to use the air force during the initial stages of a war, especially on providing air cover for the fielded army. No details are available about the role the air force played in the exercise. According to one recollection, the principles guiding the utilization of the air force were still active support, the concentrated use of air power, and close protection.[97] In due course, the fighter units performed the air patrol, air cover, and air superiority missions, while the ground attack aircraft units played the ground support role. The bomber units conducted formation bombardment and laid antitank mines, and the transport units performed airlift and airdrop missions. The exercises again exposed the backwardness of the PLAAF as well as its lack of air power strategy and doctrine.[98]

The Chinese leadership was fully aware of the shortcomings of the PLAAF and the imperative to undertake major, long-term institutional change. From the late 1970s on, Deng Xiaoping delivered repeated instructions that underscored the importance of the air force in modern warfare, given the major changes in the geopolitical situation. According to the Chinese leader, a new world war was not inevitable; instead, peace and development were the two leading trends in international affairs, and the next war would be a local, limited war fought by a professional army with modern technology. Deng pointed out that any country that went to war in the future with no air force and without air dominance would find it impossible to win, especially since both the army and navy would need air support and air cover. He wanted the air force to receive the highest priority in China's military development strategy.[99]

As a result, the PLA's share of the annual state budget dropped from a high of 18.5 percent in 1979 to about 8 percent in 1989, while the air force enjoyed a relative increase in spending. The Chinese leader, nevertheless, criticized the air force for being the most bloated force in the world. In 1985, he ordered the downsizing of the air force and the other military services by a total of one million men.[100] The initial cut involved more than 20 percent of the aviation units and almost all of the antiaircraft artillery units, which were either demobilized or downsized. By the early 1990s, the air force had trimmed almost one-third of its total force.[101]

In the meantime, Deng had become increasingly dissatisfied with the nature of the PLAAF as a defensive air force. He believed that a defense-oriented air force no longer fit in with the requirements of fighting a limited, local war under high-technology conditions. According to a PLAAF study, the Chinese leader advised the air force that "active defense itself is not necessarily limited to a defensive concept. Active defense also contains an offensive element." He further noted that "the air force's bombers are also defensive weapons" and "anyone who wants to destroy us will be subject to retaliation."[102] PLAAF literature interprets these statements as Deng Xiaoping's military thought on the air force.

In 1987, the PLAAF responded by drafting a new five-year plan to build up the air force starting in 1990. For the first time in its history, the PLAAF seemed to shift its focus to quality, not quantity. The plan was expected to lay a solid foundation by 2000 for a modern air force with Chinese characteristics (a term used to characterize a uniquely Chinese way of doing things), and its weapon systems would gradually become upgraded with both offensive and defensive capabilities (*gongfang jianbei*).[103]

However, making such aspirations and objectives a reality was hampered by bureaucratic tradition; one study comments on "[M]eetings, platitudes, studies, and reports recommending more meetings and studies."[104] The immediate problem for the PLAAF was China's lack of an aviation industry to support the new needs of the air force. The Chinese aviation industry had long focused on domestic air defense, giving priority to surface-to-air missiles; high-altitude, high-speed, all-weather interceptors; early-warning radars; and electronic countermeasure systems. Despite China's open-door policy that would enable the acquisition of some weapon systems from foreign suppliers, the Chinese leadership insisted that the PLAAF should continue to rely on China's indigenous efforts to build the air force. Deng Xiaoping once criticized those who sought foreign purchases as an expedient measure to meet the air force's needs: "How many advanced airplanes can you afford to purchase? We will become poor soon after we have bought a few airplanes." Deng urged the air force to prepare itself to continue to fight with inferior weapons for many years to come.[105]

Perhaps even more than before, the PLAAF had made no corresponding theoretical studies to support its modernization efforts. In 1988 the air force published a book titled *Kongjun zhanyigaize* (Science of Air Force Campaigns), which reiterated the concept of command of the air, based on the 1963 air force field manual. However, it specified a new direction for the development of the air force, with three eye-catching phrases: integrated resistance (*zhengti kangji*), active counterattack (*jiji fanji*), and close protection (*yanmi fanghu*).[106] The essence of these integrated air defense operations was to weld fighter aircraft, surface-based defense, and command, control, communication, and intelligence elements into an efficient defensive network. That network would fend off the enemy's air attacks, while seeking opportunities to launch counterattacks against the enemy's airfields and air forces in order to weaken its air strength so that it could wage no further attacks. Because PLAAF airfields and air bases were important targets for the enemy, the book further underlined the employment of positive measures to enhance the protection of aircraft, troops, and all types of facilities, which was critical for maintaining defensive capabilities. For the same reason, the book maintained as a general rule that the air force would only launch counterattacks at the right times to strike the enemy's key positions: namely, when enemy warning and defense systems appeared weak.

At the time, the PLAAF apparently continued to think of itself as a defensive air force and recognized China's technological disadvantage. However, it never seemed to realize that new aviation technology had widened the technology gap between China and other air powers to a shocking level.[107]

THE EMERGENCE OF A NEW OFFENSIVE AND DEFENSIVE STRATEGY

The new role that air power played in the 1991 Gulf War shattered the Chinese view of air power as primarily an air defense force. For the first time, Chinese Air Force officers watching their television screens could witness how modern air power rendered traditional air defense systems obsolete. This forced them to recognize that China had fallen behind the West not only in technology, but also in doctrinal air power thinking. Time and space were no longer the allies of those who had been confident that China's existing air defense systems could prevent any attacks deep into the nation's heartland. Serious doubts were raised about the traditional interpretation of China's defense capabilities, including the common belief that an inferior force could overcome a superior enemy. Drawing on lessons learned from Iraq's defeat in the Gulf War, the Chinese central military leadership pointed out that "a weaker force relying solely on the defensive would place itself in the position of having to receive blows," and that only by "taking active offensive operations" could the weaker seize the initiative.[108] Following such a line of thinking, PLAAF strategists also looked to the West's writing for theoretical guidance. One study specifically quoted Douhet's argument that "a weaker air force could also defeat a stronger enemy one provided it can compensate for the difference in strength by showing more intelligence, more intensity, and more violence in its offensive actions."[109] They concluded that the PLAAF must shift from a purely defensive to a combined defensive-offensive posture by adding more offensive forces.[110] All these discussions were confined within the boundaries of the traditional interpretation of people's war, taking tactically offensive action within a basically defensive strategy.

China's evolving security interests, including a potential confrontation with Taiwan, also favored consideration of augmenting the PLAAF's offensive capabilities. Since 1993 Beijing has adopted a new military strategy, placing an emphasis on fighting and winning a future regional war under high-technology conditions along China's periphery. The momentum of the independence movement in Taiwan was simultaneously viewed as an increasingly serious challenge to China's sovereignty and security.[111] The central military leadership made the proper readjustment to the air force's strategic missions, requiring it to maintain strong capabilities not only for defensive operations, but also for offensive ones. This new mission requirement included the air force's traditional tasks, such as securing air dominance over China's own airspace, supporting the army and the navy, and directing paratrooper operations, as well as carrying out independent air campaigns. It also made a new demand that the air force be able to launch attacks against the enemy's air assets on the ground in a potential local conflict along China's coast.[112] Although the PLAAF did not possess any credible offensive capabilities at the time, the changing military and security environment in the region and in the world compelled the PLAAF to envision such a transformation.

While discussions about the need to enhance the PLAAF's offensive capabilities continued, Chinese strategists had to face one cruel reality: the negative impact that the air defense–oriented strategy had exerted on the air force since its establishment. The development of avionics and precision-strike weapons had made the PLAAF even more "short-legged" (*duantui*) and "shortsighted" (*jinshi*). Even the Chinese acknowledged that their aircraft had short ranges and lacked airborne radars.[113] Using the air force in any contemplated offensive actions over the enemy's territory would be suicidal. China's indigenous defense industry offered no immediate solution to correct this weakness, because all existing programs were aimed at air defense.[114] Moreover, huge investment and many years would have been needed to create a much more sophisticated industrial base.

The only stopgap measure appeared to be foreign purchases and assistance. The West had imposed its arms embargo on China following the June 1989 events at Tiananmen Square. Thus, Beijing again turned to Russia for assistance in modernizing its air force. Beginning in 1992, China bought three batches of Su-27s—a total of seventy-four aircraft—along with their accessories. The Su-27 was one of the world's most modern aircraft and had a state-of-the-art weapon system, but had limited offensive capabilities as a fighter. The Chinese decided to purchase the Su-27 because they were impressed by its aerobatic performance.[115] In 1996 China entered into an agreement with Russia on the licensed production of two hundred Su-27s at the Chinese aircraft factory in Shenyang.[116] The existing domestic program still attached importance to the development of fighter aircraft (presumably the J-10) as an "assassin's mace" (*sashoujian*), allowing the air force to fight a local war under high-tech conditions. In other words, the ongoing debate about increasing the air force's offensive capabilities had little influence on China's traditional thinking about defensive ones.[66]

For the air force, the North Atlantic Treaty Organization (NATO) air campaign in Kosovo came as a rude awakening. While the Gulf War brought the importance of air power, long-range precision strikes, and electronic warfare (EW) to the attention of the Chinese, they still found the actual effect of air power debatable. According to a 1991 report from the PLA Military Science Academy, the Americans themselves admitted that 70 percent of their bombs did not hit their targets. Iraq's defeat in 1991 stemmed from its failure to put up effective resistance in the air or on the ground because its command systems had been destroyed, as well as its failure to organize a withdrawal because its leaders had been indecisive.[118] Allied operations in Kosovo further shocked the Chinese and drew their attention to the characteristics of the air strike because the former Yugoslavia's air defense capabilities were much better than those of Iraq. One senior air force researcher responded with alarm, noting immediately that the air strike was a "war of all dimensions" (air, sea, space, electronics, and ground), that air and space assets played a leading role throughout the war, and the control of information determined the victory.[119] This analysis represented a dramatic change in the Chinese view that any future conflict, especially one with the United States, would be dominated by long-range precision strikes preceded by intensive overhead reconnaissance and computer network attacks.

Throughout the 1990s, the PLAAF's theoretical studies on air power flourished, with the publication of two major works, *Kongjun zhanyi xue* (Science of

Air Force Campaigns) and *Kongjun zhanshu xue* (Science of Air Force Tactics),[120] along with several other individual articles.[121] Whether there was any connection between the NATO air strike against Serbia and the ongoing debate inside China about the role of the PLAAF, it seemed that Chinese thinking about air forces was not fully developed until 1999. For the first time, the air force was regarded as a strategic armed service that could play a critical role in protecting national sovereignty and security. In early March 1999 Jiang Zemin, secretary-general of the CCP and president of China, spoke about the air force's strategic objective: to transform gradually from a homeland air defense force to one that was capable of both offensive and defensive operations. He then called on the air force to "bear the brunt of, and be employed throughout the entire course" of the conflict, and "to complete certain strategic missions independently."[122]

To achieve these objectives, later that year China adopted a three-step strategy for the development of the air force over the next several decades.[123] According to China's defense white paper, China expects to lay a solid foundation for the development of the PLA into a more high-tech and more balanced network-centric joint force by 2010, to accomplish mechanization and make major progress in informatization by 2020, and to reach the goal of modernizing national defense and the armed forces by the middle of the twenty-first century.[124]

GROWING CHINESE AIR POWER AND ITS INHERENT PROBLEMS

The immediate impact of this three-step strategy on the PLAAF was a new order placed for the purchase of Su-30s from Russia. The Chinese apparently had very quickly become dissatisfied with their fleet of Su-27s, which could not fire precision weapons and had inadequate flight ranges. In late 1999 China fulfilled a $1.85 billion contract with Russia to purchase thirty-eight Su-30MKK fighter-bombers with upgraded avionics, larger weapon payloads, and air-refueling capabilities for the PLAAF. Thereafter, China placed two additional orders of thirty-eight aircraft for the air force and twenty-four Su-30 MK2s for the PLA naval aviation forces. The acquisition of these new Russian-made fighter-bombers empowered the PLAAF with real long-range strike capabilities, and has inevitably tipped the military balance in East Asia.[125]

Nevertheless, the Russian-made fighter-bombers did not considerably improve the PLAAF's precision-strike capabilities. The air force turned to the domestic aircraft industry for the JH-7 fighter-bomber, which the PLAAF had initially rejected in favor of the Su-30. The improved variant, the JH-7A, has been upgraded with two more powerful domestic-made turbofan engines and a new fire control system capable of launching precision strikes using antiradiation missiles and laser-guided bombs.[126] Recent evidence also indicates that a new EW model aircraft provides an electronic escort for the attack group. The first batch of JH-7As entered service early in 2004. The PLAAF is reportedly very enthusiastic about this new multirole fighter that, as part of a rapid replacement program, is currently allowing the PLAAF to phase out the obsolete fleet of Q-5 attack aircraft.[127]

The effort to build an offensive and defensive air force has also focused on the development of early warning and control aircraft. One lesson learned from the NATO air operations in the former Yugoslavia was that the Serbian MiG-29 was not intrinsically inferior to NATO's fighters. However, NATO's airborne warning and control system (AWACS) aircraft put the MiGs at a decisive disadvantage, while its planes cut Serbian forces off from their sources of information and prevented them from organizing an effective defense. The development of AWACS and electronic aircraft has been at the center of the Chinese effort to achieve modern air power. However, in 1999 the U.S. government successfully pressured Israel into canceling a sale of the Falcon AWACS system to China. This cancellation prompted China to gather its own talents and resources and to build its own AWACS. The Chinese AWACS was developed in a high-low combination with the KJ-2000 based on the Russian Il-76MD airframe and the KJ-200 based on the Y-8F-200 transport platform. These platforms were handed over to the PLAAF in 2005 and 2006, respectively. Simultaneously, China developed five other different types of EW aircraft, the High New (*Gaoxin*) series, as well as the Y-8 variant, which have gradually entered service since 2004.

China's air power today is not the same as it was a decade ago. When the PLAAF participated in the military parade in 1999 to celebrate the fiftieth anniversary of the founding of the People's Republic, the Russian-made Su-27 fighter was China's only third-generation warplane. Ten years later, on October 1, 2009, the Chinese Air Force flew its domestic-produced third-generation fighter, the J-10, over Tiananmen Square along with Chinese-made AWACS aircraft.[128] Recently, Chinese Air Force commander Gen. Xu Qiliang proclaimed that the PLAAF's current development strategy is to become an air force with integrated capabilities for both offensive and defensive operations in the air as well as in space (*kongtian yiti*), with a focus on the improvement of its detection and early warning, air strike, antimissile air defense, and strategic delivery capabilities.[129]

The PLAAF still faces many challenges in reaching its development goals. To understand China's claims about the development of the PLAAF and air and space power for the present and the near future, it is necessary to consider several key issues, including strategy, force structure, the officer corps, the enlisted force, unit training, logistics, and maintenance. The PLAAF's new strategic vision, calling for the development of a strategic air force with long-range capabilities, as well as integrated air and space operations employing firepower systems that incorporate advanced information technology (*xinxi huoli yiti*), is modeled on the concept of U.S. practices as the Chinese have perceived them. This strategic vision, however, differs from USAF doctrine on counterspace operations to achieve space superiority.

In the midst of its discussions on how to integrate air and space power from a broad perspective, the PLAAF continues to face constraints.[130] The most important constraint is perhaps that the PLAAF does not possess its own space assets or strategic missiles. Instead, these systems remain under the control of the General Armament Department and the Second Artillery Force, respectively. Apparently, the PLAAF has continued to lose recent debates about whether these capabilities should be placed under its control.[131] The PLAAF has no choice but to concentrate on building facilities and institutions to receive satellite services for communication, weather, navigation, and global positioning. This will enable the PLAAF to make the transition from being a traditional air force to one enabled by space-based information (communications, positioning, navigation, timing, and intelligence, surveillance, and reconnaissance) capabilities.[132]

For the past several years, China has made efforts to streamline and optimize the PLAAF's force structure. These efforts include retiring earlier generations of aircraft, reducing the number of troops, and deploying third-generation combat aircraft and ground-to-air missiles. Although the PLAAF has been modernized and its force size has been significantly reduced, it still faces substantial replacement problems. While the older J-7 and J-8 fighters remain in service, the initially purchased Su-27s and the recently assembled Chinese J-11s appear incapable of fully supporting the mission requirements of the PLAAF, which now places increased emphasis on offensive versus defensive roles.[133] The size of the Chinese Air Force and its offensive capabilities will remain limited until a significant number of J-10s and J-11Bs enter service in the next five years.[134] Even so, the PLAAF will continue to rely on upgrading second-generation aircraft to maintain a sizeable air force.

In 2005 the PLAAF established an additional transport division and one special aircraft division to enhance its long-range airlift and airborne early warning (AEW) capabilities. Russia's failure to deliver thirty-four Il-76MDs as scheduled in 2008 will keep this newly created transport division underequipped. A limited number of Y-7s and Y-8s will constitute the majority of airframes in the interim until the Chinese-designed Y-20 large transport enters service.[135] The PLAAF has also made slow progress in incorporating the newly acquired support system, such as AEW/AWACS, aerial-refueling tankers, intelligence collection platforms, and signal-jamming aircraft, all of which are necessary to increase the effectiveness of combat aircraft and warfighting capability.[136] Furthermore, the PLAAF is still described as a mixed force of aviation, ground-air defense, airborne, signal, radar, electronic countermeasures, technical reconnaissance, and chemical defense units. This mixed force structure will continue to complicate China's air and space decisions, particularly in regard to training, allocating roles and missions among services and branches, and influencing resource allocations for Chinese Air Force modernization.

The PLAAF regards the recruitment of talented people as a key to transforming the Chinese Air Force into a force able to fight high-tech wars under informatized conditions.[137] Unlike the USAF, whose officers all have college degrees, with over half holding advanced degrees, only one-third of Chinese Air Force officers are college or university graduates and only 5 percent have a master's degree.[138] The PLAAF reshuffled the officer corps of the units that had received new generation aircraft and equipment by transferring educated personnel from air force headquarters, research institutes, and universities to fill as much as 80 percent of the leadership and technical positions in these units. Although these measures will improve the quality of the force, the PLA still lacks an effective assignment system to rotate officers periodically both across and within their specialties. Likewise, the promotion of Chinese officers is still determined at the unit level, where personal relationships and departmental favoritism influence individual initiative and organizational success. Major challenges remain for the PLAAF in retaining highly educated personnel, encouraging capable officers to serve longer, finding personnel with the special expertise necessary to fill key technical positions, and recruiting young talent to join the service.[139]

As previously discussed, Chinese aviation units are transitioning from older generation aircraft to new aircraft with significantly improved capabilities. The PLAAF is also enhancing its training by incorporating new systems and methods that increase the emphasis placed on technical and tactical training in complex environments, combined arms and aircraft-type training, and joint training under mission-oriented and confrontational conditions. The PLAAF has modernized flight training and created its own "Red Flag" training base, modeled on the U.S. program at Nellis Air Force Base.[140] Despite these changes, Chinese fighter pilots only fly an average of 120–130 hours per year, while their U.S. counterparts average 250–300 hours per year. Still other discrepancies appear in training requirements. USAF fighter pilots fly around fifty hours of air refueling, AWACS command and control, dissimilar air combat training, and night training missions before they assume operational roles. By contrast, the Chinese flight-training manual requires that a pilot who is to receive air-refueling training must have at least one thousand flying hours.[141] This suggests that even though the PLAAF has adopted new pilot training guidance, the equipment, overall requirements, procedures, and methods are still not comparable to U.S. standards and quality of training.

The PLAAF has reportedly begun reorganizing its air logistics and maintenance systems to support deployed units in conducting mobile offensive operations, but many areas are still weak. Few PLAAF field stations were built to support the multiple types of aircraft deploying to their airfields. Recently, the PLAAF has adopted a plan to modernize airfields in batches, adding new equipment that can more efficiently move supplies from depots to the field. The plan includes integrated computers that can track spare parts and improved logistics and maintenance support for individual weapon systems and units as a whole.[142]

The PLAAF faces three major challenges in the areas of logistics and maintenance. The first is a shortage of personnel with the knowledge and skills to serve in a variety of positions. While the PLAAF has begun to convert some junior officer maintenance billets to noncommissioned officer billets, it is not yet clear whether these actions have helped or hindered the PLAAF's overall maintenance capabilities.[143] Second, reform is still at the initial stages of experimentation and is occurring only at local levels. This suggests that new systems are not yet standardized across the air force.[144] Third and finally, the PLAAF's limited resources in logistics support will primarily be focused on units receiving new equipment. Currently, the PLAAF enjoys the benefits of a favorable military spending policy, but budget challenges are likely. As long as the General Logistics Department continues to control military finances, a funding shortfall for the air force is inevitable in the years ahead.[145]

CONCLUSION

For the past sixty years, China has devoted energy and resources to developing a modern air force. The effort has often been frustrated by the Chinese leadership's misunderstanding of the PLAAF's actual experience in the Korean War and in homeland air defense operations during the Cold War, and by ignorance of the role that air power can play. China's military literature describes the PLAAF role in these conflicts as an unbroken string of victories and heroism. Few have dared question the legitimacy of these records and stories, including a claim that the PLAAF is the only air force in the world to have ever defeated the USAF, which have become important components of the service tradition.[146] These accounts led not only to distortion, but also to a self-delusion that perpetuated the view of Chinese air power as a defense or deterrence force.

The Soviet influence on Chinese efforts to develop their air power was undeniable. The defensive nature of the people's war doctrine determined that China had no need to develop a strategic air power capability. The Cold War strategic environment, furthermore, limited the Chinese to thinking only of the defensive role of air power. Therefore, the PLAAF long perceived itself as a homeland defense force, whose task was to maintain a limited role to support the army and navy. In constructing its forces the PLAAF concentrated on the development of campaign and tactical capabilities, with an emphasis on size over quality. While constantly criticizing the Western concept of air dominance as a capitalist military theory that overemphasized the role of weaponry in war and warfare, the Chinese maintained a pseudoscientific attitude that characterized their leadership's sporadic instructions as profound military thought on air power, and thus used those instructions as guidance. It was against such a historical and philosophical background that for more than half a century the PLAAF built its own institutional culture in terms of concepts and theory about air power and its usage, organizational structures, and procedures.

The dominant role played by air and space power in military conflicts since the 1991 Gulf War, coupled with Beijing's concern about a possible conflict in the Taiwan Strait, has driven China to adopt a new strategy for transforming the PLAAF into a modern air force. The strategy includes developing advanced aircraft and integrating them with effective support systems, conducting offensive and defensive operations against ground- and sea-based targets, and relying heavily on information-intensive systems to employ air and space power effectively.

Although the speed of Chinese air and space modernization has caused concern in the West, progress is always likely to be constrained by the technological limitations of the Chinese defense industry and by the resources needed to support modernization. Perhaps even more essential is that this air and space transformation will continue to be tempered by inherent differences in the institutional cultures of the PLA ground forces and the PLAAF. While the PLA as a whole is transforming as it introduces new advanced weapons, the real struggle it faces is against traditional concepts, older ways of doing things, outdated organizational structures, and limited funding. It will take far longer to nourish an institutional culture that enables the PLAAF to embrace both offensive and defensive capability as an independent strategic force.

Nevertheless, it is undeniable that the PLA's warfighting potential has grown in tandem with China's economy. Assuming its economy continues along a steady trajectory, China will be able to commit further resources to the more challenging aspects of the three-step strategy, such as informatization. Should these goals be realized, the United States and other powers face a genuine challenge in terms of preparing themselves to encounter increasingly capable Chinese aerospace power in the coming decade.

PART III

The following chapters investigate a selection of representative air forces in three very different regions. The Asia Pacific region, Latin America, and Continental Europe all include countries whose air forces play a considerable role within the armed forces, and offer a variety of experiences and challenges.

THE ASIA PACIFIC REGION

In chapter 7, Dr. Alan Stephens examines air power through the prism of five contrasting models—Japan, Vietnam, Indonesia, Singapore, and Australia, identifying significant continuities and discontinuities. The author notes that air power in this region was built on precisely the same foundations as in the rest of the world—reflecting Hugh Trenchard's observation that a capable air force depends on excellent recruitment and training, high operational standards, a sound economy, and a strong indigenous industrial, research, and development base. But these various air forces were also shaped by the distinctive geographic, cultural, social, and political forces in their respective countries, which in combination tended to create distinctive outcomes. For example, whereas Europe is essentially continental, the Asia Pacific area is largely maritime, often with vast operating distances that demand different approaches to strategy, technology, and force structures. Furthermore, during the twentieth century, the Asia Pacific region was acutely affected by wars of national liberation and the aftereffects of colonialism, which, in some instances, stifled the growth of air power and, in other cases, encouraged innovation. The pervasive influence of local cultures and politics has added even more nuances to an already intriguing series of case studies.

Japan's defeat in the Second World War can be traced to the rise and fall of air power, with *culture* being immensely influential. Vietnam was able to thwart France and then the United States through highly *creative responses* to air power. In Indonesia *politics* was and is the driving factor in developing air power. For the city-state of Singapore, the air force sends a *message* to other countries in the region: although Singapore has no strategic depth it is fully capable of retaliation. In Australia, more than in most countries, air power is an integrated part of the overall *defense strategy*.

Stephens uses a matrix for the analysis, mapping the strategic construct of Shape-Deter-Respond against four basic capabilities: control of the air, strike, maneuver (airlift), and intelligence, surveillance, and reconnaissance (ISR). Only a few regional air forces have the capacity to shape events, and none can compete with the implicit influence of American air power or, increasingly, with that of China, India, or Russia. However, all Asia Pacific air forces have some potential to deter and to respond, with individual capacities ranging from minimal to considerable. The credibility of each country's posture is largely defined by its ability to control national airspace and to conduct strategic strike missions. The author concludes that those countries that promote quality over quantity, avoid needless complexity, and focus on their specific circumstances will be best placed to serve their national interests through air power.

LATIN AMERICA

Dr. James S. Corum sheds light on the use of air power in Latin America, ranging from conventional warfare to counterinsurgency and counterdrug operations. He notes that the most common use of air power has been in various forms of counterinsurgency operations, and that in several cases the Latin American countries provide a good model for how smaller countries can use air power effectively to deal with terrorists, drug lords, and guerrilla fighters.

The author begins with the Mexican Revolution (1910–32) and U.S. interventions in Latin America from 1914 to 1934. Next he examines early U.S. operations against revolutionaries in Haiti and the Dominican Republic, and against Sandinista forces in Nicaragua. Corum goes into some depth in analyzing the Chaco War, fought between Bolivia and Paraguay between 1932 and 1935, to demonstrate the importance of air superiority and early uses of air power in aggressive patrolling, reconnaissance, and bombing. These actions inflicted heavy casualties on men and horses, but equally crucial was the value of airplanes in transporting troops, supplying food and weapons, and serving as early versions of air ambulances with medical evacuation teams. The author then focuses on the Peru-Ecuador war of 1941, where overwhelming air power on the Peruvian side, which included a paratroop force, ensured a rather quick and decisive victory. He next examines how most Latin American countries (with the notable exception of Argentina) supported the United States in the Second World War.

During the Cold War era, Latin American air forces served as proxy forces in covert operations where the United States sought to combat leftist forces without becoming directly involved in actual operations, such as in Guatemala in 1954, at the Bay of Pigs in 1961, and during the Sandinista war from 1979 to 1990. Various national air forces participated in conventional wars on the continent, such as the so-called "Soccer War" between El Salvador and Honduras in 1969, the better known Falklands War between Argentina and the British in 1982, and the lesser known conflict between Peru and Ecuador in 1995. Toward the end of the chapter, Corum focuses on protracted wars, emphasizing conflicts in Guatemala and El Salvador, and the insurgency and war on drugs in Colombia. He notes that the air arm of the Colombian National Police, with several dozen helicopters, is one of the largest and most modern combat air forces in Latin America, and concludes that its use of air power in counterdrug and counterinsurgency operations has proven highly effective.

CONTINENTAL EUROPE

Global Air Power comes full circle by ending with a chapter on Europe, the first continent to develop air forces for combat. In this final chapter, Dr. Christian F. Anrig offers an assessment of key issues that have dominated the debate on air power doctrine, roles, and missions in Continental Europe since the fall of the Soviet Union and the Gulf War of 1991. Operation Desert Storm was a watershed event in its application of air power. Its use of precision-guided munitions, stealth technology, and standoff missiles set the agenda for the post–Cold War period that witnessed the accession of ever more members to both the North Atlantic Treaty Organization (NATO) and the European Union (EU).

Chapter 9 pays particular attention to France, Germany, and the Netherlands, highlighting the specific challenges they face and the strategic choices they made in order to build relevant air forces. The author also provides insight into the Swedish, Polish, and Czech air forces. As in the previous case studies, the fundamental issues involved in maintaining or developing a sustainable air force reveal a range of commonalities, despite each country's unique history and culture. Within the larger framework of defense policy and defense reviews, Anrig's discussion centers on each country's experiences during the wars in which it has been involved since 1990, paying special attention to the transatlantic partnership, the debate on NATO's power projection role, and the alternative and complementary defense structures of the EU, including the EU Battle Groups.

The main challenges, according to Anrig, are mandates for "out-of-area" operations, the degree to which Continental European air forces can develop common capabilities that are equally available to both NATO and the EU, and how Continental Europe should develop air capabilities for the full spectrum of operations. These range from airlift and ISR to the classical missions of gaining and maintaining air superiority, providing close air support, performing interdiction, and even carrying out strategic bombing, as demonstrated during Operation Allied Force in 1999. Smaller European countries certainly cannot afford to cover the full spectrum of aerospace capabilities, and even medium-sized states must struggle hard to sustain credible air forces. The situation has been exacerbated by a relative overemphasis on classical air power missions at the expense of the force-enabling areas, such as deployable air bases or air-to-air refueling. The result is a continuing imbalance between the shaft and the spear. The issue, then, is the extent to which role specialization can mitigate those shortfalls and how well countries can cooperate to complement each other. One thing is clear: only by making difficult choices about procurement, force structure, and doctrine can Continental European air forces contribute operationally, as well as politically, to supporting "their natural Anglo-Saxon counterparts."

7

THE ASIA PACIFIC REGION

Alan Stephens

The Asia Pacific region is so vast and diverse that it might seem incongruous to treat it as a single entity for the purpose of analyzing air power. It is that very diversity, however, that makes the region such an instructive case study. Since air power was first used systematically almost one hundred years ago, leading exponents such as France, Germany, the United Kingdom, the United States, Russia, and Israel have tended to build on similar foundations; namely, broadly shared national interests, sound education systems, a strong industrial base, economic independence, freedom from colonization and, with the exception of Russia, a lengthy exposure to democracy.[1] By contrast, few Asia Pacific countries can claim all of those characteristics. Japan, for example, has been industrialized since the late nineteenth century, and is exceptionally homogeneous ethnically and culturally, but it was a totalitarian state utterly resistant to foreign cultures until 1945, after which it was ruled by Allied occupation forces for seven years before transitioning to democracy. And well into the second half of the twentieth century, other Asia Pacific states had to fight wars of national liberation and struggle against the aftereffects of colonialism, as a consequence of which social cohesion, industrialization, and democracy were all comparatively slow to evolve. Nor can the immutable effects of geography be ignored: whereas Europe, the Middle East, and Central Asia are largely continental, the Asia Pacific region is largely oceanic. It would seem reasonable to expect that those distinctive characteristics would have encouraged distinctive approaches to the use of air power.

That has indeed been the case. For instance, third-world Vietnam defeated both the French and the Americans without any offensive air power whatsoever, but with an admirably inventive application of the defensive; Japan has been a major air power for ninety years and has conducted some of the most spectacular long-range strikes in history; first-world nations Australia and Singapore have constructed powerful air forces well suited to their particular circumstances as an island continent and a city-state respectively; Indonesia—the world's largest Muslim nation—has seen its air force treated with indifference, even overt hostility, by its politically vigorous army; developing countries like Malaysia and Thailand continue to lay the foundations needed for modern air power; and fascist Burma and desperately poor Laos, preoccupied as they are with internal security, have tended to marginalize air capabilities as they have placed priority on their armies.

This chapter examines air power in the Asia Pacific region through the prism of five contrasting models: Japan, Vietnam, Indonesia, Singapore, and Australia, with the objective of identifying significant continuities and discontinuities. Special attention is paid to the context set by politics, culture, and geography; to competency in the traditional roles of control of the air, strike, maneuver, and information/surveillance/reconnaissance (ISR); and to professional mastery, including the ability to sustain air power.[2]

APPLYING AIR POWER

The aphorism that time spent on reconnaissance is never wasted applies equally to debate on defense and strategic studies, with due alteration for detail. In this case, time spent on establishing context is never wasted. Before the air power capabilities of a region can be sensibly discussed, it is necessary to understand both the broader forces at play, and what air power can and cannot do.

Like any form of military power, air power can only be applied in accordance with the prevailing political ethos. For example, toward the end of the Korean War in early 1953, United Nations (UN) air forces mounted an intensive bombing campaign against the dams holding the irrigation water for North Korea's rice crops. In a peasant-based economy, rice was *the* essential item for daily life, and it seemed probable that the destruction of the irrigation system would eventually bring the North Korean regime to its knees. However, international condemnation forced the UN Command to terminate the campaign. Regardless of the state of the war, the prevailing political ethos would not countenance an attempt to starve North Korea into submission. Military actions cannot be separated from politics, a truism that has been apparent more recently in the Middle East, Iraq, and Afghanistan.

Turning to the question of what air power might or might not deliver, it is helpful to begin by situating a particular nation's capabilities within the classic strategic construct of shape-deter-respond.[3] Under that construct, the order of priorities is: first, try to influence the environment in which we function—political, diplomatic, economic, social, cultural, military—toward our interests; second, if shaping is not entirely successful, seek to deter behavior that might be inimical to those interests; and third, if deterrence fails, respond as necessary anywhere along the spectrum of action, from soft sanctions at one extreme to war at the other.

The obvious model for shaping geostrategic affairs through air power is the United States, whose combination of air- and space-based information systems and weapons confers a singular ability to project air power globally at short notice. The U.S. force structure—people, aircraft, weapons, information sensors, bases, logistics, command and control networks, communications, training and education, research and development, and so on—represents a capability that can shape decision making and actions through its mere existence. American air power implicitly imposes itself on every other nation, including those of the Asia Pacific.

Shaping can also be achieved explicitly, through the application of military force. Operation Southern Watch in Iraq illustrates the point. A remarkable decade-long campaign initiated by the United States, Russia, France, and the United Kingdom (UK), Southern Watch began in August 1992 following Saddam Hussein's refusal to comply with a UN resolution, and it continued until the second invasion of Iraq by an American-led coalition in March 2003. The operation involved the successful enforcement of a no-fly zone for all Iraqi aircraft in the area south of latitude 33° north, a vast space that was dominated by combat air patrols flying primarily from Turkey and off aircraft carriers in the Persian Gulf and the Red Sea.[4] For more than ten years Southern Watch shaped the behavior of Saddam's government by controlling the movement of strategic forces and inhibiting the development of weapons of mass destruction. The shaping effect of the air blockade could have been extended if desired; for instance, by halting the movement of ships and large road vehicles.

Shaping is a complex business that generally requires resources and skills beyond the reach of most nations. Nevertheless, it is not necessary to be a superpower to exert a shaping influence. The Republic of Singapore Air Force (RSAF), for instance, has been developed over the past thirty years into a form that sends a clear message of capability and intent to any potential aggressors within Southeast Asia. Air power is only one means by which Singapore has sought to shape its geostrategic environment, but its message has been notably unambiguous.

The next stage of the strategic continuum, deterrence, while also complex, is often a more realistic ambition. A deterrent effect might be created offensively, by an intimidating attacking capability; or it might be created defensively, by protective capabilities that would make the perceived cost of an assault irrational. The RSAF again illustrates the point. At the offensive end of the deterrence spectrum the RSAF has an impressive strike force, complemented by the necessary personnel, information systems, command and control networks, and air-to-air refueling resources; at the defensive end, its excellent integrated air defense system should give any would-be attacker pause for thought.

The essence of deterrence is, as Sun Tzu observed some 2,500 years ago, to win without fighting. The Royal Australian Air Force (RAAF) illustrated this aphorism perfectly during the early phases of the UN-sanctioned operation to liberate Timor-Leste in 1999–2000, when there were concerns that extremist elements of the Indonesian Army might escalate their opposition to dangerous levels. The deployment of RAAF F-111 bombers to a base in northern Australia within range of Timor and key Indonesian targets sent a message of intent that reportedly was understood in Jakarta and that, together with the complementary deployment of U.S. warships, made it easier for wise heads to prevail.

Like everything else in life, deterrence is relative. Massive force might not be required: depending on circumstances, a section of riflemen or a few strike aircraft can constitute a credible deterrent capability.[5] It is clear from the evolving structures of a number of Asia Pacific air forces that generating a deterrent effect is one of their key objectives, a topic that is discussed in more detail below.

The ability to "respond" is at the bottom end of the strategic continuum and is the default posture of most of the world's air forces. A "response" might be made by conducting any one of a variety of missions, including airlifting troops or police to a crisis area, enforcing an air defense zone, striking at enemy targets, and so on. Even the most modest air power can respond in some fashion.

Within the construct of shape-deter-respond, air power historically has delivered four basic capabilities: control of the air; strike; ISR; and maneuver (airlift). Those capabilities contain numerous subsets such as defensive and offensive counter-air operations, maritime strike, troop lift, close air attack, maritime patrol, medevac, air-to-air refueling, resupply, interdiction, search and rescue, and airdrop. The basic roles have, however, remained constant, and constitute a valid template for assessing what a particular air force can and cannot do.

Air power capabilities can only reach their full potential if they rest on a strong foundation. Ninety years ago the head of the world's first independent air force, the (British) Royal Air Force's Sir Hugh Trenchard, defined that foundation as follows: a central flying school to set and maintain standards; research and development units for the technological edge; a cadet college to educate the future leaders; a staff college to give those leaders the finishing touches; and an apprentice scheme to train the mechanics. Trenchard also assumed the existence of an indigenous industrial base to build the aircraft and weapons. Ever since then most first-rate air forces have reflected Trenchard's model.

It is implicit in that model that only wealthy developed countries will be able to raise and sustain high-quality air forces, and generally that has been the case. Since air power was first used systematically during World War I, only a handful of countries have maintained strong, balanced air forces. But change is constant. When Trenchard formulated his blueprint there was no routine international travel, no instant media, no cellular telephones, no World Wide Web, and no ready access to a vast range of information. In short, there was no global village.[6] And it was precisely the revolutionary social and technological changes epitomized in the concept of the global village that in 2001 made possible the most devastating strategic air strike conducted since the atomic bombs were dropped on Japan more than half-a-century previously, namely, al Qaeda's attacks on New York City and the Pentagon.

Al Qaeda had none of Trenchard's building blocks. Yet for three hours it was able to assemble an extraordinarily effective air force (albeit one with a limited lifespan). Its pilots gained control of the air by deception; and they successfully attacked two of their three targets, in the process throwing the world's most powerful nation into chaos. By any measure it was an astonishing operation. The point here is that while the traditional model for developing air power remains valid, increasingly there will be opportunities for innovative thinkers to exploit emerging technologies and imaginative strategies.

Because air power is subjected to more uninformed commentary than any other form of combat power, a note on analytical method is necessary. It is far easier to look up the latest edition of *The Military Balance* and count aircraft numbers than it is to make an informed assessment of the quality of essential enabling activities. Confusing arithmetic with analysis has always been a worrisome habit, but at a time when capabilities are defined more than ever by information dominance, skilful planning, expert command and control, and the ability to sustain a rapid tempo of operations, all of which depend on high-quality people, it borders on the culpable. Anyone can buy a squadron of fifth-generation strike/fighters—the point is to be able to do something meaningful with them. Competence in enabling skills is central to the assessments drawn in this chapter.

Air power's innate qualities of reach, speed, and perspective give it an unrivaled ability to project influence globally. Nations that possess advanced air power can and will exert their authority in any region, regardless of whether they have a geographic presence. Thus, before turning to the five Asia Pacific case studies, the specter of the United States, Russia, India, and China must be addressed.

As the world's only superpower, if perhaps a fading one, the United States has had a long-standing military presence in the Asia Pacific region that it can be expected to maintain. Currently the United States enjoys a substantial advantage in aerospace power through its unrivaled combination of high-quality personnel, advanced technology, excellent command and control networks, and national industrial base. That advantage should persist for at least the next ten years. However, as the Indian and Chinese economies continue to expand, and as Russia continues to reinvent itself following the collapse of the Soviet Union, these three emerging superpowers will provide alternative sources of regional military authority and mentorship. They will affect every member of the Asia Pacific community of states to a greater or lesser extent.

The Asia Pacific's member states have vastly differing air power capabilities, but all can be examined within the construct of shape-deter-respond, and against the four basic capabilities of control of the air, strike, maneuver, and ISR. Judgments have been weighted in favor of the ability to deliver precision firepower, offensively and defensively, day and night, in all weather, a methodology that acknowledges the reality that air forces ultimately are in the business of coercion. No country has experienced that air power truism more starkly than the first case study, Japan.

AIR POWER AS CULTURE—JAPAN

The second half of the nineteenth century was a period of severe humiliation for Japan, as the chauvinistic nation was compelled by Western military power to accept a series of exploitative commercial arrangements. Realizing that radical change was needed if Japan were not to become a permanent vassal state, reformists swallowed their pride and embarked upon a program of rapid industrial modernization, known as the Meiji Restoration (after the era of the Meiji emperor, 1867–1912), which drew on the expertise of advanced economies such as Britain, France, the United States, Germany, and the Netherlands. The military-industrial results of that modernization were revealed in May 1905, when the Imperial Japanese Navy (IJN) crushed the Russian Baltic Fleet in the Tsushima Strait between Korea and Japan, a stunning triumph of the Asiatic over the European that seemed to threaten the apparent natural order.

Japan's ruling oligarchic and military elite held vaulting geostrategic ambitions. They believed (not unreasonably) that Japan was entitled to become a great power, which among other things meant emulating Western states by seizing colonies. That belief was intensified by Japan's demand for external sources of commodities, including oil, coal, iron ore, bauxite, tin, and rubber. The intellectual justification for this worldview emerged during the 1930s in the idiosyncratic notion of a "Greater East Asia Co-Prosperity Sphere," which postulated an Asia free from Western domination and guided by a benevolent Japan. Japanese strategists envisaged a sphere of influence extending from the west coast of India, down through Southeast Asia, across into the Pacific, and up to North Asia, which would provide the materials and labor a great power needed. The concept was of course a euphemism for self-serving militarism and conquest; nevertheless, it was used to justify Japan's invasions of Manchuria in 1931 and China in 1937, and the attacks of December 7 and 8, 1941, that extended World War II into the Pacific.

Sea and land power were to underwrite this expansionist policy. The navy was, in effect, the strategic service: at some stage, it would have to defeat the region's major shaping force, the U.S. Navy, in a so-called decisive battle. Control of the sea in turn would allow the army the freedom of maneuver it needed to invade and occupy new colonies. Air power, while valued, was regarded as a supporting force. In the event, however, it was to be the comprehensive failure of air power that would be the primary military cause of Japan's defeat in World War II.

Japan's foray into air power began after the battle of Tsushima, with the Japanese Naval Air Force (JNAF) being formed in 1909 and the Japanese Army Air Force (JAAF) in 1910. The roles allocated to each air arm reflected its origins. The JNAF was tasked with surface fleet protection, coastal defense, sea patrols, and anti-submarine warfare; and the JAAF with army support, especially reconnaissance and close attack.[7] Securing control of the air was a common responsibility.

Both services benefited in their formative years from the assistance of countries that would become their enemies in World War II. A French aviation mission assisted with the development of the JAAF in 1919; similarly, Japanese naval aviation owed a good deal to training provided by France, the United States, and Britain. Herbert Smith, chief designer for the Sopwith Company, visited Japan in 1923 and passed on his expertise; while in 1930 an RAF mission provided instruction in air-fighting tactics and gunnery, advice that was to prove valuable against Commonwealth air forces over Malaya and Singapore eleven years later.

Local companies such as Mitsubishi, Kawasaki, Nakajima, and Aichi initially copied their European mentors, and then started designing their own aircraft. An emphasis on warplanes suitable for control of the air and long-range strike and reconnaissance established a harmony between Japan's geostrategic ambitions and indigenous production. By the mid-1930s, the JNAF's Mitsubishi G3M3 bomber, powered by two Mitsubishi 1300 horsepower radial engines, could carry 1,800 pounds of bombs 3,800 miles. Formations of G3s made world headlines in 1937 when they conducted transoceanic bombing raids, sometimes in extremely poor weather, from bases in Japan and Formosa against Chinese targets in and around Shanghai, Nanking, and Hangchow. At the time, these were by far the longest raids flown by any bombers from any country. Also designed and built by Mitsubishi for the navy was the wonderfully maneuverable A6M Zero-sen single-engine fighter, which when it appeared in 1940 was the equal of, if not better than, any of its Western counterparts.

Rapid technological progress was accompanied by intense doctrinal debate, with early air power proponents vigorously promoting the new weapon's strategic potential.[8] Two revealing observations can be made regarding this debate.

First, prior to the invasions of Manchuria and China, opinion had been divided over the importance of control of the air. Unacceptable loss rates of bombers without fighter escorts quickly resolved the matter. Establishing control of the air became an essential precursor to all other air, land, and sea actions.

Second, as noted above, Japanese planners believed that strategic domination ultimately would be achieved through a decisive naval battle fought between "all big-gun Capital Ships."[9] At the same time, however, the potential of air power was appreciated, even if it was regarded as a supporting force. Because of the need for fleet protection, and the perceived value of harassment raids, Japan became a world leader in naval aviation. The IJN's *Hōshō* was the world's first purpose-designed aircraft carrier when it was commissioned in 1922; and by the time of the attack against Pearl Harbor on December 7, 1941, the Japanese Navy was operating a potent fleet of six large and several small carriers.

Notwithstanding scores of land battles that often were fought to the death and an exhausting naval campaign, if there were such a thing as a "decisive" battle in the Pacific, then it was fought largely by air forces. Japan's wartime rise and fall can be traced through four crucial air operations: Pearl Harbor; the sinking of HMS *Prince of Wales* and *Repulse*; the battle of Midway; and the atomic attacks on Hiroshima and Nagasaki. The rise was defined by air forces of exceptional quality, the fall by an unforgivable failure of military culture.

The JNAF's preemptive strike against the U.S. Navy (USN) at Pearl Harbor was a brilliantly conceived operation, since rivaled only by Israel's air campaign on the first day of the Six Day War in June 1967, and al Qaeda's stunning attack against the United States in September 2001. Features of Pearl Harbor that warrant attention are the attainment of air supremacy by surprise, the vast combat range the Japanese attack force achieved through the use of aircraft carriers, and the skill of the JNAF's pilots. Most of those qualities were again evident only three days later, on December 10, 1941, when formations of land-based JNAF bombers operating from Indochina shocked the world by sinking the Royal Navy's (RN's) allegedly impregnable warships, *Prince of Wales* and *Repulse*, off the east coast of Malaya. Again, the JNAF secured control of the air partly through surprise and partly through their opponent's incompetence; again, the skill of the Japanese aircrew stunned their enemies.

Yet only six months later, Japanese naval air power was crushed by a USN task force at the Battle of Midway. Fought June 4–7, 1942, Midway was remarkable for the fact that the fleets neither sighted each other nor exchanged a single shot, all of the offensive action being conducted by carrier-borne aircraft. Superior intelligence and better on-deck refueling and rearming procedures (which facilitated a faster turnaround of aircraft) eventually tipped the balance in favor of the Americans, who sank four Japanese carriers for the loss of only one of their own. As historians later recognized, the defeat at Midway destroyed Japan's ability to project power and to protect its increasingly extended lines of communication; in turn, those strategic vulnerabilities made it unlikely that Japan could win a prolonged war.

Three years later, the unopposed atomic attacks on Hiroshima and Nagasaki by U.S. Army Air Forces (USAAF) B-29 bombers exposed the utter disintegration of Japanese air power. The ferocity with which Japanese soldiers had fought as they gradually retreated toward their home islands had led U.S. intelligence to conclude that as many as half a million Allied troops and millions of Japanese soldiers and civilians would be killed if a ground invasion were mounted; consequently, in an attempt to end the war rapidly and conclusively, U.S. president Harry Truman authorized the use of atomic bombs.

By this stage, the Allies' operational and technical expertise had reached its peak, whereas Japanese air power had all but ceased to exist, other than in the self-destructive form of kamikaze suicide bombers. Numerous factors contributed to Japan's dramatic reversal of fortune, but none was more significant than culture.

Once the United States had recovered from the shock of Pearl Harbor and had placed its unrivaled industrial and training potential onto a wartime footing, Allied victory in the Pacific theater was more likely than not. Nevertheless, Japanese strategists, aircraft designers, industrialists, and airmen still won some great victories, and might well have made things far more difficult for the Allies. Their great problem was that everything they did—leadership, planning, organizational arrangements, aircraft production, training, even combat operations—was diminished by suffocating cultural prejudices. Interservice rivalry and the samurai warrior ethos were at the heart of this self-inflicted wound.

Japan had entered World War II with two separate air forces that had been created to meet the separate ambitions of the navy and the army. At no stage did the JNAF and JAAF cooperate; on the contrary, commanders refused to share ideas, plans, aircraft, facilities, maintenance, training, and intelligence. Indeed, according to one veteran, the two services "hated each other."[10] Among other things, this intransigence inhibited the technological progress that is an essential characteristic of a first-rate air power. Thus, for example, by the time the air war had reached the Japanese home islands in 1944, the JNAF was still using the Zero on the front-line, whereas the fleets of the USAAF and the USN had evolved through many iterations, culminating in two of the war's outstanding fighters, the P-51 Mustang and the F6F Hellcat.

Turning to the legacy of the samurai ethos, Japan's ruling elite may have accepted the need for technological modernization, but they were largely incapable of understanding the need for social and intellectual modernization.

Thus, in general, they refused to challenge authority and traditional practices, instead celebrating such alleged virtues as unquestioning obedience to orders, an unyielding belief in narrow elitism as opposed to broad professionalism ("the invincibility of a refined technique"), and the certainty that a spiritual conviction in victory would overcome any technological inferiority. Already superseded by the rapid advancement of mechanized warfare, the ethos was particularly ill suited to the war in the air.[11] Pilot training perfectly illustrates the problem.

At the start of the war, Japanese pilots were arriving at operational units with about eight hundred hours of flight experience, compared to two hundred in Western air forces.[12] Early combat seemed to validate this elitist approach, as highly skilled Japanese pilots swept their opponents aside. However, simple arithmetic made it obvious that, for a given amount of resources, the JNAF and JAAF would be able to train far fewer pilots than their enemies (albeit to a higher standard). And once the number of those elite pilots began to decrease, as it inexorably did once the war settled into a grim battle of attrition, the system could not produce replacements fast enough. By contrast, the Allies' philosophy of training very large numbers of very good (as opposed to elite) pilots was a crucial factor in their ability to maintain overall quality and sustain pressure.

No greater indictment of the Japanese ethos can be found than the notion of kamikaze warfare. Itself an aberration of culture, kamikaze followed a self-destructive philosophy that removed any possibility that the system might regenerate itself. When the USAAF B-29 the *Enola Gay* dropped its atomic bomb over Hiroshima on August 6, 1945, not a single Japanese fighter rose in defense of the homeland.

Shaped by the disquieting implications of World War II, Japan's contemporary political and social cultures make the nation's approach to air power in the twenty-first century one of the world's more intriguing case studies.

Any discussion of Japan's current defense culture must start with the country's constitution. Drafted after the war under the guidance of occupation forces commanded by Gen. Douglas MacArthur, the Japanese constitution is deliberately couched in pacifist terms. War is renounced, as are the acquisition of war potential and the right to behave belligerently. Article 9 specifically precludes the possession of offensive forces, a clause with clear implications for strategic air power. The right of sovereign states to self-defense is, however, recognized.

Because of that right, the constitution acknowledges the need to maintain a minimum level of armed strength. Precisely what constitutes a "minimum" is,

of course, debatable, noting that for many years Japan has probably spent more
on defense than each of China, Russia, and North Korea.[13] Similarly, although
Japan's military services have been officially titled "Self-Defense Forces" since
1954, just when self-defense becomes belligerence is also a matter of interpretation. The Israeli Air Force, for example, insists that the preemptive air strikes
that have been a feature of its operations for more than forty years are conducted
for defensive purposes.

Japan's long-term military ambitions beyond the right to resist armed aggression are difficult to interpret. It is nevertheless clear that the Japanese have
been edging toward a wider role in the Asia Pacific region. In 2005 the lower
house of parliament, the Diet, approved three controversial bills that allowed the
Japan Self-Defense Force (JSDF) to expand its outlook from an exclusive focus
on homeland defense toward an active engagement in regional security. Participation in peace operations, a greater involvement in bilateral and multilateral
exercises, and dialogues and defense exchanges soon followed. Naval refueling
operations in the Indian Ocean in support of international forces fighting in Iraq
and Afghanistan took the process a step further, to the extent that some Japanese
politicians accused their government of "remilitarization."

Those modest initiatives notwithstanding, the JSDF's main role is to deter
invasion. Should deterrence fail, the task becomes one of repelling enemy forces,
acting in concert with the United States under bilateral security arrangements.

Japan's defensive posture makes sense for more reasons than the dictates of
the constitution. Northeast Asia is an area of strategic uncertainty, with most
concern centered on North Korea and China. Japan has been worried by North
Korea's ballistic missile program ever since that erratic nation test-fired a Tae-
podong-1 missile over Japanese airspace into the Pacific Ocean in August 1998.
In mid-2006 trials of the long-range (six thousand kilometers) Taepodong-2
were followed by North Korea's first underground nuclear test. A provocative
program of weapons testing has continued, culminating in a total of eleven missile launches into the Sea of Japan in July 2009.[14]

China may not constitute an immediate threat to Japan, but its emergence
as a military superpower cannot be ignored. China's official defense policy, *National Defense in 2008,* explicitly promotes strategic capabilities such as long-
range missiles, space-based systems, strike aircraft, and precision weapons.[15] The
People's Liberation Army Air Force (PLAAF) already operates some two thousand strike/fighters (including fleets of Su-27 Flanker derivatives and locally

designed jets), long-range precision weapons, and airborne early warning and control (AEW&C) platforms (incorporating Israeli electronics).[16] A vigorous local aerospace industry builds, assembles, and modifies all of those systems and more. Those kinds of capabilities send a clear signal that China is intent on modernizing and extending the reach of its strike power. Another signal, and one that raises more concern in Japan, comes from the successful development of the Dongfeng-31 and -41 intercontinental ballistic missiles, noting that China has had nuclear warheads since the mid-1960s.

Given that context, the priorities of the Japan Air Self-Defense Force (JASDF) should be control of the air, ISR, and a (defensive) preemptive strike capability. Professional, well equipped, and supported by an excellent indigenous industrial base, the JASDF is highly capable in the first two of those roles, less so in the third.

Air defense is organized around four regional commands, with overall control exercised by a central air defense command. Almost all of the country's airspace is protected by an around-the-clock warning and surveillance system based on twenty-eight ground radar sites, thirteen Northrop-Grumman E-2C Hawkeyes, and four Boeing E-767 airborne warning and control system (AWACS). Firepower is provided by a combination of air superiority fighters and surface-to-air missiles (SAMs).

A dozen or so squadrons of fighter/interceptors are equipped with about 350 aircraft, the most advanced of which are two hundred Mitsubishi F-15s and seventy Mitsubishi F-2s (with more on order). The F-15s have been upgraded with new computers, electronic countermeasures, forward-looking infrared systems, and radar—although the latter is not an Active Electronically Scanned Array (AESA). The locally designed and built F-2s are, however, fitted with AESA radars, making them more capable against cruise-missile threats. Additional ISR is provided by about twenty RF-4Js, which are being replaced by F-15s fitted with synthetic aperture radar.

All fighter aircraft are armed with the AIM-7F/M air-to-air missile for beyond-visual-range combat, and the AIM-9L, AIM-9M, and AAM-3 for within-visual-range. Strike weapons include the widely used Pavetack and Paveway series of guided gravity bombs, and the Maverick, ASM-1, and ASM-2 short- and medium-range air-to-surface missiles. The glaring omission here is a long-range, precision, land-attack missile.

Because of concerns over regional offensive missile programs, Japan's air defense has been bolstered by land-based Patriot PAC-3 antimissile batteries, and four Aegis-class air warfare destroyers armed with SM-3 missiles. Japan may also acquire a terminal high altitude area defense system from the United States to add an extra layer to its control of the air shield.

Successive governments have fostered Japan's aerospace industry at vast expense. Proven aircraft have been built under license, and new designs have been brought into service from the drawing board. Platforms relevant to most air power roles have been manufactured or assembled, including P-3C and NP-3C maritime-patrol and reconnaissance Orions built under license by Kawasaki Heavy Industries, the indigenous ShinMaywa flying boat, and Sikorsky Blackhawk and Seahawk helicopters built by Mitsubishi. Currently the XC-2 twin-engine airlifter and the four-engine XP-1 maritime-patrol aircraft are under development by Kawasaki. Turning to higher performance jets, the subsonic T-4 trainer was designed and built by Kawasaki, and the supersonic T-2/F-1 light fighter/trainer by Mitsubishi. Mitsubishi followed up the T-2/F-1 with the technologically challenging and enormously expensive F-2 fighter. Based partly on the F-16 and incorporating about 40 percent U.S. content, the F-2's acceptance into operational service in 2000 was a major achievement. It is also noteworthy that Japan is the only country other than the United States that builds parts for the Patriot missile.

Japan's indigenous defense satellite program is equally significant. Responding to the regional ballistic missile threat, Japanese research agencies have for some years been quietly developing a national early warning and reconnaissance satellite network, described as a multipurpose information gathering satellite system. The Mitsubishi Electric Company is the prime contractor, and the National Space Development Agency is also involved. While the network will rely to some extent on American technology, it will be Japanese built, owned, and operated, and will provide an independent strategic reconnaissance capability.

These kinds of aerospace engineering achievements have major implications. Only a handful of countries are capable of designing, building, and modifying high-performance aircraft and military satellites. Should Japan ever decide to assume a more proactive military posture, its technological infrastructure would be a great asset. The ability to build and modify platforms, weapons, and sensors in-house would enable the JASDF rapidly to increase its strategic power.

A glimpse into Japan's military future may be provided when the long-running competition to acquire a new fighter, known as the F-X program, is resolved. Consistent with the national defense policy, the JASDF has been trying to acquire the world's preeminent (defensive) control of the air platform, the

F-22 fifth-generation fighter. However, it seems unlikely that the United States will release the F-22 for foreign sales. If that is the case, the logical alternative is the F-35 Joint Strike Fighter.[17] The point here is that in addition to being a leading-edge air superiority platform (albeit inferior to the F-22), the F-35 will be an exceptional strike and ISR system. Thus, should Japan choose the F-35 for air defense, it will simultaneously acquire, de facto, the fundamentals of a potent strike force. That potency would be enhanced by the JASDF'S KC-767J tankers and E-767 AWACS, which would facilitate the information superiority, networking, and extended range needed to conduct strategic strike missions. If standoff precision weapons were added to the mix, the message would be clear. An equally clear message would be transmitted should speculation that the Japanese defense ministry might sponsor a "sixth-generation" manned fighter, armed with a directed energy weapon, ever become fact.

But perhaps the challenge for the Japanese military will once again be more one of culture than of technology. Politics and the management of the economy can be used as a barometer here.

Japan's economy boomed during a sustained post–World War II "economic miracle," but has been sluggish for a decade now. Many analysts believe that the conservative Liberal Democratic Party (LDP) that dominated national politics from 1955 to 2009 had neither the temperament nor the resolve to turn things around by embracing a freer political system, more open trade policies, and a transparent banking system. Japan would not be able to shake off its cultural inhibitions, so this line of argument continued, until the industrial-age generation which rebuilt the country after World War II was replaced by information-age thinkers.[18]

The LDP's defeat in August 2009 by the socially progressive, center-left Democratic Party of Japan has been described by some commentators as "the end of the post-war political system in Japan" and the beginning of a new era.[19] The DJP has already announced that it wants a "less subservient" partnership with the United States. Refueling missions in the Indian Ocean by the JSDF in support of U.S.-led coalition forces in Afghanistan have been discontinued, and U.S. Air Force (USAF) access to Japanese bases is under review.[20] These are interesting gestures, but not much more than that. At this stage it remains unclear whether Japan's new government will challenge the postwar cultural mind-set and assume an independent, fully engaged posture in the Asia Pacific region. Until or unless it does, Japanese air power will remain circumscribed.

AIR POWER AS CREATIVE RESPONSE—VIETNAM

Few analysts would regard Vietnam as a notable air power. Yet twice within the space of twenty years Vietnam used air expertly, if unconventionally, to help defeat Western powers, first against the French at Dien Bien Phu in 1954, and then against the United States and its allies during the American war from 1962 to 1975. Because of inferior technology, Vietnamese commanders could only respond to their opponents' ostensible air power superiority. That apparent disadvantage did not, however, prevent them from demonstrating admirable initiative.

France had invaded and annexed Vietnam in 1861, following centuries of incursions by its traders, profiteers, and Christian missionaries. French rule was interrupted by World War II, when Japanese forces occupied Indochina. Unlike the British, who had sufficient good sense to realize that the end of the war also signaled the end of colonialism, France attempted to reestablish its authority following Japan's surrender on September 2, 1945. On that same day, however, the Vietnamese nationalist leader Ho Chi Minh declared the independence of the Democratic Republic of Vietnam. For the next nine years Ho's forces, known as the Viet Minh, fought to expel the French invaders. Vietnam's war of national liberation was contested mainly by ground forces, with the French constantly trying to draw their opponents into large-scale "conventional" battles, while the Viet Minh favored hit-and-run tactics. Air power played a secondary but crucial role, especially at Dien Bien Phu, where the Viet Minh's ability to win control of the air through innovative tactics and without any aircraft—indeed, without an air force—was decisive.

A succession of French commanders had struggled to come to terms with conditions in Vietnam. Constantly thwarted by their enemy's elusiveness, they vacillated tactically, uncertain whether to emphasize maneuver or position. Regardless of tactics, they relied on air to resupply deployed troops and garrisons, and to provide close attack for forces in contact. Occasionally air was vital. In the Red River Delta in late 1950 and early 1951, for example, napalm strikes by air force fighters saved the French Army from being overrun; at Nghia Lo in 1951 and Na San in 1952, the interdiction of enemy supply lines smashed dangerous Viet Minh offensives.[21]

By mid-1953 France's theater commander, Gen. Henri Navarre, had become thoroughly frustrated by his inability to contain the Viet Minh. Navarre's expeditionary forces might have been occupying the major population centers, but the Viet Minh held sway over much of the countryside. Confounded by his enemy's strategy, Navarre became increasingly eager to draw the Viet Minh into a set-piece battle that he believed his forces could dominate with firepower and aerial resupply.

Simultaneously, Navarre was concerned that the Viet Minh might make a move against the Laotian ancient capital of Luang Prabang, which was regarded as a center of gravity for Indochina. Overland access to Luang Prabang from Vietnam was dominated by the Muong Thanh Valley, so Navarre established a powerful garrison at the village of Dien Bien Phu in the center of the valley, with the dual objectives of blocking access to Luang Prabang and drawing the Viet Minh into the open. An air force of ten thousand men and some four hundred light bombers, fighters, and airlifters stood ready to support the garrison. Ho Chi Minh and his military commander, former history professor Gen. Vo Nguyen Giap, decided that the time and place were right to fight a decisive battle, and accepted the challenge.

All-out fighting in the set-piece encounter the French had been seeking began on March 13, 1954, and ended in humiliation two months later when their battlefield commander, Col. Christian de Castries, surrendered unconditionally. Conducted more in the nature of a siege than anything else, the battle nevertheless included many of the features of modern warfare: massed assaults and counterattacks, heavy firepower, desperate infantry clashes, complex trench systems, airborne offensives, aerial resupply, close air attack, bombing strikes against lines of communication, paratroop drops, armed reconnaissance patrols, attempted breakouts and reinforcements, and, on the final day, the detonation by the Viet Minh of a massive mine at the very edge of de Castries's command post. However, if any one action were to be regarded as decisive, it was the Viet Minh's unorthodox victory in the fight to control the air.

Three of de Castries's planning assumptions for victory at Dien Bien Phu were of the utmost importance. These were: that his air strike and artillery forces would give him firepower dominance; that his army would be able to hold the airstrip inside the garrison; and that, through airlift and airdrop, he would enjoy an abundance of resupply the Viet Minh could not match. In the event, all three assumptions fell victim to the French expeditionary force's inability to control the air.

The Muong Thanh Valley is surrounded by steep, heavily wooded hills. Paths are few and narrow, ravines and rivers many and awkward. All approaches are difficult for men on foot, let alone for the movement of heavy machinery and supplies. The valley's commanding heights are some ten kilometers (about six miles) from Dien Bien Phu, a distance the French believed was "superior to the useful range of any enemy artillery."[22] De Castries concluded that the Viet Minh were unlikely to have much artillery, and that even if they did manage to manhandle some batteries onto the hilltops, the guns would be small and lack range. Consequently, he expected to derive a battle-winning edge from his own artillery. Equally important to de Castries was the airstrip, which would be used for resupply and for launching close air attack sorties.

As a means of both strengthening this perceived advantage and offsetting his army's static posture, de Castries established two artillery support bases on hill outcrops known as Doc Lap and Him Lam, respectively three and two kilometers north and northeast of the airstrip; and a third, more powerful base at Hong Cum, seven kilometers south. These fire-support bases were fundamental to de Castries's plan, because they seemed to confer a battle-winning edge in what was expected to be the decisive application of firepower. As events transpired, de Castries's plan had failed on one count even before any heavy fighting began, and it failed on a second as soon as it did.

In the weeks leading up to the battle, General Giap had revealed an astute understanding of the nature of air power by using ground forces to mount what was, in effect, a preemptive control of the air campaign.[23] Sabotage and artillery attacks were conducted against French air bases throughout Indochina with the objective of weakening de Castries's air forces. Simultaneously, the Viet Minh manhandled two regiments of medium-range artillery pieces together with heavy mortars onto the hilltops surrounding Dien Bien Phu.[24] Giap was contesting control of the air through unconventional means. The instant Viet Minh guns started bombarding Dien Bien Phu's airstrip on March 13, de Castries knew his situation was dire. It quickly became worse.

Geographically separated from the main French force, difficult to support, and even more difficult to reinforce, the forward artillery bases at Doc Lap and Him Lam immediately came under intense fire, and then assault. De Castries's defensive problems were compounded by his inability to use the guns at the third forward base, Hong Cum, which were out of range. When both of the northern forward artillery bases fell inside the first forty-eight hours, the strategic calculus had shifted dramatically in the Viet Minh's favor. Giap was able to move some of his own artillery onto Doc Lap and Him Lam and bombard the main French garrison from close range, which made operations from the airstrip almost untenable. The Viet Minh had gained control of the air via preemption, sabotage, ground maneuver, and artillery fire.

Navarre attempted to regain control of the air for de Castries by ordering strikes against the Viet Minh's lines of communications into Dien Bien Phu, with the intention of stopping the flow of artillery and mortar rounds. But foreshadowing the Americans' inability twelve years later to strangle resupply along the Ho Chi Minh Trail, Viet Minh porters simply shifted their tracks a few hundred meters or approached along other lines.[25]

Weight of fire now favored the locals. French attempts to break out from Dien Bien Phu failed, as did all efforts to reinforce the besieged garrison. At one stage consideration was given in Washington to mounting a massive air strike against Viet Minh positions, even perhaps with nuclear weapons, but President Dwight Eisenhower decided that the potential costs were likely to exceed any strategic gains. Dien Bien Phu fell on May 8, 1954, following a recklessly brave frontal assault on de Castries's command post by the Viet Minh.

Historians today properly remember Dien Bien Phu as one of the twentieth century's turning points, with General Giap's victory signaling the end of French colonialism in Indochina. Few, however, appreciate that the battle was largely decided by an extraordinary campaign to control the air.

Only a decade later, Vietnam was invaded again. Following France's withdrawal, the country was divided into two "regrouping zones" on either side of the seventeenth parallel, with the North primarily Communist and the South primarily anti-Communist. Intended as a temporary line of demarcation until elections could be held, the division hardened into a territorial boundary when the United States, believing that the South would lose a fair election, obstructed the process. As the North began to assert itself, American military intervention increased, and by the early 1960s had assumed the proportions of a full-scale invasion. U.S. and South Vietnamese forces were supported by contingents from a number of other countries, most notably South Korea and Australia, while Communist forces consisted of the Viet Cong from the south, and the People's Army of Vietnam (PAVN) from the North.

Air was pivotal in the South, where for the first time maneuver via rotary-wing aircraft was an integral component of American Army doctrine. Control of the air—which the Army routinely assumed would exist—was fundamental to that doctrine. Once again the Communists could only react, having no air force in South Vietnam; once again they demonstrated impressive adaptability. Parallels with the war against France were evident, as sabotage and ground-based antiaircraft weapons became their primary means for contesting control of the air.

A sabotage mission early in the war in November 1964 set the tone by inflicting heavy damage and alarming previously complacent U.S. politicians and military commanders, who despite all the objective evidence had underestimated their opponents. Viet Cong guerrillas infiltrated to within four hundred meters (about a quarter of a mile) of the major U.S. base at Bien Hoa and fired some seventy mortar rounds at accommodation barracks and parked jet bombers. Four troops were killed and thirty wounded; and thirty jets were hit, with five being destroyed and eight severely damaged. The result was out of all proportion to the effort, and from then on raids against air bases remained an effective method by which the Communists could challenge the United States's ostensible air supremacy.

Another method was targeting helicopters. Helicopters may provide battlefield mobility, but their inefficient aerodynamic shape and high-drag rotors make them slow and noisy, undesirable characteristics for approaching hostile landing zones and disembarking troops.

In the early phases of the war the Communists' antiaircraft weapons consisted mainly of light artillery and machine guns, but as the intensity of the fighting increased, so too did the quality and caliber of their antiaircraft artillery (AAA). By the late 1960s mobile AAA regiments had proliferated in South Vietnam, and were employing visually aimed 12.7mm guns, the potent ZSU-23-4 self-propelled radar-guided AAA, and a variety of radar-directed 57mm and 85mm weapons. Often the Viet Cong located their AAA at choke points, such as mountain passes used for transit by helicopters and fixed-wing airlifters, or on high ground overlooking air bases.

The end result was that, contrary to popular belief, control of the air was contested on a daily basis in South Vietnam, and it was by no means one-sided. It might seem incongruous that no Communist fighter aircraft operated below the seventeenth parallel, that there were no classic dogfights, and that the Viet Cong did not even have an air force, but all of that is irrelevant. Any suggestion that the United States and its allies enjoyed air supremacy in South Vietnam reveals a limited understanding of the concept.

No better illustration of the campaign's intensity can be found than the U.S. Army's loss rate of helicopters. Over the course of some ten years, the U.S.

Army operated 11,827 helicopters of all types in South Vietnam. A staggering 43 percent were destroyed, 3,587 by hostile action—primarily AAA and small-arms fire—and 1,499 by other causes.[26] That loss rate would have been unsustainable for any other organization, and probably should have been for the American Army. As a senior USAF commander later noted, the effectiveness of enemy ground-based air defenses meant that "helicopters had limited utility as a combat weapons system."[27] Fixed-wing aircraft losses were less severe but also substantial. Perhaps the true character of this extraordinary fight to control the air has eluded many analysts because it took place between U.S. helicopter pilots and Vietnamese soldiers, rather than between fighter pilots and integrated air defense systems.

Before turning to the air war over North Vietnam, a battle fought at the remote U.S. Marine Corps garrison of Khe Sahn early in 1968 must be mentioned. Fourteen years after Dien Bien Phu, General Giap, still commanding the PAVN, sought to reprise his momentous victory by laying siege to Khe Sahn. Distant from supply depots and reinforcements, Khe Sahn's six thousand soldiers seemed dangerously exposed to Giap's force of thirty thousand combatants. This time, however, Giap had miscalculated the quality of his opponent's air power and was unable to win control of the air. During the seventy-seven-day siege, U.S. strike aircraft flew more than thirty thousand sorties and dropped around 110,000 tons of bombs onto enemy positions, causing as many as ten thousand casualties. Simultaneously, Khe Sahn's supply lines were kept open by C-130 Hercules, which air-dropped and air-landed around 12,500 tons of ammunition, food, and medical stores. Khe Sahn was one of those occasions where air power saved an army.

As the American invasion escalated, North Vietnam was subjected to a massive bombing campaign; ultimately, the tonnage of bombs dropped on Indochina was three times greater than that dropped on Germany during World War II. North Vietnam lacked the capability to respond in kind, either against South Vietnam or the U.S. homeland, so its leaders again resorted to their only option: namely, to respond by adopting a defensive posture.

Some effort was made to construct the classic air defense model of fighter aircraft, ground-controlled intercept radars, and coordinated command and control, but in general Vietnamese People's Air Force (VPAF) pilots flying MiG-17s, -19s, and -21s were outnumbered and outclassed. When they did join battle, they tended to imitate the hit-and-run tactics of their Viet Cong comrades, fighting only when conditions were favorable, and making single-run passes. Thus, air-to-air combat, while often highly publicized, was relatively unimportant.[28] By contrast, North Vietnam eventually assembled a formidable ground-based air defense system of SAMs, AAA, radars, and command and control networks, and the daily battle between that system and U.S. air strike packages assumed major dimensions.

The American air campaign against the North suffered from several weaknesses. Senior U.S. commanders underestimated the potency of North Vietnam's integrated air defense system; some American aircraft types and weapons were ill suited to the nature of the conflict, having been developed for nuclear war in Europe against the Soviet Union, rather than conventional war in Southeast Asia against third-world nationalists; electronic warfare systems and communications were sometimes inadequate; and coordination between the USAF and the USN was abysmal.[29]

Initially AAA constituted the greatest danger, accounting for two-thirds of U.S. losses over the North. Increasingly, however, artillery was superseded by SA-2 missile batteries. A somewhat rudimentary weapon, the supersonic SA-2 nevertheless was effective up to forty kilometers and sixty thousand feet, and was a serious threat. Missile launch rates increased rapidly, from about thirty per month in 1965 to 740 by late 1967. Kill rates averaged around one aircraft for every sixty missiles fired. American pilots developed techniques to out-maneuver the SA-2, which sometimes could be seen as soon as it was fired, and improved electronic countermeasures were introduced. Anti-radiation missiles that homed in on a SAM site's radar were an important addition to the Americans' armory; but reflecting the move/countermove that has always characterized warfighting technology, North Vietnamese defenders often chose not to activate their radars, instead firing their SA-2s visually if conditions permitted.

The air war over North Vietnam reached a climax during Operation Linebacker II, a short, fierce campaign directed by U.S. president Richard Nixon at the end of 1972, in an effort to force the Communists to the negotiating table. In only twelve days, B-52 heavy bombers flew 729 missions and dropped more than fifteen thousand tons of bombs on thirty-two industrial and military targets around Hanoi and Haiphong, badly damaging most. Fifteen B-52s were shot down, all by SA-2s. These were the most intense bombing raids conducted since

World War II, and were later credited by U.S. secretary of state Henry Kissinger with forcing the Communists to reopen cease-fire negotiations.[30]

Well before Linebacker II, the war had become deeply unpopular in the West, and the United States and its partners had started withdrawing their forces, to all intents and purposes abandoning their South Vietnamese allies. In April 1975 a PAVN tank memorably crashed through the gates of the presidential palace in Saigon (now Ho Chi Minh City), symbolizing the North's victory and reunifying the country.

Consistent with the aphorism that we learn more from defeat than from victory, the war in the air over Vietnam provides an instructive case study for Western analysts. In the North, the Communists responded to the U.S. bombing campaign by constructing a potent ground-based air defense system, and the air battle had to be re-fought and re-won every day. In the South, it is often asserted that the Communists conceded air supremacy, a conclusion that is repudiated by the U.S. Army's staggering loss-rate of helicopters. There was unquestionably a fight to control the air—it is just that it was different.

Vietnamese air power today is in a state of flux. The newly unified nation emerged from the American war with a capable point air defense system based on Soviet radars, SAMs, AAA, and fighter/interceptors; additionally, as a result of its victory, Vietnam captured large numbers of American aircraft. Since then, however, the imperative to concentrate on national development, combined with what was for decades an inefficient and backward economic system, has seen much of this hardware and its associated infrastructure deteriorate.

The VPAF continues to operate legacy Soviet-era air defense and ground attack weapons, including some 150 Su-22s and MiG-21s, to which the addition of twenty or so Su-27s and Su-30s has added a veneer of modernization. A fleet of some one hundred helicopters and fixed-wing airlifters more accurately reflects the VPAF's preoccupation with border protection, national development, and the exploitation of offshore oil and gas fields. Prickly relations arising in part from disputes over resources have inhibited collaboration with neighboring China, an emerging military superpower.

Given its history, Vietnam could be expected to conduct reasonably effective defensive air operations, but its capacity to mount and sustain offensive missions at a distance would seem questionable. Vietnam has an admirable history of responding to aggression from the air with ingenuity and resilience. Should threats emerge in the future, it would again have to call on those qualities.

AIR POWER AS POLITICS—INDONESIA

The nineteenth-century Prussian strategist Carl von Clausewitz famously described war as a "mere continuation of policy [politics] by other means." Von Clausewitz's timeless insight into the philosophy of warfare defines the relationship between the military and the state. Specifically, the sole legitimate purpose of a military force is to serve the state by applying organized violence in accordance with higher (political) policy. Among other things, this relationship explains why civilian (elected) rule of the military is a basic principle of democratic government.

Indonesia won its independence from the Netherlands in December 1949, following a four-year war of national liberation. In the years since then the country has experienced various forms of governance, in name at least. From 1949 to 1957 a system of "parliamentary democracy" was practiced under the leadership of the charismatic first president, Sukarno; and from 1957 to 1965, "guided democracy" was allegedly observed ("democracy with leadership" was Sukarno's mischievous definition of a system he still controlled). Sukarno was removed from office following a military coup in 1965, and for the next thirty-three years Indonesians lived under a system known as "*pancasila* democracy," with the government dominated by the leader of the coup, President Suharto.[31] Suharto was forced out of office in 1998 by a combination of elite dissatisfaction and widespread dissent arising from his administration's corruption, nepotism, and repression. Since then, four popularly elected presidents have overseen a remarkably successful transition to genuine democracy.

In practice, Sukarno's rule was authoritarian, and Suharto was a dictator. It is the Suharto years of pancasila democracy from 1965 to 1998 that are of most interest here, because of the context they established for Indonesian air power.

The fact that Suharto was a dictator is irrelevant to this study—over the course of history, nations have been governed more often by dictators than by democrats. The feature of his rule that warrants attention is the *nonmilitary* role the Indonesian Army *Tentara Nasional Indonesia–Angkatan Darat* (TNI–AD) assumed, in order, it was claimed, to help build the nation. Yet by the time of Suharto's demise, the TNI–AD was regarded as one of the "most corrupt and abusive" armies in the world.[32] A corollary of the army's extended period of extraordinary political and social power was the marginalization of the Indonesian Air Force *Tentara Nasional Indonesia–Angkatan Udara* (TNI–AU).

Indonesia was in crisis in 1965. Sukarno's erratic leadership had damaged the economy, created serious internal divisions, and caused tensions throughout Southeast Asia. Political instability was exacerbated by schisms between left-wing and right-wing factions in the army, and by the growth of the Indonesian Communist Party. Suharto and fellow right-wing generals seized power in a coup that may have killed as many as half a million people.

It was understandable that Suharto's so-called New Order would want to employ the army to help stabilize the situation, because it was one of the few national institutions with a degree of discipline and a countrywide presence. The issue, though, was how the army would be used or, perhaps more accurately, how it would allow itself to be used. And this was the problem for air power and, by association, for Indonesia's ability to project its national interests into the Asia Pacific region.

Dwifungsi (dual function) was the doctrine used to justify the army's intrusion into "every effort and activity of the people," including politics, ideology, economics, and culture. It was a role the TNI–AD embraced enthusiastically. Special seats were reserved in parliament for members of the armed forces, and army officers routinely served as cabinet ministers, ambassadors, governors, mayors, heads of corporations, members of the judiciary, and so on. For more than thirty years the TNI–AD and Suharto colluded to retain their hold on power. In the process they unquestionably served their own interests; whether they served those of the nation is less certain. What is certain is that under a regime of army domination and severely constrained budgets, the air force struggled.

Indonesia's ties to Moscow during the Sukarno years saw the air force accumulate a grab bag of Soviet aircraft, more by way of casual handouts than by any rational force structuring process. Soviet technical support to and interest in the TNU–AU were indifferent, and led to abysmal aircraft serviceability rates and barely competent crews. It should not have surprised anyone that the TNI–AU performed poorly during a campaign of low-level military aggression instigated by Sukarno against the proposed federation of Malaysia in the early 1960s. A nadir was reached when an attempted paradrop into southern Malaya using C-130s degenerated into a fiasco, one aircraft reportedly crashing and others dropping troops in the wrong place. An unsympathetic observer might have speculated that the army was getting the air force it deserved.

After ousting Sukarno, Suharto gradually rebuilt bridges with the West, establishing military links in the process. As part of that process, an RAAF training team visited Indonesian Air Force units in 1971 and found that skill levels were low and the infrastructure in disarray.[33] Ostensibly experienced TNI–AU pilots were well below standard, while air bases were littered with un-airworthy MiG fighters, Ilyushin bombers, Antonov transports, and assorted helicopters. Still, the RAAF's presence was an indication that matters were starting to improve. Western cooperation brought with it access to high-quality training, first-world logistics support, and advanced weapons systems.

The Indonesian government took a noteworthy initiative in 1976 when the amalgamation of several small aerospace firms led to the establishment of the *Industri Pesawat Terbang Nusantara* (IPTN). Headed by the mercurial Dr. B. J. Habibe, IPTN had an ambitious agenda, planning to build airframes, engines, and weapons systems under license, before branching into local design and construction. Light fixed- and rotary-wing transport aircraft projects were successfully managed, but organizational and financial difficulties tempered the full realization of Habibe's vision. Nevertheless, the factory, now known as the *PT Dirgantara Indonesia* (PTDI), continues to build components for major companies including Boeing, Lockheed Martin, Airbus, and Fokker, and there are plans to bring a locally produced unmanned aerial vehicle (UAV) into service. As a center of aeronautical engineering expertise, the PTDI could be used to modify TNI–AU aircraft and systems to meet the demands of short-notice defense emergencies.

Despite that progress, overt institutional inferiority and inadequate funding remained heavy burdens for the air force. The Timor-Leste crisis of 1999 exposed the broader defense policy problem this has created.

Indonesia invaded and annexed Timor-Leste in 1975, which was then subjected to a brutal occupation by the army. International pressure eventually led to a plebiscite on independence, which the citizens overwhelmingly supported. The army was bitterly opposed to the Indonesian government's decision to withdraw from Timor-Leste, and for some time open defiance seemed possible, even to the extent of unauthorized armed opposition to the landing of United Nations peacekeeping forces. In the end the TNI–AD obeyed orders, but not before many of its troops and local militia gangs—raised, armed, and sponsored by the army—had destroyed around 70 percent of Timor-Leste's infrastructure and murdered 1,500 civilians.

Had the Indonesian government decided to oppose the intervention, the UN peacekeeping force would have been at its most vulnerable during its transit to and landing at Timor-Leste: the loss of, say, a troop ship or an airlifter, with heavy casualties, might have been sufficient to end the mission before it had gathered momentum. The irony is that the TNI–AD would have played no part in any such (hypothetical) action, and the TNI–AU, which would have been the logical weapon of choice, was incapable of doing so.

Once the rancor associated with the Timor-Leste intervention had subsided, Indonesia's relations with the West resumed their upward trend. Regular exercises with the Singaporean and Australian air forces have sharpened the TNI–AU's operational skills; and education, engineering, and logistics management have also benefited. Nevertheless, the air force continues to confront serious obstacles. Technical standards remain low, with aircraft serviceability rates averaging around 40 percent, compared to 80 percent for advanced air forces. Following a succession of fatal accidents, the airworthiness of the TNI–AU's fleet has become "a matter of national concern."[34]

Adding to the air force's challenges is its complex mix of aircraft. The acquisition pendulum has constantly swung, first toward the Soviets (MiG-15s and Tu-16s), then the Americans (A-4s, F-5s, and F-16s), then to the British (Hawks), and now back to the Russians (Su-27s and Su-30s). Two Russian Su-27SK interceptor/ground attack aircraft and two Su-30MK multirole fighters were acquired in 2003; a year later, plans were announced to add another six Su-27s and two Su-30s.

It is interesting to speculate whether the acquisition of the Su-27s and Su-30s represents a reaction to the Timor-Leste crisis, when the Indonesian government realized that the subjugation of the air force has left it with little capacity to respond to security contingencies with strategic power, should it wish to do so. The Su-27 and Su-30 can be formidable aircraft when integrated into an effective warfighting system. But the TNI–AU has poor command and control, no AEW&C, inadequate weapons, and only two ancient C-130B air-refueling tankers of dubious quality; furthermore, it is traditionally weak in planning and conducting complex air operations. Consequently, it is doubtful whether the service could do much with its Flankers. If used offensively, a handful of (daylight) strikes against fixed targets would be feasible; if used defensively, point-defense over major facilities or urban areas for limited periods might be managed. As things stand, the Indonesian Air Force has a modest capacity to respond, very little capacity to deter, and none whatsoever to shape.[35]

At the same time, Indonesia's economy is recovering from the shocks of the past decade, and the country continues to progress remarkably well in its transition to democracy. As national confidence increases, politicians may wish to shift the emphasis of their defense policy away from internal security and the army and toward the region and the air force. Speculation that the government might seek to acquire around two hundred Flankers and F-16s supports this thesis, noting however that the TNI–AU will lack the personnel skills to manage such a fleet for many years.

If the TNI–AU is to flourish, it must be allowed to invest in a full suite of enabling capabilities, including ISR, command and control, advanced communications, precision weapons, and, most importantly, the kinds of people who can make those capabilities work. But while change for the better may be on the march in modern Indonesia, the dead hand of the army continues to weigh heavily on the air force. It seems unlikely that the TNI–AU will receive sufficient resources to reach its potential inside the next twenty years.

AIR POWER AS A MESSAGE—SINGAPORE

The contrast between the development of air power in Indonesia and in Singapore could scarcely be more pronounced. In large part, that contrast mirrors the respective state-building experiences. Whereas Indonesia's political history since independence has been turbulent, the Republic of Singapore's history has been ordered and single-minded. Dominated since 1959 by the first prime minister, the formidable Lee Kuan Yew, in only half a century Singapore has been transformed from a small British trading post into one of the world's wealthiest nations. That wealth is based on trade, manufacturing, advanced technology, finance, and, not least, intellectual capital. The seemingly inexorable rise of capital reserves and living standards has been accompanied by a no less impressive rise in the capabilities of the defense forces.

Singapore's leaders have always been sensitive to their challenging geostrategic circumstances. The city-state has no strategic depth—a strike/fighter can overfly the country in three minutes. Furthermore, Singapore's predominantly Chinese (77 percent) population of 4.6 million is keenly aware of its proximity to two large Muslim nations, Indonesia (240 million people) to the south, and Malaysia (26 million people, 60 percent of whom are Muslim) to the north. Relations with both have never been entirely comfortable. Singapore's armed forces have been developed with a measured efficiency that reflects those circumstances.

Independent air power was established in Singapore in 1968 with the formation of the Singapore Air Defense Command (SADC). The SADC's role was defined by its title; namely, to protect the island from air incursions. From the outset, the SADC emphasized quality. Initially the service drew heavily on the (British) Royal Air Force and the RAAF for assistance, before expanding its contacts to incorporate other first-rate air forces, including those of France, the United States, and Israel. As expertise grew, so too did the SADC's roles. In particular, the acquisition in 1973 of a fleet of A-4 Skyhawk fighter/bombers signaled the beginning of what was to become a powerful strike force. Also indicative of continual progress was the ability of the local aerospace industry to extensively modify the A-4s. Consistent with its expanding capabilities, in 1975 the SADC was renamed the RSAF.

By that time, Singapore's military strategy of defensive deterrence had become known colloquially as the "poisonous shrimp." The metaphor's meaning was clear enough: any aggressors who might be tempted to help themselves to Singapore's attractions were likely to finish up seriously ill. Many within the region perceived a degree of paranoia in Singapore's attitude; by the same token, the city-state's military professionalism generated respect.

Today's RSAF is Southeast Asia's preeminent air power: capable, well balanced, technologically advanced, and staffed by skilled personnel. Advanced weapons systems are complemented by excellent indigenous technical services, a first-class training organization, and modern command and control practices. Singapore's long-standing membership of the Five Power Defence Arrangements has been a source of valuable experience, as has regular participation in air combat exercises in the United States and Australia.[36] Additional expertise has been garnered from long-term deployments in the United States, France, and Australia. The RSAF clearly is capable of achieving its defined role: "To deter, and to win a swift and decisive victory if deterrence fails."

The RSAF's strike/fighter force of some ninety F-16s and F-5s, supported by well-organized ground-based air defense and command and control systems, would be confident of achieving control of the air against any regional competitor. Equipped with all-weather targeting pods and precision-guided weapons, the same aircraft would be equally capable in the strike role. The probable acquisition of up to one hundred F-15SG Strike Eagles and F-35 Joint Strike Fighters in the next decade will maintain the RSAF's qualitative edge, especially when linked to the service's existing proficiency in ISR, air-to-air refueling, and airborne early warning and control. Alone among the air forces of Southeast Asia, the RSAF can aspire to pursue network-centric warfare, an enabling mechanism that will greatly facilitate the ability to detect, analyze, track, and engage targets in near–real time.

Joint warfare capabilities are enhanced by substantial numbers of maneuver and close attack platforms, including CH-47 Chinook heavy-lift helicopters and AH-64D Apache combat helicopters. Singapore also has been a pioneer in the use of UAVs for ISR, and currently operates about 120. RSAF operations are underpinned by an accomplished indigenous aerospace industry, notably the Singapore Technologies Aerospace Company, while a strong cadre of highly qualified reservists, including combat-trained airline pilots, could be mobilized in a defense emergency.

If a question mark were to be placed alongside Singapore, it might concern the tendency to fast-track so-called scholars, highly educated young officers who, some critics contend, are promoted more for their natural intelligence than for any deep understanding of military affairs. The end result, it is sometimes claimed, can be a degree of naivety at the higher levels of command. The criticism is moot, but is often mentioned.

In summary, Singapore has gradually shifted its military posture from defensive deterrence to offensive deterrence. That shift is best represented by the RSAF, which has evolved from the purely responsive nature of the SADC into a force that, through its quality and structure, exerts a shaping influence in the Asia Pacific region. Singapore's "poisonous shrimp" has become a respected "porcupine."[37]

AIR POWER AS STRATEGY—AUSTRALIA

Three factors of superseding importance define Australia's geostrategic circumstances. First, the country is an island continent—that is, it is surrounded by an air-sea gap; second, its immediate area of strategic interest is vast; and third, it has a small population. It would seem logical, therefore, that Australia's military strategy should seek to exploit advanced technology to control the air-sea gap as a means of *shaping* and *deterring* in the region. The key capabilities required for this so-called defense of Australia strategy are ISR dominance, a flexible air defense system, long-range strike forces, and a high-quality personnel base. As it happens, those were precisely the characteristics demonstrated by Australian air power during World War II, the only occasion on which the nation has been exposed to a serious military threat.

Prior to the war, Australian defense policy was based on a dependent relationship with the UK. Should the distant dominion find itself in trouble, the theory was that the mother country, in the form of the RN, would appear over the horizon and save the day. Local assistance would come in the main from the

Royal Australian Navy (RAN), which during the interwar period from 1919 to 1939 received some 60 percent of all defense spending, with 30 percent going to a citizen-based army, and a mere 10 percent to the RAAF.[38]

However, when the war in the Pacific started, the RN was unable to come to the rescue, for the very good reason that it was preoccupied in the Northern Hemisphere, fighting for the UK's own survival. Despite the Australian government's prewar largesse, the RAN was manifestly incapable of dealing with any Japanese invasion fleet by itself, and the army had no role to play against what was, in the first instance, a maritime (air/sea) threat. As revealed by the actions at Pearl Harbor, Coral Sea, and Midway, and by the sinking of the *Prince of Wales* and *Repulse*, Australia's best option for controlling an air-sea gap would have been a modern air force based on reconnaissance, striking power, and defense. But at the start of the war the RAAF had a mere handful of aircraft, every one of which was obsolescent.

Confronted by reality, the Australian government dramatically reversed its prewar policy. Resources were poured into the RAAF, with expenditure increasing by a factor of 11, compared to factors of 8 for the army and 3.5 for the navy.[39] By 1945 the RAAF had become the world's fourth-largest air force, with more than 170,000 personnel and 5,600 aircraft, including the most advanced fighters and bombers.

Air was central to the war in the Pacific. Commonly referred to as an "island hopping" campaign, the Allies' strategy would be more accurately described as "airfield hopping." As islands on which airfields were located were progressively captured, the Allies' offensive air operations moved steadily closer to Japan. Critical actions included the Battles of Coral Sea and Midway in mid-1942, where U.S. air power destroyed Japan's ability to project force, thus establishing the basis of eventual victory; the Battle of the Bismarck Sea in March 1943, where USAAF and RAAF land-based aircraft devastated a large Japanese convoy, ending any possibility that Australia might be invaded; and the atomic attacks on Hiroshima and Nagasaki, which effectively ended the war without a single Allied soldier having to set foot on Japan.

The RAAF emerged from World War II with an institutional ethos that has since allowed it to remain at the forefront of regional air power. Three features have been vital. First, an excellent recruitment, education, and training system has continued to deliver high-quality people. Second, technical expertise has been sustained by a capable local aircraft industry and a very good defense scientific and technology service. Finally, operational excellence has been sustained through rigorous training, and by constant interaction with other leading air forces.

The national military strategy under which the postwar RAAF has functioned has oscillated between two main forms: expeditionary campaigns ("forward defense"), and the defense of Australia. Because the former has been the dominant model, the Australian Defense Force (ADF) has been regularly deployed on operations to places as diverse as Malta, Korea, Malaya, Thailand, Vietnam, Somalia, Timor-Leste, Iraq, and Afghanistan. The disaster of Vietnam saw the expeditionary strategy briefly superseded by the defense of Australia model, but following al Qaeda's attacks on America, alliance politics persuaded the government to commit substantial forces to the invasions of Iraq and then Afghanistan. But as those campaigns increasingly were perceived as unwinnable wars of choice, popular opinion started to favor a less aggressive posture. In particular, the notion of trying to shape and to deter in the region, and of responding only as a last (instead of a first) resort, seemed consonant with the geostrategic and ethical imperatives of the twenty-first century.

That logic was evident in the 2009 Defense White Paper, *Defending Australia in the Asia-Pacific Century: Force 2030* (DWP 2009).[40] The title unambiguously identified the ADF's geographic area of primary interest and its main purpose. While expeditionary campaigns must always remain an option, shaping and deterring in the region represent the preferred approach to national defense. Reflecting the experience of World War II, air power is the centerpiece of the associated strategy. As it happens, for the past thirty years the RAAF has been quietly assembling the necessary force structure.

The starting point is the indigenous, very long-range radar system known as "Jorn."[41] Reportedly capable of detecting targets ranging from surface ships to missiles to stealth bombers out to several thousand kilometers, the Jorn network provides 24/7 broad area coverage of the air-sea gap across Australia's vast northern reaches. The exceptional situational awareness this provides will be complemented by six Boeing-737 Wedgetail AEW&C platforms and seven high-altitude UAVs (probably Global Hawk).

DWP 2009 confirmed that air defense, strike, and additional ISR will be provided by a fleet of around one hundred F-35 Joint Strike Fighters. With its strategic reach boosted by the acquisition of five KC-30A multirole tanker transports, the F-35 will elevate the ADF's capabilities to a new level. Because the F-35s will not arrive until around 2019, twenty-four F/A-18F Super Hornets have been ordered as a bridging capability, for delivery in 2010. Strike weapons are likely to include a combination of the joint air-to-surface standoff missile, the Harpoon anti-ship missile, joint direct attack munitions, the small diameter bomb, and possibly the Norwegian Joint Strike Missile.

Mention should be made of the RAAF's approach to the human factors dimension of operating the F-35; that is, to the dimension that acknowledges that acquiring a capability involves more than simply buying hardware. Ever since Australia joined the Joint Strike Fighter program as a partner in 2002, RAAF pilots have been periodically visiting the United States to fly the F-35 simulator in order to develop warfighting tactics relevant to the aircraft's unique combination of characteristics (such as stealth, advanced networking, sensor fusion, and a distributed aperture system). Furthermore, for some two years the RAAF has had a pilot on exchange with a USAF F-22 squadron, accumulating experience on the world's only other fifth-generation fighter that will be directly transferable to the F-35.

The RAAF's capacity to position forces, weapons, and supplies over a range of distances and within a variety of topographies will be enhanced by the decision to acquire additional C-130Js and up to ten "light" fixed-wing transport aircraft (reportedly the C-27 Spartan is favored), while simultaneously retiring twelve elderly C-130Hs. Substantial additional capacity will be added by the five KC-30As, each of which can carry around 270 troops plus cargo. Battlefield maneuver will be boosted by the replacement of the army's Black Hawk and Chinook helicopters with more capable platforms.

The maritime nature of the Asia Pacific region means that most maneuver operations conducted by the ADF's air transport assets are likely to fall at the "respond" end of the strategy continuum; that is, they are unlikely to be as influential in shaping and deterring as ISR, air defense, and strike.

ISR requires special mention, because while no individual system can achieve its full potential without being integrated into the total force, ISR is the ADF's single most important technological capability. Consequently, DWP 2009's announcement that the ADF will acquire its own remote-sensing satellite, probably fitted with high-resolution, cloud-penetrating synthetic aperture radar, is enormously significant. So too is the decision to buy up to seven high-altitude, long-endurance UAVs. Space and UAVs are an essential part of the future of air power.

The new UAVs will be complemented by eight manned aircraft, almost certainly the Boeing 737–derived P-8A Poseidon, which will replace the RAAF's

AP-3C Orions. Although described as "maritime patrol" aircraft, the P-8As are likely to devote much of their effort to broader ISR tasks, including cooperation with land forces.

The final comment on Australian air power strategy must go to an intriguing topic: missile defense. At a time when the proliferation of short-, medium- and long-range missiles is one of the major features of armed conflict and international tension, Australian decision-makers have taken a hands-off approach to the problem, rejecting the idea of "a unilateral national missile defense system" and undertaking only to "review" the situation annually. But notwithstanding that declaratory policy, the ADF seems to be gradually assembling a de facto missile defense system, through the sensors and weapons associated with Jorn, the imaging satellite, AEW&C, the UAVs, the F-35, the F/A-18F, naval air warfare destroyers, and a networked command and control system. Especially noteworthy for control of the air in general and missile defense in particular is the decision to fit the air warfare destroyers with the Standard Missile-6, a top-of-the-range weapon which should have a genuine capability against ballistic and cruise missiles.

In summary, the ADF's networked system of high-quality air and space capabilities will represent an unsurpassed ISR, air defense, and precision-strike system in the Asia Pacific region. That system in turn underpins an air power–based national defense strategy.

CONCLUSION

The Asia Pacific region's distinctive geography suggests that air power's unique ability to create a strategic effect and to respond rapidly at a distance through ISR, strike, and maneuver should be central to defense policies, and that has indeed been the case. Advanced air forces from Japan, the United States, and Australia have played decisive roles in various conflicts, to the extent that Japan's early success and then ultimate defeat in World War II can be tracked through the rise and fall of its air power.

It is no coincidence that successful regional air forces have been built on the essential qualities identified by Hugh Trenchard almost one hundred years ago: excellent recruitment and training, high operational standards, a sound economy, and a vigorous indigenous industrial and research and development base. Nor is it a coincidence that in the contemporary Asia Pacific, emergent air powers such as Singapore reflect those qualities, while nations that are struggling fall short in one or more categories.

In addition to geography, air strategy in the Asia Pacific region has been strongly influenced by colonialism. Third-world Vietnam, for example, twice exploited air power with admirable creativity to defeat first-world states during its fight for national liberation. By contrast, in Indonesia, post-colonial politics and the overt domination of an intensely politicized army have prevented the air force from achieving its potential. National cultures have also been immensely influential, with Japan's experience in World War II providing a remarkable case study; moreover, it seems possible that culture, rather than economics or education or industrialization, may still inhibit the full development of Japanese air power.

Only a few Asia Pacific air forces have the capability to shape events in the region, and none can compete with the implicit influence of American air power or, increasingly, with that of China, India, and Russia. All Asia Pacific air forces have some potential to deter and to respond, with individual capacities ranging from considerable to minimal. The credibility of each state's deterrence posture will be largely defined by its ability to control national airspace, and to conduct strategic strike missions. Those countries that manage their defense resources astutely by promoting quality over quantity, avoiding needless complexity, and focusing on their specific circumstances, will be best placed to promote their national security through air power.

8

LATIN AMERICA

James S. Corum

From 1910 to the present, military aviation has played a key role in Latin American conflicts. Air power has been used in many forms, ranging from counterinsurgency to conventional war to naval air strikes in the Falklands War. Many of the air conflicts have been similar to Western conventional wars with ground support being the main role of air power. However, in some respects, the Latin American air wars have been different from other conflicts with the extensive use of air power in counter-drug operations, a major feature of recent Latin American conflicts. Air power has also played a key role in several covert operations in the region. One of the most common uses of air power has been in counterinsurgency operations, and in several cases the Latin American nations provide a good model of how smaller nations can use air power very effectively. This chapter will try to portray the Latin American military air power experience in all of its variety and under many different conditions.

AIR POWER IN THE MEXICAN REVOLUTION

The first use of military aviation in a conflict in Latin America occurred during the Mexican Revolution. With the overthrow of the Porfirio Diaz dictatorship in 1910, Mexico was plunged into a violent twenty-year revolution. Governments came to power only to find themselves immediately in conflict with other revolutionary factions.[1]

Mexican war minister Manuel Mondragón established an army air unit in April 1913 with a Curtiss aircraft that was used for reconnaissance and bombing rebel factions. The Constitutionalist faction fighting the government established its own air arm and in 1913 acquired a Martin twin-engine aircraft that was used by Gen. Alvaro Obregon's Army of the North. In 1914 the Army of the North also acquired three French Moisant single-seat planes.

As the revolution continued, the Constitutionalists, who eventually won, built up their small air force manned by a mix of Mexican and foreign mercenary pilots. The revolutionary leader Pancho Villa, who rebelled against his former Constitutionalist allies, also tried to create his own air force, acquiring two or three Wright biplanes.

Little came from these early efforts. During the period between 1913 and 1918, the aircraft available to the Mexicans were generally obsolete and in poor condition. While the factions flew small bombing raids against each other, the effects were minimal. Only with the end of World War I and the ready availability of surplus military aircraft from the major powers could the Mexican government build something resembling a capable air force. For example, in 1920, the Mexican Army bought thirteen twin-engine modern Farman F 50 bombers. The U.S. government, wanting to see stability restored to Mexico, supported the government of President Alvaro Obregon in 1923 against a coup by Gen. Victoriano Huerta. Huerta was able to convince a large part of the army to support his rebellion. To bolster the established government, the United States furnished Obregon's forces with a squadron of new DH-4 bombers, which were armed with two machine guns and could carry a 450-pound bombload. In the three months of fighting, the DH-4 bombers played a key role in suppressing the rebellion. In 1927 and again in 1929, two major coups by army factions threatened to again overthrow the government and throw Mexico back into revolutionary chaos. Again, the government, now with a capable air force of several squadrons flying American DH-4 bombers, Vought Corsair OU-2M fighters, and British Bristol F-2 fighters, employed its aviation units in an aggressive bombing campaign against the rebel forces and defeated both rebellions.

The conflict in Mexico also forced the Americans to intervene in what became the first use of U.S. military air power. After Pancho Villa's army had been defeated by the Constitutionalists in a series of battles in 1915, Villa decided he might restore his fortunes by invading the United States. In March 1916 Villa crossed the border with several hundred men and attacked the town of Columbus, New Mexico, killing seventeen Americans.[2] The American public was outraged by the act and President Woodrow Wilson sent American troops to cross the border to destroy Villa's forces.[3]

Gen. John J. Pershing was given command of a U.S. Army force that would soon exceed fifteen thousand men.[4] To support his force operating in the rugged mountain country of northern Mexico, Pershing deployed the U.S. Army's only air squadron. From March 1916 to August 1916, the First Aero Squadron, under the command of Capt. Benjamin D. Foulois, flew 540 sorties, covered over 19,000 miles, and logged over 340 flying hours in support of Pershing's troops on the ground.[5] It was an embarrassing moment for U.S. aviation history. The Curtiss biplanes were not up to the task. One of the pilots, Lt. Edgar S. Gorrell, recalled that the squadron "was in horrible shape."[6]

> The [eight] airplanes were not fit for military service, especially along the border. . . . The squadron had no machine guns, no bombs, and none of the utensils of warfare later known to World War I flyers. . . . It wasn't until about sixty days after reaching Columbus that we got our hands on about sixteen machine guns which represented, according to Captain Walsh, the Ordnance Officer, about 50 percent of the total number of army machine guns then in existence. . . . The bombs arrived in April and were but three-inch artillery shells and nobody knew how to use them. They were sent, not for use, but so that certain authorities in Washington could tell the newspapers that we were equipped with bombs.[7]

The airplanes themselves had seen extensive service and were plagued with mechanical problems. Flying in the high mountains with harsh weather conditions was exceptionally dangerous. While conducting reconnaissance, the pilots could not tell the difference between Villista soldiers and government forces. Lack of training and ordnance meant that the squadron was ineffective in the attack role. The aircraft broke down one by one and by summer the First Aero Squadron was no longer capable of operations. In the end, the U.S. Expeditionary Force chased Villa's forces all over northern Mexico, fought some small battles, and forced Villa to disperse his army into small bands to evade pursuit. Ultimately, the Americans failed to capture Villa, but he was no longer a threat to either the Mexican or U.S. governments. Despite the extraordinary efforts of the American airmen to keep their obsolete planes flying in miserable conditions, the air contribution to the campaign had been minimal.[8]

U.S. MARINE AVIATION IN LATIN AMERICAN INTERVENTIONS, 1914–34

In other regions of Latin America, the U.S. military had considerably more success in employing air power against various irregular forces during an era of U.S. intervention in the Caribbean from 1914 to 1934. The U.S. Marine Corps was the primary U.S. intervention force in these years, and the Marines served in that role in Haiti from 1915 to 1916, and again from 1919 to 1920; and in Santo Domingo from 1916 to 1917 and again in 1919.[9] For Marine aviators, fighting "bandits" in the Dominican Republic and *cacos* in Haiti was a formative experience, but Nicaragua was the proving ground for the Marines' concept of air power in small wars—a laboratory that also provided the basic formula for Marine air-ground teams in World War II and up to the present.

Marine aviation was employed to support a friendly regime in the Dominican Republic and a colonial kind of protectorate in Haiti. U.S. Marines were deployed to Haiti in 1915, and, for the next fifteen years, the Marine air units supported a Marine brigade of roughly eighty officers and twelve hundred enlisted men and an American-trained and -led *gendarmerie* of about twenty-seven hundred men.[10] U.S. forces also occupied the Dominican Republic and were later deployed to support a U.S.-friendly regime in Nicaragua, which would be the main scene of action for the U.S. forces in the region. The Marine First Air Squadron, with six Curtiss JN aircraft, deployed to the Dominican Republic in February 1919. Six more Jennies and six HS-2L flying boats of the Fourth Air Squadron began operations at Port-au-Prince, Haiti, on March 31, 1919. The First Squadron remained in the Dominican Republic until 1924 and the Fourth Squadron departed Haiti in 1934. Initially, the low-performance Curtiss Jennies served only as reconnaissance planes, but in 1919 they began to be replaced with DH-4Bs and the Marine air units soon took part in active combat, guiding patrols to make contact with the guerrillas and bombing and strafing bandit formations. However, the lack of current intelligence and reliable air-to-ground communications limited the value of the squadron in its attack role and its greatest contribution remained that of support. But even in support, the Marines on the ground recognized the new dimension that aviation brought to anti-bandit operations. It was in these operations that aviation became an integral part of the Marine Corps.[11]

The United States had been in and out of Nicaragua several times in the late nineteenth and early twentieth centuries to protect its extensive interests and to support a pro-U.S. regime. However, in the mid-1920s, the Nicaraguan government was threatened by a leftist movement led by Augusto Sandino, a Nicaraguan nationalist who had at one time lived in Mexico and served under

Pancho Villa. Sandino and his rebel forces were opposed to U.S. interests and to the U.S.-backed government. In 1927 the Marines intervened in Nicaragua to support a truce between warring factions and to try to achieve a negotiated settlement to the conflict. But Sandino broke off the negotiations and withdrew to the wilderness of northern Nicaragua to carry on the fight. President Adolfo Diaz of Nicaragua issued an appeal for additional Marines to disarm them.

The United States responded by sending a Marine brigade to Nicaragua that included the Fifth and Eleventh Marine Regiments. In January 1927 these ground forces were reinforced with Marine Air Squadron VO-1M with eight officers, eighty-one enlisted men, and six DH-4Bs. A second Marine air squadron, VO-4M, soon arrived as reinforcements.[12] By mid-1927 the Marine brigade had a strength of 178 officers and 2,725 enlisted men. The mission of the two squadrons was to support the ground forces of the Second Brigade "by providing observation aviation, ground attack aviation, and transport service."[13]

Sandino's base of operations was the remote province of *Nueva Segovia*. Because of the isolation of the area, aviation became a key support force. Air transport came into its own in Nicaragua. Thanks to aircraft, the Marines could maintain garrisons in remote areas, airdrop vital supplies, and even carry casualties back to a hospital in the rear. At first, the Marines used the DH-4s and Corsairs to fly vital supplies, but by the end of 1927 they received a TA-2 Atlantic-Fokker trimotor transport. During its first six weeks of operation, this one airplane ferried 27,000 pounds of freight and 204 passengers. The trip between Managua and the provincial capital normally required ten days to three weeks by oxcart, but the Fokker could accomplish the same mission in under two hours.[14] Additional trimotors were immediately requested and two more Fokkers arrived in 1928.[15]

Major combat operations ensued with the arrival of Marines in Sandino's base area. Thanks to the newly arrived Vought 02U-1 Corsairs, which became the primary combat plane of the campaign, the Marines were able to accurately bomb and strafe the Sandinista forces. In July 1927 Sandino's forces made a major attack on the small U.S. garrison in Ocatal and were cut to pieces by U.S. aircraft.[16]

Through the latter half of 1927 and through 1928 the Marine and Nicaraguan government ground units, supported by Marine air units and transports, chased Sandino into an even more isolated region of northern Nicaragua. Whenever Sandino's forces tried to make a stand, the Marine Corsair dive-bombers bombed them relentlessly. The Marines put patrols deep into the jungle and on mountains and supplied them by airdrop.[17] Marine aircraft dropped thousands of propaganda circulars over enemy camps.[18] In addition, the Marine aviators provided air mobility for the ground forces. In November and December 1931, guerrillas operating in the Chinandega, Leon, and Esteli provinces threatened to overwhelm the local Guardia troops. Prompt movement of additional government soldiers to key points by air transport turned the tide. The rapid moves by the government forces kept the guerrillas constantly off-balance. The guerrillas ultimately quit the area.[19]

The guerrillas were eventually exhausted by years of fighting, and Sandino agreed to a peace settlement with the government and ended the conflict. In 1933 the last of the Marine and U.S. Army intervention forces withdrew from Nicaragua.

THE CHACO WAR, 1932–35

The most dramatic use of military air power in Latin America before World War II was the Chaco War, fought between Bolivia and Paraguay between 1932 and 1935. It was a savage war with enormous casualties (sixty thousand Bolivian dead and thirty thousand Paraguayan dead) fought over the Chaco region, an arid wasteland in the heart of South America.

The conflict had its origins in Bolivia's loss of its access to the Pacific Coast to Chile in the War of the Pacific in the nineteenth century. For the next generation, the successive Bolivian governments were obsessed with obtaining access to the ocean. The Paraguayans offered Bolivia a free-port status on the Rio Paraguay, a deep river that ran from Paraguay and flowed into the Rio de la Plata and then to the Atlantic. But Bolivia would have nothing less than a port on its own territory. Having no chance to retake the old Pacific coastline, Bolivians revived ancient colonial land claims to the arid and unpopulated Chaco region—which could give Bolivia access to the Rio Paraguay and the ocean. The only problem was that the Paraguayans were in the Chaco first. What little occupation and exploitation of the Chaco that had been accomplished by the 1920s was by Paraguay. Since the Paraguayans were in possession of the key points of the Chaco, Bolivians saw invasion as their best option.

The tension over the Chaco had built up for decades. In the 1920s there had been several armed clashes in the Chaco between the national military forces. Both nations feared that war was inevitable, and through the 1920s both of these small nations strained their economies to the fullest measure to raise large armies and to equip them with the best weapons they could buy on the world market. The armed forces of both nations had been trained by European officers and understood the important role that air power had played in World War I, so Bolivia and Paraguay endeavored to provide themselves with the most modern air forces they could afford. In the arms race the Bolivians had the distinct advantage. In 1930 Bolivia had a population of three million people to Paraguay's one million, and the Bolivian economy, supported by its rich tin and silver mines, was more than three times larger than Paraguay's economy, which was based on cattle, palm oil, and cotton.

In the 1920s Paraguay strained its national budget to equip its forces and bought arms and artillery from Europe. Paraguay built up its army, navy, and aviation forces and, by 1931, had about four thousand regular armed forces with the capability to mobilize approximately sixteen thousand more.[20] Bolivia, with a much larger population than Paraguay and a steady income from its tin and silver mines, was able to buy a considerable amount of modern weaponry in the decade preceding the war, and outfitted its six thousand-man army with modern machine guns, artillery, and even some British tanks.[21]

The commander of the Bolivian army in this period was German general Hans Kundt, a retired World War I general staff officer and brigade commander who had served in Bolivia as commander of the German military mission before 1914. Kundt never familiarized himself with the Chaco or made serious plans to fight a war there. Kundt was also reluctant to rely on his Bolivian officers—and he had some very good ones, preferring to micromanage the army. Kundt was removed from his post as chief of staff and war minister and was sent into exile for his role in an attempted coup in 1930. However, it was a great stroke of luck for the Paraguayans that Kundt was brought back by the Bolivians to serve as commander in chief after the start of the war.[22]

The Paraguayan commander in the Chaco was Gen. Jose Felix Estigarribia, an exceptionally capable officer who had taken the French general staff course. In 1931 he was appointed as commander in the Chaco. He was a quiet but intense man and was popular with the troops.[23] He would prove to be one of the great military commanders in Latin American history.

Bolivia set about creating an aviation corps in 1916 and established a flight-training center in La Paz in 1925 under the command of Maj. Bernardino Bilbao Rioja, one of Bolivia's first six military pilots.[24] The Bolivian Air Corps grew into a capable air force in the 1920s.[25] In 1927 Bolivia ordered nine Vickers Vespa fighters from Britain, very capable fighter planes for their time. Between 1926 and 1927 Bolivia took delivery of six French Breguet XIX bombers, which were very sturdy and capable aircraft. In 1929 Bolivia bought four more improved models of the Breguet XIX. In 1932 the Bolivians deployed an air group of three Vickers Vespa fighters, three Breguet XIX bombers, five or six Vickers Type 143 fighters, and several Fokker CVs to the border of the Chaco.[26]

The real beginning of a military air corps in Paraguay came with the arrival of a French Air Force mission in 1926 that trained and organized a small air force equipped mostly with French fighters and bombers.[27] At the outbreak of the Chaco War, the Paraguayans possessed an air force of about twenty-five pilots and a few dozen mechanics and fitters. The Paraguayans also possessed a handful of light utility and transport aircraft as well as the training aircraft based near Asunción. The Paraguayan Potez 25 bombers were on the verge of obsolescence in 1932, but in the 1920s this rugged, dependable, and versatile aircraft had proven itself in France's colonial campaigns. The commander of the Paraguayan Air Corps was an Argentine, Lt. Col. Vicente Almonacid, who had flown with the French on the Western Front in the First World War.

THE WAR BEGINS

In June 1932 the Bolivian Army seized a small Paraguayan outpost on Lake Pinantuten. In July Paraguayan troops retook the post and the Bolivian Army countered by taking the villages of Corrales and Toledo. Bolivian forces, supported by fighter and bomber aircraft, massed for an assault on the Paraguayan fort at Boquerón, which fell at the end of the month after heavy fighting. The war quickly escalated and by August, six thousand more troops were on their way from Bolivia to reinforce the six thousand men already in the Chaco. The Paraguayans were able to match the Bolivian troop buildup and José Félix Estigarribia built up an army corps at his forward base at Isla Poí. The Paraguayans built an airstrip at Isla Poí and deployed a small force of combat aircraft—all the combat power it could muster (about ten planes)—to the battle zone.

Although the Bolivians had vastly greater forces and resources than the Paraguayans, throughout the war the Bolivian Army fought at a huge disadvantage. Bolivia mobilized its forces in the heartland of Bolivia, then transported units and supplies by train to the main Bolivian base, the city of Villa Montes in the Bolivian lowlands. From Villa Montes Bolivian soldiers faced a march of two hundred to three hundred miles through the choking dust and heat of the Chaco until they reached the front lines. The heat and lack of fodder meant that horses did not survive long in the Chaco, and both armies were soon dismounted. This meant that the basic transport in the Chaco was the motor truck, and this was in short supply in both armies. Since there were only enough trucks for supplies, soldiers had to march for weeks to reach the front line and often arrived exhausted and malnourished.

In contrast, the Paraguayans had much shorter transport routes from their population centers. The Paraguayans could use river transport up the Rio Paraguay to their main base at Puerto Casada. From Puerto Casada there was a narrow-gauge railroad into the Chaco so troops and supplies faced a relatively short march to the front. During the war these transport factors worked to Paraguay's advantage and largely negated the Bolivian advantage in manpower and material.

In September Estigarribia went on the offensive and put the Bolivian garrison of several hundred men at Boquerón under siege. The opposing air force began aggressive patrolling and reconnaissance in order to spot enemy troop movements. The Bolivians, with their larger air force, had the clear advantage in the air and, on September 8, Bolivian Vespa fighters spotted the Second Paraguayan Infantry Regiment on the road toward Boquerón and repeatedly bombed and strafed the column, inflicting heavy casualties among the men and horses. Many Paraguayan soldiers panicked and disappeared into the brush and the regiment had to be reassembled over the next few days.[28]

This example of the effectiveness of Bolivian air superiority taught the Paraguayans valuable lessons. The Paraguayans became masters of camouflage to negate Bolivian air power. The Paraguayans used the thick brush of the Chaco to camouflage movements that outflanked the Bolivians. Trails were cut under the trees and the Paraguayans wove the branches of the trees together over a road to minimize the aerial signature. Paraguayan artillery was the prime target of the Bolivian air patrols, but the guns were especially hard to spot if camouflaged and carefully dug in. The Paraguayan efforts paid off, and Bolivian air patrols subsequently failed to give the Bolivian high command a clear picture of Estigarribia's offensive operations.[29]

The Paraguayans were able to block the attempts to relieve the Boquerón garrison and it surrendered on September 10.[30] This Bolivian defeat at the outset of the war gave the Paraguayans confidence and shook Bolivian morale. The Bolivian Air Corps commander insisted that his forces should attack the Paraguayan capital at Asunción to break the Paraguayan morale. It would have been easy, as the Bolivian air units at the Muñoz and Ballivían airfields were within easy range of Puerto Casada and Asunción, and both cities had few antiaircraft defenses. But the Bolivian High Command canceled the plan because bombing Asunción would have been likely to cause an international outcry against Bolivia.[31] Instead, the Bolivians bombed the Paraguayan base at Puerto Casada. The air raids provoked a strong reaction from the Argentine government, as many Argentines lived in Puerto Casada and ran the railroad into the Chaco (and supported the Paraguayan war effort). After the bombing of Puerto Casada, Argentina told the Bolivian government that it would not accept casualties among its civilians and dropped hints that it might enter the war on the side of Paraguay if attacks persisted. Bolivia subsequently called off the air attacks.

During the fighting of 1932 both air forces conducted aggressive reconnaissance and patrolling and during the Boquerón campaign both sides carried out numerous attacks in support of the ground troops.[32] Air-to-air fighting was relatively rare. Still, conditions in the Chaco were bad enough to inflict a steady attrition. Both air forces lost planes to accidents on the rough Chaco airfields and obtaining replacement aircraft became a priority for both governments.[33] The Bolivians ordered twenty Curtiss Wright C14R Osprey two-seater reconnaissance and light bomber biplanes from the United States and planes began arriving in January 1933. The Osprey had one forward and one rear .30 caliber machine gun and could carry up to 260 pounds of bombs in various configurations. It was a very capable ground attack aircraft for its day and the Bolivians used it as a fighter-bomber throughout the war. Also ordered from the Curtiss Corporation were Hawk II and Sea Hawk fighter planes. Four were delivered in 1932, two in 1933, and three in 1934. This single-seat fighter was among the best fighters of its day and it had an armament of two forward-firing machine guns and a bomb load of 220 pounds. It was also used as a fighter-bomber.[34] An important addition to Bolivian air power was the purchase of three Junkers Ju 52 trimotor transports from Germany. This transport could operate in the roughest conditions and could carry a load of three tons. The first arrived in October 1932 and the other two in 1933, and the Junkers became the backbone of the Bolivian air transport service.[35]

The Paraguayans ordered five Fiat CR 20bis biplane fighters from Italy to replace the French Wibault fighters, whose engines tended to overheat. The Fiat fighters were capable aircraft, but were generally rated as inferior to the

Bolivians' Curtiss Hawk fighters. Thus, the Bolivians were able to maintain the qualitative edge in aircraft throughout the war.

Finding replacement planes was made difficult by a League of Nation and United States embargo on arms sales to both combatants.[36] However, Bolivia and Paraguay showed considerable ingenuity in evading international controls and importing enough aircraft to keep their air forces flying. Bolivia relied upon support from Chile, which had bought the license to assemble Curtiss aircraft, including the Curtiss Falcon. Chile ignored the League of Nations embargo and sold Falcons to Bolivia and these replaced the Vickers Ospreys lost to combat and accidents.[37] Curtiss Hawks and Sea Hawks, Bolivia's best fighter planes, were also acquired through Chile.

Under the embargo, France impounded nineteen aircraft ordered by Paraguay. However, the Paraguayans showed as much ingenuity as the Bolivians in obtaining aircraft. Seven Potez 25s sold by France to Estonia were mysteriously rerouted and shipped to Paraguay.[38] During the war, Uruguay and Argentina connived to support the Paraguayan aircraft purchases, and trainers, transports, and liaison aircraft were obtained by Paraguay through Argentine sources.

When Kundt took over the Bolivian Army in late 1932 he ordered a series of fruitless frontal assaults against well-dug-in Paraguayan forces at Nanawa. Through the first half of 1933 the Bolivians maintained the offensive against the Paraguayan defenses at Nanawa. Although the Bolivians had the advantage in artillery and used their aircraft to bomb the Paraguayan front lines, the Bolivian Air Corps failed in the mission to find and knock out the Paraguayan artillery because it was almost impossible to find the well-camouflaged Paraguayan gun positions deployed in the thick bush. The Bolivian main assault of July 1933 was a bloody failure with over 2,000 Bolivians killed for the loss of only 149 Paraguayan dead and 400 wounded.[39]

After defeating the Bolivian offensive at Nanawa, Estigarribia looked to strike a major blow. In November 1933 Paraguayan air and ground patrols spotted some major gaps in the Bolivian lines at Campo Via that offered the Paraguayans a chance for a bold move. Estigarribia quietly concentrated his forces, and on December 3 unleashed a double-envelopment maneuver that surrounded the Bolivian Fourth and Ninth Divisions. General Kundt reacted slowly to the crisis. Bolivian reconnaissance pilots gave accurate reports about the strength and locations of the Paraguayan troop movements that would have allowed the Bolivians to react, but Kundt rejected the reports as alarmist and inaccurate because he was convinced that the Paraguayans were not ready to conduct operations on a broad front.[40] Kundt's failure to react caused Bolivia's greatest defeat. On December 11, the two surrounded Bolivian divisions surrendered. Over twenty-six hundred soldiers were killed and approximately seventy-five hundred taken prisoner. At one stroke, most of the Bolivian combat forces had been destroyed. Kundt was removed as the commander in chief of Bolivia's forces and the Bolivian Army went into a headlong retreat. The Campo Via victory provided the Paraguayans with a vast amount of captured material and allowed Estigarribia to maintain the offensive.[41]

With Kundt gone the Bolivian morale and effectiveness improved. As the Paraguayan logistics lines lengthened, their advance slowed and the Bolivians established a defensive line in the central Chaco. In May 1934 the Bolivians counterattacked and trapped the Paraguayan Second Division at Cañada Strongest. The Paraguayans fought their way out of the pocket, but in the process five hundred died and fifteen hundred were taken prisoner.

Estigarribia looked for a means to again outmaneuver the Bolivians and found it in November 1934 when his surprise flank attacks surrounded the Bolivian Reserve Corps. Two thousand Bolivian troops died and four thousand prisoners were taken. Only two thousand fought their way out of the pocket to safety. The Bolivian Air Corps performed brilliantly and prevented the destruction of the retreating Cavalry Corps with a series of relentless and effective air attacks that slowed down the pursuing Paraguayan forces and allowed the army corps to retreat in good order.[42]

By early 1935 the Bolivian Army had withdrawn from the Chaco and had fallen back to Villa Montes. The Paraguayans maintained the pressure on the Bolivians and crossed the Rio Parapiti into Bolivia in April 1935. In a series of savage counterattacks very effectively supported by the Bolivian Air Corps, the Bolivians threw the Paraguayans back across the river. By this point, both nations were exhausted. An armistice was signed on June 14, 1935, ending the war. Subsequent negotiations recognized Paraguay's claim to the Chaco.[43]

Assessing the Air War

Aircraft and pilot attrition during the war was high for both sides, and operational accidents accounted for most deaths.[44] During the war, Bolivia operated between fifty-seven and sixty-two combat aircraft and twenty-two trainers and transport aircraft. Paraguay operated thirty-two combat aircraft and twenty-three training and transport aircraft. According to the official Paraguayan figures, Paraguay lost eight aircraft in combat during the war and Bolivia lost ten.[45] The primary cause of combat losses was ground fire.

Air power played a major role in the Chaco War and both air forces demonstrated tactical competence and considerable ingenuity. Both sides employed aircraft mostly in reconnaissance and ground attack roles, although some long-range bombing raids were also carried out against enemy air bases and supply depots.[46] While the fighter and bomber squadrons of the two air forces acquired most of the glory, the transport and utility aircraft of both sides played a very important role. The lack of infrastructure in the Chaco made air transport a necessity for transporting supplies, and both sides used a variety of transport and light aircraft to support their forces. Bolivia had a fairly impressive air transport capability and it was used to the fullest throughout the war to transport ammunition, fuel, and medicine to the front. During the war, the Ju 52s alone carried over forty-four hundred tons of cargo to the front. Paraguay also pressed into service all of its transport and utility aircraft. Both nations used aircraft to transport the sick and wounded from front-line airfields to hospitals in the rear. The Ju 52s transported an estimated forty thousand wounded and sick Bolivian troops to the rear and the Paraguayan transports were also used as air ambulances. Many thousands of lives were saved by a well-conceived air evacuation system.[47]

THE PERU-ECUADOR WAR OF 1941

Since the nineteenth century Peru and Ecuador have fought conflicts over the southern border. The area most in contention was the eastern corner of Ecuador in the Amazon basin. After years of tension, in 1941 Ecuador occupied the disputed town of Zarumilla in order to maintain its access to the Amazon region. Peru was far stronger and much better prepared for war, and in this minor Ecuadorian military action the Peruvians saw the justification to initiate major hostilities.

From the start there was no doubt as to the outcome. Ecuador had only eighteen hundred troops in the contested border region and Peru had thirteen thousand men ready to move on the Ecuadorian border. Ecuador had no air force, whereas Peru had a capable small air force consisting of a fighter squadron of North American NA-50 fighter planes and a bomber squadron of Caproni 310 bombers. In addition, an Italian military mission in Peru had trained a Peruvian paratroop unit, the first such unit in Latin America.

The war opened with skirmishing along the Zaramilla River on July 5, and by July 6 the Peruvian Air Force was engaged in close support operations all along the front. The Ecuadorians attempted to counterattack, but the Peruvians were able to advance steadily under cover of their air force. In addition to bombing the Ecuadorians, the Peruvian air units conducted leaflet drops over the Ecuadorian capitol urging the civilians to surrender. The ability of the Peruvians to fly unopposed over their capital unnerved the Ecuadorians.

On July 27, 1941, the Peruvians carried out a bold stroke by dropping a paratroop company at the vital port city of Puerto Bolivar. The Ecuadorians were completely unprepared for such a move, as it was the first combat airborne operation in the Western Hemisphere. The Peruvians secured the city and held it until the advancing ground forces could enter almost unopposed. Four days later the Ecuadorians agreed to a cease-fire. During the three-week war Ecuador lost the whole coastal province of El Oro and some towns in the Andean province of Loja. Although the overwhelming force of the Peruvians almost guaranteed victory, thanks to their air force and a well-conducted airborne operation the war was a very quick and decisive one.[48]

LATIN AMERICAN AIR POWER IN WORLD WAR II

During World War II, the United States provided arms, aircraft, training, and support to allied nations in Latin America. Mexico and Brazil, as active combatants, received the most extensive aid and training. The initial effort was organized around the mission of fighting the German U-boat menace in the Caribbean and Atlantic. Later, the United States carried out a large-scale program to build up the Latin American allies (virtually all of Latin America aside from Argentina) and, by 1945 the Latin American nations were generally equipped and trained on the U.S. model. Between 1941 and 1945 the Americans provided over two thousand aircraft under the Lend Lease program to Latin America.[49]

The anti-submarine campaign of 1941–42 was the beginning of the U.S.–Latin American security partnership. When America was thrust into the war the Caribbean area became a vital part of the Allied war effort. The region was a source of oil to America and Europe and it contained some of the busiest shipping lanes in the world. However, the region was virtually undefended. In 1941 the United States set up a Caribbean defense headquarters and built and expanded air bases in Florida, Puerto Rico, Panama, and the British colonies of St. Lucia, Antigua, Trinidad, and Jamaica. In 1941 there were fewer than 250 planes available for the defense of the region, and these were mostly short-range obsolete fighters of limited effectiveness in the anti-submarine war.

As the undersea war heated up and the shipping losses became catastrophic, the Latin American nations joined in the effort to combat the U-boat threat.[50] The regional military response was based on the Rio Conference agreements of January 1942. Aircraft were the best anti-submarine weapons and several Latin American nations made airfields available to the Americans. Virtually every nation in Latin America cooperated with the United States by providing overflight rights and airfields for American use. The need for common defense overruled nationalist and anti-American sentiment, even in Mexico. The United States dispatched B-18 medium bombers, A-20 light bombers, and patrol planes to Latin American bases in January 1942. Guatemala allowed the Americans to base planes at Guatemala City. Mexico was reluctant to allow Americans on its soil, but when German U-boats sank two Mexican ships in May 1942, Mexico declared war on Germany and allowed the Americans to operate anti-submarine patrols from the airfield at Cozumel. That same month, the Mexican Air Force began anti-submarine patrols against U-boats and five U.S. Navy patrol bombers were transferred to Mexico to help in the anti-submarine campaign.

In the desperate days of 1942, anything that could fly was pressed into service to hunt U-boats. In July 1942 a Mexican Air Force patrol flying modified T-6 trainers attacked and damaged a U-boat off the coast of Tampico. Even the small Cuban Air Force flew patrol missions with light planes. When they spotted U-boats they tracked them and radioed the position to an American bomber unit that could attack with heavy depth charges. Five U.S. Navy anti-submarine patrol squadrons were also based in northern Brazil to cover the mid-Atlantic shipping lanes that came under attack.[51] Due largely to the urgency of the anti-submarine campaign, the United States expanded its plans to provide aircraft to Latin America.

After their initial successes, the Germans found it more and more difficult to operate in Caribbean waters. Although few U-boats were confirmed sunk, the constant air surveillance and air attack drove the Germans to waters where the kills were easier. The Caribbean anti-submarine campaign was the first instance of U.S.–Latin American direct military cooperation.[52] This cooperation between the United States and all the Latin American nations, excepting pro-Axis Argentina, became the basis for a series of close relationships and defense agreements after the Second World War. In the meantime, the United States cooperated closely with Mexico and Brazil to help those nations stand up air units that could serve with Allied forces in combat abroad.

Brazil made the largest contribution to the Allied cause of all the Latin American nations. In 1943 the Brazilian government decided to stand up a division-sized expeditionary force to serve with the Allied forces in Germany. The Brazilian force was sent to Italy in 1944 and served with the U.S. Fifth Army. An air unit of the expeditionary force was stood up in December 1943 as the First Fighter Group of the Brazilian Air Force. The unit was sent to the United States for advanced training and was equipped with the American P–47D fighter. The First Fighter Group, a squadron-sized force with more than forty pilots and three hundred ground crew, arrived in Italy in October 1944 and served as part of the American 350th Fighter Group. The First Fighter Group saw little air-to-air combat as the Luftwaffe fighter force had been effectively destroyed in Italy. Like many other Allied fighter units in Italy, the First Fighter Group became a close support and ground attack unit. The Germans had a large and very capable antiaircraft force in Italy and the ground attack operations of the First Fighter Group were often very hazardous. Pilot and aircraft attrition throughout the campaign was fairly heavy as the Brazilian fighter unit flew constant missions attacking German transport and supply points. When the Allies went on the final offensive to overrun northern Italy in 1945 the First Fighter Group earned the respect of their American colleagues by inflicting heavy casualties on the retreating Germans.[53]

When Mexico went to war in 1942 it was reluctant to do more than to counter the German U-boat threat. However, as the war progressed, the Mexican government decided that it was in the best interest of Mexico to make a more substantial contribution in the form of creating an air squadron that would serve in active combat with the Americans. In March 1944 the Mexican Air Force formed the 201st Fighter Squadron, a unit of 42 officers and 249 enlisted men, and in June 1944 dispatched the unit for training in the United States. After ten months of training in the United States, the 201st Squadron was deployed to serve with the Fifth U.S. Air Force fighting in the Philippines. The 201st Squadron, nicknamed the "Aztec Eagles," was equipped with the rugged P–47 fighter-bomber and was assigned to serve with the American Fifty-eighth Fighter Group supporting the ground campaign against the Japanese forces in Luzon. Through May and June 1945 the 201st carried out close air strikes on Japanese ground troops and carried out its missions effectively. In its brief time in combat the 201st lost several pilots killed in combat and operational accidents.[54]

POST–WORLD WAR II COVERT AIR OPERATIONS

In the post–World War II era the United States was primarily interested in keeping Latin America free of Communist influence. Since the old style of military intervention was no longer acceptable to international opinion, a more subtle form of intervention was developed in which the Central Intelligence Agency (CIA) would covertly equip and support anti-Communist factions to overthrow leftist regimes. No American forces were directly involved and the covert approach allowed U.S. interests to be furthered at low cost and with less risk of international or public resistance.

Guatemala, 1954

In the early 1950s it looked as if Guatemala was ripe for a Communist takeover. The leftist president of Guatemala, Jacobo Arbenz, had nationalized American property and was close to the Guatemalan Communist Party. It was reported that he was obtaining weapons from the Soviet bloc. In 1954 President Dwight Eisenhower approved a CIA plan to overthrow President Arbenz. The CIA financed a force of rightist Guatemalans that was set up in Nicaragua under former army colonel Castillo Armas, a political enemy of Arbenz. Part of the rebel force included a small air force of war-surplus aircraft, including four B-26 bombers, three P-47 fighters, and two C-47 transports. American and Nationalist Chinese pilots flew the planes.

On June 18, 1954, Armas's small rebel army invaded Guatemala from Honduras. The rebel air force bombed military installations in Guatemala City and dropped leaflets over the capital. The rebel air force, which was superior to the regular Guatemalan Air Force, played a key role in demoralizing the Arbenz supporters and convincing the Guatemalan military to offer only minor resistance to the rebels. There were a few skirmishes and a few more air raids and Arbenz resigned the presidency and left the country on June 27.[55] It was one of the most successful CIA operations of the era. While there was some Latin American outcry against the obvious U.S. involvement in the coup, it was relatively minor. A pro-Communist regime had been overthrown and replaced by a pro-American regime at a very small cost.

The Bay of Pigs, 1961

When Fidel Castro came to power in Cuba in 1959 the United States had serious grounds for alarm. While Arbenz's relationship with the Communists was tenuous, Castro instituted a full-scale dictatorship in Cuba and openly allied himself with the Soviet bloc. Hundreds of thousands of refugees flooded out of Cuba and a large resistance movement was organized inside Cuba. The Eisenhower administration began planning to overthrow Castro's regime with a covert campaign just like the one that had proven so successful in Guatemala. A force of eighteen hundred Cuban exiles was organized and trained secretly in Nicaragua to carry out an invasion of Cuba. The invasion was expected to ignite a national rebellion against Castro and lead to an overthrow of the regime.

A key part of the plan was the creation of a small covert air force for the exile Cubans. Twenty surplus B-26 bombers and some C-46 transports were obtained by the CIA and a group of former U.S. military pilots and aircrew who had flown in B-26s were recruited. The key mission of the covert air force was to destroy the Cuban Air Force, now called the Revolutionary Air Force (FAR), by surprise attacks on its main airfields. The CIA planes would be painted in FAR markings to give the impression that some Cuban pilots had rebelled. Once the FAR was destroyed, the exiles would have the freedom to land at the Bay of Pigs on the south coast of Cuba.

When the Kennedy administration came to power in early 1961 it was alarmed by the large arms shipments that the Soviet Union had poured into Cuba. The Cubans had not only received tanks and artillery, but were also awaiting MiG-15 and MiG-17 jet fighters from the Soviet bloc. Kennedy believed the time had come to stop Castro and ordered the operation to proceed.

The whole operation was one of the worst planned and executed military operations ever carried out by the United States. The intelligence assessments that Cuba was eager to overthrow Castro were wildly overoptimistic. The efficiency of Castro's army and secret police was underrated. Castro had armed and equipped a force of two hundred thousand regular troops and militia with light and heavy weapons supplied by the Soviet Union and was ready to oppose any U.S.-led invasion.[56] Castro also had a fairly capable air force that included eight T-33 fighters, twelve B-26 bombers, twelve Sea Fury FB 11 fighters, ten C-47 transports, and eight helicopters.[57]

The operation began on April 15, 1961, when six of the exile air force B-26s attacked the three main FAR bases. The attack went badly wrong. The strike destroyed five of the FAR planes and damaged others while two of the exile planes were lost. The attack failed to deliver a knockout blow to the FAR, which was now alerted to be ready for invasion. With Castro alerted to an attack he put his own plans into motion, and had his secret police immediately arrest one hundred thousand likely political opponents all across Cuba. In one blow, the resistance movement that was expected to rise up was crippled.[58]

Rather than call off the invasion, the CIA force, carried by five civilian cargo ships, landed at the Bay of Pigs the next day. That day, Castro's air force attacked and the FAR's Sea Furies sank the SS *Houston*, one of the invasion ships. Later that day the SS *Rio Escondido* was sunk by Castro's air force for the loss of one Sea Fury.

The exile air force persisted with its operations and the C-46 transports successfully dropped a small paratroop company to support the beach landings. B-26s tried to provide close support, but suffered further losses from Castro's fighter planes. The FAR's successful air attacks had so disrupted the landings that the Cuban Army was given time to mobilize forces and pen the invaders into a small beachhead. Air attacks forced the surviving exile ships out to sea and in four days the exile brigade was forced to surrender.

THE SANDINISTA WAR, 1979–90

After the Anastasio Somoza dictatorship in Nicaragua was overthrown by a broad national coalition in 1979 the Marxist elements of the revolution, called the Sandinistas, seized power, created a pro-Soviet one-party state and began suppressing all the other political parties in the country. By 1980 a new civil war had broken out as anti-Marxist Nicaraguans organized themselves inside the country and in neighboring Costa Rica and Honduras.

After Ronald Reagan assumed the U.S. presidency in 1981, the Americans began a program of covert financial and military support to the opponents of the Marxist Daniel Ortega regime in Honduras. A strong U.S. ally in the region, Honduras allowed the anti-Marxist forces, known as Contras, to establish bases in Nicaragua where the United States and other Latin American nations provided aid, arms, and training. Once trained and armed, rebel groups would infiltrate back across the rugged mountainous borders of Nicaragua and would establish themselves in the villages and wage a guerrilla war against the Sandinista government.

In the mid-1980s the CIA operated a variety of light civilian transport aircraft in support of the Contras. The covert planes would drop supplies inside Nicaragua for the Contras. Although the Sandinista Air Force made a major effort to interdict these flights, only one was ever shot down.

For their part, the Sandinistas were provided with a modern air force by the Soviet Union and employed a force of Hind 25 heavy-attack helicopters as its main strike force against the Contra rebels. After a decade of struggle, the two sides were locked in a stalemate. The Contras could not hope to mount a conventional offensive in the face of the Sandinista air power. On the other hand, the Contra forces remained in the field and largely controlled the rural areas of northern Nicaragua. As a result of years of negotiations the Sandinistas allowed a free election, which they promptly lost by a landslide. The conflict ended with a centrist democratic government taking power.[59]

POST–WORLD WAR II CONVENTIONAL WARS

EL SALVADOR–HONDURAS WAR, 1969

El Salvador and Honduras have long been rivals. Through the 1960s three hundred thousand Salvadoran peasants had left their overpopulated country and settled in less-populated Honduras. Along the border there were more than twenty villages claimed by both nations. In 1969 Honduras took action to try to return many of the illegally settled Salvadorans and tensions flared up between the countries. After a series of riots at Salvadoran/Honduran soccer games, the nations broke diplomatic ties and on July 10, 1969, the Salvadoran Air Force (*Fuerza Aerea Salvadorena* or FAS), attacked the main Honduran air base near Tegucigalpa, destroying and damaging several aircraft. At the same time twelve thousand Salvadoran troops invaded Honduras. The FAS, equipped with ten F-51s and five B-26s, flew in support of the army.

The Salvadorans, with their larger and better equipped army, drove the Hondurans back. But the Hondurans had a capable air force with twelve F4U Corsairs and eight transports and this force struck back at the Salvadorans on July 15, bombing the main Salvadoran airfield at Ilopango and the national oil refinery. Air-to-air combat ensued, and the Hondurans were victorious, as Honduran Corsairs shot down two Salvadoran F-51s.

Although the Salvadorans continued their advance the Honduran air strikes demonstrated that the Hondurans were ready to strike major targets inside El Salvador. On July 18 the Organization of American States called for a cease-fire and both nations accepted. Over the next few weeks the Salvadorans withdrew behind the pre-war border and El Salvador and Honduras settled the border issues through diplomacy.[60]

Air Power in the Falklands War

The Falklands/Malvinas War of 1982 is a notable one for airmen. The war pitted two modern and capable air forces and naval air arms against each other. The decisive battle that determined the fate of the islands was fought in the air with the ground war largely a sideshow. Indeed, if the Royal Navy had not been able to fend off Argentine air attacks, no British landings on the Falklands would have been possible.

The Falklands had been a problem ever since Britain had seized the islands in the 1830s.[61] Negotiations between Argentina and Britain were in progress in 1982 but the Argentine military junta feared that Britain would garrison the islands. Seeing a window of opportunity to act before the British sent a significant force to the Falklands, the junta ordered the islands to be seized. On April 1, 1982, Argentine troops landed on the Falklands. Argentina expected the British to quietly cede sovereignty of the islands, but instead, Britain decided to mobilize its forces to retake the small, isolated colony.

The Argentine junta had been so sure that Britain would accept its fait accompli that no plans had been made to defend the islands or repel a British task force. With a powerful British task force assembling to retake the islands, the Argentines had only a few weeks to cobble together a force and create a plan to defend the Falklands. The Argentines created a theater command under Vice Adm. Juan Lombardo for command of Argentine naval units and the Falklands garrison, which amounted to two brigades and over ten thousand soldiers by the end of April. The Argentines also set up a special force called the Southern Air Force (*Fuerza Aerea Sur*, or FAS) under the command of air force Brig. Gen. Ernesto Horacio Crespo, a highly experienced pilot and commander, and he was given the mission of attacking the British fleet.[62]

Argentina had a large, relatively modern air force. The *Fuerza Aerea Argentina* (FAA) possessed first-rate front-line aircraft including Mirage III interceptors. The naval air arm was acquiring a squadron of Super Étendard fighters from France. The FAA had Israeli-made Mirage 5 fighters (called Daggers), Mach 2 aircraft effective in both the air-to-air and strike roles. The primary attack aircraft of both the FAA and the navy were several dozen A-4 Skyhawks bought from the U.S. Navy in 1972.[63] The FAA possessed eight Canberra bombers, a small transport force, and several squadrons of IA-58 Pucaras. The Pucara was a twin-engine turboprop attack aircraft designed and manufactured in Argentina. It could mount a 30mm cannon and a variety of bombs. It had the advantage of being able to operate from small, rough airstrips. The naval air arm had Aero-macchi 339 jet trainers, a small aircraft configured as a light strike fighter. The pilots of the FAA and naval air arm were well trained and the services had a good infrastructure and ground units that could effectively repair and maintain aircraft.[64] However, the FAA had been designed to fight Argentina's main enemy—Chile. With this in mind, the force was trained for short-range operations and for supporting ground troops. Only a few units of the Argentine Navy were trained to fight a naval campaign.

In a war fought at long range the Argentines were at a tremendous disadvantage. The small airfield at Port Stanley was the only hard-surface airfield in the Falklands and its short length, 4,500 feet, meant that transport planes such as the C-130s could land and operate, but the FAA's high-performance jets could not. So Argentina's best air units were all based on the Argentine mainland—with the four main bases all between five and six hundred miles from the Falklands. This meant that the Argentines would operate at the most extreme limits of their endurance. Planes sent to attack the British fleet or support ground forces in the Falklands would have a long and stressful flight, and then only five minutes over the target area before they had to return home or run out of fuel.

Although aerial refueling was the answer, the FAA had only two tankers (KC-130s) to serve the whole air force and navy. This was the result of concentrating all its military planning on neighboring Chile, an anticipated conflict with short ranges. Air Force and Navy A-4 Skyhawks were equipped for aerial refueling, but the Mirage IIIs and Daggers were not, so this reduced their ability to attack or serve as fighter cover. The Argentineans also lacked long-range reconnaissance aircraft. Another deficiency was armament. The FAA's primary air-to-air weapon was the French-made Matra 530 infrared air-to-air missile, which was far inferior to the U.S.-made AIM-9L Sidewinder that was carried by the UK Harriers.[65] The British AIM-9 had a wider field of vision (90–120 degrees) and a much more sensitive infrared seeker that could lock on to the friction heat generated by the front of the enemy aircraft in flight. Unlike the Argentines, Harrier pilots did not have to get behind their opponents to fire a deadly shot; they could even engage the AIM-9s coming head-on.[66]

With only a few weeks before the British task force arrived General Crespo prepared his aerial strike force. The Skyhawks and Daggers made simulated bombing runs against the Argentine Navy destroyers while the warships simulated missile defense and made evasive maneuvers. The FAS was deployed to four air bases within range of the Falklands.[67] In all, the FAS had approximately 110 combat aircraft (including the Pucaras) based on the Argentine mainland with an additional twelve naval strike aircraft available.

The Argentines sent ten thousand troops to the Falklands, with the largest force (seven thousand men) on East Falkland Island near Port Stanley. Because of the British naval blockade, the whole force was dependent upon an FAA airlift. The FAA had seven C-130s and a few Fokker F-27 twin-engine light transports. To reinforce the effort, every national airline aircraft that was capable of landing at Port Stanley was pressed into service to ferry the troops and equipment to the islands. The FAA transport force performed extremely well and the airlift effort of the FAA to support the forces in the Falklands lasted until the last day of the campaign.

A sizeable air force was deployed to the Falklands to serve under the command of Army Gen. Mario Menendez—and not under the command of the FAS. Nineteen army, navy, and air force helicopters were sent to the Falklands to serve in the reconnaissance and trooplift roles.[68] For strike operations, twenty-four Pucaras were ordered to the islands and the navy air arm sent six Aeromacchi 339 light strike aircraft and six T-34B Mentors.

The opening shots of the campaign were fired on May 1, 1982, when the first wave of the British invasion force arrived in Falklands waters. The British task force, under the command of Adm. John Woodward, was built around two light carriers (HMS *Hermes* and HMS *Invincible*), over twenty destroyers and frigates, and a host of troopships and support vessels carrying a British brigade with full equipment. The carriers each had a complement of Royal Navy Harrier jets and helicopters. The British first wave consisted of sixty-five ships protected by modern antiaircraft missile systems.[69] However, the primary British strike and defense weapon was the force of twenty-one Harrier jet Vertical Take-off and Landing (VTOL) aircraft. The small Royal Navy Harrier force would be bolstered by a further fourteen RAF Harriers coming later on two large cargo vessels that had been modified with flight decks for VTOL Harrier operations. Four more Harriers were flown from Ascension Island, with numerous tanker refuelings, to reinforce the British late in the campaign.[70] The Harrier was more modern than anything the FAA flew and, although it had a short range, it could fly as a Combat Air Patrol (CAP) over the fleet for forty minutes to an hour. This was a significant time advantage over the Argentine attackers who only had a few minutes to find their targets and engage the enemy. The Royal Navy maintained a CAP of two Harriers equipped with extremely lethal AIM 9L Sidewinder air-to-air missiles over the fleet. The small number of Harriers available for fleet defense meant that Argentina's best hope for success was to attack the fleet while the Harriers were diverted or on deck refueling.

On May 1, RAF Vulcan bombers, flying from the British base at Ascension Island thousands of miles away, bombed the Port Stanley airfield, and Harriers struck the Goose Green and Port Stanley airfields.[71] The FAS, now alerted to the British fleet, sent waves of strike aircraft to attack the British ships. The *Fuerza Aerea Sur* could not send a large strike force to overwhelm the British air defenses because of range and bomb-load restrictions. To carry a one-ton bombload for six hundred nautical miles, Skyhawks needed aerial refueling. With only two air tankers the FAS could send only small flights, usually four aircraft at a time. Each flight had to be carefully planned to make the refueling rendezvous.[72]

On the first day almost all of the FAS strike forces went into action. British warships shelling Port Stanley were attacked by a flight of Daggers dropping bombs and strafing with cannon. The Argentine pilots reported that one ship had been heavily damaged and that two others had received varying degrees of damage. The Argentines claimed a triumph in damaging three ships and shooting down at least five Harriers. Heartened by their perceived success despite the loss of five aircraft and others damaged, the FAS prepared to mount more strikes.[73] In reality the British had lost no planes and suffered only minor damage to one ship. Because the FAS had no means of conducting a battle-damage assessment they had to rely upon pilot and antiaircraft gunner reports—which consistently overestimated the effect of both the Argentine defense and air strikes. Thus, the Argentines were largely in the dark about the true condition of their enemy through the whole campaign.

On the other hand, the British had the advantage of good reconnaissance capabilities. One must also assume that the United States provided its ally with satellite intelligence photos of Argentine airbases that allowed the British to count and identify enemy aircraft on mainland runways. Because the FAS lacked long-range reconnaissance assets the Argentines had few means to locate the British ships. The lack of reconnaissance assets, combined with the often poor weather, meant that approximately one-third of the Argentine aircraft sent to strike the British in the course of the campaign returned without making contact.

On the second day of the campaign, May 2, the Argentine cruiser *General Belgrano* was sunk by the British nuclear submarine HMS *Conqueror*. After this action, the Argentine Navy remained in port and the Argentine Navy's aircraft carrier, which might have been a potent threat to the British Fleet, was not used.

Argentina's most formidable weapon was the Exocet missile. The FAS had five Super Étendard fighters configured to carry the airborne Exocet missile.[74] The radar-guided Exocet could be fired at long range, nearly thirty miles, and carried a 950-pound warhead. It streaked along at almost Mach-1 speed at a very low level—just over the wavetops—and it was difficult to shoot down. If it struck its target, the result was usually devastating. Fortunately for the British, only five of the airborne Exocets had arrived from France before the Common Market nations and NATO arms embargo kicked in. Argentina could not obtain any more.

On May 4 an Argentine plane identified what it believed was British carrier HMS *Hermes* east of Port Stanley. Two Super Étendards, each armed with one Exocet, took off for the long flight. The Étendards mistakenly picked up the destroyer HMS *Sheffield* and fired both Exocets at long range. One Exocet found the target, crippled the *Sheffield*, and caused heavy casualties. The *Sheffield* was later abandoned and sank.

From May 1 to May 20, the British task force carried out a systematic campaign of bombing and shelling Argentine installations and forces in the Falklands. The British lost three Harriers from combat and accidents.[75] On May 21, the main British ground force landings were in progress and that day the Argentines sent virtually the whole air strength of the FAS to attack the British ships, about seventy-five aircraft. Flying in flights of four, Argentine Skyhawks and Daggers dropped to a one-hundred-foot altitude for the last one hundred miles to San Carlos Bay. High hills screened Argentine aircraft from detection until the last moment when they popped up over the hills and began the bomb runs. The frigate HMS *Ardent* was damaged in an early attack and sunk by a second Argentine attack. Four other ships were damaged but the Argentines paid a fearful price. The British downed nine FAS aircraft (five Daggers and four Skyhawks).

May 25, Argentina's Independence Day, saw a major air effort by the FAS. One Exocet strike aimed at the HMS *Invincible* went astray; the missile's radar was diverted by large amounts of chaff and locked on the cargo ship *Atlantic Conveyor*. The *Atlantic Conveyor* was hit and later sank. The FAS lost three aircraft in the morning. In the afternoon Argentine Skyhawks sank the destroyer HMS *Coventry*. May 25 was the worst day for the British in the campaign. Yet most of the two British ground force brigades were ashore and ready to mount the final offensive against the Argentine ground forces.

The twenty-four FAA Pucaras based in the Falklands were whittled down by strikes on the Port Stanley airfield and in air-to-air combat. But the remaining planes at Port Stanley and the FAS were still ready to attack the British fleet whenever the weather was clear. On June 8, the troopships *Sir Galahad* and *Sir Tristram* were unloading troops of the Welsh Guards at the port of Fitzroy when five Daggers and five Skyhawks hit both ships, which caught fire and were abandoned. Later that afternoon Skyhawks caught the landing vessel *LCU F4* sailing and sank it. However, the Harrier CAP caught up with the Skyhawks and promptly shot down three with Sidewinders.

The British ground units began their offensive on May 28, when they surrounded and forced the surrender of the Argentine garrison at Darwin. During the last two weeks of the campaign the Harriers carried out numerous close air support missions to help the British troops. The British methodically rolled up the Argentine Army until the last forces were cornered in a perimeter around Port Stanley by June 8. With no hope of relief, General Menendez surrendered with over eight thousand men at Port Stanley on June 14. The British had won the war.

Although the British won, the record of the FAS in the Falklands War was also impressive. The only Argentine senior commander who demonstrated real competence and professionalism in the Falklands War was the FAS commander General Crespo. Given the technological inferiority of the Argentine Air Force and naval air arm, the problems of range, the lack of tankers, and the lack of reconnaissance assets, Crespo did very well with the forces available. The FAS inflicted considerable damage upon the British. Indeed, most British losses in the war came not from the ten-thousand-man Argentine army force, but from the hundred-plus FAS aircraft.[76] The FAS paid a very heavy price and during the course of operations the FAS lost 41 percent of its aircraft to combat and operational accidents. Despite this heavy attrition rate the morale and fighting spirit of the FAS never broke.[77]

Peru versus Ecuador, 1995

The border between Peru and Ecuador had been a point of dispute for decades when war erupted in 1995. Through the 1960s and 1970s, Peru had worked to become a South American superpower, building up large armed forces equipped primarily with heavy weapons from the Soviet bloc. In the 1960s and 1970s, the

Peruvian Air Force (FAP) acquired a squadron of supersonic Mirage 5s from France. In the 1970s, Peru bought thirty-six American Cessna A-37s and, in 1974, Peru signed an arms deal with Moscow and purchased thirty-two Sukhoi Su-22 ground attack planes. In the early 1980s Peru acquired Mi-8 utility helicopters and Mi-25 attack helicopters from Soviet allies. In addition to all these purchases, Peru had a fleet of Canberra bombers and a squadron of Mirage 2000P fighters.

Although the Peruvian forces looked impressive on paper, an economic meltdown in the 1980s had undermined the armed forces. From 1988, Peru lacked the funds to maintain its large air force and operational rates of the air force inventory dropped to 50 percent, or lower, for some models. The ground forces also deteriorated during this period.

In contrast, the Ecuadorians had managed their economy better and had built up their air and ground forces in the 1970s and 1980s. The Ecuadorians bought the Mirage F1J fighter from France and the Jaguar fighter from the UK. In the 1980s Ecuador bought twelve Kfir C-2 supersonic fighters from Israel. A variety of helicopters were bought, mostly from the United States. While the Ecuadorian Air Force was smaller than Peru's, the Ecuadorians were able to maintain their planes and training levels at a much higher standard.

In early 1995 Ecuador was prepared for war and Peru was not. A skirmish on the border in the contested Cenepa Valley in the Amazon basin on January 9 developed into a larger battle. Within days both sides reinforced their forces and by late January both nations were in full mobilization mode. Battalion-sized battles were fought in the isolated Cenepa Valley, with both sides carrying out close air strikes in support of their ground forces. The Ecuadorians were well equipped with modern antiaircraft missiles and their Kfir interceptors made operations hazardous for the Peruvian Air Force. During a major battle on February 10, the Ecuadorians shot down a Peruvian A-37 and two Su-22s for no losses. On February 13, the Peruvians lost two helicopters to Ecuadorian ground fire. By this point, it was obvious that the Peruvians were outmatched on the ground and in the air. On February 17 both sides agreed to a cease-fire and to have peacekeepers occupy the disputed region. At the final tally of the air war, one Ecuadorian aircraft had been damaged, while the Peruvians had lost at least four helicopters and five jets (one Canberra, one A-37, three Su-22s) and had several damaged aircraft. Ecuador emerged the clear winner and its more effective air power had made the difference.[78]

PROTRACTED WARS—AIR POWER IN LATIN AMERICAN INSURGENCIES

The period from the 1960s to the 1990s saw several major insurgencies in South and Central America. In every case, air power played a central role in the counterinsurgency strategy of the government.

The use of air power in all these conflicts follows common characteristics. While people think of air power primarily in the strike role, it was the ability of air power to move troops and supplies quickly around the country that made the greatest impact. The infrastructure in rural Central and South America is poorly developed, so air transport is a necessity for any military force.

The wars since the 1960s have been largely helicopter wars. The ability to move a company, or even a platoon, rapidly to a threatened region makes the difference between victory and defeat for government forces. Aerial reconnaissance is also a key part of the counterinsurgency strategy. The ability to put rural and remote areas under regular surveillance makes the rebels disperse their forces and operate at night and under cover.

Guatemala

The three-decade-long civil war in Guatemala had its roots in long-standing social and economic problems exacerbated by a tradition of dictatorial governments. From 1963 until 1996, Guatemala fought a counterinsurgency campaign against Marxist rural movements.[79]

In the early 1960s, Guatemala had a small air force of a few hundred men and a few dozen aircraft composed of World War II surplus trainers, P-51 Mustang fighters, B-26 bombers, and some C-47 transports.[80] The United States, concerned about the growth of Communist influence in Latin America after Castro came to power in Cuba, was ready to send aid and advisers to Guatemala when a rural Marxist insurgency began in 1963. At that time Guatemala had a small but capable army supported by a small air force, the *Fuerza Aerea Guatemala* (FAG), equipped with surplus U.S. F-51s and B-26s. The United States also supplied four T-33 jet trainers as part of a comprehensive modernization program for the FAG. The T-33s, based on the F-80 fighter that had served as a fighter-bomber in the Korean War, were a suitable aircraft for a low-level insurgency and they were put into service as attack aircraft.[81]

By 1965 the insurgency had expanded and now covered several regions. The United States delivered more T-33s as well as armed Sikorsky UH-19B helicopters, the first military helicopters in Central America. These provided the

Guatemalan Army with an air assault capability.[82] The Sikorskys were armed with .30 caliber machine guns and 2.7-inch rocket pods for gunship support.[83] By 1966 a rearmed and now well-trained Guatemalan Army was ready to undertake major operations against the insurgents. Under guidance from U.S. advisers, the FAG reorganized its combat aircraft into Special Warfare Composite Squadrons composed of two to three F-51s, one to two T-33s, a B-26, a UH-19B, and a pair of C-47 transports.[84] The Guatemalan Air Force initiated a large-scale bombing campaign against rebel-held areas in addition to supporting the army's ground operations.

By 1967 Guatemalan armed forces had acquired UH-1 helicopters from the United States and had become a formidable force. From 1966 to 1968, the enlarged army and police forces conducted an all-out offensive against the rebel forces based in the northeast. The Guatemalan forces attacked the rebel infrastructure and demolished villages providing support; aerial bombing and napalm were used. By 1968 the rebellion had been largely stamped out at the cost of an estimated eight thousand rebels and civilians killed.[85]

Guatemala was largely quiet during the early 1970s with only a low level of insurgency in a few areas. However, the economic crises of the 1970s combined to make the conditions of the rural population worse than ever.[86] Leftist rebel groups found a welcome home in the overwhelmingly Indian western highlands. In the late 1970s the insurgency was reborn as a mostly Indian rebellion organized around four Maoist-oriented rebel groups, the largest group of insurgents known as the Guerrilla Army of the Poor.[87]

The Guatemalan generals replied to guerrilla activity in the highlands with brute force methods. As the insurgency grew, the U.S. government, under the Carter administration, became critical of the Guatemalan junta for its appalling human rights record. To forestall the embarrassment of a cutoff of U.S. aid, in 1977 the Guatemalans declared they would forgo all U.S. military assistance rather than accept reforms the United States was imposing on them. Guatemala would go it alone and handle the insurgency in its own way.[88]

Rejecting U.S. aid was especially hard on the Guatemalan Air Force, which was almost completely equipped with U.S.-made aircraft provided under military aid programs.[89] Guatemala had a fairly capable small air force of about 650 officers, noncommissioned officers (NCOs), and enlisted men, with its principal combat aircraft an A-37 squadron. The FAG constituted five squadrons: a fighter-bomber squadron, a reconnaissance squadron with light observation aircraft, two transport squadrons, and a helicopter squadron. The air force operated out of four major bases. The FAG's mission was to support the army that had grown to about twenty-seven thousand men by the early 1980s.[90]

Ending the U.S. aid program meant that when the FAG lost three A-37s to accidents in the 1970s, they could not be replaced.[91] As aircraft wore out Guatemala searched the world aviation market for affordable aircraft to replace losses and retirements. A deal was made in 1979 to import twelve Swiss Pilatus PC-7 turboprop trainers, which could also be modified as very capable counter-insurgency fighters.[92] Guatemala was able to reinforce its small helicopter fleet in 1981–82 by purchasing Bell 206B helicopters and Aroespatiale Alouettes. Four Fokker F-27 light transport planes were acquired in 1982.[93]

The air force saw considerable combat. Rebel villages were bombed. If the rebels engaged the army in battle, the army could retaliate with a flight of A-37s or attack helicopters. Between 1981 and 1983 the Arava and C-47 transports of the air force dropped paratroop units to seize territory deep in rebel country. Fighting was sustained but on a low level with squad and platoon actions being the norm. A shortage of spare parts caused by the U.S. military aid cutoff meant that no more than 50 percent of the air force was operational at any time.

One means of maximizing the military capabilities of Guatemala was to en-gage all of the nation's civilian pilots and aircraft in a program to support the war effort. In 1982 civilian pilots were enlisted in a force called the *Commando Especial Reserva Aerea*. The one to two hundred civilian pilots were given the rank of lieutenant in the air force reserve, a bit of military instruction, and issued uniforms and sidearms. Aircraft owners provided aircraft, usually light single- or twin-engine Piper, Beech, and Cessna planes, for a few days at a time. The air force reserve took over much of the routine, non-combat flying operations of the air force, which allowed the small FAG to concentrate on combat operations. The light civilian aircraft flew observation missions and personnel and supplies to small rural airstrips. It was an efficient and simple means of moving people and getting mail and supplies to isolated garrisons in a country where road move-ment was routinely ambushed. The mobilization of the nation's civil air fleet and pilots was an imaginative stroke and generally regarded as very successful.[94]

Gen. Rios Montt was relieved of the presidency after only sixteen months in power, but the strategy that he had mapped out continued from 1983 to 1986. The western highlands of Guatemala, the center of the revolt, were systemati-cally cleared and largely depopulated.

For its small size and lack of modern equipment, the Guatemalan Air Force played an important role in the war.[95] The psychological effect of air power played an important role in demoralizing the insurgents. In the memoirs of one of the guerrilla leaders, he speaks of helicopters and aircraft being constantly over his band and aircraft "thundering" over isolated villages.[96] At one point, he speaks of "the sky dark with helicopters and military planes" and "hundreds of soldiers parachuting into Indian villages."[97] Of course, the Guatemalan Air Force never had the capability to provide anything resembling constant air coverage, nor could it have ever dropped more than 150 paratroopers.

The insurgency ended in 1996 with the now-democratic Guatemalan government agreeing to support political reforms and economic aid for the rural poor. For their part, the rebels put down their arms and agreed to work peacefully within the Guatemalan system.

EL SALVADOR

The civil war in El Salvador was rooted in long-term dissatisfaction with the military dictatorship of that country. A 1979 coup promised quick democratic reforms, but when these did not pan out, a coalition of Marxist and leftist groups united as the Farabundo Martí National Liberation Front (FMLN) and, in 1980, initiated a major insurgency.[98]

In January 1981 Ronald Reagan became the president of the United States, and his administration was willing to provide large amounts of economic and military aid to bolster El Salvador. Thus began a commitment by the United States that would last until the end of the war in 1992.[99] The war in El Salvador saw significant involvement by the United States in the form of military and economic aid, advisers, and training.[100] Almost a quarter of the U.S. military aid was provided to the FAS.[101] Indeed, the FAS played a major role in the eventual success by government forces.[102]

In 1981 the FAS was a small force of under a thousand men consisting of a small paratroop battalion, a security force, a small antiaircraft unit, and four small flying squadrons with a grand total of sixty-seven mostly obsolete aircraft.[103] The FAS had its primary air base at Ilopango on the outskirts of the capital, and there was a smaller base at San Miguel in the eastern part of the country. In 1982 a rotary-wing operating base was established at La Union.

By 1981 the ten thousand rebels of the FMLN Alliance held large areas of El Salvador's fourteen provinces (called departments).[104] The rebels could put large, well-armed forces into the field. Cuba and Nicaragua supplied weapons, and light aircraft brought arms and supplies into El Salvador at night from Nicaragua, using small landing strips set up for crop dusters.[105] Early in the war, the El Salvadoran Armed Forces (ESAF) tried to conduct sweeps in company and battalion strength and these tactics worked to the benefit of the rebels, who could pick an engagement with company-strength government units and then ambush the reinforcing column. Whole companies of the Salvadoran Army were annihilated.

With the decision to aid the Salvadorans, the U.S. Southern Command helped the government of El Salvador to craft a national strategic plan.[106] U.S. policy emphasized reform and economic development. The military strategy increased the size of the ESAF and oriented it to counterinsurgency. Between 1980 and 1984, the ESAF more than tripled in size from twelve to forty-two thousand troops.[107] The ESAF was provided with modern weapons and equipment and the U.S. Army trained special "hunter" light infantry battalions. These light battalions would patrol aggressively and keep the rebel columns under pressure. Air power had a major role in the national strategy and the air force was to be enlarged and improved with the primary emphasis on building up a large and capable helicopter force that could lift infantry companies and battalions for offensive operations and also provide helicopter gunship support. This type of mobility could provide a rapid reaction force to block and pin down rebel columns that engaged the ground troops.

The United States provided a total of $48,920,000 in military equipment sales, military equipment credits, and military aid to El Salvador in 1981.[108] In 1982 the military assistance and sales program for El Salvador grew to $82,501,000 with another $2,002,000 for the international military education and training program (officer and NCO training).[109] A steady stream of new aircraft for the FAS flowed south throughout the conflict.[110] Due to strong opposition from liberals in the U.S. Congress, the administration imposed a strict limit of fifty-five military personnel that could be assigned to the U.S. Military Group in El Salvador.[111] FAS personnel had to be trained outside their country in the United States or at the Inter-American Air Force Academy at Albrook Field in Panama.

From 1981 to 1984 the FAS was limited in its ability to provide effective support to the army because of the lack of training in coordinating air-ground operations. The FAS was also essentially a daytime air force with a minimal ability to operate at night. In January 1982, it seemed that the FAS had been delivered a crippling blow when a hundred rebel commandos infiltrated the main FAS base at Ilopango and destroyed fourteen aircraft and damaged five more. In fact, it was a blessing for the Salvadorans as the worn-out 1950s French Ouragan fighters destroyed by the rebels were replaced by A-37s provided by the United States. The A-37s were far more capable aircraft for a counterinsurgency war. O-2 reconnaissance aircraft were also provided as well as UH-1H helicopters to replace the losses.[112]

By 1984 the U.S. military aid program began to pay off in terms of increased effectiveness of the government forces. The Salvadoran Army now outnumbered the rebels four to one and the new hunter battalions could use a more aggressive strategy and take the war to the rebels. The FAS had also been strengthened, was better trained, and was now equipped with enough helicopters to carry out airmobile operations with the army.

In 1984 and 1985 government forces gained the initiative. Air power in the form of the A-37 fighters, helicopter gunships, and helicopter lift played a major role in the government's success.[113] Helicopter gunship missions were increased by three or four times their previous rate of operations during March to May 1984.[114] During 1984 U.S. military assistance enabled the FAS to increase its helicopter inventory from nineteen at the start of the year to forty-six by year's end.[115] The most effective single air unit in the war was the FAS's five medevac helicopters, which were coupled with improved medical care for the Salvadoran Army made possible through U.S. aid. The availability of rapid medevac significantly improved the morale of the army and made it noticeably more aggressive on the battlefield.[116]

According to former FMLN leaders, the FAS played a major role in turning the initiative over to the government. The U.S.-supplied O-2 light reconnaissance planes covered the country thoroughly and the rebels could no longer operate in large units.[117] Rebel forces had to stay on the move. Still, the rebels adapted to the risk of aerial attack and after the FAS proved its ability to successfully insert company-sized forces to deal with FMLN attacks, the FMLN learned to spot likely helicopter landing zones and prepare them for ambush.[118]

U.S. military aid built up the FAS infrastructure and, by the mid-1980s, Ilopango had become a well-equipped air base.[119] In 1985 and early 1986 the FAS aircraft and helicopters supported several large army offensives that finally reduced some of the FMLN's major strongholds.[120] FMLN leaders credited the greater air mobility of the army in the mid-1980s with causing "a very significant turn in the war."[121] By 1987 the FAS was a force of twenty-five hundred men with an airborne battalion, five airplane squadrons, and a large helicopter force.[122]

Progress in El Salvador's internal political situation was made after the free elections of the mid-1980s, which saw reformer José Napoleon Duarte assume the presidency. Human rights abuses by the armed forces were curbed and rural aid increased. By 1990 the nation was exhausted by more than a decade of war and both sides agreed to peace talks in 1990. A cease-fire was agreed to in 1991, and peace accords were signed between the government and the FMLN in early 1992. The war was ended by a compromise solution that worked very much on the government's terms.

COLOMBIA: INSURGENCY AND THE DRUG WAR

Colombia offers one of the most complex instances of insurgency because the insurgency in Colombia is closely connected with the drug war. Since the early 1980s the U.S. military and State Department have coordinated military, police, and intelligence efforts with several South and Central American nations in order to reduce and interdict the flow of drugs, primarily cocaine, that are produced in South America.

For twenty-five years the United States has partnered most closely with Colombia, the country that has been most damaged by the illegal drug trade. Peru, Ecuador, Brazil, and several Central American nations have also cooperated. In Colombia the U.S. State Department has operated a small contract air force that includes surveillance aircraft to find the centers of cocaine growing and processing and crop dusting planes that spray herbicide on the crops when they are found. The military involvement in the drug war is large and, on the U.S. side, primarily consists of aerial surveillance assets employed to spot small aircraft and boats shipping drugs through the Caribbean region. Latin American nations have also taken a strong role in the surveillance and interdiction campaign.[123] In the 1990s Peru had an extremely aggressive policy of shooting down unregistered suspected drug aircraft, with over one hundred shootdowns to the credit of the Peruvian Air Force. Brazil has modified one of its highly successful jet-transport designs and created a highly sophisticated aerial intelligence platform, the Embraer 145, and has employed that plane in surveillance operations to spot drug-growing areas. Such a plane and its use in cooperation with other Latin American powers symbolize a new status for Brazil as a leading Latin American power.[124]

Armed helicopters are commonly deployed by the U.S. Coast Guard to stop and seize vessels running drugs through international waters. The tactics are constantly changing, as drug traffickers move their operations quickly. As soon as one area becomes heavily patrolled, they find a hole through the coverage. In 2000, in response to better interdiction efforts in the Caribbean, the drug traffickers moved their transport routes to the Central American coast, where there was very thin police and military coverage.[125] At the time of writing, the campaign against drug trafficking continues, with a significant daily international air effort involved.

Colombia has been plagued by several insurgencies by Marxist groups since the 1960s. The most significant one has been the Revolutionary Armed Forces of Colombia (FARC), a rural-based Marxist group that, by the year 2000, was able to field twenty thousand well-armed insurgents organized into large units. However, in the 1970s and 1980s, the main problem for the Colombians was not the insurgency but the growing drug trade carried out by the huge cartels in Cali and Medellin. In the late 1980s and early 1990s the power of the cartels was broken, but the drug trade continued with smaller groups taking over the trafficking. At this time the FARC took over much of the cocaine processing and trafficking, even forcing farmers to grow coca in the areas they controlled. With vast sums coming in from the drug trade, the FARC was able to buy sophisticated weapons, and special items such as night-vision devices, and had the financial resources to pay its soldiers well. Although the drug trade proved a boon for FARC's logistics, the organization over time transformed itself from being a revolutionary movement for the people into just another vast crime cartel.

For political reasons, the U.S. aid to the Colombian armed forces is very constrained. But the United States is much freer to provide aid in the form of counterdrug assistance. Thus, a great part of the U.S. aid to Colombia since the 1980s has gone to support the Colombian National Police's special counterdrug battalions. These battalions are equipped like military units and have been supplied with a variety of U.S. helicopters, including more than thirty of the very modern UH-60 helicopters for trooplift.[126] The air arm of the Colombian National Police, centered on several dozen helicopters, makes this force one of the largest and most modern combat air forces in Latin America. Since the drug trade supports the FARC insurgents, there is little practical difference between counterdrug operations and counterinsurgency.

Although Colombia has one of the larger and more professional armed forces of South America, the military made little progress in the 1990s as Colombia became progressively more unstable. By 2000 there were several thousand murders a year by FARC, mostly of civilians, and hundreds of kidnappings. The FARC was able to take on army battalions and beat them. The government under President Andrés Pastrana pursued a policy of negotiations with the FARC and allowed the FARC its own zone of control in central Colombia. In 2000 there was a major step forward when the U.S. Congress passed Plan Colombia, which provided $1.3 billion in aid to the Colombian police and military. With the aid, the Colombians were able to purchase new aircraft and modernize their forces.

Plan Colombia provided a sound basis for the government to take a more aggressive policy and the government developed a comprehensive strategy to fight the rebels. In 2002 the war reached a major turning point when President Pastrana canceled the peace talks and moved to restore government authority in the sanctuary zone after FARC guerrillas hijacked a domestic airliner and murdered a senator on the flight.

An enlarged and better equipped army, police, and air force went on the offensive in February 2002. Air force fighters and helicopter gunships struck eighty-five strategic points within the FARC zone as the offensive commenced on February 20.[127] The army carried out joint operations against FARC strongholds in north-central Colombia with several battalions of troops employed.[128] Air power played a key role in the government offensive. The Colombian Air Force's Kfir fighter-bombers had been recently configured to use precision munitions, and known rebel bases across northern Colombia were attacked with precision-guided munitions (PGMs). The strategy of using PGMs has been very effective in taking out FARC leaders, and also minimizing the chance of collateral damage and casualties.

Colombian forces since 2002 have followed sound counterinsurgency practices of securing and supporting the rural population, while also aggressively seeking out FARC units with heliborne troops. The Colombian forces have, since 2002, shown a high degree of effectiveness in mounting large battalion plus operations, supported by air power.

In 2002 President Alvaro Uribe succeeded President Pastrana. Uribe favored a hard line against the insurgents who killed thirty-five hundred Colombians yearly. He won with a landslide of 52.8 percent, with only 31.8 percent for his closest challenger.[129] Uribe followed Plan Colombia with Plan Patriota, which increased taxes to pay for an increase in the police and military forces. Since 2002 the Colombian government has reduced the FARC threat significantly. From twenty thousand armed troops in 2002, the FARC was reduced to five thousand in 2008. The murder and kidnapping rates have dropped to 25 percent of their 2000 totals.

Air power has been a key element in the increasingly successful campaign against Colombian insurgents. The Colombian Air Force is organized for counterinsurgency, with nine thousand personnel in 2010 organized into six regional combat commands. Each combat command contains a task force of three to four squadrons. Each command has a mix of fixed-wing strike aircraft and helicopters.

The Colombian Air Force's main strike force consists of a squadron of Kfir C.7 fighters and a squadron of Mirage 5 fighters. These planes are able to use PGMs. Several squadrons are equipped with the American A-37, but as these wear out they are being replaced with the highly effective Embraer turboprop A-29 Supertucano, made in Brazil.[130]

Colombia also has a small naval air arm. However, as noted, the Colombian Police's air arm constitutes an air force in itself. Between the police and air force, the Colombians have more than 100 fixed-wing strike aircraft and more than 250 modern helicopters, including UH-60s and refurbished Mi-8 helicopters. These helicopters have the power to lift troops and cargo into the mountainous terrain that covers most of Colombia. At the time of this writing, Colombian air power remains a major part in the ongoing campaign against the insurgents. In many respects, Colombia represents a high point in the ability of Latin American air forces to conduct sustained operations effectively.

CONCLUSION

For almost one hundred years the air forces of Latin America have employed air power, very often with decisive effects in war and very often in highly innovative ways. It should be remembered that Bolivia and Paraguay were among the first nations to employ aerial medical evacuation on a large scale. Peru was the first nation in the Western Hemisphere to field a paratroop force and use it successfully in combat. Indeed, any small nation today could learn some useful lessons from Ecuador's successful air power and air defense tactics that essentially won the 1995 conflict with Peru.

In the field of counterinsurgency the Latin American nations have used air power most successfully, and in that aspect of warfare, nations such as Colombia can serve as a model in the employment of air power as a key weapon to break a major insurgency. In another air power mission today, that of countering drug shipments, Brazil is continuing the tradition of innovation with its development of highly effective military surveillance aircraft based on civilian transports. Given the history and geography of Latin America, and the ongoing political and economic instability that fuels insurgencies and hostility between regional states, it is certain that we will see the continued conventional and unconventional conflicts in the region. If history is any guide, we can also expect that air power will be play a central role in future Latin American conflicts.

9

CONTINENTAL EUROPE

Christian F. Anrig

At the dawn of the twentieth century, continental European states were at the forefront of the development of air power.[1] The Italian invasion of Libya in 1911 saw the first use of aircraft in combat, and France soon took the lead in the development of aircraft engines. Together with Germany, it wielded the most potent air arm at the dawn of World War I.[2] After the abrogation of the Treaty of Versailles, German air power was again on the rise, with the Luftwaffe playing a crucial role in enabling the sweeping Blitzkrieg campaigns.

During World War II, however, the United States and the United Kingdom emerged as the leading nations in the application of air power, particularly in the field of strategic bombing. Economic and industrial realities after 1945 further reinforced that the United States would dominate air power throughout the entire Cold War period. American air power dominance became particularly apparent in Desert Storm, coinciding with the end of the Cold War. Despite a transatlantic air power capability gap, continental European air forces remain significant. Some of them have been regularly deployed to out-of-area operations in the Balkans, Afghanistan, and elsewhere.

This chapter specifically scrutinizes national responses to Desert Storm, which, in conjunction with the end of the Cold War, proved to be a major catalyst for the evolution of Western air power. It goes on to examine Western air campaigns and the underlying defense-political transformation throughout the 1990s. Thereafter, the chapter highlights the specific challenges for European nations to maintain relevant air forces, which requires strategic choices. Specifically, former Warsaw Pact air forces had to make far-reaching decisions and to undergo radical force transformations. Finally, recent and current European operations are examined in order to assess European transformational efforts.

SCOPE OF CHAPTER

There are a number of common denominators that provide the basis for Western European culture. These include Christianity, the Renaissance from the fourteenth to the seventeenth century, the Enlightenment of the eighteenth century, and the ripple effects of the French Revolution that laid the basis for modern European nation-states. Yet in spite of these common denominators and recent supranational developments, such as the creation of the European Union (EU), continental Europe does not form a homogeneous bloc.

As of 2010 twenty-seven nations are members of the EU. Additionally, there are several non-EU member states that possess an air force or at least air power assets. Since it is not feasible to cover every continental European air force in the required depth, this chapter looks at a selection of major continental European air powers. It particularly scrutinizes the *Armée de l'air* (French Air Force), the *Luftwaffe* (German Air Force), and the *Koninklijke Luchtmacht* (Royal Netherlands Air Force). The study of these three air forces highlights the impact of different defense policies upon air power. At the one end of the scale is the Netherlands with its transatlantic orientation and with its strong emphasis upon the North Atlantic Treaty Organization (NATO) as its most important security pillar. At the other end is France, which only recently reentered NATO's integrated military structure. Germany highlights the reciprocal relationship between internal politics and air power. Besides these three air forces, the chapter analyzes some aspects of the Swedish, the Polish, and the Czech air forces. The first is a nonaligned air force. The latter two represent two former Warsaw Pact air forces that have undergone far-reaching transformation processes.

THE COLD WAR LEGACY

Throughout the Cold War, continental Europe was partitioned by the Iron Curtain, splitting the Continent into a democratic west and a Communist east. The result was nearly half a century of deterrence postures between NATO and the Warsaw Pact.

Since Western powers faced a significant numerical disparity in conventional forces, NATO nations agreed to shift the emphasis of their defense to nuclear weapons. This move was underlined by NATO's third strategic concept, which was issued in its final form on May 23, 1957. Specifically, the concept advocated massive retaliation as a key element. The underlying principle of massive retaliation was that NATO could not afford to engage in limited wars with the Warsaw Pact, in which the opponent could bring to bear its numerical conventional superiority.[3]

Massive retaliation meant that so-called tactical aircraft were ancillary to the nuclear task or were co-opted into this task. As a result, nuclear predominance ossified conventional air power thinking and doctrine.

NATO's doctrine of massive retaliation was soon called into question. First, European decision makers started to doubt whether an American president could afford to sacrifice an American city for a European city. Second, the Soviet Union's nuclear arsenal increased considerably, eroding NATO's competitive advantage in nuclear deterrence. NATO responded by adapting its doctrine. NATO's fourth strategic concept in its final version was issued on January 16, 1968, providing the basis for the doctrine of Flexible Response. Embracing flexibility and escalation, the doctrine identified three generic response modes to Warsaw Pact aggression: direct defense, deliberate escalation, and general nuclear response. Direct defense specifically aimed at defeating aggression on the level at which the enemy would choose to fight, thereby attempting to limit a potential conflict.[4]

Yet NATO's leading military leaders, particularly U.S. Army Gen. Bernard W. Rogers, the eighth Supreme Allied Commander Europe, highlighted the Western alliance's inadequate conventional means to implement the doctrine of Flexible Response. As a reaction to the perceived imbalance of power in conventional forces, he advocated the so-called Follow-on Forces Attack (FOFA) concept in the early 1980s. It aimed at countering an incursion by striking deep against Warsaw Pact second- and third-echelon forces.[5] In essence, FOFA attempted to draw upon superior Western technology by emphasizing characteristics such as increased range, accuracy, and mobility of modern weapons systems in order to exploit the depth of the battlefield. As such, NATO air power was at its core. Yet throughout the 1980s, Western European nations were reluctant to make the required investments in advanced sensors, platforms, and deep-strike weapons, which led a well-respected British air power practitioner and commentator to conclude: "Once again doctrine had far outrun the capability to implement it. Happily, on this occasion the potential opposition believed it."[6]

Like their NATO counterparts, the Warsaw Pact militaries emphasized the importance of offensive counter-air operations, particularly airfield attack. Yet the air superiority battle was ancillary to the combined arms offensive.[7] As such, Soviet and Warsaw Pact air power doctrine was profoundly influenced by the Soviet Union's experience in the Great Patriotic War, emphasizing massive air support on the decisive axis to facilitate breakthroughs and exploitations.[8]

European countries with a policy of nonalignment or neutrality developed their own distinct approaches to air power during the Cold War era. Among them, Sweden, with its national aircraft industry, provides an impressive example of how a relatively small country was able to push technological boundaries to maintain a potent air arm almost independently.

With the dissolution of the Soviet Union in 1991, a new period began that was initially characterized by dialogue and cooperation between NATO and former Warsaw Pact countries, finally leading to full NATO membership of former foes. Against the backdrop of international crises, particularly the Balkan civil wars, NATO became a prominent military player in crisis management. Developments in the 1990s and beyond were also characterized by the development of the EU's security and defense policy. Since then, the EU has taken on responsibility for a number of crisis management and stabilization operations, primarily on behalf of the United Nations.

REACTIONS TO DESERT STORM

Despite various European contributions, the air effort during Operation Desert Storm was one-sided. The U.S. services logged 85.4 percent out of the total of 118,661 fixed-wing aircraft sorties. In contrast, the United Kingdom, France, and Italy flew 6.7 percent, with the United Kingdom contributing more than two-thirds of the European share.[9]

Two main enablers of Operation Desert Storm were precision-guided munitions, which offset the need for mass attacks to achieve a high probability of success, and stealth aircraft, which provided access to high-threat environments. Other innovations included the E-8 Joint Surveillance Target Attack Radar System, which is, in effect, an Airborne Warning and Control System (AWACS) analogue for ground surveillance. Capable of viewing the entire Kuwaiti theater of operations in a single orbit, the aircraft proved vital in thwarting the Iraqi attack and potential follow-on attacks against the Saudi coastal town of Al Khafji in the early stages of the air campaign.[10] In essence, the destruction of Iraqi armor revealed a new relationship between air and land power. Though air power could not hold ground, it denied it to Iraqi ground forces.[11]

Western air power could essentially draw upon the innovations that had occurred throughout the 1980s against the backdrop of FOFA. Given that only the United States Air Force (USAF) had made real efforts to implement the concept, it came as no surprise that basically all innovations of Desert Storm hinged upon American air power. French and British air assets, for their part, could only contribute in a limited manner to the precision-strike campaign. Desert Storm turned out to be a catalyst for the evolution of European air power in the post–Cold War era, regardless of direct air force involvement in the campaign.

FRANCE

Operation Desert Storm proved to be a watershed for French defense policy in general and for the French Air Force in particular. Politically, French participation in Operation Desert Storm aroused tensions within the republic's leadership, as there were deep concerns about Franco-Arab relations. At the insistence of the French defense minister Jean-Pierre Chevènement, French combat aircraft were not based together with American forces in order to avoid close linkage with the coalition. Moreover, the defense minister wanted to relegate French operations to purely defensive actions. In the course of the campaign, he opposed strikes against targets on Iraqi territory, but was overruled by President François Mitterrand. On January 27, 1991, Chevènement resigned from his post.[12] Political reluctance to subordinate French forces to the joint force commander during the force buildup led to a military marginalization of the French Air Force contingent.

As a consequence of the altered geostrategic environment and the experience of Desert Storm, France gradually abandoned this isolationist stance. French authorities began to realize that in order to make a meaningful contribution to theater wars, the French armed forces needed to be fully interoperable.

From the very beginning, the French Air Force was a key player in the air campaigns over the Balkans and was able to integrate itself swiftly into the American-led air campaign over Afghanistan in early 2002. This change in paradigm is also reflected in France's involvement in cooperative endeavors that aim at harmonizing European air power. For instance, together with the Royal Air Force, the French Air Force established the Franco-British European Air Group with the aim of improving interoperability. The organization was later expanded into the European Air Group to include the air forces of Belgium, Germany, Italy, the Netherlands, and Spain.[13]

In particular, the Gulf War air campaign had a major impact upon the renewal of conventional forces. One French commentator argued that the shortfalls encountered by the French Air Force during Operation Desert Storm raised serious concerns about previous procurement priorities. For instance, French aircraft lacked appropriate identification friend or foe (IFF) systems, adequate self-protection suites, and navigation systems. Moreover, the French Air Force did not possess sophisticated command and control systems, which proved indispensable for an air campaign of the scale and complexity of Desert Storm.[14] The relatively poor quality of conventional assets had been primarily caused by France's nuclear doctrine. The preeminence of this doctrine largely rendered conventional forces secondary.[15] In the wake of Desert Storm and with the end of the Cold War, France's 1994 Defense White Book put a new emphasis upon conventional forces and conceived of their primary employment outside the nuclear context.[16] Two years later, the most far-reaching defense reform took place after Jacques Chirac had become president. Besides modernizing the conventional component of France's defense architecture, Chirac's reform also abolished national service.[17] Shortcomings in the light of Operation Desert Storm were supposed to be the particular driving factor behind the reform.[18]

The computerized command and control system SCCOA (*système de commandement et de conduite des opérations aériennes*) became an important pillar in the French Air Force's move toward enhanced and interoperable conventional forces. The deployable component of SCCOA can support an air component command headquarters in deployed operations.[19] Enhanced interoperability on a technical level has enabled the French Air Force to take on a lead role on the international stage. As such, the French Air Force was in charge of the NATO Response Force (NRF) air component during the second half of 2005.[20] Despite France's non-integration into NATO's integrated military command structure at the time, the French Air Force was the first European air force to provide an air command infrastructure on a national basis to the NRF.

THE NETHERLANDS

During the 1990–91 Gulf crisis, the Royal Netherlands Air Force intended to contribute actively to the coalition against Saddam Hussein. Yet due to the concentration of extremely large numbers of aircraft in the Gulf region, all hosting options for Dutch F-16s turned out to be exhausted.[21] While the Royal Netherlands Air Force could not immediately contribute to Operation Desert Storm, it did so indirectly. Two Dutch Patriot missile batteries were dispatched to Diyarbakir for protection against potential Iraqi Scud missiles. This deployment was soon to be reinforced with two further Hawk missile squadrons and Stinger units. The Royal Netherlands Air Force also deployed a Patriot missile battery to Israel.[22]

The Netherlands' immediate reaction to Desert Storm was to draw lessons from the fighter deployment to the Gulf area that nearly took place. On a strategic level, efforts were made to speed up the decision-making process for any future crisis management operations. On an operational level, squadrons were geared up for rapid deployment.[23] In the remainder of the 1990s and beyond, Dutch decision makers proved indeed capable of swiftly committing air units to operations across the spectrum of military force, and the Royal Netherlands Air Force took on a significant role in the ensuing Balkan air campaigns.

It can, therefore, be concluded that Desert Storm was not so much about a paradigm change, as in the case of the French Air Force, but a matter of recalibrating already existing Dutch defense capabilities. However, it is interesting to note that with regard to doctrine development and the acquisition of an air-to-ground precision-strike capability, Operations Deny Flight and Deliberate Force proved to be more critical for the Royal Netherlands Air Force than Desert Storm. Hence, there appears to be a difference between just observing an operation and fully taking part in it.

The first national air power doctrine was promulgated after major air operations over Bosnia and Herzegovina. Against the backdrop of operations in the Balkans and elsewhere, a conceptual grasp of air power doctrine was deemed essential in order to translate limited political goals into military operations.[24] In his foreword to the first Royal Netherlands Air Force air power doctrine of 1996, the commander in chief established an explicit link between the employment of the Royal Netherlands Air Force in deployed operations during the early 1990s and the need for adequate air power doctrine.[25] The Balkans air campaigns gave an impetus to Dutch doctrine development. Moreover, only after Operation Deliberate Force was it decided to acquire a precision ground attack capability by procuring AGM-65G Maverick air-to-ground missiles,[26] though Desert Storm most clearly revealed the key role of precision-guided munitions in modern air warfare.

GERMANY

In the midst of Germany's reunification, Iraq's occupation of Kuwait surprised already-busy German authorities. A direct German military involvement in the Gulf crisis could have provoked the Soviet Union to stop supporting German reunification and as such was categorically ruled out. Moreover, there seemed to be no constitutional foundation for German involvement, and the popular repudiation of any direct military involvement. As a result, Germany's main contribution to the Gulf War turned out to be significant financial support.[27] While the Federal Republic avoided becoming directly involved in Operation Desert Storm, a NATO request to contribute to Turkey's defense against potential Iraqi incursions was received in Bonn. Eventually, the German decision makers hesitantly agreed to dispatch eighteen Alpha Jet light attack aircraft. The United States and Turkey were anything but pleased by this lukewarm response. As a reaction, Germany reinforced its air force units in Turkey by deploying eleven air defense batteries.[28]

Germany's and the German Air Force's reactions to Desert Storm can be characterized by discomfort, both politically and militarily. Uneasiness with offensive air power doctrine in deployed operations led to heated political debates regarding the proper interpretation of the German Constitution, as will be further examined below. Militarily, this uneasiness led to a gap in doctrinal thought throughout the 1990s. Immediately after Desert Storm, on March 22, 1991, the German Air Force published its first Air Force Service Instruction, *Luft-waffendienstvorschrift* LDv 100/1, which corresponds to an air power doctrine. The document was doctrinally still anchored in the later stages of the Cold War. For instance, air power's role in alliance defense, particularly in the context of Flexible Response, was dealt with in depth.[29]

In contrast, the Royal Air Force, which almost simultaneously published the first edition of Air Publication (AP) 3000 *Royal Air Force Air Power Doctrine*, soon deemed a revision of AP 3000 necessary in order to reflect lessons learned from Desert Storm. In 1993, the Royal Air Force published the second edition of AP 3000. In its foreword, the chief of the air staff highlighted the importance of recent operational experience for the doctrine development process: "Since AP 3000's first edition was published, the Royal Air Force's air power doctrine has been reviewed and refined in the light of operational experience in the Gulf and elsewhere."[30] Yet no successful attempt at revising air power doctrine was made within the German Air Force in order to reflect recent lessons from Desert

Storm. This doctrinal gap was mirrored in a lack of a broader doctrinal debate in the form of publications, quite in contrast to the Cold War era, when there was a lively doctrinal and conceptual debate on air power's role in the defense of alliance territory.

Only after a radical defense reform was the doctrinal process in Germany revived. Germany's defense reform began with a reform proposal approved by the Social Democratic and Green coalition government in June 2000.[31] Yet due to the continuing imbalance between available means and operational requirements and in the wake of September 11, the newly appointed Social Democratic defense minister Peter Struck undertook a far-reaching review of the reform, which was reflected in the 2003 Defense Policy Guidelines.[32] These guidelines can be understood as a fundamental paradigm shift toward a more proactive defense policy. This transformational process increased the need to anchor the German Air Force doctrinally. As a consequence, the chief of the air staff ordered the revision of German air power doctrine. The revised LDv 100/1 was issued in April 2009, though not released to the wider public. In essence, developments at the strategic level had a direct impact upon doctrine development.

In terms of equipment, particularly in the area of precision-guided air-to-ground munitions, the German Air Force belatedly started to implement the lessons of Desert Storm. A precision-strike capability by means of laser-guided bombs was only acquired in 2001.[33] This delay can be related to two further issues: a strained budget situation in the wake of German reunification and the primary defensive mission of the German Armed Forces. As regards the former, the 1994 Defense White Book announced that any major equipment programs for the main defense forces would be deferred to the period after 2000.[34] As regards the latter, Russia's military potential remained a significant factor in Germany's threat perception throughout the 1990s. Accordingly, territorial defense continued to be the constitutional bedrock for the German Armed Forces, determining their size and structure.[35] This, in turn, led to a continuation of Cold War concepts. For instance, unlike the Royal Air Force, which disposed of its submunitions dispenser system partly as a consequence of Desert Storm,[36] the German Air Force retained its equivalent system. The submunitions dispenser system MW-1 remained the main armament of the German Tornado fleet throughout the 1990s.[37]

As in the case of doctrine development, the German Air Force made significant steps in the light of Germany's defense reform. As will be further analyzed, a considerable number of cruise missiles were procured, and, more importantly, the German Air Force acquired autonomous operational command and control structures. Throughout the Cold War, the German Air Force hinged almost entirely upon NATO command and control facilities. Yet with an increasing German involvement in deployed operations, a national capacity for command and control became necessary. In late 2001, the German Air Force set up its Air Operations Command, Kommando Operative Führung Luftstreitkräfte. The command was primarily designed as a national nucleus for multinational air component command headquarters in the framework of EU or NATO operations.[38]

THE BALKAN AIR CAMPAIGNS

In the wake of Desert Storm, the United States and Europe were confronted with internal conflicts in the Balkans. For many European air forces, the Balkans air campaigns were the first true baptism of fire since World War II. As such, they were normative determinants for the development of European air power. As analyzed above, they specifically influenced the Royal Netherlands Air Force. In Germany, they ignited a heated debate among political decision makers and the wider public on the use of force, which culminated in the German Air Force's full-fledged participation in Operation Allied Force.

The predominant ethnic conflict in the Balkans was the Bosnian Civil War from 1992 to 1995. In response to this conflict and on behalf of the United Nations (UN), various European nations deployed ground contingents and NATO conducted its first air campaign. Operation Deny Flight commenced on April 12, 1993. Its primary purpose was to enforce a no-fly zone over Bosnia by means of round-the-clock combat air patrols.[39] The mission spectrum was soon extended to include air-to-ground strikes. Yet due to the narrow rules of engagement, the parties in the civil war could execute air space violations with near impunity,[40] and a total of only four close air support missions were authorized in the course of two years.[41]

A consensus for sustained air strikes in response to Bosnian Serb aggression only began to form in the second half of 1995. In particular, the massacre at Srebrenica in July 1995 constituted a watershed in the European attitude toward the employment of military force. The immediate event triggering a more robust air campaign was the shelling of a Sarajevo marketplace on August 28, 1995. Operation Deliberate Force was launched early on August 30.[42] Intertwined with diplomatic efforts, the campaign was halted for negotiations. After these faltered, the bombing was resumed.[43] On September 14, after approximately two weeks of bombing, the Bosnian Serbs agreed to UN terms, which caused air strikes to be suspended permanently.[44]

Deliberate Force was a light air campaign. A mere 3,515 sorties, including 750 strike sorties, were flown by the alliance's air forces,[45] and slightly more than 1,000 air-to-ground munitions were released.[46] This low number of munitions was coupled with careful targeting. The air component commander, Lt. Gen. Michael Ryan, considered the avoidance of collateral damage to be of strategic importance.[47] Targeting focused on the mobility and command infrastructure of the Bosnian Serb army, without going beyond the immediate area of the adversary's ground operations.[48] The campaign required a total of up to three hundred aircraft, among these approximately twenty air-to-air refueling aircraft.[49] In terms of sorties, the U.S. services accomplished by far the most (66 percent), followed by the United Kingdom (10 percent), and France (8 percent), with Dutch, German, Italian, Spanish, Turkish, and common NATO aircraft flying the remainder.[50]

Deliberate Force was embedded into a comprehensive approach, producing the November 1995 Dayton Accords, which brought stability to Bosnia and Herzegovina. The air campaign could also draw upon air-ground synergies that were lacking during Operation Allied Force in 1999. A heavy French, British, and Dutch multinational brigade was deployed to Bosnia in mid-1995. According to the French general commanding the brigade, artillery fire paralyzed Bosnian Serb military movements around Sarajevo and produced synergistic effects with air power.[51] Probably most importantly, a Bosnian and Croat ground offensive underpinned the air strikes of August and September 1995.[52]

Between March 24 and June 9, 1999, NATO embarked upon its largest air campaign over the Balkans, the goal being to stop the suppression of the Albanian majority in Kosovo. Why Milosevic gave in is still a controversial issue. A plethora of factors were identified, with air power underpinning all the other factors: the increasing threat of a NATO ground intervention, declining support from Russia, NATO's cohesion as an alliance, and diplomatic interventions. In its destructive effect, Operation Allied Force was much larger than its predecessors. Approximately 23,000 munitions were released, of which 35 percent were precision guided, including 329 cruise missiles.[53] Unlike in Operation Deliberate Force, however, these munitions were not exclusively aimed at the military mobility and command infrastructure. The air campaign went beyond the confines of Kosovo and laid waste to large parts of the Serbian infrastructure, including oil refineries along the Danube.

EUROPEAN DISTINCTIVENESS

Air operations over the Balkans were significantly influenced by the political context. In general, continental European constituencies were not at ease with the use of military force. Therefore, the application of force had to be clearly related to humanitarian purposes. Given the prevailing political sensitivity, the Balkan air campaigns were only gradually escalated. This was in direct contrast to Operation Desert Storm, where several target sets were attacked simultaneously. Taking account of the delicate situation, European politicians and militaries alike were reluctant to go down the path of an all-out war. The situation in Germany and to a lesser degree in the Netherlands aptly highlights the reality of this politically fragile environment. For instance, after a Dutch F-16 had downed a MiG-29 on the first night of Operation Allied Force, the Netherlands chief of defense Adm. Lukas Kroon was concerned about too much publicity of this incident.[54]

During the first half of the 1990s, the participation of German AWACS personnel in Operation Deny Flight aroused staunch political opposition within Germany. Both the junior government party and the major opposition party appealed to the Federal Constitutional Court.[55] In particular, there was disagreement on how to interpret Article 24, Paragraph 2, of the constitution. The specific article states that the German Federation can participate in collective security systems for the preservation of peace.[56] Previously, this article had been interpreted to limit German combat operations to the collective defense of the alliance. On April 8, 1994, however, the constitutional court issued its verdict, declaring that German participation in Operation Deny Flight did not violate the constitution.[57] The main ruling of the court, following on July 12, 1994, stated that the Federal Republic could assume full responsibility as a member of a collective security system, including armed out-of-area operations. However, parliamentary approval was deemed a necessary prerequisite.[58] These verdicts paved the way for the German Air Force's gradual integration into Western air campaigns over the Balkans.

Despite this new interpretation of the German constitution, the foreign and defense ministers made it clear at the time that Germany would continue to adhere to a policy of restraint and that missions outside Europe would remain an exception. However, pressure upon German decision makers to make a substantial contribution to Operation Deny Flight increased. In particular, NATO authorities expressed a clear interest in Germany's deploying Tornado ECRs, especially designed for suppression of enemy air defenses (SEAD).[59]

On June 30, 1995, the German parliament voted in favor of a more robust military commitment in the context of the Bosnian civil war. The core of the German contribution encompassed eight Tornado ECRs and six reconnaissance Tornados.[60] Yet nationally imposed rules of engagement provided a very narrow margin for German aircraft to operate. Their employment was essentially restricted to supporting defensive actions of the multinational brigade, previously deployed by the French, British, and Dutch to Bosnia. As a consequence, German Tornados could not directly participate in the UN-mandated air campaign Deliberate Force.[61] Yet indirectly, the presence of German SEAD Tornados over Bosnia inhibited hostile surface-to-air missile (SAM) activity.

German politicians' reluctance to employ offensive air power did not imply that they were averse to exposing aircrews to significant risks. The German Air Force, together with the USAF, the Royal Air Force, the French Air Force, and the Canadian Air Force, constantly contributed to the air bridge to Sarajevo from late 1992 to early 1996,[62] despite the fact that German aircraft were shot at on several occasions.[63] At least equally dangerous was airdropping relief supplies in eastern Bosnia from March 1993 to August 1994.[64]

In the medium term, the massacre at Srebrenica proved to have a considerable impact upon the German public and the political decision-making process. For instance, the slogan "never war again from German soil" was replaced by the slogan "never Auschwitz again, never Srebrenica again." However, this shift in public opinion did not yet embrace combat operations across the spectrum of military force.[65]

In the late 1990s, the Federal Republic was confronted with the dilemma of contributing to a NATO air campaign against the backdrop of a humanitarian disaster, but without a UN mandate. Almost paradoxically, a Social Democratic and Green coalition government committed the German Air Force to take part fully in Operation Allied Force.[66] A total of ten Tornado ECRs and four reconnaissance Tornados were deployed.[67] These aircraft accounted for 1.37 percent of the allied aircraft fleet, which, in the course of the campaign, was increased to 1,022 assets, and for 1.33 percent of allied sorties.[68] Nevertheless, the German

Tornado fleet released a significant number of high-speed antiradiation missiles, 236 out of a total of 743 expended throughout the entire campaign.[69]

The political significance of German participation in Operation Allied Force was more significant than Germany's military contribution. For the first time, the German Air Force participated as an equal partner in a peace enforcement operation. As regards German public opinion, popular support was solid, amounting to approximately two-thirds. Yet German participation in a high-intensity ground war scenario would clearly have been declined.[70] Hence, air power proved instrumental in overcoming German compunctions.

From an Anglo-American point of view, political concerns on the continent seemed to inhibit air power's potential. In particular, France was singled out as a scapegoat for gradualism in Operation Allied Force. In a hearing of the Senate Committee on Armed Services in October 1999, the American air component commander Lt. Gen. Michael Short argued that France should not have been allowed to restrict American aviators, who had borne the main bulk of the air campaign over Serbia and Kosovo.[71] In fact, the issue was more complex, as there was also an intra-American dispute. The Supreme Allied Commander Europe, Gen. Wesley Clark, and Lt. Gen. Michael Short disagreed on the most effective target sets. Whilst the former identified Serbian ground forces in Kosovo as the center of gravity of the campaign, the latter put a premium upon leadership-related targets in Belgrade.[72]

Besides President Clinton, only Tony Blair and Jacques Chirac were in a position to veto possible targets. The French president expressed deep concern about striking both targets that might cause significant collateral damage and targets in Montenegro, including SAM sites and an airfield,[73] in an attempt to limit the conflict. Accordingly, he vetoed for a time the bombing of two television towers and key bridges across the Danube in Belgrade. This, among others, led to the above American accusations. In their study on the Kosovo conflict, Daalder and O'Hanlon point out that the strategy of gradual escalation also came out of Washington. American authorities anticipated a limited and short air campaign. This belief is reflected in the relatively modest air armada that was within striking range of Serbia on the eve of Operation Allied Force.[74] In essence, the debate involved a generic issue about gradualism versus shock and awe, or, in USAF parlance, parallel warfare. Despite Chirac's efforts to limit the conflict, the French Air Force carried a heavy burden. In terms of precision-guided air-to-ground munitions, French airmen released more than double the amount that British forces dropped.[75]

A genuine European air power doctrine would have to embrace the realities of Europe's fragile political environment. Therefore, it would have to conceptualize the gradualist approach, even if this runs contrary to the widely accepted American view of air power doctrine, which places an emphasis on striking several target sets simultaneously in order to supposedly paralyze the enemy's system.

Gradualism and political restraints are not negative things per se. Deliberate Force is an example of how a gradualist joint campaign effectively achieved its goals through concentration on the mobility and command infrastructure of the Bosnian Serb army and in conjunction with the other tools of grand strategy. Regarding Operation Allied Force, it is generally accepted that Russian diplomacy played an important role in convincing Milosevic to agree to a G8 plan. It is doubtful whether the Russians would have been willing to exert pressure on the Serbs if NATO had embarked upon an all-out air campaign from the very beginning.

THE EUROPEAN DEFENSE ARCHITECTURE

In order to understand modern European air power, it is necessary to analyze the evolution of the European defense architecture in the post–Cold War era. Crisis management in the Balkans and elsewhere shaped NATO's strategic outlook. Besides the North Atlantic Alliance, the EU has emerged as an important security actor.

NATO AFTER THE COLD WAR

NATO responded to the end of the Cold War by developing a new strategic concept, which was completed in November 1991. The concept put a premium upon dialogue and cooperation with former adversaries. It also provided for major changes in NATO's defense posture, such as reductions in the size of military forces on the one hand, and improvements in their mobility, flexibility, and adaptability on the other.[76]

Cooperation with non-NATO states was sought through the Partnership for Peace (PfP) program, which was launched at the January 1994 Brussels summit. On a strategic level, PfP has aimed at enhancing Europe's security and stability. On an operational level, the goal has been to improve interoperability.[77] Originally relegated to the lower spectrum of military force, the PfP mission spectrum was extended in 1997 to include peace enforcement operations.[78]

Cooperation with former Warsaw Pact states was gradually increased. At the 1997 Madrid summit, NATO issued accession invitations to the Czech Republic, Hungary, and Poland. These countries became formal members of the alliance in early 1999. Two further enlargement rounds have taken place since then. In 2004, Slovakia, Slovenia, Romania, Bulgaria, Estonia, Latvia, and Lithuania formally entered the alliance, and, in 2009, Croatia and Albania became NATO members.

Parallel to NATO's enlargement, power projection became an increasingly important focus for the alliance. Whereas collective defense alongside deterrence had provided the fulcrum during the Cold War era, the emphasis shifted toward crisis management operations. The strategic concept adopted at the 1999 Washington summit underlined NATO's willingness to support UN-mandated out-of-area operations.[79]

In the wake of September 11, NATO's military transformation experienced a further impetus at the Prague summit in 2002, when the alliance launched the NRF project. The NRF is a quickly deployable and technologically advanced joint force comprising up to twenty-five thousand troops and support personnel.[80] It is designed for worldwide employment across the spectrum of military operations: from evacuation operations and deployment as a mere show of force to Article 5 (collective defense) operations.[81] The NRF basically serves two purposes: to improve NATO power-projection capability and to accelerate NATO's military transformation.

THE EMERGING EUROPEAN SECURITY AND DEFENSE POLICY

In parallel to NATO's reorientation, EU members agreed on the development of a Common Foreign and Security Policy (CFSP) in 1992. Five years later, the European Council incorporated the Western European Union's Petersberg tasks into the EU's CFSP.[82] These include humanitarian and rescue tasks, peacekeeping, and peace enforcement.

Despite declarations of broad political intentions, the EU was far from having concrete military tools at its disposal. In particular, the EU lacked adequate means to deal with the crises in the Balkans. As such, it was essential to improve European military capabilities if the EU wanted to take on strategic responsibilities. This realization led Great Britain and France, the two critical European military actors, to reconcile opposing views. In particular, the United Kingdom, which had never been a keen supporter of autonomous European military capabilities, became apprehensive that Europe's military powerlessness might imperil the very foundation of the Atlantic partnership. The rapprochement between

France and the United Kingdom led to the bilateral Saint Malo Declaration in late 1998.[83] The political leaders of both countries jointly declared that "the Union must have the capacity for autonomous action, backed up by credible military forces, the means to decide to use them, and a readiness to do so, in order to respond to international crises."[84]

Almost immediately, the wider European community embraced the principles of the Saint Malo Declaration. The June 1999 EU summit in Cologne was a significant milestone toward a European Security and Defense Policy (ESDP) as an integral part of the EU's CFSP. In particular, the apparent European shortcomings during Operation Allied Force proved to be an important catalyst for making rapid progress in common European defense matters.[85]

In essence, NATO and the ESDP complement each other. While the former guarantees a strong transatlantic link and has continued to be the sole framework for collective defense in Europe, the latter is suited to respond to requests by the UN secretary-general against the backdrop of crises in the developing world. In many cases, the ESDP is the natural choice for crisis management and stabilization operations, as EU-led operations are in certain theaters regarded as more benign than potential NATO operations.[86] Yet while the ESDP provides a suitable framework for peace support operations, it is less suited for major theater wars.

Critics argue that the EU is producing irrelevant redundancies in relation to NATO. Yet these supposed redundancies are predominantly of a conceptual and not of a physical nature. As such, there are no separate EU troops. Since both NATO and the EU largely draw on the same troops and assets, there is a single set of forces for both institutions. Only in the domain of command and control are there some redundancies, such as a civil-military planning cell within the European Union Military Staff. Yet these redundancies come at a relatively minimal cost. In order to secure freedom of action, a host of potential options must be made available. The current environment is characterized by the formula "the mission defines the coalition," as expressed by the former U.S. secretary of defense Donald Rumsfeld.

CHALLENGE TO MAINTAIN RELEVANT AIR POWER

Desert Storm manifested the United States' uncontested air power dominance. In the wake of the Gulf air campaign, European air forces realized that, in order to remain interoperable, far-reaching reforms would be necessary. By the late

1990s, however, many of the required measures had not yet been implemented. As a consequence, Operation Allied Force again highlighted a glaring transatlantic capability gap, and once again the United States shouldered the largest burden. While some Europeans could provide niche capabilities, such as SEAD, or were able to deliver precision-guided munitions, many European allies could not effectively cooperate with U.S. services,[87] which contributed 59 percent of all allied aircraft involved in the air campaign and released over 80 percent of the expended munitions.[88]

Using Allied Force as a benchmark for European air power, a British commentator argued in 2000 that the five hundred all-weather fighter-bombers that the United Kingdom, France, Germany, and Italy could field at the time needed to be increased by about 50 percent.[89] Moreover, discrepancies were particularly glaring in the areas of air-to-air refueling and airborne standoff jamming. With regard to the former, U.S. services provided approximately 90 percent. With regard to the latter, European allies entirely depended upon American capabilities.[90]

Both NATO and the EU have launched initiatives in order to address the European air power shortfalls. In the course of Operation Allied Force, NATO member states agreed upon the so-called Defense Capabilities Initiative. The initiative aimed at improving the alliance's effectiveness in deployed operations.[91] In parallel, the EU launched initiatives such as the Helsinki Headline Goal in December 1999 and the Headline Goal 2010 in 2004.[92] Identified shortfalls were generally identical to those identified by the NATO initiatives. In the following subsections, European responses to the capability gaps in the domains of precision strike and air mobility are analyzed. Moreover, the benefits of role specialization and doctrinal innovations are highlighted.

PRECISION STRIKE

In the early stages of Operation Allied Force, U.S. services and the Royal Navy launched a total of 329 cruise missiles. In contrast, the continental Europeans were not in a position to contribute to the cruise-missile campaign. Since then, the Germans and French have been acquiring over one thousand missiles combined.[93] The French Air Force has also been introducing an innovative modular air-to-ground armament, the *armement air sol modulaire* (AASM). It is essentially a rocket-propelled all-weather precision-guided bomb. In April 2008, French

Rafale fighter-bombers engaged Taliban positions with AASMs for the first time, with target coordinates provided by a Canadian forward air controller.[94]

Moreover, modern fighter-bombers, such as the Rafale or Eurofighter Typhoon, are capable of carrying a standard air-to-ground weapons load of four to six precision-guided munitions, whereas in Allied Force strike aircraft such as the Tornado, the Mirage 2000D, or the F-16 carried a standard precision-weapon load of only two laser-guided bombs. This results in significantly enhanced precision firepower per aircraft, which challenges the previously cited argument that Europe would have to increase its five hundred all-weather fighter-bombers by 50 percent. In conclusion, European air forces have been addressing the lack of precision firepower effectively. This is quite in contrast to the force-enabling areas, which results in a continuing imbalance between the shaft and the spear.

Air Mobility

Throughout the Cold War, European armed forces were supposed to fight "in place." With the end of the bipolar East–West confrontation, the sudden move toward out-of-area operations produced significant shortages in airlift capacities. This situation was not significantly improved throughout the 1990s. Even at the end of the decade, Western European air forces did not own one single military wide-body transport aircraft capable of lifting a main battle tank. In 2001 the Royal Air Force was the first European air force to bridge this gap by first leasing and later procuring C-17 Globemasters.[95] Yet a sustained and robust European solution can only be expected through the commissioning of the A400M transport aircraft, Europe's latest military transport aircraft developed by Airbus. The program is currently experiencing serious delays, with the earliest possible deliveries not expected to take place before late 2012.[96] Germany and France together plan to acquire 110 A400M aircraft.[97]

The problem of insufficient airlift capacities is compounded by a shortfall in deployable ground infrastructure. Against the backdrop of the NRF air component, the French Air Force and the Royal Air Force have so far been the only air forces to provide deployable air bases on a national basis.[98]

Air-to-air refueling is another area where Europe lacks adequate resources, as was revealed in Operation Allied Force. In 2006 the USAF, the U.S. Navy, and the U.S. Marine Corps operated a total of more than 650 tanker aircraft. In contrast, various European air forces combined could muster approximately seventy aircraft, with the United Kingdom and France providing the bulk of them.[99] For the foreseeable future, this gap will remain. In renewing their aging tanker fleets, the United Kingdom and France are not striving to expand their fleets but to maintain current capacity levels.[100]

EUROPEAN AIRLIFT COOPERATION

European deficiencies are often presented as inextricably linked to tight defense budgets. It is unrealistic to assume that European defense budgets would be increased significantly in the coming years. As a consequence, potential remedies lie in reallocating existing defense resources and in cooperative approaches. Other alternatives include role specialization as well as doctrinal and operational responses.

In the wake of the Kosovo air campaign, European air forces have embarked upon a number of complementary initiatives aiming at rationalization through cooperation. Since 2001, European Air Group air forces have jointly coordinated their airlift; first, through the European Airlift Coordination Cell based in Eindhoven, which further evolved into the European Airlift Center (EAC) in mid-2004.[101] Subsequently, on July 1, 2007, the EAC merged with the Sealift Coordination Center to become the Movement Coordination Center Europe in Eindhoven, which made it possible to draw joint synergies.[102]

In addition to these organizational efforts, European states chartered and jointly procured strategic transport aircraft. Currently, there are two complementary initiatives. The first initiative is the so-called Strategic Airlift Interim Solution (SALIS), under which a multinational consortium of sixteen countries, led by Germany, charters Antonov An-124–100 transport aircraft. SALIS has been operational since March 2006. The SALIS initiative helps to bridge the European capability gap until the commissioning of the European A400M military transport aircraft. SALIS is equally available to both NATO and EU operations, underlining the strategic partnership between the two alliance frameworks.[103] NATO's Strategic Airlift Capability initiative, for its part, is an example of shared ownership. Ten NATO countries plus two PfP nations, Finland and Sweden, jointly procured three C-17 Globemasters, the first of which arrived at its home airbase in Papa, Hungary, on July 27, 2009.[104]

The most far-reaching project in the domain of European airlift integration is the European Air Transport Command (EATC) initiative. In May 2007, Germany, France, Belgium, and the Netherlands signed an agreement on the establishment of the EATC, and in February 2010, the defense ministers of the four countries declared that the multinational command, headed by a German major general, would begin operations in Eindhoven, Netherlands, in the second half of 2010.[105] This multilateral approach provides each of these countries with an unprecedented surge capability. As such, the EATC will significantly strengthen Europe's rapid reaction crisis management capabilities.

ROLE SPECIALIZATION

In contrast to pooling assets, role specialization has proven to be more politically sensitive. It has been perceived to supposedly limit national freedom of action. Nations are reluctant to become reliant on other allies for particular niche capabilities; yet, due to a lack of resources, role specialization has been taking place by default. In particular, small nations have to limit themselves to a very narrow bandwidth of capabilities, which generally include interceptors, a tactical precision-strike and reconnaissance capability, and a limited airlift capability. Role specialization in certain key areas particularly in the force-enabling areas, such as deployable ground infrastructure, would allow small- and medium-sized states to gain leverage within an alliance or coalition. According to the same logic, reluctance to go down the path of role specialization contributes to the European imbalance between the shaft and the spear. In general, smaller countries, which have to make hard choices due to limited defense resources, emphasize offensive over force-enabling capabilities.

DOCTRINAL AND OPERATIONAL RESPONSES

Doctrinal and operational responses can have a force-multiplying effect. In particular, they allow European air forces to rely more on human-centric approaches to air power than on hardware and software. During Operation Deny Flight, the American air component commander announced that all missions had to be escorted by dedicated SEAD aircraft. Yet some European allies expressed their doubts about this (from their vantage point) far from flexible way of operating.[106] Instead of over-reliance on SEAD and standoff jamming, Dutch pilots preferred to make their flight patterns as unpredictable as possible.[107] It was basically an argument on whether to rely on rigid force packages with standoff jammers and dedicated SEAD assets or to draw upon air power's flexibility.

The Royal Netherlands Air Force displayed other distinct national approaches in the air campaigns over Bosnia. Since the mid-1980s, the Royal Netherlands Air Force had been pioneering the so-called swing-role concept. In contrast, other European F-16 users were employing the aircraft in fixed roles only.[108] All Dutch F-16 pilots were trained in both the air-to-air and air-to-ground role.[109] Due to this concept and its inherent flexibility, Dutch F-16s could be retasked from an air defense to an attack mission while in the air, providing the combined air operations center with some extra leeway.[110]

Accordingly, size is not necessarily proportional to effectiveness. Thus far, size has not played a significant role in generating relevant European air power. For Operation Allied Force, the German contingent contained ten dedicated SEAD Tornados. This small contingent released approximately one-third of the total amount of high-speed antiradiation missiles expended throughout the entire campaign.

STRATEGIC CHOICES

The development and production of cutting-edge combat aircraft depend upon an advanced and solid industrial base. Only a few nations actually fulfill these prerequisites and are also willing to expend the required resources. For these reasons, international cooperation has always been an important consideration in European aircraft development. The United Kingdom, Germany, and Italy joined together to develop and produce the most important European all-weather strike aircraft of the Cold War, the Tornado. Currently, the most important European combat aircraft program is the Eurofighter Typhoon. From a German vantage point, there is more to industrial cooperation than just cost savings, as it has contributed significantly to closer alliance relations and has, therefore, enhanced European and transatlantic security.[111]

France's move toward strategic independence has been reflected by the country's approach to acquiring aerospace assets. The French aerospace industry develops and produces not only domestic combat aircraft, but also a coherent range of subsystems, such as laser designator pods as well as air-to-air and air-to-ground armaments. However, strategic independence has its price. Very often, domestic programs are more costly than buying off the shelf, and delays of national or European development programs have direct corollaries for operational effectiveness. In particular, the French air mobility fleet has been plagued by chronic overstretch. While the Royal Air Force started to lease four C-17 Globemasters in 2001, the French Air Force as of 2009 was not expecting first deliveries of the A400M transport aircraft until late 2012.

Among the smaller European countries, Sweden stands out. Throughout the Cold War era and beyond, the Swedish aerospace industry has proved its capacity to produce very sophisticated aircraft, such as the Draken, the Viggen, or the Gripen, Sweden's latest fighter-bomber. A substantial part of Gripen components, however, have been procured from abroad, which helped to break the cost spiral. As such, the Swedish aerospace company Saab considers systems integration as one of its core competencies.[112]

The Netherlands is at the other end of the scale. Buying off the shelf has been facilitated by the country's strong transatlantic orientation. It has allowed the procurement of advanced American aircraft at low unit costs without excessive delays. The Netherlands procurement policy puts a premium upon meeting the Royal Netherlands Air Force's operational needs. Yet buying off the shelf has not contributed to European strategic autonomy. Moreover, the Netherlands and other European countries with a similar procurement policy have indirectly benefited from a domestic European aerospace industry. In the absence of such an industry, American suppliers might have had no incentive to offer their products at competitive prices.[113]

While buying off the shelf can generate effective air power at relatively low costs, such an approach, if pursued by major European countries, paradoxically threatens to widen the transatlantic capability gap. Some of the larger constituencies might be less inclined to spend significant sums on assets not produced domestically that can only be used effectively in conjunction with U.S. forces. A domestic European aerospace industry is of strategic importance. Accordingly, a British commentator argued, "it would be difficult to envisage any credible European security policy in the second century of air power without a credible aerospace industrial base to sustain it."[114]

The acquisition of modern fighter-bombers has far-reaching corollaries. As such, it requires or reflects profound strategic choices. These choices are aptly exemplified by the European involvement in the Joint Strike Fighter (JSF) project and European space programs.

THE JOINT STRIKE FIGHTER

The American-led F-35 Lightning II, or JSF, offers a hybrid between buying off the shelf and being fully involved in a multilateral aircraft program: hierarchical partnership. Apart from the United Kingdom, four European nations are committed to the American-led F-35 project. As of mid-2009, governments planned to buy the following quantities: Italy 131, the Netherlands 85, Norway 56, and Denmark 48 aircraft.[115] It is interesting to note that the two largest European JSF customers, the United Kingdom and Italy, are also Eurofighter Typhoon core nations. The involvement of these nations in the JSF program cannot be reduced to one single factor. Yet both countries' transatlantic outlook, which, in the case of the United Kingdom, is underpinned by the so-called special relationship with the United States, might have been among the predominant factors. Other factors most probably include access to cutting-edge American technology or interoperability with the USAF and other U.S. services.

Unlike European cooperative programs, however, the JSF program does not offer truly equal partnership. Besides the United States as the lead nation, there are different levels of partnership. With the United Kingdom being the only level I partner and the Netherlands and Italy being the only level II partners, the remaining participating nations are relegated to level III partnership.[116] Accordingly, against the backdrop of a potential deal with Norway, a Saab official argued: "What differentiates the Gripen deal is that it offers real and equal partnerships in the ongoing development of a real project. There will be no restricted black boxes or secret source codes."[117]

Despite the hierarchical structure of the JSF partnership, Italy and the Netherlands have significant industrial stakes in the project. For instance, Italy plans to build a final assembly and checkout facility, which is likely to become a maintenance, repair, and overhaul-and-upgrade hub for European JSF operators.[118] Lockheed Martin has identified the Dutch company Fokker Elmo as the main supplier of wiring harnesses.[119]

Finally, on November 20, 2008, Norway eliminated the next-generation Gripen from its fighter competition and settled for the JSF. Norwegian officials argued that the Gripen would not only be unable to fully meet Norway's requirements, but that the aircraft would also be more expensive to acquire and operate than the JSF. Saab reacted by accusing Norwegian authorities of "incomplete, or even faulty, analysis."[120] In particular, Saab's chief executive challenged the Norwegian evaluators' decision, arguing that they had more than trebled his company's guaranteed bid price, which resulted in a figure higher than the estimated figure for the JSF. Moreover, Saab conducted air-to-air combat simulations, thereby supposedly putting paid to Lockheed's marketing claims of the JSF's air-to-air combat superiority.[121] The Norwegian fighter contest has a precedent. In the 1970s, the F-16, which at the time lacked an all-weather capability, was favored over the Swedish Viggen, specifically suited for the Nordic climate.[122]

However correct or faulty the evaluation results in the Norwegian fighter contest were, the debate conveys that the procurement of modern combat aircraft has strategic implications that exceed the technical and combat-tactical level. It could be argued that Norway's choice also reflects its national security strategy, which puts a premium upon securing the United States' commitment to Europe's Nordic region.

Norway also has had an excellent experience with its F-16 fleet. The aircraft's operational service life was originally estimated at twenty years. Yet through multinational upgrade programs together with Belgium, Denmark, and the Netherlands, the service life could be extended by ten to fifteen years.[123] Moreover, technological developments in the post–Cold War era, combined with precision-guided munitions, turned relatively small fighter-bombers such as the F-16 into truly multirole platforms with outstanding weapons-delivery accuracy. In contrast, complex aircraft, such as the F-111 or the Tornado, were required for all-weather precision strike during the Cold War. A repeat of the F-16 success story might be expected with the JSF.

Like Norway, Denmark has put a premium upon strong transatlantic relations. The country deliberately opted out of defense endeavors within the framework of the EU. NATO and close cooperation with the United States are regarded as the central security pillars. As such, there are, besides the Swedish Gripen, two American fighters amongst the potential candidates in the Danish F-16 replacement contest.[124]

AEROSPACE POWER

Current British air power doctrine, AP 3000 *British Air and Space Doctrine,* quite explicitly underlines the United Kingdom's dependency upon U.S. space assets.[125] In contrast, French authorities embarked upon ambitious military satellite programs after Desert Storm. The first truly capable French military observation satellite, Helios I, was launched in July 1995, and it quickly exposed the strategic implications of European dependence upon U.S. satellite intelligence. In September 1996, Washington justified air strikes against Iraq by claims of large-scale Iraqi troop movements against Iraqi Kurds. France disputed these American claims on the grounds that Helios I did not observe any of the claimed troop movements. Discrepancies of this kind subsequently contributed to a widening rift between the Anglo-American and the French policies on Iraq.[126]

Germany has also been taking significant steps to establish a national strategic reconnaissance capability. In late 2006 the first of five SAR (synthetic aperture radar)-Lupe satellites was launched. In addition to this national capability, an agreement to secure a satellite data exchange between Germany and France was implemented in early 2008. Project MUSIS is supposed to further synchronize national European satellite programs with the aim of establishing a multinational strategic reconnaissance capability. Participating states are France, Germany, Italy, Spain, Belgium, and Greece.[127] Yet countries that have traditionally put a particularly strong emphasis upon the transatlantic alliance have generally opted out of European military space programs. For instance, the Netherlands declined participation in the French Helios II satellite surveillance program.[128]

FORMER WARSAW PACT AIR FORCES

After the Cold War, many former Warsaw Pact states embarked upon a profound transformation in order to bring their militaries up to NATO standards. This has been a challenging task, as interoperability requires not only hardware commonality, but also considerable changes in training and concepts. In the case of Poland, the country's new transatlantic strategic outlook is reflected in the Polish Air Force's transformational efforts.

THE POLISH AIR FORCE

In 1990 the Polish Air Force was among the largest in the Warsaw Pact, comprising in excess of five hundred jet aircraft.[129] By 2009, a decade after Poland's entry into NATO, this figure was reduced to 128 combat capable aircraft, including a mix of F-16Cs and Soviet aircraft, such as the MiG-29.[130] The first of forty-eight F-16s was received in November 2006. It is one of the most visible signs of Poland's military transformation toward a military that puts a particular premium upon interoperability with the United States. Yet the rapid transformation also involved a number of significant difficulties, such as an initial shortage of trained personnel to operate the Polish F-16 fleet, which led to a number of malfunctions.[131]

Poland also aggressively modernized its transport fleet. On March 24, 2009, two months after its last AN-26 was withdrawn from service, the Polish Air Force received the first of five U.S. surplus C-130E Hercules transports. The refurbished aircraft are expected to remain operational for the next twenty years, supporting Polish military deployments through a range of tasks, including medical evacuation and rescue missions. Poland is also a member of NATO's Strategic Airlift Capability that can draw upon jointly acquired C-17 Globemasters based in Hungary.[132] In an attempt to extend its capabilities profile, the Polish Air Force is considering procuring two tanker-transports that would support deployed F-16 operations. With the Boeing KC-767 being one of the prime contenders, the choice might be American again.[133]

American preponderance relates not only to the procurement of hardware, but also to training and education, a domain where the Polish Air Force seeks particularly close alignment with the USAF. Polish F-16 pilots have been trained in the United States,[134] and a significant number of Polish mid-career officers have been attending the Air Command and Staff College and the Air War College at Maxwell Air Force Base, Alabama. Annual U.S. Air National Guard and Air Reserve deployments to Poland supplement these close ties in order to conduct joint F-16 training.[135]

Yet it is interesting to note that, for real operations, the Polish Air Force has so far relied upon Soviet equipment. As such, the Poles deployed MiG-29s for air policing duties to the Baltic States from March to July 2008.[136] NATO's 2004 expansion round included Estonia, Latvia, and Lithuania, none of which possessed adequate assets to collectively control their airspace. As a consequence, air policing and quick reaction alert (QRA) duties fell upon the wider NATO community. This particular NATO commitment is expected to last until around 2018.[137]

THE CZECH AIR FORCE

Like the Polish Air Force, the Czech Air Force has undergone a far-reaching transformation process from a former Warsaw Pact air force to a NATO air force. The most visible sign of the Czech Air Force's transformation is its equipment. Yet unlike their Polish counterparts, the Czechs opted for a European fighter aircraft. The Czech Air Force started operating the JAS 39 Gripen in 2005 under a ten-year lease deal brokered via Sweden's Defense Material Administration.[138] The Czech Gripen fleet consists of fourteen fighters, which are used both for training and frontline purposes, with two aircraft plus one reserve held at permanent QRA status within the Czech Republic. Apart from participating in multinational exercises in Belgium and France,[139] the Czech Gripen contingent also hosted Air National Guard F-16s, maintaining good relations with both the USAF and European air forces.[140]

From May to September 2009, the small Czech Gripen fleet was particularly heavily tasked with two simultaneous QRA commitments. Besides national QRA duties, four Czech Gripen fighters deployed to the Baltic States, where they were held at high readiness against the backdrop of NATO's air policing mission. Given the limited number of airframes, the Gripen contingent is currently unable to expand its mission spectrum to include roles such as air-to-ground attack. As a result, the Czech Air Force focuses on the air-to-air domain and intends to hone its skills by participating in a number of integrated exercises. With an eye on a combined Czech-Swedish deployment across the Atlantic for a Red Flag exercise in the United States, air-to-air refueling was identified as an important enabling ability to be included in the Czech program.[141] Overall, Czech Gripen operations are testimony to a successful, albeit limited, force transformation.

TWO KINDS OF OPERATIONS

The Balkan air campaigns, under a NATO banner and partly on behalf of the UN, were a formative experience, as they represented for many European air forces the first real combat operations since World War II. These days, European nations engage primarily in two types of operations: NATO or EU operations. Accordingly, a division of labor is gradually becoming apparent between the two cooperative security bodies. While the EU focuses on rapid reaction interventions and stabilization operations in the developing world and in the Balkans, NATO's International Security Assistance Force (ISAF) in Afghanistan conducts a support operation against the backdrop of the war on terror. Simultaneously, European air forces have contributed to Operation Enduring Freedom, which runs in parallel to ISAF.

Continental European Air Powers and the War on Terror

Despite alliance frictions during and in the wake of Operation Allied Force, the French Air Force was the first European air force to engage targets in Afghanistan by means of fighter-bombers. In order to provide French decision makers with autonomous intelligence, the offensive phase was preceded by a reconnaissance phase. The deployment of attack aircraft was two-pronged. On the one hand, French Super Étendards from the aircraft carrier *Charles de Gaulle* had been flying missions over Afghanistan since December 2001. Together with Italian Navy Harriers operating off the carrier *Garibaldi,* these aircraft supported U.S. Navy fighter-bombers circling over Kandahar and Tora Bora. On the other hand, six

Mirage 2000Ds together with two tanker aircraft were deployed over six thousand kilometers from France to Manas, Kyrgyzstan, on February 27, 2002. On March 2 these aircraft, alongside sixteen French Navy Super Étendards, took part in Operation Anaconda.[142] American air component commander Lt. Gen. T. Michael Moseley argued that, given the ferocity of the fighting on the ground, he had to involve the French Mirage aircraft immediately. The French detachment was the first to be based at Manas, and Moseley acknowledged France's role in establishing a new front for operations over Afghanistan. In June 2002, U.S. Marine Corps F/A-18D Hornets joined the French Mirage aircraft in Manas.[143] Over a protracted period of time, French aircrews covered vast distances from Manas to Afghanistan in each sortie.

After this initial deployment, the French Air Force has continued to deploy combat aircraft to Central Asia, for both Operation Enduring Freedom and ISAF. In August 2007 the French Air Force Central Asia detachment redeployed from Dushanbe, Tajikistan, to Kandahar, in order to reduce transit time and increase NATO's firepower in the southern Afghan provinces. The detachment has comprised a total of six combat aircraft, normally combining three Mirage 2000Ds with either three Mirage F1s or three Rafales. The latter and French Navy Rafales, operating from the aircraft carrier *Charles de Gaulle,* saw action over Afghanistan in March 2007 for the first time.[144]

Earlier operations out of Manas, Kyrgyzstan, highlighted the move toward European cooperation when, on October 1, 2002, the French Mirage 2000D detachment was replaced by a European F-16 detachment.[145] This combined detachment consisted of eighteen F-16s from Denmark, the Netherlands, and Norway, supported by a Dutch KDC-10 tanker aircraft. The European Airlift Coordination Cell based in the Netherlands supported the transition.[146]

Despite some legal and procedural obstacles, the multinational F-16 deployment was effective. In order to further improve cooperation between the European F-16 users, Lt. Gen. Dick L. Berlijn, then commander in chief of the Royal Netherlands Air Force, put forward the initiative of a combined European expeditionary F-16 wing. The defense ministers of Belgium, Denmark, the Netherlands, Norway, and Portugal signed the memorandum of understanding for establishing the European Participating Air Forces' Expeditionary Air Wing (EPAF EAW) during the NATO summit in Istanbul on June 28, 2004. The participating air forces intend to make optimum use of complementary assets in out-of-area operations in order to increase efficiency.[147] Through their combined commitment, EPAF nations as a group can deliver more robust and sustainable force packages than autonomous national efforts would allow. Since their deployment to Manas as part of Operation Enduring Freedom, European EPAF EAW F-16 users have continued to support ISAF operations in Afghanistan.

The ability to engage at the higher end of the spectrum of military force had become an important pillar of Dutch foreign and defense policy. In a speech given in early 2004, the Dutch defense minister underlined that allied solidarity must be apparent not only from a country's military capabilities, but also from its willingness to share risks. He specifically argued that politicians must have the courage to shoulder responsibility for operations entailing fatal casualties in order to make a significant contribution to stability and security.[148]

Unlike France or the European F-16 users, Germany has been reluctant to deploy offensive air power to Afghanistan. Nevertheless, the German Air Force played a vital role in running the Kabul airport and dispatched reconnaissance Tornados in support of ISAF.[149]

EU OPERATIONS

Rapid reaction interventions throughout the Cold War and in the 1990s were predominantly conducted within national frameworks. Against the backdrop of an emerging ESDP, these operations assumed a multinational dimension under the aegis of the EU. The former colonial powers France and Great Britain have accumulated the most experience. In the post–Cold War era, Operation Palliser was the most prominent British rapid reaction intervention in Africa. In 2000, the operation decisively contributed to the stabilization of civil-war-torn Sierra Leone with minimal, but rapid, use of force. This section looks at an examination of the French experience, which, together with the British experience, has served as a blueprint for European transformation efforts.

On many occasions, the French armed forces intervened in Africa in order to stabilize hotspots or evacuate Western citizens. Providing air mobility and close air support, French Air Force units have proved indispensable for mission success. Probably the most significant French intervention in Africa during the 1990s was Operation Turquoise, lasting from June 22 to August 22, 1994. Its goal was to stop genocide in Rwanda and establish a safe haven. Given Rwanda's landlocked location and distance from France—more than eight thousand kilometers or five thousand miles—rapid deployment could only be executed by airlift. Due to European strategic airlift shortfalls, the air bridge was considerably augmented by chartered Russian wide-body transport aircraft. A total of three thousand personnel and seven hundred vehicles and additional cargo were moved during the operation. Moreover, the twelve hundred French frontline troops were supported by twelve combat aircraft.[150] Other major joint interventions were conducted by the French armed forces in Central Africa (1996), in the Congo (1997), and in the Ivory Coast (beginning in late 2002). On each occasion, inter- and intra-theater airlift proved to be a vital key to success.

In the light of the emerging ESDP, French decision makers shifted from autonomous toward multilateral operations. The EU's first autonomous military operation outside Europe took place in the southern provinces of the Democratic Republic of Congo, lasting from June 6 to September 1, 2003. The EU conducted Operation Artemis on behalf of the UN in order to respond to the killing of approximately four hundred civilians by civil war factions in the city of Bunia. Out of 1,860 troops, France provided 1,660.[151] The main combat element consisted of 150 French and approximately 75 Swedish special operations forces. The deployment of these Swedish forces is an interesting aspect, particularly given the country's legacy of neutrality. During the operation, French Mirage aircraft conducted close air support and reconnaissance missions,[152] providing the small ground contingent with superior firepower. Successful rapid reaction interventions draw upon air-ground synergies in order to make up for limited deployed troops and assets.

The United Kingdom and France again shaped European defense policy when the EU Battle Group concept was, for the first time, raised at the Franco-British summit in Le Touquet on February 4, 2003.[153] The concept became a key aspect of Headline Goal 2010, ratified by the EU member states in 2004.[154] After prolonged interventions in Bosnia and elsewhere, the EU Battle Group concept represents a move toward more robust, but timely, limited operations, primarily based on Chapter VII of the UN Charter, which does not require the consent of the warring parties.

The EU Battle Group concept has accelerated the force transformation process of smaller European countries. At the Council of Defense Ministers in Brussels, on November 22, 2004, it was declared that Sweden, Finland, Norway, and Estonia intended to establish a multinational battle group, with Sweden as the lead nation. Out of fifteen hundred troops, Sweden contributes approximately eleven hundred. The Nordic Battle Group's first standby lasted from January 1 to June 30, 2008.[155] Though being a land-centric concept, it had a significant impact upon the Swedish Air Force's move toward deployed operations. During the standby period, eight NATO interoperable JAS 39 C Gripen combat aircraft provided part of an air component.[156] This rapid reaction Gripen unit represented a preliminary point of culmination in a shift from an air force that was almost exclusively geared toward autonomous territorial defense to an air force that can effectively operate in an international framework.

After Operation Artemis, European Union Force (EUFOR) RD Congo was the second major military operation conducted by the EU in the Democratic Republic of Congo, this time in 2006. Unlike Artemis, EUFOR was not a rapid reaction intervention, but a stabilization operation, with the aim of providing security during the election in conjunction with other civilian EU crisis management cells. Besides a number of European countries dispatching airlift assets, Belgium deployed unmanned aerial vehicles, while the French Air Force made a vital contribution by making available forward-deployed bases and providing a contingent of Mirage F1s, which reportedly had a deterrent effect at potential trouble spots.[157] France's military experience on the African continent was also important for the joint and multinational Operation EUFOR Tchad/RCA. Lasting from January 2008 through March 2009, the joint EU operation took place against the backdrop of the crisis in Darfur. A key objective was to protect civilians, particularly refugees, in Chad and in the Central African Republic.[158]

CONCLUSION

In the latter half of the Cold War, continental European air powers were an integral part of NATO's Flexible Response doctrine. Since air power effectively brought to bear superior Western technology, it was a key element of NATO's efforts to meet the Warsaw Pact's numerical superiority in conventional forces. Though it went unnoticed in the east, doctrinal aspirations exceeded available continental European means.

Coinciding with the end of the Cold War, Desert Storm offered a very vivid demonstration of modern air power. According to their national circumstances, the three air forces examined reacted differently to the air war in the Gulf. Yet in the short and long terms, each of the air forces started to implement the lessons learned from modern air warfare. This is clearly reflected in their approaches to deployed operations, procurement, and doctrine.

For the French Air Force, Desert Storm brought about a paradigm shift in terms of interoperability, international integration, and a new emphasis upon advanced conventional weapon systems. It can be concluded that, among the three air forces examined, Desert Storm had the biggest impact upon the French Air Force. Naturally, there is a difference between "being there" and "watching an operation." Yet from a sociopolitical point of view, the employment of French air power across the spectrum of military force was nothing extraordinary, as the French armed forces regularly engaged in deployed operations on the African continent throughout the 1980s.

On a political level, the Netherlands swiftly implemented lessons from the Dutch F-16 fighter-bombers that were nearly deployed to the Gulf area in 1990–91. On a military level, however, as regards doctrine development and the acquisition of an air-to-ground precision capability, Operations Deny Flight and Deliberate Force were more defining, being the first air campaigns for the Royal Netherlands Air Force. Political willingness to employ the air force across the spectrum of military force became a significant leverage for Dutch air power. The Royal Netherlands Air Force is a paradigm of smaller nations making significant contributions to multinational operations out of proportion to their size, if there is political resolve combined with adequate military capabilities. Yet even in the case of the Netherlands, an overtly aggressive stance in the use of offensive air power had to be avoided. In order not to arouse public concerns, the Dutch supreme commander deliberately imposed a low profile upon the Royal Netherlands Air Force's performance in Operation Allied Force. This was particularly the case after a Dutch F-16 downed a Serb MiG-29 in the first night of operations: the only European air-to-air kill in major air campaigns throughout the 1990s.

German authorities, for their part, felt uneasy with offensive operations and continued to emphasize the territorial defense of the alliance throughout the 1990s. This led to a gap in doctrinal innovation within the German Air Force, which was overcome with the defense reform starting in 2000 and culminating in 2003. Given its historical legacy, Germany has been constrained in employing military power in real operations. Besides air mobility, the German Air Force has been relegated to SEAD and reconnaissance. Strike aircraft have so far not been employed in real operations. Since there is still a dichotomy between operational potential and political acceptance, the German Air Force's wide-ranging capabilities are unlikely to be fully translated into effective operational output in deployed operations. This collides with Germany's strong emphasis upon NATO, as alliance solidarity might be questioned by its main allies. Nevertheless, German aircrews have been exposed to significant risks entailing potential fatalities, as demonstrated by the German Air Force's outstanding role in the air supply efforts to Sarajevo and eastern Bosnia during the first half of the 1990s.

Germany's uneasiness with offensive air power is, to various degrees, shared by other European partners, as was demonstrated throughout the Balkan air campaigns. While European partners preferred gradual bombing campaigns that were closely intertwined with diplomatic efforts, American airmen put a premium upon a parallel warfare strategy, striking at several target sets simultaneously.

Maintaining relevant air power has become increasingly challenging, particularly against the backdrop of a seemingly never-ending cost spiral. Each aircraft generation has been considerably more expensive in real terms than the one it has replaced. As a result, smaller European states cannot afford the full panoply of aerospace capabilities. Role specialization would mitigate European air power shortfalls, as it would help to address the imbalance between offensive air power and force enablers. So far, this option has not really been exploited, as it supposedly reduces a nation's political discretion.

Cooperation amongst the larger European air forces is also gaining in importance. The Royal Air Force, the French Air Force, and the German Air Force together—or any two of these combined—cover a significant range of aerospace power capabilities. This core of capabilities is coherent and would allow smaller European air forces to plug in and play. By providing niche capabilities, they could reinforce the existing capabilities of these larger air forces and thereby contribute to more robust and sustainable force packages. Consequently, role specialization should not happen by default, as is currently the case owing to limited resources, but as a result of a deliberate and focused policy. While it is unrealistic to assume that all alliance partners would contribute to a particular operation, be it in the framework of NATO or the EU, it is realistic to expect that any two of the larger European air forces combined with a number of smaller air forces would commit themselves. If defense resources are appropriately allocated, such a combination of air forces is likely to cover a balanced range of capabilities for deployed operations.

Not only are planners confronted with the challenge of appropriately allocating defense resources to cover a balanced spectrum of air power, but they also have to take account of the strategic corollaries accompanying aerospace procurement programs. In this regard, air power is truly strategic as it requires far-reaching choices. While buying American assets off the shelf can generate effective air power at comparably low costs, such an approach can undermine the foundations of a European aerospace industry. Some former Warsaw Pact countries have sought close alignment with the United States. In particular, Poland considers a strong transatlantic link as the ultimate guarantee of its integrity. This strategic outlook is reflected in the Polish Air Force's procurement policy.

Another area that requires strategic choices is space. It is basically a choice between European autonomy and access to, as well as dependence on, advanced American space assets. While the United Kingdom's special relationship with the United States has been beneficial to British air power, it has led to complacency in the domain of space power. In contrast, some continental European countries, particularly those that put a premium upon a strong European defense pillar, have developed their own capabilities and have been grasping space power faster. They regard space power as a core strategic competency. Moreover, complete dependence upon U.S. space assets is not a real alternative for continental European states, as they do not have access to American intelligence to the same degree as their Anglo-Saxon counterparts. This reportedly also applies to countries such as Denmark with its pronounced transatlantic orientation. The Five Eyes (Australia, Canada, New Zealand, the United Kingdom, and the United States) share a privileged partnership.

These strategic choices are closely related to the European defense architecture. While NATO was the uncontested Western European defense pillar throughout the Cold War, the ESDP offers a complementary alternative to NATO in the post–Cold War era. With the ESDP providing an effective framework for joint and comprehensive stabilization operations in the developing world, NATO has taken on a prominent role in the war on terror, and it is still the sole guarantor of collective defense. In general, it is not a question of NATO versus the ESDP, but a matter of emphasis. Countries that put a premium upon a strong transatlantic partnership prioritize NATO operations; countries that aspire to develop a more autonomous European defense put an emphasis upon ESDP operations.

Out of area, European countries are not capable of autonomously dealing with major contingencies. In these circumstances, it is "inconceivable that European governments would act independently of the U.S."[159] However, the EU needs to be capable of autonomously undertaking peace support operations across the spectrum of military force, with a particular focus upon joint early entry operations in failed states. After severe disagreements over Iraq in early 2003, it is difficult to imagine that the EU could, in a timely manner, have drawn upon NATO resources, meaning American assets, for Operation Artemis in mid-2003.

Rapid reaction operations in failed states require solid strategic airlift; intelligence, surveillance, and reconnaissance (ISR); and close air support, the last one providing escalation dominance. Once a robust number of A400M aircraft have been commissioned in the European Air Transport Command, which offers assured availability for EU operations, Europe will have unprecedented surge capacities at its disposal. In the domain of ISR, the various European satellite programs represent a major quantum leap.

Overall, continental Europe places a lesser emphasis upon air power than does the United States. Despite the lesser emphasis, continental Europe has been developing air power capabilities that have the potential of taking on a more prominent role in combined operations. Against the backdrop of a decrease in American air assets, European contributions to deployed operations might become more relevant, not only from a political point of view, but also from an operational one. Continental European air forces remain natural allies of their Anglo-Saxon counterparts.

THE FUTURE OF AIR POWER

David A. Deptula

"The world stands on the threshold of the 'aeronautical era.' During this epoch the destinies of all people will be controlled through the air."

—Brig. Gen. William "Billy" Mitchell, *Winged Defense*, 1925[1]

The advent of the aircraft and its use as both an instrument of commerce and a weapon of warfare have transformed mankind. Indeed, air power now permeates all modern societies and military organizations, having become as indispensable as it is decisive.

The seven decades that followed Brigadier General Mitchell's declaration of the "aeronautical era" demonstrated his prescience. From the skies over Europe and the Pacific in World War II, to the air battles over Korea, Vietnam, the Middle East, the Balkans, the Falklands, and South Asia—air power has meant the difference between victory and defeat.

Air power has also exerted a profound influence in the realm of commerce and, to some extent, in establishing national identity. As of the spring of 2010 over twenty-seven hundred private and national airlines operated globally, transporting an estimated 11 million passengers and 120 million tons of cargo every day.[2] Even with the widespread availability of the Internet and fast video processing, which have enabled teleconferencing and electronic mail for the past twenty years, the commercial air carrier market in the United States has experienced only an estimated 8 percent reduction in traffic.[3] Being there in person remains the preferred way to conduct business, transmit time-sensitive information, and learn about people and places.

The challenges encountered during Mitchell's "aeronautical era" have been profound. At the height of the combined bombing campaign in Europe during World War II, only 20 percent of the weapons dropped hit within a mile of their intended target.[4] Flying in adverse weather or at night was only marginally possible until the development of on-board instrumentation, and the widespread use of radar for precision approaches, traffic control, and early alerting and warning of impending air attack. Air-to-air combat was still a visual range affair to be executed with aimed, but imprecise, gun and cannon fire. In 1929 an innovative demonstration of in-flight refueling showed promise of extending aircraft range, but it would take another thirty years for force-extension in the form of dedicated air-to-air refueling aircraft to arrive in any usable form.[5] The same delay can be seen in air mobility: despite early, costly attempts at instituting airmail service in the 1920s, it took the pressures of a world war in the 1940s to drive the development of specially designed cargo-transport aircraft that subsequently gave rise to mass commercial passenger transit.

Today, air combat is conducted with aircraft forward and, in the case of remotely piloted aircraft, with the vast majority of operators thousands of miles away. Cargo can be air-dropped with satellite-guided precision anywhere on the globe. Weapons—both air-to-air and air-to-ground—can be delivered well beyond visual range with an accuracy measured in centimeters, during the day, at night, and in all weather, in some cases, even with automatic target recognition and engagement. Our aircraft are becoming less the specialized platforms associated with the monikers "fighter," "bomber," "cargo," or "intelligence, surveillance, and reconnaissance (ISR)," and instead are flying multi-role nodes in a connected commercial enterprise or in a contest between belligerents. The question that confronts us in this era is: what's next?

TECHNOLOGY: TRENDS AND THREATS

Despite the tremendous successes achieved in the "aeronautical era" and air power's demonstrated ability to deliver weapons, information, and resources with precision, air power still has its detractors. We are entering an era of disconnects and ironies with respect to the realities of air power past and present. This situation, if not carefully analyzed and studied in the rigorous lights of practical experience and a disciplined methodology, could lead us to false conclusions and a dangerous security future. Thus, it is important to review the trends currently influencing the development of air power at the micro (tactical), meso (operational), and macro (strategic) levels in the context of both today's air power capabilities and the challenges that air power will face in the future.

At the micro level, three broad trends will have an impact on air power. First, computing, sensing, and data compression technologies are advancing in ways that will radically change how we conduct operations in the atmosphere. Second, the ability to transmit data using innovative means—for instance, transmitting only the parts of a video or radar picture that have changed—enables airmen today, commercial or military, to stay far more connected to activity in other domains than their predecessors were. Third, capabilities that in the past required a specialized and tailored platform can now be combined on one aircraft.

At the meso level, our previous operational model for winning battles, which linked air power to massive, mechanized surface maneuver forces, has been supplemented by one that links air power to small elements of special operations forces or intelligence sources, capable of achieving similar effects with fewer resources. This has been made possible by the emergence of a reliable capability to deliver effects that historically required mass to compensate for a lack of precision, and of the ability to find, fix, track, and engage rapidly and precisely, at all times and under all conditions, on a global level. Together these advances have moved us into an era where desired effects from, through, and in the air domain have achieved unsurpassed prominence in meeting security objectives. From force deployment, ISR, rapid aeromedical movement, and communications, to precision and rapid force application, air power has become essential.

Finally, at the macro level, we must recognize the extent to which nations have become reliant on rapid global power projection and mobility—all made possible by air power. The United States did not become a global air power nation overnight—that capability resulted from more than sixty years of careful and thoughtful investment. However, increased reliance on and use of its airborne arm are rapidly aging U.S. aircraft systems.[6] At present rates of aircraft employment, the United States faces increasing risk that its ability to support its global responsibilities by air will erode.[7] Allies are already expressing concern over this situation. For example, in Australia's latest *Defence White Paper*, published in 2009, the country's defense minister notes that "[t]he biggest changes to our outlook . . . have been . . . the beginning of the end of the so-called unipolar moment; the almost two-decade-long period in which the pre-eminence of our principal ally, the United States, was without question."[8]

Time and technology wait for no one—especially airmen. As informative as it may be to study the trends that influence air power, those trends must be viewed in conjunction with the emerging threats to yield usable guidance for the future. As the previous chapters have outlined, air power has emerged as one of the most vital and flexible tools a nation can employ in pursuit of defense and security. This has not gone unobserved by those seeking to gain influence on the world's stage. Nations around the world continue to adopt air power in all its forms—both commercial and military. As of this writing, no less than eighteen advanced fighters and bombers are in production or design worldwide.[9]

Technology has dramatically improved not only aircraft, but also the sensors used to find, fix, track, and target to create effects. Some forty different over-the-horizon radars worldwide now enable vastly improved long-range detection and classification of aircraft over roughly two-thirds of the globe.[10] The power, reach, and fidelity of other radars have also greatly increased. At the same time, technology has led to ever more powerful weapons, including surface-to-air missiles capable of reaching their targets over hundreds of miles—even to space—and air-to-air missiles with ranges of more than a hundred miles. We can safely assume that the range and reach of these systems will continue to increase over time.[11] Of special interest are the ways and means being developed to prevent nations from employing their air power effectively; ballistic and cruise missiles with the range and precision to target and deny both land- and sea-based air power already exist today.[12]

As we become ever more connected by airborne networks, we are also becoming more and more vulnerable to possible infiltration, disruption, and displacement of those networks. Threats in the cyber domain can significantly affect operations and linkages across all domains—land, sea, space, and air alike.[13] Additionally, we are witnessing the dawn of a new era in non-kinetic high-energy weapons that have the potential to render modern electronics and sensors useless. Microwave transmitters and active electronically scanned array radars, enabled by efficient, high-powered generators, are changing the way we define target destruction. A target may appear intact, but the electronics that control its functioning may be corrupted, rendering the sensor or weapon useless. These emerging capabilities may well define our operational challenges in the future far more than the mechanisms of kinetic destruction to which we have become accustomed.

IMPLICATIONS FOR THE FUTURE

The context of the past, especially the immediate past, is too often used to define a future that reassuringly reinforces the lessons of the most recent success or failure. From Colin Gray to George Friedman, the best strategists have tried—so far unsuccessfully—to draw attention to a fundamental rule about the future: it is almost impossible to predict by extrapolating directly from the past. Thus, the most useful approach to charting possible futures involves focusing on the trends and the threats emerging from those trends rather than making comfortable, and often devastatingly incorrect, predictions.[14]

It has become commonplace in our time to draw distinctions between regular and irregular warfare, and either to elevate or underestimate the challenges of violent extremism or the rise of long-dormant but "once-great" powers. Heated as the debate is, it is academic at best, perhaps saying more about us than about the challenges we face. At its worst, the debate creates a feeding frenzy over policy, funding, and force structures that often results in mismatched capabilities between belligerents.[15]

Air Power Environments

Heeding the strategists' warning and using the context defined here and in the preceding chapters of this book, airmen can identify three environments that will dominate the development, employment, and success of air power in the future: permissive, contested, and denied. Since the latter half of the twentieth century, air power nations have had the luxury of operating in a permissive air environment. The skies over Iraq and Afghanistan are present-day examples: while the environment on the surface is far from permissive, there is virtually no contest for control of the air.

In parallel with this current success comes the tendency to assume it will prevail in the future, yet trends identified in this book may rapidly change this environment into a contested one. And, of course, as those trends create new realities, the threats they spawn will lead us toward a denied environment—one in which the legacy systems of today would find it very hard to survive, much less operate successfully.

The Response: Each Aircraft a System

The challenges presented by these environments will drive airmen to unprecedented forms of innovation. To overcome past threats, combat aircraft required altitude, speed, maneuverability, precision, and eventually low observability, and these became their dominant characteristics. In the future, the ability to integrate the various characteristics and functions performed by a variety of legacy aircraft into one "system" will define improved air power capability.

Remotely piloted aircraft and advanced fifth-generation ISR/strike aircraft already provide glimpses of this evolution in air power. These aircraft represent the leading edge of air power's future—not because there is or is not a pilot on board, but because the increased integration of multiple capabilities on one aircraft enables compression of the "find-fix-finish" equation in both time and space. No longer must a nation wait for a tertiary system to assess effects before taking appropriate follow-on action. The integrated capability greatly compresses the time required for successful closing of the "observe, orient, decide, and act" loop.

As we move into the future it will be vital to understand the critical relationship between ISR and kinetic/non-kinetic attack capabilities. This kind of capability-based perspective will become ever more essential in an era of constrained defense resources. Therefore, while we will still build dedicated ISR platforms, we must incorporate ISR capabilities into all our platforms—air, space, sea, land, and cyberspace. Doing so will require adjusting concepts and processes that govern how we allocate, plan, and employ these systems.

Traditional nomenclature associated with aircraft limits the understanding of air power's potential in this regard. Today, a single aircraft can already perform the functions of ISR, strike, close air support, electronic warfare, strategic attack, and others. These capabilities will only improve with the maturation of concepts for modular aircraft and fractionated systems that are minimally affected if parts of the systems are removed. For example, the F-22 and F-35 are not merely "fighters": they are F-, A-, B-, E-, EA-, RC 22s and 35s. In addition to having a vast array of kinetic and non-kinetic attack capabilities, they function as flying ISR sensors that will allow the United States and its allies to conduct network-centric warfare inside adversary battlespace from the first moments of any conflict. In fact, they could be viewed as "all-weather-day-night-penetrating-sensing-kinetic-and-non-kinetic-processing-communcation-action" assets.

In the future, therefore, we will judge the value of aircraft in terms of their ability to sense and communicate, as well as by how well they perform in their traditional roles. Think of this approach as the "observer effect" extended to modern air power operations: Today, in the skies over Iraq and Afghanistan, more eyes and ears are collecting and analyzing data to inform actions than at any other time in the history of manned flight.[16] The simple act of observation causes adversaries to react by immediately changing their activities.

AIR POWER: A SEAMLESS WEB

Airmen are charting a course toward dealing with the resulting challenges by projecting power without projecting vulnerability, creating an asymmetric advantage from the air without the enduring legacy of footprints, and maintaining a focus on a future increasingly dominated by contested and denied environments. This vision does not signify or imply the end of manned flight. Neither now nor on the distant horizon is there a technology that can replicate the 360-degree spherical situational awareness and judgment of a pilot flying an aircraft—either hands-on or remotely. But the integration of twenty-first century technologies into modern-day aircraft will drive a new, multidimensional concept of air power as a seamless web of elements in a vast multi-role/multi-mission system of systems. That complex system will be employed as a cohesive, networked, survivable whole, rather than as specialized, mission-specific platforms operating in series. Realization of this vision will be the airman's counter to emerging threats, and will extend the enduring capability of air power to compress operations in the third dimension.[17]

CONCLUSION

Today, a decade into the twenty-first century, Mitchell's "aeronautical era" endures, and offers even greater potential for both warfare and commerce. As in Mitchell's time, a vision of a future as yet undetermined will in large part be defined by the exploitation of air power in all its forms.

While Mitchell could not have imagined the technological breakthroughs that enable today's integration among air, space, and cyberspace capabilities, he would readily have understood the underlying concept. The success or failure of competition—military or commercial—will be determined by how well a nation can seamlessly integrate air power across permissive, contested, and denied environments, rapidly synchronizing multiple aerospace missions and functions across the domains of air, land, sea, space, and cyberspace ahead of both competitors and adversaries. The coming decades, too, will present severe challenges and high costs. However, no great nation can afford the far greater price of stagnation or failure to adapt or embrace the potential of air power in this, our modern information age.

NOTES

INTRODUCTION

1. Winston S. Churchill, *The Second World War: The Gathering Storm* (New York: Mariner Books, 1948), 100.
2. F-22 Raptor, www.fa22raptor.com, accessed April 22, 2010.
3. F-35 Joint Strike Fighter Program, www.jsf.mil, accessed April 22, 2010.
4. Carl A. Spaatz, quoted in John A. Warden III, "The Enemy as a System," *Airpower Journal* 9, no. 1 (Spring 1995): 41–55.
5. Antulio J. Echevarria, Jr., "American Operational Art, 1917–2008," in *The Evolution of Operational Art: From Napoleon to the Present*, eds. John Andreas Olsen and Martin van Creveld (Oxford: Oxford University Press, 2011).
6. Edward N. Luttwak, *Strategy: The Logic of War and Peace*, revised and enlarged edition (Cambridge, MA: Belknap Press of Harvard University Press, 2001), 200–201.

CHAPTER 1. BRITISH AIR POWER

1. Maj. J. D. Fullerton, Royal Engineers (later Colonel Fullerton of the Royal Flying Corps), Chicago Exposition, 1893, *Operation of the Division of Military Engineering*, reprinted in *Flight*, December 20, 1917, 1343.
2. H. G. Wells, *War in the Air* (London: Bell and Son, 1908), chap. 8, 1.
3. Ibid., chap. 6, 6.
4. Dr. David Clarke, "Scareships over Britain: The Airship Wave of 1909," *Fortean Studies* 6 (1999): 39–63.
5. *Flight*, August 7, 1909.
6. House of Commons Debates [hereafter HC], August 2, 1909, vol. 8 (HC 02/8), chaps. 1564–1617; and *Hansard*, Parliamentary Archives, www.Parliament.uk.
7. HC 02/8, c1565.
8. HC 02/8, c1571.
9. *Hansard*, April 5, 1909, vol. 3, c716.
10. HC 02/8, c1582.

11. Ibid.

12. HC 02/8, c1607.

13. Conference on the Laws of War, The Hague, October 18, 1907. *Annex to Regulations Respecting Law and Custom of War on Land*, Section II, chap. 1, art. 25.

14. HC 02/8, c1583.

15. HC 02/8, c1600.

16. HC 02/8, c1614.

17. *The Aeroplane*, June 8, 1911, and July 20, 1911.

18. War Office Memorandum CD 6067, April 12, 1912, printed in *The Aeroplane*, April 18, 1912.

19. R. A. Mason, *Air Power: A Centennial Appraisal* (London: Brassey's, 1994), 15–18.

20. *Flight*, June 26, 1914, 686.

21. J. H. Morrow, Jr., *The Great War in the Air* (Washington, DC: Smithsonian Institution Press, 1993).

22. RFC HQ Memo, September 22, 1916, RAF Air Historical Branch, AIR/1/522/16/12/5.

23. Cited by Hilary St. George Saunders, *Per Ardua: The Rise of British Air Power, 1911–1935* (London: Oxford University Press, 1944), 207.

24. "Report of the Prime Minister's Committee on the Air Organisation and Home Defence against Air Raids," August 17, 1917; and H. A. Jones, *The War in the Air* (Oxford: Clarendon Press, 1934), Appendices 8–14.

25. See the author's *Air Power: A Centennial Appraisal,* for early evidence of Fullerton's vision, and T. D. Biddle, *Rhetoric and Reality in Air Warfare* (Princeton, NJ: Princeton University Press, 2002), for details of Tiverton's and others' staff work.

26. "The Second Report on the Prime Minister's Committee on Air Organisation and Home Defence against Air Raids, August 17, 1917," reprinted in E. Emme, *The Impact of Air Power* (Princeton, NJ: Van Nostrand, 1959), 35.

27. Ibid.

28. Ibid.

29. Ibid., 37.

30. Ibid., 33.

31. Memorandum by W. S. Churchill, "Munitions Possibilities of 1918," Section 4, October 21, 1917, cited in ibid., 38.

32. Brigadier Newall, "The Scientific and Methodical Attack of Vital Industries," May 27, 1918, in AIR 1/460/15/312/101, cited in Biddle, *Rhetoric and Reality in Air Warfare*, 35–36.

33. Letter signed off by W. A. Robinson, Air Ministry, Strand WC 2, 11555/1918, to the Secretary, War Cabinet, 2 Whitehall Gardens, SW1, May 13, 1918.

34. Memorandum on the Bombing of Germany by GOC Independent Force, Royal Air Force, to Secretary of State for Air, June 23, 1918.

35. Trenchard to Weir, July 2, 1918, I F G./79.

36. Annexure to Process Verbal, Third Session Inter-Allied Aviation Committee,

Versailles, and July 21 and 22, 1918, 40/A/3, from the Secretary, British Section, Supreme War Council, Versailles, to Weir and Trenchard, July 23, 1918, part I, item 3, part II, item 5, University of Birmingham Collection.

37. "Memorandum on the Subjects for Discussion Proposed by the French Representative for the Third Session of the Inter-Allied Aviation Committee by GOC Independent Force," RAF at Ochey, July 9, 1918, University of Birmingham Collection.

38. Maj. Gen. Sir H. M. Trenchard, KCB, DSO, Commanding the Independent Force, Royal Air Force, "Final Dispatch to the Secretary of State for the Royal Air Force," dated January 1, 1919, published in the *Tenth Supplement to the London Gazette*, Tuesday, December 31, 1918, 135.

39. Maj. Gen. F. H. Sykes, "Memorandum by the Chief of the Air Staff on Air Power Requirements of the Empire," Air Ministry, December 9, 1918, Part I, 1, University of Birmingham Collection.

40. F. H. Sykes, *From Many Angles* (London: Harrap, 1942), 265.

41. W. Raleigh and H. Jones, *War in the Air*, vol. 6 (Oxford: Clarendon, 1937), 160.

42. *Hansard*, vol. 126, col. 1590, March 11, 1920.

43. August 15, 1919, cab. 23/15/616A.

44. Note by Secretary of State for Air Winston S. Churchill to "An Outline of the Scheme for the Permanent Organization of the Royal Air Force," Cmd. 467, December 11, 1919.

45. W. S. Churchill, *The World Crisis: The Aftermath* (London: T. Butterworth, 1929), 52.

46. M. Gilbert, *Winston S. Churchill*, vol. 4 (London: Heinemann, 1975), 202; *Hansard*, vol. 123, cols. 101–102, December 15, 1919.

47. Cmd. 467, para. 3.

48. Cab. 27/71/FC 31(2), cited in H. Montgomery Hyde, *British Air Policy between the Wars* (London: Heinemann, 1976), 91–92.

49. Cited in Gilbert, *Winston S. Churchill*, vol. IV, 217.

50. Memorandum "Mesopotamia" to the Cabinet by the Secretary of State for the Colonies, August 4, 1921, University of Birmingham Collection.

51. P. A. Towle, *Pilots and Rebels* (London: Brassey's, 1989).

52. Interview notes taken by Samuel Hoare, later Lord Templewood, November 1, 1922, cited in Lord Templewood, *Empire of the Air* (London: Collins, 1957), 36.

53. Cmd. 467, para. 1.

54. Cmd. 1581, December 14, 1921.

55. See John Ferris, "The Theory of a 'French Air Menace': Anglo-French Relations and the British Home Defence Air Force Programmes of 1921–25," *Journal of Strategic Studies* 10 (1987): 61–83, for details of Trenchard's contribution to perceptions of a French air threat.

56. CP 32(23), June 20, 1923.

57. Air Staff Memorandum, July 19, 1923, Air 5/416.

58. Cab. 23/46, July 31, 1923.

59. AIR 8/79.

60. Biddle, *Rhetoric and Reality in Air Warfare*, 82–84.
61. Neville Parton, "The Development of Early RAF Doctrine," *The Journal of Military History* 72 (October 2008): 1162.
62. Ibid., 1155–1177.
63. Ibid., 1172.
64. Ibid., 1173–1177.
65. Published as CD 64. Confidential Air Staff Memorandum No. 43, S.28279, "The War Aim of the Royal Air Force," 1929.
66. Ibid., 3.
67. Ibid., 5–6.
68. John Ferris, "Achieving Air Ascendancy: Challenge and Response in British Strategic Air Defence, 1915–40," in *Air Power History: Turning Points from Kitty Hawk to Kosovo*, eds. Sebastian Cox and Peter Gray (London: Frank Cass, 2002), 21–50.
69. *Hansard*, November 10, 1932, vol. 270, col. 632.
70. Ferris, "Achieving Air Ascendancy," 25.
71. Ibid., 39–41.
72. H. E. Wimperis, November 12, 1934, cited in Montgomery Hyde, *British Air Policy between the Wars*, 324.
73. Quoted by Air Chief Marshal Sir Christopher Foxley-Norris in the obituary of Air Marshal Sir Arthur MacDonald, *The Independent*, September 9, 1996.
74. Correspondence with the author, September 13, 1972.
75. Randall T. Wakelam, *The Science of Bombing: Operational Research in RAF Bomber Command* (Toronto: University of Toronto Press, 2009), 15.
76. M. Dean, *The Royal Air Force and Two World Wars* (London: Cassell, 1979), 49.
77. JPC Paper 155, p. 146, cited in Biddle, *Rhetoric and Reality in Air Warfare*, 115.
78. Sir John Slessor, *The Central Blue* (London: Cassell, 1956), 169–170.
79. AIR 2/2767, cited in Montgomery Hyde, *British Air Policy between the Wars*, 406.
80. AIR 8/226, cited in ibid., 408.
81. Ibid., 409.
82. Minutes of Chief of the Air Staff Meeting, January 18, 1938, cited in ibid., 413.
83. Dean, *The Royal Air Force and Two World Wars*, 72, 101.
84. Squadron Leader Brian Armstrong, "Through a Glass Darkly: The Royal Air Force and the Lessons of the Spanish Civil War," *Air Power Review* 12, no. 1 (2009).
85. Ibid., 49–50.
86. Biddle, *Rhetoric and Reality in Air Warfare*, 118.
87. Peter H. Oppenheimer, "Luftwaffe Lessons from the Spanish Civil War," *Journal of the Institute for Historical Review* 7, no. 2 (Summer 1986).
88. Biddle, *Rhetoric and Reality in Air Warfare*, 122.
89. Wakelam, *The Science of Bombing*, 16.
90. Stuart Hadaway, "The Development of RAF Bombs, 1919–1939," *RAF Historical Society Journal* 40 (2009).
91. *The Aeroplane*, April 12, 1922.
92. Cited in *The Aeroplane*, December 13, 1922, 448.
93. Andover Records SW/3/a-d/1938.

94. Marshal of the Royal Air Force Sir Thomas Pike, letter to author, May 25, 1972.
95. Wing Commander F. J. W. Mellersh, "Air Armament, Training and Development," RAF Staff College, Andover, May 11, 1939, Air Historical Branch A/11/6.
96. Ibid., 53.
97. Letter to the author from Air Chief Marshal Sir Alfred Earle, June 4, 1972.
98. R. A. Mason, "Royal Air Force Staff College, 1922–1972," MOD Bracknell, 1972.
99. Slessor, *The Central Blue*, 203–206.
100. Deputy Director of Operations to Deputy Chief of the Air Staff, September 1937, cited in Biddle, *Rhetoric and Reality in Air Warfare*, 116.
101. J. C. Slessor, *Air Power and Armies* (Oxford: Oxford University Press, 1936).
102. Slessor to Trenchard, October 28, 1939, Trenchard Papers, IV/54/94.
103. Alastair Horne, *To Lose a Battle: France 1940* (London: Macmillan, 1969); and S. W. Peach, "Air Power and the Fall of France," in *Air Power History*.
104. "Role of the RAF in War and the Strategic Use of Air Power, Part Nine, Operations in France, September 1939–June 1940," Air Ministry London, March 1944.
105. Wann-Woodhall Report, September 1940, cited by Wing Commander H. Smyth, *From Coningham to Coningham-Keyes, Royal Air Force Historical Society Journal* 44 (2009): 29.
106. Brad Gladman, "Logistics Doctrine and the Impact of War: the Royal Air Force's Experience in the Second World War," in *Air Power History*, 198–206.
107. Air Chief Marshal Sir Frederick Rosier, "How the Joint System Worked (1)," in *The End of the Beginning*, Bracknell Paper no. 3, Royal Air Force Historical Society (March 1992): 27.
108. Smyth, 35–36.
109. Middle East (Army and RAF) Directive on Direct Air Support (1941).
110. Brad Gladman, "The Development of Tactical Air Doctrine in North Africa, 1940–1943," in *The End of the Beginning*.
111. Vincent Orange, "The Commanders and the Command System," in *The End of the Beginning*.
112. Ibid., 44.
113. Alfred Price, "The Rocket Firing Typhoons in Normandy," *RAF Historical Society Journal* 45 (2009).
114. "Proposal for the Conduct of Air Warfare against Britain," Luftwaffe Operations Staff (Intelligence), November 22, 1939, translated by Air Ministry AHB 6, June 21, 1947.
115. Alfred Jodl, Chief of Operations, June 30, cited in W. L. Shirer, *The Rise and Fall of the Third Reich: A History of Nazi Germany* (New York: Simon & Schuster, 1960), 758–760.
116. Ibid., 753.
117. Horst Boog, "Eagle Attack," *The Sunday Telegraph*, June 17, 1990.
118. Ibid., 6.
119. Ibid., 7.

120. R. Hough and D. Richards, *The Battle of Britain: The Jubilee History* (London: Hodder and Staughton, 1989), 112.
121. Mason, *Air Power*, 51–53.
122. Luftwaffe 8th Abteilung Study, "The Course of the Air War Against England," July 7, 1944, translated by Air Ministry AHB 6, May 20, 1947, 6.
123. Ibid., 6.
124. Ibid., 7.
125. Führer Conferences on Naval Affairs, 1940, 101, cited in Shirer, 773.
126. Richard Overy, *The Air War, 1939–1945* (Washington, DC: Potomac, 2005), 32–34.
127. Horst Boog, "The Policy, Command and Direction of the Luftwaffe in World War II," *RAF Historical Society Journal* 41 (2008): 67–85.
128. AIR 14/249, August 22, 1939, and June 5, 1940, cited in Biddle, *Rhetoric and Reality in Air Warfare*, 182–183.
129. Portal to Air Ministry, July 16, 1940, cited in C. Webster and N. Frankland, *The Strategic Air Offensive against Germany, 1939–1945*, vol. 1 (London: HMSO, 1961), 150.
130. Churchill to Beaverbrook, cited in Biddle, *Rhetoric and Reality in Air Warfare*, 88.
131. Ibid., 186–187.
132. Ibid., 189 and note 51.
133. Chief of Air Staff Meeting, June 2, 1941, cited in ibid., 194.
134. R. V. Jones, "The Intelligence War," *RAF Historical Society Journal* 41 (2008): 19.
135. Wakelam, *The Science of Bombing*, 23.
136. C. Webster and N. Frankland, *The Strategic Air Offensive against Germany, 1939–1945*, vol. 4 (London: HMSO, 1961), 143–148.
137. Henry Probert, *Bomber Harris: His Life and Times* (London: Greenhill, 2001).
138. Wakelam, *The Science of Bombing*, 72–75.
139. Webster and Frankland, *The Strategic Air Offensive against Germany, 1939–1945*, vol. 4, 154.
140. Ibid., 275.
141. Biddle, *Rhetoric and Reality in Air Warfare*, 229–230.
142. Eaker to Arnold, November 16, 1943, cited in ibid., 225.
143. Wakelam, *The Science of Bombing*, 138–156.
144. Interview with author, Bracknell, July 1977, Joint Services Staff College Library Archive.
145. Mason, *Air Power*, 54–57.
146. J. S. Cox, "The Years of Development," *RAF Historical Journal* 31 (2004): 129–142.
147. A. V. M. Bufton, ACAS (Int), cited in ibid., 132.
148. Sir Arthur Harris, interview with the author, July 1977.
149. R. Overy, "The Bombing of Germany in World War II: A Reappraisal," *RAF Historical Journal* 31 (2004): 151–159.
150. Ibid., 158.
151. Sir Philip Joubert de la Ferté, *The Third Service* (London: Thames and Hudson, 1955), 153.

152. "Digest of the Group Discussions," *Seek and Sink*, Bracknell Paper no. 2, RAF Historical Society (1992), 81.
153. E. Thomas, "Intelligence," in *Seek and Sink*.
154. "Digest of the Group Discussions," in *Seek and Sink*, 74–79.
155. G. Till, *Air Power and the Royal Navy* (London: Jane's, 1979), 175.
156. JCS 1725/1, May 1, 1947.
157. H. Wynn, *RAF Nuclear Deterrent Forces* (London: HMSO, 1997), 2.
158. Prime Minister memo, August 29, 1945, cited in ibid., 8.
159. Ibid., 24.
160. M. Gowing, *Independence and Deterrence*, vol. 1 (London: Macmillan, 1974), 185.
161. M. J. Armitage and R. A. Mason, *Air Power in the Nuclear Age* (London: Macmillan, 1984), 62–69.
162. NATO Defence Committee DC6/1, December 1, 1949, para. 5.
163. A. Beaufre, *NATO and Europe* (London: Faber and Faber, 1967), 51.
164. H. Probert, "The Royal Air Force and the Korean War," *RAF Historical Society Journal* 21 (2000): 64.
165. Armitage and Mason, *Air Power in the Nuclear Age*, 217–218.
166. Ibid., 219–220.
167. AHB IIH/272/3/40A, cited in Wynn, *RAF Nuclear Deterrent Forces*, 131.
168. Cmd. 9391, MOD London, February 1955, paras. 19, 23.
169. T. C. G. James, "The Impact of the Sandys Defence Policy on the RAF," *RAF Historical Society Journal* 41 (2008): 45–65.
170. Cmd. 124, *Defence: Outline of Future Policy* (London: HMSO, April 1957).
171. Ibid., para. 13.
172. Ibid., paras. 61, 62.
173. *Royal Air Force Manual of Operations, AP 1300*, 4th ed. (London: Air Ministry, March 1957).
174. Ibid., 2–3.
175. Ibid., 8–10.
176. Wynn, *RAF Nuclear Deterrent Forces*, 318.
177. Ibid., 275.
178. Ibid., 276.
179. Chief of the Air Staff, 1971, July 25, 1958, cited in ibid., 278.
180. Sir Frank Cooper, "The Direction of Air Force Policy in the 1950s and 1960s," *RAF Historical Society Journal* 11 (1992): 22.
181. *Hansard*, June 10, 1976, vol. 912, c. 1705.
182. Chief of the Air Staff to Wing Commander R. A. Mason, Annex A to Chief of the Air Staff 91111, November 2, 1976.
183. Armitage and Mason, *Air Power in the Nuclear Age*, 2–3.
184. Air Vice-Marshal J. W. Price, "The Alert Measures Committee," *RAF Historical Society Journal* 30 (2003), 14.
185. Maj. Gen. Julian Thompson, "Harrier Operations," *RAF Historical Society Journal* 30 (2003): 115.
186. Air Vice-Marshal George Chesworth, Chief of Staff to Air Commander, "The Higher Direction of War: The Falklands—National Command," *RAF Historical Society Journal* 16 (1995).

187. S. Cox and S. Ritchie, "The Gulf War and UK Air Power Doctrine and Practice," in *Air Power History*, 290.
188. Air Chief Marshal Sir William Wratten, "An RAF Commander's View of Cooperation in the 1990–91 Gulf War," *RAF Historical Society Journal* 32 (2004): 118.
189. Ibid.
190. Air Marshal Ian MacFadyen, "Tactical Command—Theatre/Wing: The Gulf War," *RAF Historical Society Journal* 16 (1996): 59.
191. Ibid., 57.
192. Wratten, "An RAF Commander's View of Cooperation in the 1990–91 Gulf War," 119.
193. Group Capt. Glen Torpy, "Tactical Command—Squadron: The Gulf War," *RAF Historical Society Journal* 16 (1996): 79.
194. "Discussion," *RAF Historical Society Journal* 16 (1996): 44.
195. R. Hallion, ed., *Air Power Confronts an Unstable World* (Washington, DC: Brassey's, 1997).
196. Statistics from a briefing by Lt. Gen. Michael Ryan to the February 1996 Corona South meeting, Orlando, Florida.
197. NAC Memorandum to UN Secretary-General, MCM-KAD—August 4, 1993.
198. *Statement on the Defence Estimates 1994*, Command 2550 (London: HMSO, 1994), 33, 43.
199. UK Secretary of State for Defence Malcolm Rifkind, London, January 14, 1993 (press conference transcript).
200. BBC Radio interview, August 3, 1993, reported in the *Daily Telegraph* (London), August 4, 1993.
201. *Daily Telegraph* (London), August 5, 1995.
202. R. A. Mason, "Operations in Search of a Title: Air Power in Operations Other Than War," in *Air Power Confronts an Unstable World*.
203. *Hansard*, vol. 259, cols. 453–534, May 4, 1995.
204. House of Commons Defence Committee (HCDC), *Lessons of Kosovo*, vol. 1 (London: HMSO, October 2000), para. 24.
205. Ibid., Appendix 1, UNSCR 1199, 4.
206. Ibid., 32–33.
207. Ibid., Appendix 1, UNSCR 1203, 3.
208. UK Defence Minister George Robertson, cited in HCDC Report, 69.
209. John E. Peters, Stuart Johnson, Nora Bensahel, Timothy Liston, and Traci Williams, *European Contributions to Operation Allied Force: Implications for Transatlantic Cooperation*, Project Air Force (Santa Monica, CA: RAND Corporation, 2001).
210. HCDC Report, 133.
211. Ibid., 134.
212. Royal Air Force Commanders' Briefing Note, CBN/24/02, DCC (RAF), September 5, 2002.
213. Air Vice-Marshal Glen Torpy, evidence to the House of Commons Defence Committee, in *Lessons of Iraq, Third Report of Session 2003–2004*, vol. 1, HC57-1 (London: The Stationary Office Ltd., March 16, 2004), 34.

214. Ministry of Defence, *Operations in Iraq: First Reflections* (London: UK MOD, July 2003), 3.
215. *The Strategic Defence Review*, CM 3999 (London: HMSO, July 1998): 21–30.
216. Ibid., 11.
217. Benjamin Lambeth, "The RAF Contribution to Operation Iraqi Freedom," unpublished report prepared for United States Central Command Air Forces (Santa Monica, CA: RAND, November 2005), University of Birmingham Collection.
218. HC 57-1, 99–103.
219. Lambeth, "The RAF Contribution to Operation Iraqi Freedom," 26–27.
220. Capt. Doug Beattie, MC, *An Ordinary Soldier* (London: Simon & Schuster, 2009).
221. Charles Haddon-Cave QC, The Nimrod Review, HC 1025 (London: Her Majesty's Stationery Office [HSMO], October 28, 2009).
222. Air Chief Marshal Sir Stephen Dalton, "Dominant Air Power in the Information Age," address to International Institute for Strategic Studies, London, February 15, 2010.
223. *The Strategic Defence and Security Review*, CM 7948 (London: HSMO, October 2010): 25–27.
224. *AP 3000 British Air and Space Power Doctrine*, 4th ed. (London: Air Staff, 2009).
225. Ibid., 14.

CHAPTER 2. U.S. AIR POWER

1. Military aviation has been the subject of numerous sweeping histories and compendiums. The author recommends the following for their reliability and discernment: John Andreas Olsen, ed., *A History of Air Warfare* (Washington, DC: Potomac Books, 2010); Walter J. Boyne, *The Influence of Air Power upon History* (Gretna, LA: Pelican Publishing Co., 2003); Stephen Budiansky, *Air Power: From Kitty Hawk to Gulf War II* (New York: Viking, 2003); Sebastian Cox and Group Capt. Peter Gray, eds., *Air Power History: Turning Points from Kitty Hawk to Kosovo* (London: Frank Cass, 2002); Alan Stephens, ed., *The War in the Air, 1914–1994* (Maxwell AFB, AL: Air University Press, 2001); and Air Vice-Marshal Tony Mason, *Air Power: A Centennial Appraisal* (London: Brassey's, 1994). Older works still of value are Robin Higham, *Air Power: A Concise History* (New York: St. Martin's Press, 1972); and (particularly) Eugene M. Emme, ed., *The Impact of Air Power: National Security and World Politics* (Princeton, NJ: Van Nostrand, 1959).
2. Frederick Stansbury Haydon, *Aeronautics in the Union and Confederate Armies* (Baltimore, MD: Johns Hopkins University Press, 1941); Tom D. Crouch, *The Eagle Aloft: Two Centuries of the Balloon in America* (Washington, DC: Smithsonian Institution Press, 1983); and Juliette A. Hennessy, *The United States Army Air Arm, April 1861 to April 1927* (Washington, DC: Office of Air Force History, 1985).
3. Exhibited today in Paris's Musée des Arts et Métiers; see Charles H. Gibbs-Smith, *Clément Ader: His Flight-Claims and His Place in History* (London: HSMO, 1968); for a diametrically opposite view, see Pierre Lissarrague,

Clément Ader: Inventeur d'avions (Toulouse: Bibliothèque historique Privat, 1990). Ader was more successful as a theorist on air power; see his *L'aviation militaire* (Paris: Berger-Levrault, Éditeurs, 1911).

4. Final Report of War Department, in *Langley Memoir of Mechanical Flight*, Charles M. Manley, ed., pt. II, a study in the *Smithsonian Contributions to Knowledge* series 27, no. 3 (1911).

5. Robert L. Lawson, *The History of U.S. Naval Air Power* (New York: Military Press, 1985), 10.

6. Archibald D. Turnbull and Clifford L. Lord, *History of United States Naval Aviation* (New York: Arno Press, 1972), 12–20.

7. See, for example, Herbert A. Johnson, *Wingless Eagle: U.S. Army Aviation through World War I* (Chapel Hill: University of North Carolina Press, 2001).

8. Relative force structure figures from: Ernst von Hoeppner, *Deutschlands Krieg in der Luft* (Leipzig: Verlag von K. F. Koehler, 1921), 7; Von Hardesty, "Early Flight in Russia," in *Russian Aviation and Air Power in the Twentieth Century*, eds. Robin Higham, John T. Greenwood, and Von Hardesty (London: Frank Cass, 1998), 22; Paul-Louis Weller, "L'aviation française de reconnaissance," in *L'aéronautique pendant la guerre mondiale, 1914–1918*, ed. Maurice de Brunoff (Paris: Maurice de Brunoff, 1919), 63; Charles Christienne and Gen. Pierre Lissarrague, *A History of French Military Aviation* (Washington, DC: Smithsonian Institution Press, 1986), 59; and Jerome Hunsaker, "Forty Years of Aeronautical Research," *Smithsonian Report for 1955* (Washington, DC: Smithsonian Institution, 1956), 243 (though his European figures are clearly in error).

9. U.S. War Department, *Final Report of Gen. John J. Pershing* (Washington, DC: GPO, 1919); *Aircraft Production: Hearings before the Subcommittee of the Committee on Military Affairs* [hereafter *Aircraft Production Hearings*] vol. 1, 65th Cong., 2d sess. (Washington, DC: GPO, 1918); H. Snowden Marshall, E. H. Wells, and Gavin McNab, *Final Report of the Committee on Aircraft Investigation* [May 2, 1918], 6, in Benedict Crowell Papers Collection, box 1, fol. 6, Kelvin Smith Library, Case Western Reserve University, Cleveland, OH; and B. Crowell and R. F. Wilson, *The Armies of Industry*, vol. 1 (New Haven, CT: Yale University Press, 1921), 346.

10. For Mitchell's pursuit of air power and Trenchard's influence, see Andrew Boyle, *Trenchard: Man of Vision* (London: Collins, 1962), 298–299; Lt. Col. Mark A. Clodfelter, "Molding Airpower Convictions: Development and Legacy of William Mitchell's Strategic Thought," in *The Paths to Heaven: The Evolution of Airpower Theory*, ed. Col. Philip S. Meilinger (Maxwell AFB, AL: Air University Press, 1997), 79–114; Brig. Gen. Alfred F. Hurley, *Billy Mitchell: Crusader for Air Power* (Bloomington: Indiana University Press, 1975); and Douglas Waller, *A Question of Loyalty: Gen. Billy Mitchell and the Court-Martial That Gripped the Nation* (New York: HarperCollins, 2004).

11. Battle Orders No. 3, HQ Air Service First Army, September 13, 1918, in *The U.S. Air Service in World War I*, ed. Maurer Maurer, vol. 3 (Washington, DC: Office of Air Force History, 1979), 365–367.

12. "Report of American Aviation Mission," in *Reorganization of the Army: Hearings before the Subcommittee of the Committee on Military Affairs*, 66th Cong., 1st sess. (Washington, DC: GPO, 1919), 199–205.

13. Baker testimony, *Hearings before the Committee on Military Affairs on a United Air Service*, 66th Cong., 2d sess. (Washington, DC: GPO, 1919), 395.

14. Brig. Gen. William Mitchell, *Our Air Force* (New York: E. P. Dutton & Company, 1921), 183–184, 193, 199, 222–223; see also his "America in the Air," *National Geographic Magazine* 39, no. 3 (March 1921): 347; Joint Army and Navy Board, *Report of the Joint Board on the Results of Aviation and Ordnance Tests Held during June and July, 1921*, in U.S. Congress, *Congressional Record*, 67th Cong., 1st sess. (August 20, 1921), 8625–8626; and A. Lincoln, "The USN and the Rise of the Doctrine of Air Power," *Military Affairs* 15, no. 3 (Autumn 1951).

15. *Report of the Select Committee of Inquiry into Operations of the United States Air Services*, 68th Cong., 1st sess. (Washington, DC: GPO, 1925); President Calvin Coolidge to Curtis D. Wilbur [SecNav] and Dwight F. Davis [SecWar], Sept. 12, 1925, in U.S. Senate, Morrow Board, *Aircraft in National Defense*, Senate document no. 18, 69th Cong., 1st sess. (Washington, DC: GPO, 1925), 85–86 [hereafter *Morrow Report*]; U.S. Congress, House, *Aircraft: Hearings before the President's Aircraft Board, October 14 and 15, 1925*, vol. 5, 69th Cong., 1st sess. (Washington, DC: GPO, 1925); and U.S. Congress, House, *Department of Defense and Unification of Air Service: Hearings before the Committee on Military Affairs, January 19 to March 9, 1926*, 69th Cong., 1st sess. (Washington, DC: GPO, 1926).

16. For examples, see Rear Adm. Bradley Fiske to Rear Adm. William Sims, November 30, 1920, quoted in William M. McBride, "Challenging a Strategic Paradigm," *The Journal of Strategic Studies* 14, no. 1 (March 1991): 75; "General Mitchell's Publicity Propaganda Running Contrary to Policy Approved by President Harding," n.d., RG 255, box 230, ser. 45-1, Mitchell folder, National Archives and Records Administration, Archives II.

17. AIR 5/244, Group Capt. Malcolm Christie to Director of Air Ministry Air Intelligence, A. A. Nos. 269 and 582, May 26, 1923, and May 9, 1924, National Archives, Kew, England. For a perspective on the RNAS, the RAF, and the debt of the latter to the former, see Christina J. M. Goulter, "The Royal Naval Air Service: A Very Modern Force," in *Air Power History*, eds. Cox and Gray, 51–65.

18. For example, see Walter Raleigh, *The War in the Air*, vol. 1 (Oxford: Clarendon Press, 1922), 206–211, 478–489; Hilary St. George Saunders, *Per Ardua: The Rise of British Air Power, 1911–1939* (London: Oxford University Press, 1945), 211–224.

19. From aviation appropriation and expenditures enumerated in the various *Annual Reports* of the Navy Bureau of Aeronautics and the Army Secretary of War, 1925–1932; Rudolf Modley and Thomas J. Cawley, eds., *Aviation Facts and Figures, 1953* (Washington, DC: Lincoln Press, Inc., for the Aircraft Industries Association of America, Inc., 1953), Table 5-1, 102.

20. I thank Professor Eric Grove, professor of naval history and director of the Centre for International Security and War Studies, University of Salford, for insights on the attitude of RNAS airmen toward the RFC.

21. As noted by Lt. Col. Mark Clodfelter in his "Molding Airpower Convictions: Development and Legacy of Mitchell's Strategic Thought," in *The Paths to Heaven*, ed. Col. Phillip S. Meilinger (Maxwell, AFB: Air University Press, 1997), 107.

22. A. Lincoln, "The USN and the Rise of the Doctrine of Air Power," *Military Affairs* 15, no. 3 (Autumn 1951): 146–147; Barry Watts and Williamson Murray, "Military Innovation," in *Military Innovation in the Interwar Period*, eds. Williamson Murray and Allan R. Millett, rev. ed. (Cambridge: Cambridge University Press, 2005), 395–396; for Eisenhower's crucial role in the creation of the air force, see Herman S. Wolk, *Reflections on Air Force Independence* (Washington, DC: Air Force History and Museums Program, 2007), 75, 79–81.

23. *Report of the Select Committee of Inquiry into Operations of the United States Air Services,* 68th Cong., 1st sess. (Washington, DC: GPO, 1925), 4–5, 17–18; *Hearings before the Select Committee of Inquiry into Operations of the United States Air Services, Parts 2,* 68th Cong. (Washington, DC: GPO, 1925), 617, 663; Robert Frank Futrell, *Ideas, Concepts, Doctrine: Basic Thinking in the United States Air Force,* vol. 1: *1907–1960* (Maxwell AFB, AL: Air University Press, 1989), 42–44; and James P. Tate, *The Army and Its Air Corps: Army Policy toward Aviation, 1919–1941* (Maxwell AFB, AL: Air University Press, 1998), 19–20.

24. *Report of the Select Committee of Inquiry into Operations of the United States Air Services,* 68th Cong., 1st sess. (Washington, DC: GPO, 1925), 5–7; *Aircraft: Hearings before the President's Aircraft Board, October 14 and 15, 1925,* vol. 5, 69th Cong., 1st sess. (Washington, DC: GPO, 1925), 1457–1461.

25. Louis Breguet, "L'aviation aux Etats-Unis," *L'Aérophile* 33, no. 23–24 (December 15, 1925): 356–361.

26. *Department of Defense and Unification of Air Service: Hearings before the Committee on Military Affairs, January 19 to March 9, 1926,* 69th Cong., 1st sess. (Washington, DC: GPO, 1926), Tables "Aircraft Strength United States Army Air Service: Status of Airplanes on Hand, June 30, 1925," and "United States Naval Airplane Strength, June 30, 1925," 116–117. These remarkably low figures are supported by a letter by John J. Ide of the NACA to Dwight Morrow, November 4, 1925, in the Ide Papers, box 1, fol. 4, Archives of the National Air and Space Museum, Smithsonian Institution, Washington, DC.

27. "Sees Hope for U.S. Aviation," *Chicago Tribune,* April 27, 1927, a clipping from the W. B. Robertson Scrapbook Collection, vol. 2, Missouri History Museum Library and Research Center, St. Louis, MO.

28. *Morrow Report,* 99–115; United States Navy, *Report of the Secretary of the Navy, 1926* (Washington, DC: GPO, 1926), 596–598; United States War Department, *Report of the Secretary of War, 1926* (Washington, DC: GPO, 1926), 33–35; *Report of the Secretary of War, 1927* (Washington, DC: GPO,

1927), 41; and Nick A. Komons, *Bonfires to Beacons: Federal Civil Aviation Policy under the Air Commerce Act, 1926–1938* (Washington, DC: GPO, 1978), 78, 88.

29. For this technical turnaround, see Peter W. Brooks, *The Modern Airliner: Its Origins and Development* (London: Putnam, 1961); and Ronald Miller and David Sawers, *The Technical Development of Modern Aviation* (New York: Praeger Publishers, 1970). For interwar flying boat development, see Capt. Richard C. Knott, USN, *The American Flying Boat: An Illustrated History* (Annapolis, MD: Naval Institute Press, 1979), 73–113.

30. For an example of disparagement regarding American airliner development, see R. H. Mayo (Technical Adviser) to George Woods Humphery (General Manager), September 6, 1933, "Atlantic Services, 1930–35," AW/1/6161 Pt. 1, Imperial Airways Papers, British Airways Archives, Hatton Cross, England. For MacRobertson's race and unease over its outcome, see "The Empire Air Routes," *Flight*, November 1, 1934; AIR 2/1668, *Daily Express*, November 16, 1934, clipping; Air Ministry, "Minutes of Meeting Held in Room 415, on Friday, November 23, 1934, at 11:30 Hours," National Archives, Kew; and *Hansard*, HL Deb (November 21, 1934), vol. 95, cc. 33–72.

31. For example, see Capt. A. R. Ladd, "An Estimate of the Powers and Limitations of Commercial Aviation, in Relation to National Defense" (Ft. Leavenworth, KS: Command and General Staff School, 1932), Combined Arms Research Library, Ft. Leavenworth, KS.

32. Lt. Col. William C. Sherman, "Tentative Manual for the Employment of Air Service" (HQ 1st Army, 1919), reprinted in U.S. Air Service, *Air Service Information Circular* 1, no. 76 (June 30, 1920), and reprinted subsequently as Document no. 55 in *The U.S. Air Service in World War I*, ed. Maurer Maurer, vol. 2 (Washington, DC: Office of Air Force History, 1978), quote is from 313.

33. USAAS, Training Regulation 440-15, *Fundamental Principles for the Employment of the Air Service* (Washington, DC: War Department, January 26, 1926), was revised once, in 1935, but that, of course, was before the wars in Spain and China. For more on doctrinal and organizational development in this time period, see Chase C. Mooney and Martha E. Layman, *Organization of Military Aeronautics, 1907–1935 (Congressional and War Department Action)*, no. 25 of the *Army Air Forces Historical Studies* series (Washington, DC: HQ AAF Assistant Chief of Air Staff/Intelligence, Historical Division, December 1944), 78–103; Thomas H. Greer, *The Development of Air Doctrine in the Army Air Arm, 1917–1941*, USAF Historical Study 89 (Maxwell AFB, AL: USAF Historical Division, Research Studies Institute, Air University, 1985), 40–41; Futrell, *Ideas, Concepts, Doctrine*, vol. 1, 63; and James A. Mowbray, "Air Force Doctrine Problems: 1926–Present," *Airpower Journal* 9, no. 4 (Winter 1995).

34. For a survey of interwar carrier development and thought, see Geoffrey Till, "Adopting the Aircraft Carrier: The British, American, and Japanese Case Studies," in *Military Innovation in the Interwar Period*, eds. Murray and Millett, 191–226.

35. Robert T. Finney, *History of the Air Corps Tactical School, 1920–1940* (Washington, DC: Center for Air Force History, 1992), 64, 67–69. See also James L. Cate, "Development of United States Air Doctrine, 1917–41," *Air University Quarterly Review* 1, no. 3 (Winter 1947); and Stephen L. McFarland, *America's Pursuit of Precision Bombing, 1910–1945* (Tuscaloosa: University of Alabama Press, 1995), 40–44, 68–104.

36. For examples, see Capt. St. Clair Streett, "What Principles Should Govern the Strategical Employment of the Air Force, with Particular Consideration to the Most Suitable Objectives?" (Ft. Leavenworth, KS: Command and General Staff School, 1934), 30–36; and Capt. Robert Olds, "A Critical Analysis of Air Force Action in Coastal Air Defense, by Phases" (Ft. Leavenworth, KS: Command and General Staff School, 1935), 1–2, 36–37, copies of both are in the Combined Arms Research Library, Ft. Leavenworth, KS.

37. Though many air advocates regarded the outcome as an attempt to defuse growing agitation for a separate air service.

38. Lt. Col. Barry D. Watts, *The Foundation of U.S. Air Doctrine: The Problem of Friction in War* (Maxwell AFB, AL: Air University Press, December 1984), 18–23. For the memoirs of its principal architect, see Haywood S. Hansell, Jr., *The Air Plan That Defeated Hitler* (Atlanta, GA: Higgins-McArthur/Longino and Porter, 1972). The Army Air Corps became the Army Air Forces in 1941, remaining so until the formation of the U.S. Air Force in 1947. See also Herman S. Wolk, *Cataclysm: General Hap Arnold and the Defeat of Japan* (Denton: University of North Texas Press, 2010), 15–55.

39. Gen. Henry H. Arnold, *Global Mission* (New York: Harper & Brothers, Publishers, 1949), 179–180. For Roosevelt's role as an air advocate, see Jeffery S. Underwood, *The Wings of Democracy: The Influence of Air Power on the Roosevelt Administration, 1933–1941* (College Station: Texas A&M University Press, 1991), 60–72, 123–141.

40. Turnbull and Lord, *History of United States Naval Aviation*, 259–281; Charles M. Melhorn, *Two-Block Fox: The Rise of the Aircraft Carrier, 1911–1929* (Annapolis, MD: Naval Institute Press, 1974), 87–115; Thomas Wildenberg, *All the Factors of Victory: Admiral Joseph Mason Reeves and the Origins of Carrier Airpower* (Washington, DC: Potomac Books, 2003), 136–199; and Mark R. Peattie, *Sunburst: The Rise of Japanese Naval Air Power, 1909–1941* (London: Chatham Publishing, 2001), 52–76.

41. Data from USN Chief of Naval Operations, "Schedule of Naval Aeronautic Organizations as of April 30, 1945," OP-31-B2-RSM (SC) A-21-1 (May 7, 1945), Naval Historical Center Archives, Washington Navy Yard, Washington, DC.

42. For example, post-1934 American fighter aircraft, such as the Seversky P-35, Curtiss P-36, Brewster F2A, and Grumman F4F, were generally as technically advanced as their foreign contemporaries (despite some myths to the contrary), and thus their successors, with the sole exception of the Lockheed P-38 (first flight 1939), effectively constituted extrapolations of existing American design practice rather than new departures in technology and design concept. This reflected the generally high standard of American aircraft design practice at the time.

43. Arnold, *Global Mission*, 197. See also H. Duncan Hall, *North American Supply* (London: HMSO, 1955), 59–71, 216–218, 429.

44. Report of M. A.Valencia, "Special Report on Spanish Government Air Force," February 21, 1937, in *Spanish War: Reports and Articles*, vol. 1, ed. Lt. Col. Byron Q. Jones (Washington, DC: War Plans Division, Army War College, December 27, 1940), 14, a typescript compilation in the Library of the U.S. Army Military History Institute, U.S. Army War College, Carlisle Barracks, Pennsylvania. I thank Judith Meck for her assistance in locating this reference. This observation, incidentally, is consistent with similar reactions of ground forces exposed to air attack from the onset of battlefield air attack. For a telling later example, see Maj. Gen. Frido von Senger und Etterlin, *Neither Fear Nor Hope* (New York: E. P. Dutton, 1964), 224. For an overview, see Group Capt. Andrew P. N. Lambert, *The Psychology of Air Power*, Whitehall Paper Series (London: Royal United Services Institute for Defence Studies, 1995).

45. Hansell, Jr., *Air Plan That Defeated Hitler*, 55.

46. Years later, Martin-Marietta chairman Norman R. Augustine would memorably dub this the "Yes-but-would–it-have-worked-in-theory?" school of thought.

47. Col. Thomas A. Cardwell III, USAF, *Airland Combat: An Organization for Joint Warfare* (Maxwell AFB, AL: Air University Press, December 1992), 13.

48. Interview of Maj. Gen. Ross E. Rowell, USMC, Oct. 24, 1946, U.S. Marine Corps Historical Center, Archives Section, Washington Navy Yard. I thank Danny Crawford for making this available. See also Robert Sherrod, *History of Marine Corps Aviation in World War II* (Washington, DC: Combat Forces Press, 1952), 1–53; Richard P. Hallion, *Strike from the Sky: The History of Battlefield Air Attack, 1911–1945* (Washington, DC: Smithsonian Institution Press, 1989), 71–74; and Gary C. Cox, "Beyond the Battle Line: U.S. Air Attack Theory and Doctrine, 1919–1941" (Maxwell AFB, AL: Air University, School of Advanced Airpower Studies, June 1995), 6–30.

49. Mecozzi was arguably the most thoughtful and perceptive of interwar theorists about the value and conduct of battlefield air support operations, though, like most others, he believed it required a specialized ground attack aircraft; see his "Origini e svilupo dell'aviazione d'assalto," *Revista Aeronautica* 11, no. 2 (February 1935).

50. For Guadalajara, see M. A. Paris, Report 23, 461-W, "Spain (Combat Aviation): Subject–Major Military Operations, Air Combat Operations, Operations in Spain" (June 2, 1937), in Jones, *Spanish War*, vol. 1, 1–4; and Ramón Garriga, *Guadalajara y sus consecuencias* (Madrid: G. Del Toro, 1974), 109–180.

51. As reflected in various Command and General Staff School papers, such as Capt. Frederick I. Eglin, "Is Attack Aviation Necessary as a Distinct Type of Aviation?" (1932); Capt. Lloyd L. Harvey, "Attack Aviation" (1932); and Capt. C. Y. Banfill, "The Use of Attack Aviation in Support of a Decisive Ground Attack" (1936), copies of which are in the Combined Arms Research Library, Ft. Leavenworth, KS.

52. The major exceptions being (1) carrier task forces, which operated specialized attackers but usually under their own air cover of fighters; and (2) Russian Front air operations. On the Russian Front, from early 1943, conditions of

relative air parity meant both the Luftwaffe and the VVS could secure local air superiority over a portion of the front if they chose to do so, enabling them to operate specialized ground attack aircraft with relative impunity. Yet, over time, even these two expressed an eventual preference for the fighter-bomber. After the war, the VVS quickly abandoned the specialized attack airplane, not acquiring another until it followed America's 1970s adaptation of the A-10 design concept with an attack aircraft of its own, the Su-25.

53. For example, the Douglas SBD, though an eminent weapon for naval warfare, was adopted by the AAC as the A-24, and it proved unsuited for routine combat operations where friendly fighter cover was weak or nonexistent. The Curtiss SB2C Helldiver could only serve with the AAF and RAF in training and utility roles. The Vultee A-35 Vengeance never saw combat service with the AAF, but served with the RAF in Burma with some limited success. The North American A-36 Invader, a dive-bomber modification of the P-51A fighter, proved so unsuitable that squadrons wired its dive brakes shut and operated it as a conventional fighter-bomber, with much greater success.

54. War Dept., FM 1-5, *Employment of Aviation in the Army* (Washington, DC: GPO, 1940), 22, 42; and War Dept., FM 1-10, *Tactics and Technique of Air Attack* (Washington, DC: GPO, 1940), 1154–1219.

55. USMC, *Small Wars Manual* (Washington, DC: GPO, 1940), section 9–23.

56. War Dept., FM 31–35, *Aviation in Support of Ground Forces* (Washington, DC: GPO, 1942).

57. War Dept., FM 100–20, *Command and Employment of Air Power* (Washington, DC: GPO, 1943); and B. L. Montgomery, *Some Notes on High Command in War* (Tripoli: HQ 8th Army, January 1943), copy in the U.S. Army Military History Institute Library, which was drafted by Coningham. See also Brad Gladman, "The Development of Tactical Air Doctrine in North Africa, 1940–43," in *Air Power History*, ed. Cox and Gray, 188–206; and David Ian Hall, *Strategy for Victory: The Development of British Tactical Air Power, 1919–1943* (Westport, CT: Praeger Security International, 2008), 104–145.

58. FM 100-20, 1. It has been suggested such a screaming full-cap header was justified by the need to emphasize doctrinal change. Upon reflection, I remain convinced it reflects a permeating anxiety indicative of the years of frustration air power theorists and practitioners had spent toiling in a wilderness of doctrinally constrained thinking.

59. FM 100-20, 1, 10–11.

60. For example, see Lt. Gen. Omar N. Bradley and the Air Effects Committee of the 12th Army Group, *Effect of Air Power on Military Operations: Western Europe*, a report prepared for the United States Strategic Bombing Survey (HQ 12th Army Group Air Effects Committee, n.d.), 180. For one commander's reaction to air attack, see Lt. Gen. Fritz Bayerlein, "Panzer-Lehr Division (July 24–25, p. 44)," Manuscript A-902 (Historical Division, HQ U.S. Army Europe, n.d.), copy in the U.S. Army Military History Institute Library.

61. Quote from David N. Spires, *Air Power for Patton's Army: The XIX Tactical Air Command in the Second World War* (Washington, DC: Air Force History and Museums Program, 2002), 312.

62. Watts, *The Foundation of U.S. Air Doctrine*, 62–72.

63. Quoted in Futrell, *Ideas, Concepts, Doctrine*, vol. 1, 153 (emphasis in original).

64. James Doolittle with Carroll V. Glines, *I Could Never Be So Lucky Again: An Autobiography by General James H. "Jimmy" Doolittle* (New York: Bantam Books, 1991), 380.

65. See Walt W. Rostow, *Pre-Invasion Bombing Strategy: General Eisenhower's Decision of March 25, 1944* (Austin: University of Texas Press, 1981), 3–6, 66–87.

66. Franklin D'Olier et al., "Summary Report (European War)," in *The United States Strategic Bombing Surveys (European War) (Pacific War)*, a reprint of the European and Pacific summary reports from the United States Strategic Bombing Survey (Maxwell AFB, AL: Air University Press, October 1987), 37–38.

67. James A. Winnefeld and Dana J. Johnson, *Joint Air Operations: Pursuit of Unity in Command and Control, 1942–1991* (Annapolis, MD: Naval Institute Press and RAND, 1993), 13–38.

68. Eliot A. Cohen, "The Unsheltering Sky," *The New Republic* 204, no. 6 (February 11, 1991): 24.

69. Ronald H. Spector, *Eagle against the Sun: The American War with Japan* (New York: Free Press, 1985), 485.

70. Quoted in Col. Mike Worden, *Rise of the Fighter Generals: The Problem of Air Force Leadership, 1945–1982* (Maxwell AFB, AL: Air University Press, March 1998), 17; see also D'Olier et al., "Summary Report (Pacific War)," in *The United States Strategic Bombing Surveys (European War) (Pacific War)*, 110.

71. Chase C. Mooney, *Organization of Military Aeronautics, 1935–1945 (Executive, Congressional, and War Department Action)*, no. 46 of the *Army Air Forces Historical Studies* series (Washington, DC: HQ AAF Historical Office, April 1946), 50.

72. Marshall to Stimson, May 16, 1944, quoted in Maurice Matloff, *Strategic Planning for Coalition Warfare, 1943–1944* (Washington, DC: U.S. Army Center of Military History, 1990), 411. Marshall included as well the "Soviet numerical preponderance," and the "high quality of our ground combat units."

73. John S. D. Eisenhower, *Strictly Personal* (Garden City, NY: Doubleday, 1974), 72.

74. Statement of General Dwight D. Eisenhower, in U.S. Congress, Senate, *Department of Armed Forces and Military Security: Hearings on S.84 and S.1482*, 79th Cong., 1st sess. (Washington, DC: GPO, 1945).

75. David McCullough, *Truman* (New York: Simon & Schuster, 1992), 566.

76. Edwin L. Williams, Jr., *Legislative History of the AAF and the USAF, 1941–1951*, no. 84 in the *USAF Historical Studies* series (Maxwell AFB, AL: Air University, USAF Historical Division, Research Studies Institute, September 1955), 47.

77. Quote from Thomas K. Finletter et al., *Survival in the Air Age: A Report by the President's Air Policy Commission* (Washington, DC: President's Air Policy Commission, January 1, 1948), 8 (emphasis added).

78. Ibid., 24.

79. Futrell, *Ideas, Concepts, Doctrine*, vol. 1, 207–208.

80. General H. H. Arnold to Theodore von Kármán, November 7, 1944, office files of the Air Force Scientific Advisory Board, Headquarters Air Force, Pentagon, Washington, DC; ibid., 205.

81. For example, Theodore von Kármán, "Where We Stand: First Report to General of the Army H. H. Arnold on Long Range Research Problems of the AIR FORCES with a Review of German Plans and Developments," August 22, 1945, vol. II-1, copy #13, including Hsue-shen Tsien, "Reports on the Recent Aeronautical Developments of Several Selected Fields in Germany and Switzerland," July 1945; copy in General Henry H. Arnold Papers, Microfilm Reel 194, Manuscript Division, Library of Congress. USAAF, "German Aircraft, New and Projected Types" (1946), box 568, A-1A/Germ/1945 file, Archives of the National Museum of the USAF, Wright-Patterson AFB, Ohio.

82. USAF, *History of AAF Participation in* Project Paperclip, *September 1946– April 1948 (Exploitation of German Scientists)*, vol. 2, study no. 215 (Wright-Patterson AFB, OH: Air Materiel Command, 1948), 17–21, copy in "Paperclip" file, #11723, National Aeronautics and Space Administration Historical Division, Washington, DC; "BuAer Exploitation of German and Austrian Scientists," *Naval Aviation Confidential Bulletin* no. 4–47 (October 1947), 16–17, copy in archives of the U.S. Naval Historical Center, Washington Navy Yard; and Clarence G. Lasby, *Project Paperclip: German Scientists and the Cold War* (New York: Athenaeum, 1971), passim.

83. For example, USAAF, "German Aircraft, New and Projected Types" (1946); USAF, *History of AAF Participation in Project Paperclip, September 1946–April 1948 (Exploitation of German Scientists)*; "BuAer Exploitation of German and Austrian Scientists," 16–17; and Lasby, *Project Paperclip* .

84. For example, see Theodore von Kármán et al., *Research and Development in the United States Air Force: Report of a Special Committee of the Scientific Advisory Board to the Chief of Staff, USAF* (Washington, DC: SAB, 1949), sections III-7 to VIII-5, copy in the office files of the Air Force Scientific Advisory Board, HQ USAF, Pentagon. I thank Mark Lewis, the Air Force Chief Scientist, for alerting me to this document.

85. Their reign lasted through Vietnam; see Worden, *Rise of the Fighter Generals*, 37–39.

86. Walton S. Moody, *Building a Strategic Air Force* (Washington, DC: Air Force History and Museums Program, 1996), 224–229.

87. Memorandum, Forrestal to the service secretaries, "re Functions of the Armed Forces and the Joint Chiefs of Staff," April 21, 1948; and Truman to Forrestal, "Re: Functions of the Armed Forces and the Joint Chiefs of Staff," April 21, 1948, both reprinted as Document 7 in *The United States Air Force Basic Documents on Roles and Missions*, ed. Richard I. Wolfe (Washington, DC: Office of Air Force History, 1987), 151–166. See also R. Earl McClendon, *Unification of the Armed Forces: Administrative and Legislative Developments, 1945–1949* (Maxwell AFB, AL: Air University, USAF Historical Division, Research Studies Institute, 1952), 72–73; and Futrell, *Ideas, Concepts, Doctrine*, vol. 1, 236–237.

88. Forrestal, heavily stressed by it all, suffered a breakdown after leaving office and

subsequently jumped or fell to his death while undergoing treatment at a naval hospital outside Washington. See Walter Millis, ed., in collaboration with E. S. Duffield, *The Forrestal Diaries* (New York: Viking Press, 1951), 543–555.

89. Viscount Templewood [Sir Samuel Hoare], *Empire of the Air: The Advent of the Air Age, 1922–1929* (London: Collins, 1957), 54–71.

90. Keith D. McFarland and David L. Roll, *Louis Johnson and the Arming of America: The Roosevelt and Truman Years* (Bloomington: Indiana University Press, 2005), 168–187; Col. Phillip S. Meilinger, USAF, "The Admirals' Revolt of 1949: Lessons for Today," *Parameters* 19, no. 3 (September 1989): 81–96; and Jeffrey G. Barlow, *From Hot War to Cold: The U.S. Navy and National Security Affairs, 1945–1955* (Stanford, CA: Stanford University Press, 2009), 211–239.

91. Marcelle Size Knaack, *Post–World War II Bombers, 1945–1973*, vol. 2 of the *Encyclopedia of U.S. Air Force Aircraft and Missile Systems* (Washington, DC: Office of Air Force History, 1988), 3–51; and Norman Polmar et al., *Aircraft Carriers: A History of Carrier Aviation and Its Influence on World Events*, vol. 2: *1946–2006* (Washington, DC: Potomac Books, 2008), 45–51. For the perspective of one senior naval officer on the B-36, see Stephen Jurika, Jr., ed., *From Pearl Harbor to Vietnam: The Memoirs of Admiral Arthur W. Radford*, Hoover Institution Publication 221 (Stanford, CA: Hoover Institution Press, 1980), 159–217.

92. Lt. Gen. Omar N. Bradley to Gen. H. H. Arnold, September 1944, in the "Bradley, World War II Correspondence, 1942–45" file, "Correspondence with Major Historical Figures, 1936–1960" box, General of the Army Omar N. Bradley Papers, Archives of the U.S. Army Military History Institute.

93. John Schlight, "Elwood R. Quesada: TacAir Comes of Age," in *Makers of the United States Air Force*, ed. John L. Frisbee (Washington, DC: Office of Air Force History, 1987), 198–203; for Quesada's perspective on tactical air power, see his "Tactical Air Power," *Air University Quarterly Review* 1, no. 4 (Spring 1948).

94. This is discussed in detail in Futrell, *Ideas, Concepts, Doctrine*, vol. 1, 306–308, 365–378; and Cardwell, *Airland Combat*, 16–18.

95. Arthur W. Connor, Jr., "The Armor Debacle in Korea, 1950: Implications for Today," *Parameters* 22, no. 3 (Summer 1992): 73.

96. Maj. Roger F. Kropf, "The U.S. Air Force in Korea: Problems That Hindered the Effectiveness of Air Power," *Airpower Journal* 4, no. 1 (Spring 1990): 30–46; see also Wayne Thompson and Bernard C. Nalty, *Within Limits: The U.S. Air Force and the Korean War* (Washington, DC: Air Force History and Museums Program, 1996).

97. USAF Evaluation Group, "An Evaluation of the Effectiveness of the United States Air Force in Korea" (January 1951), book 2, vol. 1, 227–228, copy in the Air Force Historical Research Agency archives, Maxwell AFB, AL. This quote is also reprinted in Robert Frank Futrell, *The United States Air Force in Korea, 1950–1953* (New York: Duell, Sloan, and Pearce, 1961), 146.

98. Robert Debs Heinl, *Victory at High Tide: The Inchon-Seoul Campaign* (Philadelphia: J. B. Lippincott Co., 1968).

99. MacArthur quote in Futrell, *The United States Air Force in Korea*, 341; for combat cargo, see William M. Leary, *Anything, Anywhere, Anytime: Combat Cargo in the Korean War* (Washington, DC: Air Force History and Museums Program, 2000).

100. Thomas C. Hone, "Korea," in *Case Studies in the Achievement of Air Superiority*, ed. Benjamin F. Cooling (Washington, DC: Air Force History and Museums Program, 1994), 453–504; William T. Y'Blood, *MiG Alley: The Fight for Air Superiority* (Washington, DC: Air Force History and Museums Program, 2000); Xiaoming Zhang, *Red Wings over the Yalu: China, the Soviet Union, and the Air War in Korea* (College Station: Texas A&M University Press, 2002); and Kenneth P. Werrell, *Sabres over MiG Alley: The F-86 and the Battle for Air Superiority in Korea* (Annapolis, MD: Naval Institute Press, 2005).

101. Lt. Commander Anthony T. DeSmet, USN, "Effects of Doctrine and Experience on Close Air Support Operations in Korea," Report AU/ACSC/054 /2000-04 (Maxwell AFB, AL: Air Command and Staff College, April 2000); and "The Air War in Korea: A Statistical Portrait of the USAF in the First Hot Conflict of the Cold War," *Air Force Magazine* 79, no. 4 (April 1996): 66.

102. Eduard Mark, *Aerial Interdiction: Air Power and the Land Battle in Three American Wars* (Washington, DC: Center for Air Force History, 1994), 265–319.

103. Walton S. Moody, *Steadfast and Courageous: FEAF Bomber Command and the Air War in Korea, 1950–1953* (Washington, DC: Air Force History and Museums Program, 2000).

104. See, for example, Col. James T. Stewart, ed., *Airpower: The Decisive Force in Korea* (Princeton, NJ: D. Van Nostrand Co., 1957), a series of essays that had first appeared in the *Air University Quarterly Review*.

105. Winnefeld and Johnson, *Joint Air Operations*, 60.

106. See William T. Y'Blood, *Down in the Weeds: Close Air Support in Korea* (Washington, DC: Air Force History and Museums Program, 2003).

107. For a contemporary view, see William A. Gunn, *A Study of the Effectiveness of Air Support Operations in Korea*, FEC Project no. 2 (Tokyo: Operations Research Office, General Headquarters, Far East Command, January 1952), U.S. Army Military History Institute Library. See also Michael Lewis, "Lt. Gen. Ned Almond, USA: A Ground Commander's Conflicting View with Airmen over CAS Doctrine and Employment" (Maxwell AFB, AL: School of Advanced Airpower Studies, August 1997); Michael J. Chandler, "General Otto P. Weyland, USAF: Close Air Support in the Korean War" (Maxwell AFB, AL: School of Advanced Air and Space Studies, Air University, June 2003); and DeSmet, *Effects of Doctrine and Experience on Close Air Support Operations in Korea*.

108. Cardwell, *Airland Combat*, 18–22; Winnefeld and Johnson, *Joint Air Operations*, 39–62. For often-varying army, navy, and air force perspectives, see Roy E. Appleman, *South to the Naktong, North to the Yalu*, a volume in the *United States Army in the Korean War* series (Washington, DC: Office of the Chief of Military History, 1961); Futrell, *The United States Air Force in Korea*; James A. Field, *History of United States Naval Operations in Korea*

(Washington, DC: USN, 1962); and Lt. Col. Pat Meid and Major James M. Yingling, *U.S. Marine Operations in Korea, 1950–1953* (Washington, DC: USMC, 1972).

109. Ralph D. Bald, *Air Force Participation in Joint Army–Air Force Training Exercises, 1951–1954*, no. 129 in the *USAF Historical Studies* series (Maxwell AFB, AL: Air University, USAF Historical Division, Research Studies Institute, September 1957), 132. The exercises were Southern Pine, Snow Fall, Long Horn, Cold Spot, and Tacair 54-7.

110. See Futrell, *Ideas, Concepts, Doctrine*, vol. 1, 379–395; for a far more negative interpretation of AF doctrinal development in this time period, see Mowbray, "Air Force Doctrine Problems."

111. For example, heavy attack (the Douglas A3D, subsequently the A-3), light jet attack (the Douglas A4D, the A-4), all-weather attack (the Grumman A2F, the A-6), air superiority (the Vought F8U, the F-8), and fleet air defense and swing-role attack (the McDonnell F4H, the F-4). For the navy's Korean lessons learned, see Richard P. Hallion, *The Naval Air War in Korea* (Baltimore, MD: Nautical and Aviation Publishing Company of America, 1986).

112. Marcelle Size Knaack, *Post–World War II Fighters, 1945–1973*, vol. 1 of the *Encyclopedia of U.S. Air Force Aircraft and Missile Systems* (Washington, DC: Office of Air Force History, 1978), 38–133. Not all were satisfactory, and some were quite problem-ridden; see Richard P. Hallion, "A Troubling Past: Air Force Fighter Acquisition since 1945," *Airpower Journal* 9, no. 4 (Winter 1990): 4–23. For fighter pilots' perspectives, see Jack Broughton, *Rupert Red Two: A Fighter Pilot's Life* (St. Paul, MN: Zenith Press, 2007); and James Salter, *Gods of Tin: The Flying Years*, ed. Jessica Benton and William Benton (Washington, DC: Shoemaker & Hoard, 2004).

113. The quote is from Ivan A. Getting, "Vignettes of Continental Defense" (January 18, 1982), 4, a privately printed memoir sent to me by the late Dr. Getting on August 17, 1990. See also Capt. Kenneth Schaffel, *The Emerging Shield: The Air Force and the Evolution of Continental Air Defense, 1945–1960* (Washington, DC: Office of Air Force History, 1991), a thorough study that stands as a testament to the intellect and capabilities of a gifted young officer-historian who sadly died in an accident before it was published.

114. See, for example, Gen. Thomas S. Power, "SAC and the Ballistic Missile," *Air University Quarterly Review* 9, no. 4 (Winter 1957–58): 2–29.

115. Lt. Col. William L. Anderson, "Officers and Missiles," *Air University Quarterly Review* 10, no. 1 (Spring 1958): 80, and accompanying "Future Operator Requirements" chart. Regarding "the coming of astro power," see the statement of Gen. Bernard Schriever, "The USAF Reports to Congress," in the same issue, 55.

116. For the roots and highlights of scientific and technological transformation within the air force, see Nick A. Komons, *Science and the Air Force: A History of the Air Force Office of Scientific Research* (Arlington, VA: Historical Division, USAF Office of Aerospace Research, 1966); Thomas A. Sturm, *The USAF Scientific Advisory Board: Its First Twenty Years, 1944–1964* (Washington, DC:

Office of Air Force History, 1986); Michael H. Gorn, *Harnessing the Genie: Science and Technology Forecasting for the Air Force, 1944–1986* (Washington, DC: Office of Air Force History, 1988); Jacob Neufeld, George M. Watson, Jr., and David Chenoweth, eds., *Technology and the Air Force: A Retrospective Assessment* (Washington, DC: Air Force History and Museums Program, 1997); Dwayne A. Day, *Lightning Rod: A History of the Air Force Chief Scientist's Office* (Washington, DC: Office of the Air Force Chief Scientist, 2000); and Stephen B. Johnson, *The United States Air Force and the Culture of Innovation, 1945–1965* (Washington, DC: Air Force History and Museums Program, 2002).

117. For SIOP development from the perspective of its originator, see Glenn A. Kent, with David Ochmanek, Michael Spirtas, and Bruce R. Pirnie, *Thinking about America's Defense: An Analytical Memoir* (Santa Monica, CA: RAND Corporation, 2008), 22–42.

118. This interest accelerated as conditions deteriorated at the end of the 1950s and Vietnam loomed into prominence, resulting in detailed counterinsurgency studies by the Concepts Division of the Aerospace Studies Institute at the Air University at Maxwell AFB. Examples include the following Concepts Division studies: *The Accomplishments of Airpower in the Malayan Emergency, 1948–1960* (May 1963); *The Employment of Airpower in the Greek Guerrilla War, 1947–1949* (December 1964); and *Guerrilla Warfare and Airpower in Algeria, 1954–1960* (March 1965). All of these are in the archives of the Air Force Historical Research Agency, Maxwell AFB, AL.

119. For aerial refueling, see Richard K. Smith, *Seventy-Five Years of Inflight Refueling: Highlights, 1923–1998* (Washington, DC: Air Force History and Museums Program, 1998). For a background of expeditionary forces, see Richard G. Davis, *Anatomy of a Reform: The Expeditionary Aerospace Force* (Washington, DC: Air Force History and Museums Program, 2003). For a case study, see Jacob Van Staaveren, *Air Operations in the Taiwan Crisis of 1958* (Washington, DC: USAF Historical Division Liaison Office, November 1962).

120. For the USAF and ballistic missiles, see Lt. Col. Kenneth F. Gantz, *The United States Air Force Report on the Ballistic Missile: Its Technology, Logistics, and Strategy* (Garden City, NY: Doubleday, 1958), a compendium of essays reprinted from *Air University Quarterly Review;* Edmund Beard, *Developing the ICBM: A Study in Bureaucratic Politics* (New York: Columbia University Press, 1976); Jacob Neufeld, *Ballistic Missiles in the United States Air Force, 1945–1960* (Washington, DC: Office of Air Force History, 1990); and Neil Sheehan, *A Fiery Peace in a Cold War: Bernard Schriever and the Ultimate Weapon* (New York: Random House, 2009).

121. Eduard Mark, *Defending the West: The United States Air Force and European Security, 1946–1998* (Washington, DC: Air Force History and Museums Program, 1998), 17 (emphasis added).

122. Ibid., 22.

123. Ibid., 26.

124. Ibid., 28–29; Richard P. Hallion, "The USAF and NATO: From the Berlin Airlift to the Balkans," a lecture in commemoration of the fiftieth anniversary of NATO, Aerospace Power Breakfast Series, DFI International, Army and Navy Club, Washington, DC, April 15, 1999.

125. For the crisis in general and Air Force U-2 operations, see Norman Polmar and John D. Gresham, *Defcon-2: Standing on the Brink of Nuclear War during the Cuban Missile Crisis* (Hoboken, NJ: John Wiley & Sons, Inc., 2006).

126. Gary P. Myers, ed., *Executive Summary of the Missiles in Cuba, 1962: The Role of SAC Intelligence* (Offutt AFB, NE: Strategic Joint Intelligence Center, 1992), 14–15. As Myers acknowledges, this summary is based on an earlier one by Major Sanders Laubenthal, USAF, that appeared in 1983 in the *SAC Intelligence Quarterly.*

127. Myers, ed., *Executive Summary,* 15. See also McCone memorandum, document no. 91, and CIA memorandum, document no. 104, in *CIA Documents on the Cuban Missile Crisis, 1962,* ed. Mary S. McAuliffe (Washington, DC: History Staff, Central Intelligence Agency, October 1992), 305, 342.

128. SAC's involvement in the Cuban Missile Crisis is also briefly mentioned in Bruce M. Bailey, *"We See All": A History of the 55th Strategic Reconnaissance Wing, 1947–1967* (Privately printed, 55th ELINT Association, 1982), 111, copy in the library of the Air Force History Support Office, Bolling AFB, Washington, DC.

129. Indeed, in 1964, before the onset of bombing operations over North Vietnam, Undersecretary of State George Ball had written, "We are considering air action against [North Vietnam] as the means to a limited objective—*the improvement of our bargaining position with the North Vietnamese.*" See George W. Ball, "How Valid Are the Assumptions Underlying Our Viet-Nam Policies?" *Atlantic Monthly* 230 (October 5, 1964), 38. As well, Secretary of the Air Force Harold Brown stated in 1969, "air power was one of our principal bargaining counters"; see Harold Brown, "Airpower in Vietnam," *Air University Review* 20, 3 (May–June 1969). These issues are examined more completely in R. P. Hallion, "Air Power in Peripheral Conflict: The Cases of Korea and Vietnam," a paper presented at an air power symposium held at the École militaire, Paris, June 10, 1999, and reprinted in *Aviation militaire: Survol d'un siècle,* Général de brigade aérienne Hugues Silvestre de Sacy et al., eds. (Paris: Service historique de l'armée de l'air and the École militaire, 1999), 225–242.

130. Lt. General Michael C. Short, "An Airman's Lessons from Kosovo," in *From Manoeuvre Warfare to Kosovo?* ed. John Andreas Olsen (Trondheim: Royal Norwegian Air Force Academy, 2001), 288.

131. Quoted in Kenneth W. Condit, *The Joint Chiefs of Staff and National Policy, 1955–56,* vol. 6 of *History of the Joint Chiefs of Staff* (Washington, DC: Historical Office, Joint Staff, 1992), 9 (emphasis added).

132. As evidenced by Knaack, *Fighters,* passim.

133. Col. H. R. McMaster, *Dereliction of Duty: Lyndon Johnson, Robert McNamara, the Joint Chiefs of Staff, and the Lies That Led to Vietnam* (New York: HarperCollins Publishers, 1997), is the best account. McNamara's own views

of security and Vietnam can be found in his *The Essence of Security: Reflections in Office* (New York: Harper & Row, Publishers, 1968); and his remarkably (even by the standards of most memoirs) indulgent and self-serving autobiography (with Brian Van De Mark), *In Retrospect: The Tragedy and Lessons of Vietnam* (New York: Times Books, 1995).

134. Robert F. Coulam, *Illusions of Choice: The F-111 and the Problem of Weapons Acquisition Reform* (Princeton, NJ: Princeton University Press, 1977); Robert J. Art, *The TFX Decision: McNamara and the Military* (Boston, MA: Little, Brown, 1968); and Richard P. Hallion, *The Evolution of Commonality in Fighter and Attack Aircraft Development and Usage* (Edwards AFB, CA: Air Force Flight Test Center History Office, November 1985).

135. U.S. Congress, Senate, *TFX Control Investigation*, vols. 1–10, 88th Cong., 1st sess. (Washington, DC: GPO, 1963–64); U.S. Congress, Senate, *TFX Contract Investigation* (Second Series), vols. 1–3, 91st Cong., 2d sess. (Washington, DC: GPO, 1970); and U.S. Congress, Senate, *TFX Contract Investigation*, Report no. 91–1496, 91st Cong., 2d sess. (Washington, DC: GPO, 1970). For a program summary, see Knaack, *Fighters*, 223–262.

136. Glenn E. Bugos, "Testing the F-4 Phantom II: Engineering Practice in the Development of American Military Aircraft, 1954–1969," (PhD dissertation, University of Pennsylvania, 1988); Major Richard G. Head, "Decision-Making on the A-7 Attack Aircraft Program" (PhD dissertation, Syracuse University, 1971); and Knaack, *Fighters*, 265–285.

137. Bruce K. Holloway, "Air Superiority in Tactical Air Warfare," *Air University Review* 19, no. 3 (March–April 1968): 9 (emphasis added).

138. R. Frank Futrell et al., *Aces and Aerial Victories: The United States Air Force in Southeast Asia, 1965–1973* (Washington, DC: Office of Air Force History, 1976), 155–160. The U.S. Navy had a higher 5.6 victories per friendly fighter loss, reflecting better training and tactics at that time; see John B. Nichols and Barrett Tillman, *On Yankee Station: The Naval Air War over Vietnam* (Annapolis, MD: Naval Institute Press, 1987), 75–79. For both services, Vietnam taught the importance of realistic training, leading to programs such as Top Gun and Red Flag.

139. For example, dissimilar air combat training: RAAF Sabres flew as MiG-17 surrogates against F-4 and F-105 squadrons, an important contribution to the combat readiness of USAF forces. See Alan Stephens, *Going Solo: The Royal Australian Air Force, 1946–1971* (Canberra: Australian Government Publishing Service, 1995), 278–307. The South Vietnamese Air Force is examined in General William W. Momyer, *The Vietnamese Air Force, 1951–1975: An Analysis of Its Role in Combat* (Washington, DC: GPO in association with Air University, 1975), which details its accomplishments, made in the face of deficiencies in command and organization.

140. McMaster, *Dereliction of Duty*, 333.

141. For two differing interpretations, see Lt. Col. Mark Clodfelter, *The Limits of Air Power: The American Bombing of North Vietnam* (New York: Free Press, 1989); and Wayne Thompson, *To Hanoi and Back: The USAF and*

North Vietnam, 1966–1973 (Washington, DC: Smithsonian Institution Press, 2000). See also Karl J. Eschmann, *Linebacker* (New York: Ivy Books, 1989). For technology, see Gerald K. Hendricks, "The Emphasis on Limited War and Its Impact on Research and Development in the USAF," Student Report no. 82 (Washington, DC: Industrial College of the Armed Forces, April 1, 1988): 38–42; David R. Mets, *The Quest for a Surgical Strike: The USAF and Laser Guided Bombs* (Eglin AFB, CA: Air Force Systems Command Armament Division, 1987); Ron Westrum and Howard A. Wilcox, "Sidewinder," *American Heritage of Invention & Technology* 2, no. 3 (Fall 1989): 56–63; and Memo, JSC to SecDef, "Status of Guided Missile Projects" (March 15, 1950), reprinted as document no. 13 in Wolfe, *The United States Air Force Basic Documents on Roles and Missions,* 213–218.

142. For examples of VC and NVA reflections, see statements of Xuan Vu, Trinh Duc, Bui Van Tai, Huong Van Ba, and Nguyen Van Thanh in David Chanoff and Doan Van Toai, *Portrait of the Enemy* (New York: Random House, 1986).

143. Winnefeld and Johnson, *Joint Air Operations,* 63.

144. The best short account of the Vietnam air war remains Col. Alan L. Gropman, "The Air War in Vietnam, 1961–1973," in *War in the Third Dimension: Essays in Contemporary Air Power,* ed. Air Vice-Marshal R. A. Mason (London: Brassey's Defence Publishers, 1986), 33–58. The most comprehensive summary of the air war is Thomas C. Thayer, ed., *The Air War,* vol. 5 of *A Systems Analysis View of the Vietnam War, 1965–1972* (Washington, DC: Southeast Asia Intelligence Division, OASD(SA)RP, February 18, 1975), document ADA051611, Defense Technical Information Center, Defense Logistics Agency.

145. Quoted in John J. Sbrega, "Southeast Asia," in *Case Studies in Close Air Support,* ed. Benjamin F. Cooling (Washington, DC: Air Force History and Museums Program, 1991), 469.

146. McGiffert quote from John A. Doglione et al., *Air Power and the 1972 Spring Invasion* (Washington, DC: GPO in association with Air University, 1976), 103; Chanoff and Doan Van Toai, *Portrait of the Enemy,* 109, 154, 171, 185; see also Jeffrey J. Clarke, *Advice and Support: The Final Years, 1965–1973* (Washington, DC: U.S. Army Center of Military History, 1988), 486.

147. Douglas Kinnard, *The War Managers* (Hanover, NH: University Press of New England, 1977), 63.

148. Even by the most cautious analysis, Linebacker II was decisive; see, for example, Herman L. Gilster, *The Air War in Southeast Asia: Case Studies of Selected Campaigns* (Maxwell AFB, AL: Air University Press, October 1993), 75–115; for the spring invasion, see Doglione and also David K. Mann, *The 1972 Invasion of Military Region I: Fall of Quang Tri and Defense of Hue* (Hickam AFB, HI: HQ Pacific Air Forces, Directorate of Operations Analysis, CHECO/Corona Harvest Division, March 15, 1973).

149. I believe that the Nixon administration, largely through air power, created circumstances whereby South Vietnam could have prevailed as an independent nation after 1973. Its collapse in 1975 reflected far more the loss of political

will and the refusal of Congress to fund continued aid (the political fallout of the Watergate affair and the resignation of Richard Nixon) than any innate military weakness. Indeed, in retrospect, America left South Vietnam in 1973 in a condition analogous to how it had withdrawn from Korea in 1953. An American failure then to continue to support South Korean independence would have doomed it to collapse by early 1955 just as surely as it did doom South Vietnam two decades later.

150. Meir quote from John T. Correll, *The Air Force and the Cold War* (Arlington, VA: Air Force Association, September 2005), 29. Airlift was the only means of resupplying Israel during the war; the first ship of an emergency sealift only arrived after fighting had already ceased.

151. Yehuda Weinraub, "The Israel Air Force and the Air Land Battle," *Israeli Defense Forces Journal* 3, no. 3 (Summer 1986): 22–30. Over nineteen days of combat, Israel lost 109 aircraft, representing 35 percent of its prewar air combat strength. B-52 losses are covered in Brig. Gen. James R. McCarthy and Lt. Col. George B. Allison, *Linebacker II: A View from the Rock*, ed. Col. Robert E. Rayfield (Washington, DC: GPO in association with Air University, 1979), esp. 50–89, 171–172.

152. See Benjamin S. Lambeth, *The Transformation of American Air Power* (Ithaca, NY: Cornell University Press, 2000).

153. The leadership shift, to 1982, is detailed in the previously cited Worden, *Rise of the Fighter Generals.*

154. For example, SAC's last three commanders—Generals Welch, Chain, and Butler—were all Vietnam F-4 pilots.

155. A resurrection hinted at over a decade earlier in Soviet doctrinal writings; for example, see Marshal V. D. Sokolovskii, ed., *Soviet Military Strategy*, trans. and ed. Herbert S. Dinerstein, Leon Gouré, and Thomas W. Wolfe (Englewood Cliffs, NJ: RAND in association with Prentice-Hall, Inc., 1963), 413; and Kenneth R. Whiting, "The Past and Present of Soviet Military Doctrine," *Air University Quarterly Review* 11, no. 1 (Spring 1959): 38–60. See also Central Intelligence Agency, "Military Thought (USSR): The Balance of Forces of Opposing Sides in Aviation," *Intelligence Information Special Report* (12 October 1976), a translation of a secret 1968 Soviet essay redacted and released by the CIA as Report HR 70-14, approved for public release on January 16, 2006, and available at www.cia.gov.

156. The A-10 began as an effort to replace the aging and rapidly attriting propeller-driven Douglas A-1 Skyraider "for handling the close-support mission of a South Vietnam environment." See Scientific Advisory Board, "Report of the USAF Scientific Advisory Board Aerospace Vehicles Panel," May 1967, 1, copy in the office files of the Air Force Scientific Advisory Board, HQ USAF, Pentagon.

157. ATP-27A/B and ATP-33; see David J. Stein, *The Development of NATO Tactical Air Doctrine*, Report R-3385-AF (Santa Monica, CA: RAND, December 1987), 14–35. For the complex nature of the German issue, see Michael E. Thompson, *Political and Military Components of Air Force Doctrine*

in the Federal Republic of Germany and Their Implications for NATO Defense Policy Analysis, P-7557-RGS (Santa Monica, CA: RAND, September 1987).

158. See Neufeld, Watson, and Chenoweth, *Technology and the Air Force*; Day, *Lightning Rod*; and Johnson, *USAF and Culture of Innovation*.

159. Frederick L. Frostic, "Quality versus Quantity in Tactical Fighter Forces," *Journal of Defense Research* 13, no. 3 (Fall 1981: 308. I thank Mr. Frostic, the former Assistant Secretary of Defense for Requirements and Plans, for alerting me to this study.

160. For the thirty-one initiatives, see Richard G. Davis, *The 31 Initiatives: A Study in Air Force–Army Cooperation* (Washington, DC: Office of Air Force History, 1987); for the army perspective in this period, see John L. Romjue, *From Active Defense to AirLand Battle: The Development of Army Doctrine, 1973–1982* (Washington, DC: GPO, 1984); the relevant army doctrine was FM 100-5, issued in 1976.

161. See, for example, Harold R. Winton, "Partnership and Tension: The Army and Air Force between Vietnam and Desert Shield," *Parameters: The U.S. Army War College Quarterly* 26, no. 1 (Spring 1996): 100–119, who concludes, "From 1973 to 1990 the Army and the Air Force formed a solid partnership centered around the Army's ability to execute its AirLand Battle doctrine with Air Force support. The strength of this partnership was evident in extensive biservice training, doctrinal publications, and programmatic cooperation."

162. See, for example, Lt. Gen. Michael J. Dugan, "Air Power: Concentration, Responsiveness, and the Operational Art," *Military Affairs* 69, no. 7 (July 1989): 21.

163. To the frustration of Britain's senior civil and military leadership; see Margaret Thatcher, *The Downing Street Years* (London: HarperCollins Publishers, 1993), 216–236; Michael Heseltine, *Life in the Jungle: My Autobiography* (London: Hodder and Stoughton, 2000), 241–243; and Adm. Sandy Woodward, with Patrick Robinson, *One Hundred Days: The Memoirs of the Falklands Battle Group Commander* (London: Fontana, 1992), 1–22, 236–292; for the RAF side, see Group Capt. Timothy Garden, "Technology Lessons from the Falklands Conflict," in *Low-Intensity Conflict and Modern Technology*, ed. David J. Dean (Maxwell AFB, AL: Air University Press, 1986), 113–121.

164. Benjamin S. Lambeth, "Moscow's Lessons from the 1982 Lebanon Air War," in *War in the Third Dimension*, 127–148; CIC Matthew M. Hurley, "The Bekaa Valley Air Battle, June 1982: Lessons Mislearned?" *Airpower Journal* 3, no. 4 (Winter 1989): 60–70.

165. As noted by Col. Alexander Orlov, "The Soviet Air Forces in the Afghan War," in de Sacy et al., eds., *Aviation militaire*, 243–247, who notes (244–245), "Participation of helicopters in air support operations was cut down and in 1987 it was terminated in order to exclude losses. . . . ground forces could successfully solve their tasks only when air support was provided."

166. In 1990, shortly after the publication of the *Global Reach–Global Power* strategic planning framework, a much more respectable, reliable, and useful compilation of such material appeared, edited by a noted air power authority; see Lt. Col. Charles M. Westenhoff, *Military Air Power: The CADRE Digest*

of Air Power Opinions and Thoughts (Maxwell AFB, AL: Air University Press, October 1990).

167. For example, see Brig. Gen. Benjamin G. Holzman, "Basic Research for National Survival," *Air University Quarterly Review* 12, no. 1 (Spring 1960): 28–52, quote from 52.

168. For example, Major Robert C. Ehrhart's thoughtful "Some Thoughts on Air Force Doctrine," *Air University Review* 29, no. 2 (March–April 1980): 29–38.

169. Richard P. Hallion, "Doctrine, Technology, and Air Warfare: A Late Twentieth-Century Perspective," *Airpower Journal* 1, no. 2 (Fall 1987): 16–27 (emphasis in original). The perception of air force doctrine in the post-Gulf years (when one senior commander, for example, memorably informed a gathering of juniors that "doctrine is bullshit") is discussed in Barry D. Watts's almost identically titled "Doctrine, Technology, and War," a lecture given at the Air & Space Doctrinal Symposium, Air University, Maxwell AFB, April 30–May 1, 1996, www.airpower.au.af.mil/airchronicles/cc/watts.html.

170. And it is arguably worse today, particularly when extended to the cyber revolution, space, and the ever-contentious issue of joint and combined command and control. As one retired air force leader remarked recently to the author, "When I was a major, our doctrine was four pages long and it actually made some sense and was useful."

171. The cartoonish 1979 AFM 1-1 is dissected in Williamson Murray, "A Tale of Two Doctrines: The Luftwaffe's 'Conduct of the Air War' and the USAF's Manual 1-1," *Journal of Strategic Studies* 6, no. 4 (December 1983): 84–93. For a discussion of these and the general progression of air force doctrinal thinking and process in the 1970s and 1980s, see Futrell, *Ideas, Concepts, Doctrine,* vol. 2, 722–745.

172. Col. John A. Warden III, *The Air Campaign: Planning for Combat* (Washington, DC: National Defense University Press, 1988); for Warden's place in relation to the pantheon of air power theorists, see David R. Mets, *The Air Campaign: John Warden and the Classical Airpower Theorists* (Maxwell AFB, AL: Air University Press, 1999).

173. Which it eventually was. It was issued in March 1992 with a brief 20-page summary volume and a 308-page second volume of supporting thought and commentary written by Col. Dennis Drew. AFM 1-1 came out after the war; for an assessment of it in light of the Desert Storm experience, see Lt. Col. Kurt A. Cichowski, *Aerospace Doctrine Matures through a Storm: An Analysis of the New AFM 1-1* (Maxwell AFB, AL: School of Advanced Airpower Studies, Air University, 1992).

174. For a typical example, see Robert Coram, "The Case against the Air Force," *Washington Monthly* 19, no. 7–8 (July–August 1987): 17–24.

175. One senior commander stated that the USAF had but two purposes: continental air defense and "supporting the U.S. Army in battle." (Personal recollection from serving on the Secretary of the Air Force's Staff Group.)

176. Donald B. Rice, *The Air Force and U.S. National Security: Global Reach–Global Power* (Washington, DC: Office of the Secretary of the Air Force, June 1991).

177. Personal recollection. Deptula later became a key Gulf War air campaign planner, and subsequently the service's most outspoken advocate for air power. When this essay was written, he was the Air Staff A2, with responsibilities for overseeing the service's intelligence, surveillance, and reconnaissance activities. For a useful history of the evolution of the document and the interplay of various parties, see Major Barbara J. Faulkenberry, *"Global Reach–Global Power": Air Force Strategic Vision, Past and Future* (Maxwell AFB, AL: Air University, June 1995).

178. *Global Reach–Global Power*, 15.

179. From a conversation with an individual who was present when the officer made the remark, but who wishes both his identity and that of the senior officer to remain anonymous.

180. Cheney had a reputation for hot-tempered decision-making, dating back to his days in the Ford White House; earlier, he had rebuked Air Force Chief of Staff Gen. Larry Welch over a meeting on the Hill; fortunately Welch survived. Dugan, whose comments were reported in the *Washington Post*, was, alas, not so fortunate. See also Robert S. Dudney, "The Lavelle Syndrome," *Air Force Magazine* 93, no. 9 (September 2010).

181. As joint-force air commander, Horner's previously cited memoir is particularly useful for conveying his approach to the war, as is a briefing he presented on May 8, 1991, to a Congressional meeting of the Business Executives for National Security. This Desert Storm briefing is at www.airforce-magazine. com under the "Keeper File."

182. For example, see Joel Achenbach, "The Experts in Retreat," *Washington Post*, February 28, 1991.

183. Fred Frostic, *Air Campaign against the Iraqi Army in the Kuwaiti Theater of Operations*, Report MR-357-AF (Santa Monica, CA: RAND, 1994), 63–64. The psychological effects of Gulf War attacks are examined in detail in Group Capt. Andrew P. N. Lambert, *The Psychology of Air Power* (London: Royal United Services Institute for Defence Studies, 1994), 64–69. For the KTO air campaign, see Perry D. Jamieson, *Lucrative Targets: The U.S. Air Force in the Kuwaiti Theater of Operations* (Washington, DC: Air Force History and Museums Program, 2001).

184. For example, Frostic, *Air Campaign against the Iraqi Army*, 46, notes, "Casualties from air attacks . . . were relatively low. By dispersing and staying away from their vehicles and weapons, [Iraqi troops] could survive."

185. It also stimulated the navy, as Korea had earlier, to undertake fundamental and transformational changes in its planning, operations, and aircraft usage to make naval aviation much more effective in both deep-strike and littoral operations; see Report, Capt. Steven U. Ramsdell, USN, to Dean Allard, Director, Naval Historical Center, May 14, 1991, copy in the Naval Historical Center Archives, Washington Navy Yard, Washington, DC, which dissected the service's Gulf War performance; and Rear Adm. Riley D. Mixon, USN, "Where We Must Do Better," Naval Institute *Proceedings* 118, no. 8 (August 1991): 39. Mixon was commander of Task Force 155 operating in the Red Sea.

186. Charles Krauthammer, "Holiday from History," February 14, 2003, at http://www.aish.com/ci/s/48900002.html.

187. For a review of these, see A. Timothy Warnock, *Short of War: Major USAF Contingency Operations* (Maxwell AFB, AL: Air Force History and Museums Program in association with Air University Press, 2000); see also General John P. Jumper, "Rapidly Deploying Aerospace Power: Lessons from Allied Force," *Aerospace Power Journal* 13, no. 4 (Winter 1999): 4–10.

188. Donald B. Rice, *Toward the Future: Global Reach–Global Power: U.S. Air Force White Papers, 1989–1992* (Washington, DC: Headquarters Air Force, 1992); and Merrill A. McPeak, *Selected Works, 1990–1994* (Maxwell AFB, AL: Air University Press, August 1995). As the Air Force Historian in this period, I worked closely with both men, this proximity influencing my judgment of them.

189. For example, see the extension of the original GR–GP Paper, *Global Reach–Global Power: The Evolving Air Force Contribution to National Security* (Washington, DC: Office of the Secretary of the Air Force, December 1992).

190. Of this reorganization, only the latter has drawn continued questioning as to its value, primarily because it recalls the failure of the original Air Materiel Command to adequately address research and development issues at the very onset of the service, a failure that drove the emergence of Air Research and Development Command, which evolved into Air Force Systems Command, and the establishment of Air Force Logistics Command.

191. This is my personal assessment, but it is also clearly evident in his writings and testimony. Unfairly caricatured by some at the time for his interest in uniforms, heraldry, and health food, McPeak was a perceptive student of air power and air organization; his 1985 essay "TACAIR Missions and the Fire Support Coordination Line," *Air University Review* 36, no. 6 (September–October 1985): 65–72, anticipated FSCL issues encountered in Desert Storm subsequently. For McPeak on planning, see Capt. Stephen N. Whiting and Capt. Thomas K. Dale, *Air Force Long-Range Planning: Institutionalization Revisited* (Washington, DC: Air Force History Support Office in association with the Office of the Special Assistant to the Chief of Staff for Long-Range Planning, April 12, 1996), 43–49. For *Spacecast 2020*, see Lt. Gen. Jay W. Kelley et al., *Spacecast 2020*, vol. 1 (Maxwell AFB, AL: Air University Press, June 1994), v–xxix, 1–22. See also Faulkenberry, *Global Reach–Global Power*, 35–36.

192. Most notably Great Britain and Australia, via the RAF Director of Air Studies at the Royal Air Force Staff College, Bracknell, and the RAAF Air Power Studies Centre at Canberra. Each strengthened and expanded their doctrinal and air power study efforts, producing a wide range of influential papers and holding a series of international conferences and seminars of lasting significance. But many other nations did as well, including those of the former Warsaw Pact.

193. In fact, evident as early as the question-and-answer session following General McPeak's Desert Storm briefing to the news media held at the Pentagon on March 15, 1991. For the transcript of that briefing and questions, see Merrill A. McPeak, *Selected Works*, 15–50, esp. 47.

194. Personal recollection. For the record, the report came out with the expression.
195. Army and Navy Joint Board, *Report of the Joint Board on the Results of Aviation and Ordnance Tests Held During June and July, 1921*, in U.S. Congress, *Congressional Record*, 67th Cong., 1st sess. (August 20, 1921), 8625–8626; for the 1992 meeting, see Grant T. Hammond, *Paths to Extinction: The U.S. Air Force in 2025*, vol. 4 of the *Air Force 2025* study (Maxwell AFB, AL: Air University, August 1996), 2.
196. For example, see MacGregor Knox and Williamson Murray, eds., *The Dynamics of Military Revolution, 1300–2050* (Cambridge: Cambridge University Press, 2001), a useful series of case studies; and Thomas A. Keaney and Eliot A. Cohen, *Gulf War Air Power Survey Summary Report* (Washington, DC: GPO, 1993), 251, which concludes, "The ingredients for a transformation of war may well have become visible in the Gulf War, but if a revolution is to occur someone will have to make it."
197. During Desert Storm, for example, twenty-eight Americans were killed, and another ninety-seven injured by a Scud missile that was fired at Dhahran. An investigation revealed that it was the only unengaged Scud of the war, having been fired on Dhahran at a time when, by chance, a Patriot battery was down for maintenance. It is highly unlikely the same result would have occurred had the air force owned the theater missile defense, as its concept of operations, built around robust redundancy (not present, obviously, in this case), would have enabled the Scud to have been engaged.
198. Merrill A. McPeak, "Leave the Flying to Us," *Washington Post*, June 5, 2003. It might be added that Sullivan's decision made even less sense given the immense costs associated with operating the hugely bloated numbers of gas-turbine–powered tanks the service possessed, despite, after 1989, any sort of threat that would have justified such a disproportionate force. For Sullivan's views on the nature of future war, see General Gordon R. Sullivan and Lt. Col. James M. Dubik, *Land Warfare in the 21st Century* (Carlisle Barracks, PA: U.S. Army War College Strategic Studies Institute, February 1993). In one very important respect, Sullivan and Dubik were prescient: the value of small, high-technology special operations forces "equipped with secure satellite communications, laser designators, and position guidance systems" (xxiii). SOF teams thus equipped were able to call down precise CAS in Afghan operations, routinely cuing JDAM assaults from B-52s circling at high altitude, as discussed later.
199. McPeak, "Leave the Flying to Us."
200. For example, Winnefeld and Johnson, *Joint Air Operations*; Christopher Bowie, Fred Frostic, Kevin Lewis, John Lund, David Ochmanek, and Philip Propper, *The New Calculus: Analyzing Airpower's Changing Role in Joint Theater Campaigns* (Santa Monica, CA: RAND, 1993); Col. Phillip S. Meilinger, *Ten Propositions on Air Power* (Washington, DC: Air Force History and Museums Program, 1995); David E. Thaler and David A. Shlapak, *Perspectives on Theater Air Campaign Planning* (Santa Monica, CA: RAND, 1995); Col. David A. Deptula, *Firing for Effect: Change in the Nature of Warfare* (Arlington, VA: Aerospace Education Foundation, August 1995); Alan Vick, David T.

Orletsky, John Bordeaux, and David A. Shlapak, *Enhancing Air Power's Contribution against Light Infantry Targets* (Santa Monica, CA: RAND, 1996); Daniel Gouré and Christopher M. Szara, eds., *Air and Space Power in the New Millennium* (Washington, DC: Center for Strategic and International Studies, 1997); Earl H. Tilford, Jr., *National Defense into the 21st Century: Defining the Issues—A Special Report from the Strategic Studies Institute* (Carlisle Barracks, PA: U.S. Army War College Strategic Studies Institute, June 6, 1997); Oliver Fritz, Renee Lajoie, and Barry Bleichman, *A 21st Century Air Force: U.S. Global Engagement and the Use of Air and Space Power* (Washington, DC: DFI International, July 1997); and *Air Power Confronts an Unstable World,* ed. Richard P. Hallion (London: Brassey's, 1997). Numerous similar studies were prepared abroad, particularly, as noted earlier, in Great Britain and Australia.

201. John P. White et al., *Directions for Defense: Report of the Commission on Roles and Missions of the Armed Forces* (Washington, DC: Department of Defense, May 24, 1995), ES-4.

202. General John Shalikashvili, *Joint Vision 2010–America's Military: Preparing for Tomorrow* (Washington, DC: Office of the Chairman of the Joint Chiefs of Staff, July 1996).

203. HQ USAF, *Global Engagement: A Vision for the 21st Century Air Force* (Washington, DC: HQ USAF, November 21, 1996), passim. Regarding their longevity, in *Vision 2020* in 2009, *Global Engagement*'s six core competencies were now termed "distinctive capabilities," but remained: air and space superiority (replacing "aerospace" superiority), global attack, rapid global mobility, precision engagement, information superiority, and agile combat support.

204. See Hammond, *Paths to Extinction,* vol. 4 of *Air Force 2025,* 1.

205. The bottom-up review ended inconsequentially, subsequently replaced by the first QDR, undertaken by Aspin's successor, William Cohen. See John T. Correll, *In the Wake of the QDR: The Quadrennial Defense Review and Its Consequences* (Arlington, VA: Air Force Association, September 2006), 7–8.

206. Particularly the denial of Air Force AC-130 gunship support, which surviving participants universally agreed would have made a tremendous difference in survivability. See Letter and Report, Senators John Warner and Carl Levin to Senators Strom Thurmond and Sam Nunn, Committee on Armed Services, "Review of the Circumstances Surrounding the Ranger Raid on October 3–4, 1993 in Mogadishu, Somalia," September 29, 1995, 28–33, 49; and Mark Bowden, *Black Hawk Down: A Story of Modern War* (New York: Atlantic Monthly Press, 1999), 335, 340–341.

207. Quote from Air Force Doctrine Center, "Historical Evidence Database," (Maxwell AFB, AL: AFDC, 1999).

208. Quoted in Fritz, Lajoie, and Bleichman, *A 21st Century Air Force,* 20.

209. Gene H. McCall and Maj. Gen. John A. Corder, eds., *New World Vistas: Air and Space Power for the 21st Century,* 14 vols. (Washington, DC: Air Force Scientific Advisory Board, 1996). For a survey of the air force in this time period, see William T. Y'Blood, "From the Deserts to the Mountains," and "Metamorphosis: The Air Force Approaches the Next Century," in *Winged Shield,*

Winged Sword: A History of the United States Air Force, vol. 2, ed. Bernard C. Nalty (Washington, DC: Air Force History and Museums Program, 1997), 441–554.

210. Personal recollection. For the AEF, subsequently discussed, see General Ronald R. Fogleman, "Air Power and the American Way of War," a speech before the Air Force Association Air Warfare Symposium, Orlando, Florida, February 15, 1996. See also Bradley Graham, "In Carrier's Absence, U.S. Sending 18 Jet Fighters to Bahrain," *Washington Post,* October 24, 1995.

211. The two volumes were Meilinger, *The Paths to Heaven,* and Col. Bruce M. DeBlois, ed., *Beyond the Paths of Heaven: The Emergence of Space Power Thought* (Maxwell AFB, AL: Air University Press, September 1999). Recognizing the increasing role of space in national security affairs, SAAS became the School of Advanced Air and Space Studies (SAASS) in 2002.

212. For a sampling of subjects in papers the author found useful (in addition to other SAAS and AU Papers cited previously in this work), see the following SAAS and AU student Papers: Maj. Michael A. Longoria, "A Historical View of Air Policing Doctrine" (May 1992); Maj. Robert J. Hamilton, "Green and Blue in the Wild Blue" (June 1993); Lt. Col. Jerome V. Martin, "Victory from Above" (June 1994); Gary C. Cox, "Beyond the Battle Line" (June 1995); Lt. Col. Michael Fricano, "The Evolution of Airlift Doctrine and Organization" (April 1996); Major James R. Ayers, "Military Operations Other Than War in the New World Order" (May 1996); William G. Reese III, "The Doctrine Gap" (June 1996); Lt. Col. Carl R. Pivarsky, Jr., "Airpower in the Context of a Dysfunctional Joint Doctrine" (February 1997); Linda E. Torrens, "The Future of NATO's Tactical Air Doctrine" (June 1996); Major Steven L. Kwast, "Convergence or Divergence: The Relationship between Space Doctrine and Air Force Doctrine" (March 1997); Lt. Col. Milton H. Johnson, "The Impact of Doctrine on Air Force Roles and Missions" (April 1997); Lt. Col. Max D. Shaevitz, "Airpower in the Next Millennium" (April 1997); Kent Laughbaum, "Synchronizing Airpower and Firepower in the Deep Battle" (June 1997); Major Eric S. Mathewson, "The Impact of the Air Corps Tactical School on the Development of Strategic Doctrine" (March 1998); John R. Carter, Jr., "Airpower and the Cult of the Offensive" (April 1998); Major Philip M. Senna, "The JFACC and Small-Scale Contingency Operations" (April 1998); Maj. Kevin J. Fowler, "Avoiding the Seam—An Analytical Framework for Deep Attack" (June 1998); Richard F. Walker, "Facing the Future: A Doctrine for Air Control in Limited Conflicts" (June 1998); Peter A. Lee, "Power Projection: A Comparison of the Aerospace Expeditionary Force and the Carrier Battle Group" (April 1999); Lt. Col. Jay M. Vittori, "Fighting Fires with Fire" (April 1999); Maj. James P. Cashin and Maj. Jeffrey D. Spencer, "Space and Air Force: Rhetoric or Reality?" (April 1999); Maj. Michael J. Moran, "An Evolving Doctrine: Force Application from Space" (March 2000); John K. McMullen, "The United States Strategic Bombing Survey and Air Force Doctrine" (June 2001); Maj. Scot B. Gere, "Struggling to Target" (April 2002); Maj. Terrance J. McCaffrey III, "What Happened to BAI?" (June 2002); Lt. Col.

Daniel F. Baltrusaitis, "Centralized Control with Decentralized Execution" (February 2003); Maj. Lynn I. Scheel, "Long-Range Strike—Concepts and Doctrinal Implications of Future Airpower Capabilities" (June 2003); Scott E. Caine, "Incorporating Effects-Based Operations into Air Force Doctrine" (June 2003); Lt. Col. Charles Q. Brown, Jr., "Developing Doctrine for the Future Joint Force" (April 2004); Maj. Michael G. Koscheski, "Counterforce Attack" (June 2006); and Robert R. Powell, "Quenching the Phoenix" (June 2008).

213. An old thesis in new bottles widely trumpeted following the publication of Robert A. Pape's *Bombing to Win: Air Power and Coercion in War* (Ithaca, NY: Cornell University Press, 1996); for a cogent counter, see Lt. Col. Mark J. Conversino, "The Changed Nature of Strategic Air Attack," *Parameters: The U.S. Army War College Quarterly* 27, 4 (Winter 1997–98), 28–41. The increasing enthusiasm for air power debate carried over into doctrinal development, then drawing greater attention than at any previous time in the service's history. In the fall of 1991, Lt. Gen. Charles Boyd, the commander of Air University (and former Deputy Assistant Chief of Staff for Plans and Operations), had strongly endorsed a revised AFM 1-1 that, in its rationale for air and space power, went considerably beyond previous editions of the manual, setting a tone for subsequent doctrinal pursuit. See Lt. Gen. Charles G. Boyd and Lt. Col. Charles M. Westenhoff, "Air Power Thinking: 'Request Unrestricted Climb,'" *Airpower Journal* 5, 3 (Fall 1991), 4–15.

214. Carl Bergquist, "LeMay Center Realigns Doctrine Development," Air Force News Service News Release, March 6, 2009. The following is a listing, current as of this essay, of the current Air Force Doctrine Documents, their date of authority, and the corresponding DoD Joint Doctrine Publication ("Joint Pubs" or JP): AFDD 1 *Air Force Basic Doctrine* (November 17, 2003), JP 1 *Joint Warfare of the Armed Forces of the United States*; AFDD 1-1 *Leadership and Force Development* (February 18, 2006), JP 0-2 *Unified Action Armed Forces*; AFDD 1-2 *Air Force Glossary* (January 11, 2007), JP 1-02 *DoD Dictionary of Military and Associated Terms*; AFDD 2 *Operations and Organization* (April 3, 2007), JP 0-2, JP 3-0 *Joint Operations*, JP-3-30 *Command and Control for Joint Air Operations*, and JP 3-33 *Joint Task Force Headquarters*; AFDD 2-1 *Air Warfare* (January 22, 2000), JP 3-0; AFDD 2-1.1 *Counterair Operations* (October 1, 2008), JP 3-01 *Countering Air and Missile Threats*; AFDD 2-1.2 *Strategic Attack* (June 12, 2007), JP 3-0; AFDD 2-1.4 *Countersea Operations* (September 15, 2005), JP 3-30; AFDD 2-1.6 *Personnel Recovery Operations* (June 1, 2005), JP 3-50 *National Search and Rescue Manual*, JP 3-50.2 *Doctrine for Joint Combat Search and Rescue*; AFDD 2-1.7 *Airspace Control in the Combat Zone* (July 13, 2005), JP 3-30, JP 3-52 *Joint Doctrine for Airspace Control in the Combat Zone*; AFDD 2-1.9 *Targeting* (June 8, 2006), JP 3-60 *Joint Doctrine for Targeting*; AFDD 2-1.8 *Counter-Chemical, Biological, Radiological, and Nuclear Operations* (January 26, 2007), JP 3-40 *Joint Doctrine for Combating Weapons of Mass Destruction*; AFDD 2-10 *Homeland Operations* (March 21, 2006), JP 3-26 *Joint Doctrine for Homeland Security*; and AFDD 2-12 *Nuclear Operations* (May 7, 2009) [no JP counterpart].

215. Exemplified by the battle over control of joint fires, and the contentious issue of the location and coordination of the FSCL. See Lambeth, *Transformation of American Air Power*, 286–296.

216. See his rationale to air force members, published in the *Air Force Times*, August 11, 1997, as well as a transcript of an interview of General Fogleman by Richard H. Kohn, reprinted in "The Early Retirement of Gen. Ronald R. Fogleman, Chief of Staff, United States Air Force," *Aerospace Power Journal* 15, no. 1 (Spring 2001): 6–23. Fogleman retained his ties to the service, participating frequently on senior-level panels and study efforts, and continuing to champion air and space power in various venues.

217. Readers should note the following discussion is simply my personal view from observing it all unfold, including being present when he personally informed the Air Staff "two letters" he was retiring.

218. Quotes from the previously cited Kohn interview. He was particularly dismayed by a lack of accountability revealed in the aftermath of two horrific events: the shootdown of two Black Hawk helicopters by F-15s in a notorious friendly fire incident over northern Iraq in April 1994, and the crash of a B-52 from reckless maneuvering during an air show practice in June of that year. As bad as the former had been, the latter incident, stemming from a dismaying pattern of increasingly irresponsible airmanship, could have been catastrophic if it had occurred during the show itself, as the aircraft would have impacted in the midst of thousands of base visitors. It should be noted (and as was evident personally to the author at the time) that McPeak and ACC commander general Mike Loh were equally outraged by the two incidents, and moved swiftly to prevent any similar occurrence.

219. And this continues to be official documentary, report, and briefing practice, as of 2010.

220. Again, this is, of course, my own interpretation.

221. See Faulkenberry, *Global Reach–Global Power*, 42.

222. To be reassembled and reworked yet again, as subsequently discussed.

223. General Sir Michael Rose, *Fighting for Peace: Bosnia, 1994* (London: Harvill Press, 1998), 204; Rose commanded the UN force into early 1995, and his book is highly skeptical of air power.

224. Boutros Boutros-Ghali's intentions were often difficult to fathom, spawning an Air Staff witticism that while the air force had the stealthy B-2A, the UN had the stealthy B-2G. Boutros Boutros-Ghali, personal recollection.

225. See Col. Robert C. Owen, ed., *Deliberate Force: A Case Study in Effective Air Campaigning* (Maxwell AFB, AL: Air University Press, January 2000), for a detailed examination of various aspects of this air campaign.

226. Transcript of statement by Richard Holbrooke to Elizabeth Farnsworth, *Newshour with Jim Lehrer*, PBS Television, February 21, 1996. Weapons data is from Lt. General Ralph E. Eberhart, AF/XO, "Airpower: An Airman's Perspective," a briefing presented to the Air Staff, April 1996, Slide 129; and Lt. Gen. Michael E. Ryan, Commander, Air Forces Southern Europe, briefing on Operation Deliberate Force, Air Force Corona South meeting, Orlando,

Florida, February 1996, copies of these two briefings are in the archives of the Air Force Historical Studies Office, Bolling AFB, Washington, DC. For a detailed examination of precision-guided munition (PGM) use, see Richard P. Hallion, "Precision Guided Munitions and the New Era in Warfare," *Air Power History* 43, no. 3 (Fall 1996): 4–21.

227. For the Welch case, see Andrew Rosenthal, "Cheney Rebukes Air Force Chief for Arms Talk with Legislators," *New York Times*, March 25, 1989. In retrospect, the rebuke was unjustified, unwarranted, and inappropriately given, as Welch was acting well within his responsibilities and duties as Chief of Staff.

228. Personal recollections.

229. Introductory remarks to Davis, *Anatomy of a Reform*, iii.

230. Brig. Gen. William R. Looney III, "The Air Expeditionary Force: Taking the Air Force into the Twenty-First Century," *Airpower Journal* 10, no. 4 (Winter 1996): 4.

231. Fogleman, "Air Power and the American Way of War." See also Graham, "In Carrier's Absence, U.S. Sending 18 Jet Fighters to Bahrain." Bruce Roehm, CSAF Operations Group, regarding this first deployment. See also Daniel R. Mortensen, ed., *The Air Expeditionary Force in Perspective*, CADRE Research Paper 2003-01 (Maxwell AFB, AL: Air University, 2003), 3–4. Air force officials took pains to point out that AEFs were not an attempt by the air force to "replace the carrier," but rather were a means of supplementing it. For a comparison of the AEF and naval carrier battle groups, see Major Peter A. Lee, *Power Projection: A Comparison of the Aerospace Expeditionary Force and the Carrier Battle Group* (Maxwell AFB, AL: Air University, 1999).

232. Personal recollections; the author was a member of the group, which met over several months, and, as part of its outreach, prepared an air dominance study, "Control of the Air: The Enduring Requirement" (Washington, DC: Air Force History and Museums Program, 1999).

233. Rebecca Grant, *The Kosovo Campaign: Aerospace Power Made It Work* (Arlington, VA: Air Force Association, September 1999), is an excellent overall survey.

234. UN airpower skepticism is related in Clark's memoir *Waging Modern War: Bosnia, Kosovo, and the Future of Combat* (New York: PublicAffairs, 2001), 125–126, 195–196, 200–205; for Clark and targeting issues, see Lt. Gen. Michael C. Short, "An Airman's Lessons from Kosovo," in *From Manoeuvre Warfare to Kosovo?*; and Benjamin S. Lambeth, *NATO's Air War for Kosovo: A Strategic and Operational Assessment* (Santa Monica, CA: RAND, 1998), xix–xx et seq.

235. Bruce R. Nardulli, Walter L. Perry, Bruce Pirnie, John Gordon IV, and John G. McGinn, *Disjointed War: Military Operations in Kosovo, 1999*, Report MR-1406 (Santa Monica, CA: RAND, 2002), Fig. 4.5, "Task Force Hawk Helicopter Self-Deployment," 72, and related text. See also Clark, *Waging Modern War*, 258, 425; and John Gordon IV, Bruce R. Nardulli, and Walter L. Perry, "The Operational Challenges of Task Force Hawk," *Joint Force Quarterly* 29 (Autumn–Winter 2001–2002): 52–57.

236. Genaro J. Dellarocco, "Force Projection Research and Development: The Key Enabler for Army Transformation," in *Army Transformation: A View from the*

U.S. Army War College, ed. Williamson Murray (Carlisle Barracks, PA: U.S. Army War College Strategic Studies Institute, July 2001), 225.

237. Rory Carroll, "'I'm Not Right in the Head Now': A Conscript's War," *The Observer*, London, Sunday, June 20, 1999, 17. In fact, of course, Allied intelligence was far less omniscient than the Serbian veteran believed, but this quote speaks to the great psychological intimidation and perspective distortion induced by air attack.

238. Writing in the *Daily Telegraph* (London), June 6, 1999. RAND's Benjamin Lambeth is more cautious in fully ascribing credit to air power alone; see his *NATO's Air War for Kosovo*, xiv, where he notes, "One should be wary of any intimation that NATO's use of air power produced a successful result for the alliance without any significant contribution by other factors," though this view certainly mirrored most commentary of the time, as Lambeth amply notes on 220–221.

239. Quoted in Lambeth, *NATO's Air War for Kosovo*, 222.

240. Ibid., 224; Clark, *Waging Modern War*, 201–202, relates his astonishment at learning President Clinton was personally approving targets.

241. Transcript of statement of General Jumper at the Colloquy "Operation *Allied Force*: Strategy, Execution, Implications," The Eaker Institute for Aerospace Concepts, Washington, DC, August 16, 1999.

242. Short, "An Airman's Lessons," in *From Manoeuvre Warfare to Kosovo?*, 257–258. For a detailed examination of the campaign, see Stephen T. Hosmer, *The Conflict over Kosovo: Why Milosevic Decided to Settle when He Did* (Santa Monica, CA: RAND, 2001).

243. U.S. Department of Defense, *Joint Vision 2010–America's Military: Preparing for Tomorrow* (Washington, DC: GPO, June 2000), 1.

244. Secretary F. Whitten Peters and General Michael E. Ryan, *Vision 2020* (Washington, DC: HQ Air Force, 2000), passim.

245. Lt. Col. Paul D. Berg, "Global Vigilance, Reach, and Power," *Air & Space Power Journal* 22, no. 4 (Winter 2008).

246. For example, *Vision 2020*'s mission was the somewhat detached and cerebral "to defend the United States and protect its interests through aerospace power." Its 2008 mission was the more visceral "to fly, fight, and win . . . in air, space, and cyberspace." Its vision remained "Global Vigilance, Reach, and Power." In 2009, *Vision 2020*'s six core competencies were now termed "distinctive capabilities," but remained basically the same: air and space superiority (replacing *Vision 2010*'s "aerospace" superiority, a term which, though appropriate, was strangely not beloved in the Jumper-era Air Force), global attack, rapid global mobility, precision engagement, information superiority, and agile combat support. In 2009 these "distinctive capabilities" flowed from three core competencies, which were development of airmen, technology-to-warfighting, and integrating operations. The service's core values remained the Fogleman-legacy "Integrity First, Service before Self, and Excellence in All We Do." See USAF, *United States Air Force Mission* (2009), http://www.af.mil/vision.

247. Personal recollection.

248. Benjamin S. Lambeth, *Air Power against Terror: America's Conduct of Operation Enduring Freedom* (Santa Monica, CA: RAND National Defense Research Institute, 2005), 49.

249. *The Global War on Terrorism: The First 100 Days* (Washington, DC: Coalition Information Center, December 2001), 8; the sixteen nations figure is from a transcript of a joint press conference held by President George W. Bush and General Tommy Franks, commander, Central Command at the Prairie Chapel Ranch, Crawford, Texas, December 28, 2001.

250. For a sampling of thought, see Daryl G. Press, "The Myth of Air Power in the Persian Gulf War and the Future of Warfare," *International Security* 26, no. 2 (Fall 2001): 43; Robert A. Pape, "The Wrong Battle Plan," *Washington Post,* October 19, 2001; and John T. Correll, "Blood and Thunder," *Air Force Magazine* 85, no. 1 (January 2002): 2.

251. Rebecca Grant, *The Afghan Air War* (Arlington, VA: Air Force Association, September 2002), 21.

252. William Branigan, "Jubilant Afghan Fighters Set Their Sights on Kabul," *Washington Post,* January 8, 2002.

253. Edward Cody, "Taliban's 'Hide-and-Wait' Strategy Failed," *Washington Post,* December 23, 2001.

254. Karl Vick, "Rout in Desert Marked Turning Point," *Washington Post,* December 31, 2001.

255. Karl Vick, "For U.S., Attack on Kandahar Was a Victory on Two Fronts," *Washington Post,* December 26, 2001.

256. Fareed Zakaria, "Face the Facts: Bombing Works," *Newsweek,* December 3, 2001.

257. Examined in Richard B. Andres, Craig Wills, and Thomas E. Griffith, "Winning with Allies: The Strategic Value of the Afghan Model," *International Security* 30, no. 3 (Winter 2005–2006): 124–160.

258. Richard Cohen, ". . . And Now to Iraq," *Washington Post,* November 30, 2001.

259. "Close air support and other airstrikes were constrained by the lack of anticipatory pre-battle joint planning," in Richard L. Kugler, Michael Baranick, and Hans Binnendijk, *Operation Anaconda: Lessons for Joint Operations* (Washington, DC: Center for Technology and National Security Policy, National Defense University, 2009), viii. See also Grant, *Afghan Air War,* 26–27.

260. Quote from Benjamin Lambeth, "Operation Enduring Freedom, 2001," in Olsen, ed., *History of Air Warfare,* 268. For Hagenbeck's version of events, see Robert H. McElroy and Patrecia Slayden Hollis, "Afghanistan: Fire Support for Operation Anaconda–An Interview with Major General Franklin L. Hagenbeck," *Field Artillery* (September–October 2002), 5–9. Hagenbeck was subsequently promoted to lieutenant general and appointed Superintendent of the United States Military Academy at West Point in 2006.

261. For good surveys, see Rebecca Grant, *Gulf War II: Air and Space Power Led the Way* (Arlington, VA: Air Force Association, September 2003); Catherine Dale, *Operation Iraqi Freedom: Strategies, Approaches, Results, and Issues for Congress* (Washington, DC: Congressional Research Service, March 28, 2008);

and Kristin F. Lynch, John G. Drew, Robert S. Tripp, and C. Robert Roll, Jr., *Supporting Air and Space Expeditionary Forces: Lessons from Operation Iraqi Freedom,* Report MG-193 (Santa Monica, CA: RAND, 2005). For a more critical view of the air campaign, see Williamson Murray, "Operation Iraqi Freedom, 2003," in *A History of Air Warfare,* 279–296, who notes that the air campaign constituted "a throwback to the earliest days of air power when the most significant contribution that aircraft made was to the ground campaign" (279).

262. John J. Fialka and Andy Pasztor, "Grim Calculus: If Mideast War Erupts, Air Power Will Hold Key to U.S. Casualties," *New York Times,* November 15, 1990; and Harry G. Summers, Jr., "It Was a Lot of Hot Air," *Washington Times,* September 20, 1990.

263. For example, see Keir A. Lieber and Daryl G. Press, "The Nukes We Need: Preserving the American Deterrent," *Foreign Affairs* 88, no. 6 (November–December 2009): 39–51; and J. Johnson, Christopher J. Bowie, and Robert P. Haffa, *Triad, Dyad, Monad? Shaping the U.S. Nuclear Force for the Future,* Mitchell Paper 5 (Arlington, VA: Air Force Association, December 2009); and Air Force Studies Board, *Future Air Force Needs for Survivability* (Washington, DC: National Academies Press, 2006).

264. Ann Roosevelt, "USJFCOM Drops Effects Based Operations Terms and Concepts," *Defense Daily* (August 19, 2008). So ingrained is this thought that recently no less than the commander of Joint Forces Command, Gen. James Mattis, USMC, dropped reference to effects-based operations (and related terms) in joint-military professional education, a barely disguised swipe at air power proponents and planners who now accept it (fortunately) as an important targeting concept in modern war.

265. Personal recollection. The RC-12 Horned Owl demonstration subsequently led to the MC-12 Liberty tactical ISR program, introduced into Iraqi operations in 2009, and used to discover weapons caches and improvised explosive devices. See Senior Airman Wes Carter, "MC-12 Aircrews Complete 2,000th Combat Sortie," HQ 332d Air Expeditionary Wing Public Affairs, Joint Base Balad, Iraq, 24 March 2010, http://www.af.mil/news/story.asp?id=123196397, accessed March 25, 2010.

266. See Robert S. Dudney, "The Lavelle Syndrome," *Air Force Magazine* 93, no. 9 (September 2010).

CHAPTER 3. ISRAELI AIR POWER

1. For the Israeli Air Force's role during the War of Independence and the period between the War of Independence and the Sinai War, see the Israeli Air Force, *Dapey Shchakim: The Air Force in the Independence War* (Tel Aviv: Ministry of Defense Publication, 1994) (Hebrew version); Yitzhak Shtaygman, ed., *Operation "Kadesh": The Israeli Air Force from 1950–1956: Buildup and Operation* (Tel Aviv: Ministry of Defense, 1986); Ezer Weizman, *On Eagles' Wings: The Personal Story of the Leading Commander of the Israeli Air Force* (New York: Macmillan, 1976); Yaakov Erez and Ilan Kfir, eds., *IDF in its Corps: Army and Security Encyclopedia* (Tel Aviv: Revivim Publishing, 1981) (Hebrew version);

Ehud Yonay, *No Margin for Error: The Making of the Israeli Air Force* (New York: Pantheon Books, 1993); and the Israeli Air Force official website: www.iaf.org.il.

2. All the figures in the chapter, particularly numerical figures (number of sorties, downed aircraft, and order of battle), are taken from the Israeli Air Force's official history. Valuable information can be found on the Israeli Air Force official website.

3. Regarding Israel's military doctrine, see Israel Tal, *National Security: The Israeli Experience* (Westport, CT: Praeger, 2000). On the American perspective of this doctrine, see Eliot Cohen, Michael Eisenstadt, and Andrew Bacevich, *Knives, Tanks, and Missiles: Israel's Security Revolution* (Washington, DC: Washington Institute for Near East Policy, 1998).

4. For a discussion of the Israeli Air Force in the Sinai War, see references in endnote 1. Also see Moshe Dayan, *Diary of the Sinai Campaign* (New York: Schocken Books, 1967); and Motti Havakuk, "The Development of the Israeli Air Force in Light of the Sinai War Lessons," in *When the Engines Roared: 50th Anniversary of the Sinai War*, ed. Haggai Golan and Shaul Shay (Tel Aviv: Ma'arachot Publishing, 2006), 307–332 (Hebrew version).

5. For a discussion of the Israeli Air Force during this period, see Weizman, *On Eagles' Wings*; Yoash Tsidon-Chatto, *By Day, By Night: Through Haze and Fog* (Tel Aviv: Ma'ariv, 1995) (Hebrew version); Yonay, *No Margin for Error*; Ran (Peker) Ronen, *Hawk in the Sky* (Tel Aviv: Yedioth Aharonoth, 2002) (Hebrew version); and the Israeli Air Force official website.

6. See Avi Cohen, ed., *Defending Water Resources: The Policy of Using Air Raids on the Israel-Syria Border, 1956–1967* (Tel Aviv: Ministry of Defense, 1992) (Hebrew version).

7. For an in-depth discussion of the Israeli Air Force during the Six Day War, see the Israeli Air Force, *Dapey Shchakim: The Air Force in the Six Days War* (Tel Aviv: Ministry of Defense Publication, 1992) (Hebrew version); Danny Shalom, *Like a Bolt out of the Blue* (Rishon Le-Zion: BAVIR-Aviation Publications, 2002) (Hebrew version); Yossi Abudi and Ze'ev Lachish, "Operation 'MOKED': The Destruction of the Arab's Air Forces in the Six Days War," in *Battlefield: Decisive Battles in the Land of Israel*, ed. Aryeh Shmuelevich (Tel Aviv: Ministry of Defense Publishing, 2007) (Hebrew version); and Ronen, *Hawk in the Sky.* Also see Michael B. Oren, *Six Days of War: June 1967 and the Making of the Modern Middle East* (New York: Oxford University Press, 2002).

8. On the Israeli Air Force in the War of Attrition, see the Israeli Air Force, *Dapey Shchakim: The Air Force in the War of Attrition* (Tel Aviv: Ministry of Defense Publication, 1995) (Hebrew version); and the Israeli Air Force official website. See also Dima Adamsky, "The 'Seventh Day' of the Six Day War: The Soviet Intervention in the War of Attrition (1969–1970)," in *The Soviet Union and the June 1967 Six Day War*, eds. Yaacov Ro'i and Boris Morozov (Palo Alto, CA: Stanford University Press, 2008); and Dan Schueftan, *Attrition: Egypt's Post-War Political Strategy, 1967–1970* (Tel Aviv: Ministry of Defense Publication, 1989) (Hebrew version).

9. On the Israeli Air Force during the Yom Kippur War, see the Israeli Air Force, *Dapey Shchakim: The Air Force in the Yom Kippur War* (Tel Aviv: Ministry of Defense Publication, 1993) (Hebrew version); Benjamin Peled, *Days of Reckoning* (Tel Aviv: Modan Publishing, 2004) (Hebrew version); Yonay, *No Margin for Error*; Ronen, *Hawk in the Sky*; Weizman, *On Eagles' Wings*; Shmuel Gordon, *Thirty Hours in October* (Tel Aviv: Maarive, 2008) (Hebrew version); Ido Embar, "The Israeli Air Force in the Yom Kippur War Land Campaign," in *The Crossing That Wasn't: The Yom Kippur War, the Crossing Conception, the Fire Line*, ed. Uri Milstein (Israel: Survival Research Institute, 1992) (Hebrew version); and the Israeli Air Force official website. Also see Chaim Herzog, *The War of Atonement: The Inside Story of the Yom Kippur War* (London: Greenhill Books/Lionel Leventhal, 2003).

10. See Anwar Sadat, *In Search of Identity: An Autobiography* (New York: Harper & Row, 1978); and Saad El Shazly, *The Crossing of the Suez* (San Francisco, CA: American Mideast Research, 1980).

11. Ezer Weizman, *On Eagles' Wings: The Personal Story of the Leading Commander of the Israeli Air Force* (New York: Macmillan Publishing, 1976), 329.

12. On the processes in the Israeli Air Force following the Yom Kippur War, see Iftach Spector, *Loud and Clear* (Tel Aviv: Yediot Aharonoth, 2008) (Hebrew version); David Ivry, "The SAM's Destruction in the First Lebanon War," *Ma'arachot* 413 (July 2007), 68–71; Ronen, *Hawk in the Sky*; Erez, *IDF in its Corps*; Yonay, *No Margin for Error*; Peled, *Days of Reckoning*; and the Israeli Air Force official website.

13. On the Israeli Air Force during the first Lebanon war, see Ivry, "The SAM's Destruction in the First Lebanon War." Also see Ze'ev Schiff and Ehud Ya'ari, *Israel's Lebanon War* (New York: Simon & Schuster, 1984).

14. On the first Intifada, see Ze'ev Schiff and Ehud Ya'ari, *Intifada: The Palestinian Uprising: Israel's Third Front* (New York: Simon & Schuster, 1990).

15. On the military-society relationship in Israel, see Uri Ben-Eliezer, "Post-Modern Armies and the Question of Peace and War: The Israeli Defense Forces in the 'New Times,'" *International Journal of Middle East Studies*, 36, no. 1 (2004): 49–70; Uri Ben-Eliezer, "The Civil Society and the Military Society in Israel," *Palestine-Israel Journal* 12, no. 1 (2005): 49–55; Gabriel Ben-Dor, "Civil-Military Relations in Israel in the Mid-1990s," in *Independence: The First 50 Years*, ed. Anita Shapira (Jerusalem: Shazar Center Publishing, 1998), 471–486 (Hebrew version); and Stewart Cohen, *The IDF and Israeli Society: A Renewed Analysis* (Tel Aviv: BESA Center for Strategic Studies, Bar Ilan University, 2001). On the influence of these relationships on the use of military power in Lebanon from a ground force officer's point of view, see Moshe Tamir, *The War without a Name* (Tel Aviv: Ministry of Defense Publishers, 2006) (Hebrew version).

16. For a discussion of the Israeli Air Force during the second Intifada, see the Israeli Air Force official website. Also see Amos Harel and Avi Isacharoff, *The Seventh War* (Tel Aviv: Yedioth Aharonoth, 2004) (Hebrew version).

17. See Winograd Commission of Inquiry, *Interim Report*, 2007; Winograd Commission of Inquiry, *Final Report*, 2008. Also see Amos Harel and Avi Issacharof, *34 Days: Israel, Hezbollah, and the War in Lebanon* (New York: Palgrave Macmillan, 2008); Amir Rapaport, *Friendly Fire: How We Defeated Ourselves in the Second Lebanon War* (Tel Aviv: Ma'ariv, 2007) (Hebrew version); Ofer Shelah and Yoav Limor, *Captives of Lebanon: The Truth on the Second Lebanon War* (Tel Aviv: Yedioth Ahronoth and Hemed, 2007) (Hebrew version); Avi Kober, "The Second Lebanon War," perspective paper no. 22 (The Begin-Sadat Center for Strategic Studies, September 28, 2006); Avi Kober, "The Israel Defense Forces in the Second Lebanon War: Why the Poor Performance?" *The Journal of Strategic Studies* 31, no. 1 (February 2008): 3–40; and David Makovsky and Jeffrey White, *Lessons and Implications of the Israel-Hizballah War: A Preliminary Assessment*, policy focus no. 60 (Washington, DC: Washington Institute for Near East Policy, October 2006).

18. On Hezbollah, see Augustus Richard Norton, *Hezbollah: A Short History* (Princeton, NJ: Princeton University Press, 2007); and Judith Palmer Harik, *Hezbollah: The Changing Face of Terrorism* (London: I.B. Tauris, 2005).

19. See Eyal Zisser, "The Battle for Lebanon: Lebanon and Syria in the Wake of the War," in *The Second Lebanon War: Strategic Perspectives*, eds. Shlomo Brom and Meir Elran (Tel Aviv: Institute for National Security Studies, 2007); and Edward N. Luttwak, "Misreading the Lebanon War," *Jerusalem Post*, August 20, 2006.

20. See Ron Tira, *The Limitations of Standoff Firepower-Based Operations: On Standoff Warfare, Maneuver, and Decision*, memorandum no. 89 (Tel Aviv: Institute for National Security Studies, March 2007); Kober, *The Second Lebanon War*; Kober, "The Israel Defense Forces in the Second Lebanon War," 3–40; Efraim Inbar, "How Israel Bungled the Second Lebanon War," *Middle East Quarterly Magazine* (June 22, 2007); Amir Kulick, "Hizbollah vs. the IDF: The Operational Dimension," *Strategic Assessment* 9, no. 3 (Tel Aviv: The Institute for National Security Studies, November 2006); Shai Feldman, "The Hezbollah-Israel War: A Preliminary Assessment," *Middle East Brief* no. 10 (Waltham, MA: Crown Center for Middle East Studies, Brandeis University, September 2006), http://www.brandeis.edu/crown/publications/meb/MEB10.pdf; and Sarah Kreps, "The 2006 Lebanon War: Lessons Learned," *Parameters* (Spring 2007), 72–84.

21. See Azar Gat, "Isolationism, Appeasement, Containment, and Limited War: Western Strategic Policy from the Modern to the 'Postmodern' Era," in *War in a Changing World*, eds. Zeev Maoz and Azar Gat (Ann Arbor, MI: University of Michigan Press, 2001); and Edward N. Luttwak, "Post-Heroic Warfare," *Foreign Affairs* 74, no. 3 (May/June 1995): 109–122.

CHAPTER 4. SOVIET-RUSSIAN AIR POWER

1. The term "Soviet-Russian" forces and air forces has been used in the chapter to indicate the continuity that has been maintained within the military through

the breakup of the Soviet Union and the formation of independent nations. There are occasions when only the "Soviet" or "Russian" Air Force is referred to, depending on whether the nation was the Soviet Union or Russia at that time.

2. K. Subrahmanyam, "Introduction," in *Defending India*, ed. Jaswant Singh (Chennai, India: Macmillan India Ltd., 1999), xix.

3. Sanu Kainikara, *Diminishing Relevance: Emergent Air Forces at the Cross Roads*, paper no. 10 (Canberra: Aerospace Centre, 2003), 3–4.

4. Christopher Donnelly, *Red Banner: The Soviet Military System in Peace and War* (Coulsdon, UK: Jane's Information Group, 1988), 36.

5. Lionel Kochan and Richard Abraham, *The Making of Modern Russia*, 2d ed. (Hammondsworth, UK: Penguin Books Ltd., 1983), 22.

6. Russian War Ministry's Annual Report, 1908.

7. Robin Higham and Jacob W. Kipp, eds., *Soviet Aviation and Air Power: A Historical View* (London: Brassey's Publishers Ltd., 1978), 1.

8. Charles H. Gibbs-Smith, *The Aeroplane: An Historical Survey* (London: HMSO, 1960), 189.

9. "History of the Aeroplane: The Century Before," http://www.wright-brothers.org/, accessed September 30, 2009.

10. Gibbs-Smith, *The Aeroplane*, 200.

11. David R. Jones, "The Beginnings of Russian Air Power, 1907–1922," in *Soviet Aviation and Air Power*, 16–17.

12. Ibid., 20.

13. N. N. Golovin, *The Russian Campaign of 1914* (Fort Leavenworth, KS: The Command and General Staff School Press, 1933), 7.

14. Robert A. Kilmarx, *A History of Soviet Air Power* (London: Faber and Faber Ltd., 1962), 31.

15. Col. B. Simakov, "Soviet Air Force in the Years of Foreign Intervention and Civil War," *Vestnik Vozdushnogo Flota* no. 7 (July 1952): 75–85.

16. Jones, "The Beginnings of Russian Air Power, 1907–1922," 18–19.

17. Neil M. Heyman, "NEP and the Industrialization to 1928," in *Soviet Aviation and Air Power*, 41.

18. Ibid., 40.

19. Ibid., 4–42.

20. Kilmarx, *A History of Soviet Air Power*, 171.

21. Alexander Boyd, *The Soviet Air Force* (London: Macdonald and Jane's, 1977), 109.

22. Department of the Army, *Russian Combat Methods in World War II*, pamphlet no. 20–230 (Washington, DC: U.S. Government, November 1950): 8–12.

23. Adolf Galland, *The First and the Last*, trans. Mervin Savill (New York: Henry Holt and Co., 1954), 80.

24. This division of the war into three distinct phases has been adopted from John T. Greenwood, "The Great Patriotic War, 1941–1945," in *Soviet Aviation and Air Power*, 69–136.

25. Chris Bishop, ed., *The Encyclopaedia of 20th Century Air Warfare* (Leicester, UK: Silverdale Books, 2001), 95.

26. Asher Lee, *The Soviet Air Force* (London: Duckworth, 1952), 123.

27. A number of reports that have been translated from the original by the Parallel History Project and published in English are clearly indicative of this. See http://isn.ethz.ch/php, accessed October 4, 2009.

28. Raymond S. Dawson, *The Decision to Aid Russia* (Chapel Hill: University of North Carolina Press, 1959), 21–57.

29. General Augustin Guillaume, "Soviet Arms and Soviet Power," *Infantry Journal Press* (1949), 170.

30. Greenwood, *The Great Patriotic War*, 84.

31. General Franz Halder, *The Halder Diaries*, vol. 7 (Washington, DC: Infantry Journal Press, 1950), 36.

32. Kilmarx, *A History of Soviet Air Power*, 183.

33. Ibid., 186.

34. Col. E. K. Fedorov, *The Red Army* (London: Cobbett Publishing Company, 1944), 32–43.

35. Kilmarx, *A History of Soviet Air Power*, 185.

36. Fedorov, *The Red Army*, 33.

37. Boyd, *The Soviet Air Force*, 127–139.

38. Quoted in Greenwood, *The Great Patriotic War*, 133.

39. Royce D. Zant, "Soviet National Strategy," National Security Report (Washington, DC: Library of Congress, 1989), 6, http://www.globalsecurity.org/military/library/report/1989/ZRD.html, accessed October 5, 2009.

40. Ernest May, John Steinbruner, and Thomas Wolfe, *History of Strategic Arms Competition, 1945–1972*, ed. Alfred Goldberg (Washington, DC: Historical Office, Office of the Secretary of Defense, March 1981, declassified with deletions, December 1990), 634.

41. Col. Gen. A. Gastilovich, "Military Thought: Theory of Military Art Needs Review," quoted in *CIA Documents of the Cuban Missile Crisis, 1962* (Washington, DC: History Staff, Center for the Study of Intelligence, Central Intelligence Agency, declassified October 1992), 21.

42. Kathryn Weathersby, *Soviet Aims in Korea and the Origins of the Korean War, 1945–1950: New Evidence from Russian Archives*, working paper no. 8 (Washington, DC: Cold War International History Project, Woodrow Wilson Center for Scholars), 5.

43. Major Robert B. Greenough, "Communist Lessons from the Korean Air War," *Air University Quarterly Review* (Winter 1952/53): 25.

44. Kilmarx, *A History of Soviet Air Power*, 236–237.

45. Anthony Robinson, ed., *Aerial Warfare* (London: Orbis Publishing, 1982), 264.

46. Pawel Monat, "Russians in Korea," *Life*, June 27, 1960, 76.

47. Col. James T. Stewart, ed., *Airpower: The Decisive Force in Korea* (Princeton, NJ: D. Van Nostrand Company, 1957), 131

48. William Green, "The Development of Jet Fighters and Fighter Bombers," in

The Soviet Air and Rocket Forces, ed. Asher Lee (New York: Frederick A. Praeger, 1959), 139–140.

49. Bishop, ed., *The Encyclopaedia of 20th Century Air Warfare,* 414.
50. Sanu Kainikara, "Sukhoi's Formidable Flanker Family," *Asia-Pacific Defence Reporter* (April/May 1999): 69.
51. Ibid., 171.
52. Lynn Hansen, "The Resurgence of Soviet Frontal Aviation," *Strategic Review* (Fall 1978): 73–74.
53. Raymond L. Garthoff, "Introduction," in Marshal V. D. Sokolovsky, *Military Strategy: Soviet Doctrine and Concepts* (London: Pall Mall Press, 1963), xi.
54. Ibid., 232.
55. William Koenig and Peter Scofield, *Soviet Military Power* (Greenwich, CT: Bison Books Corp., 1983), 111.
56. Jacob W. Kipp, "Soviet Tactical Aviation in the Postwar Period," *Aerospace Power Journal* (Spring 1988): 21–26.
57. Kilmarx, *A History of Soviet Air Power,* 118.
58. V. I. Lenin, *Works,* vol. 33, 1941–51 (Moscow: State Political Publishing House, 1951), 74.
59. Joseph P. Mastro, "The Lessons of World War II and the Cold War," in *Soviet Aviation and Air Power,* 204–207.
60. Nikita Khrushchev, *Khrushchev Remembers: The Last Testament* (New York: Bantam Books, 1976), 250–262.
61. Oleg Penkovsky, *The Penkovsky Papers* (Garden City, NY: Doubleday, 1965), 253–257.
62. Kipp, *Soviet Tactical Aviation,* 21–26.
63. N. Semenov, "Gaining Supremacy in the Air," *Voennaia Mysel* no. 4 (April 1968), as translated in Joseph Douglas, Jr., and Amoretta M. Hoeber, "Selected Readings from Military Thought, 1963–73, Vol. 5," *Studies in Communist Affairs,* part I (Washington, DC: GPO, 1982), 203.
64. Koenig and Scofield, *Soviet Military Power,* 17.
65. William E. Odom, *The Collapse of the Soviet Military* (New Haven, CT: Yale University Press, 1998), 82–83.
66. Barnett R. Rubin, *The Fragmentation of Afghanistan* (New Haven, CT: Yale University Press, 1995), 20.
67. Sanu Kainikara, *Red Air: Politics in Russian Air Power* (Boca Raton, FL: Universal Publishers, 2007), 244.
68. Bill Sweetman, "Sukhoi Su-25 Frogfoot," *International Defence Review* no. 11 (November 1985): 1760–1761.
69. Ibid.
70. Russell G. Breighner, "Air Defence Forces," in *Soviet Armed Forces Review Annual, 1982–1983,* ed. David Jones (Gulf Breeze, FL: Academic International Press, 1984), 160–165.
71. Sanu Kainikara, "MiG-29 Fulcrum—An Agile Fighter," *Asia-Pacific Defence Reporter* (February/March 1999): 13–16.

72. Sanu Kainikara, *Papers on Air Power* (Canberra: Air Power Development Centre, 2007), 83–87.

73. Ibid., 85.

74. Archie Brown, *The Gorbachev Factor* (New York: Oxford University Press, 1996), 115–120.

75. Harriet Fast Scott and William F. Scott, *Soviet Military Doctrine: Continuity, Formulation and Dissemination* (London: Westview Press, 1988), 262.

76. Richard Pipes, *Communism: A Brief History* (London: Weidenfield and Nicholson, 2001), 33–42.

77. Scott and Scott, *Soviet Military Doctrine*, 263.

78. Christian Nünlist, "Cold War Generals: The Warsaw Pact Committee of Defence Ministers, 1969–1990" (Washington, DC: The Parallel History Project on NATO and the Warsaw Pact, 1993), 13–17.

79. Margaret Roth, "Soviet Troop Cuts Could Vastly Change East-West Relations," *The Navy Times*, December 19, 1988, 14.

80. Raymond L. Garthoff, "New Thinking and Soviet Military Doctrine," in *Soviet Military Doctrine from Lenin to Gorbachev (1915–1991)*, ed. Willard C. Frank, Jr., and Philip S. Gillet (Westport, CT: Greenwood Press, 1992), 196.

81. Raymond L. Garthoff, "New Thinking and Soviet Military Doctrine," *The Washington Quarterly* 11, no. 3 (Summer 1988): 131–142.

82. Garthoff, "New Thinking and Soviet Military Doctrine," in *Soviet Military Doctrine from Lenin to Gorbachev (1915–1991)*, 201–203.

83. Kent D. Lee, "Implementing Defensive Doctrine: The Role of Soviet Military Science," in ibid., 272.

84. Ibid., 272–274.

85. Kainikara, *Red Air*, 272–274.

86. Mary C. Fitzgerald, "The New 'Aerospace War' in Soviet Military Thought," in *The Soviet Military and the Future*, eds. Stephen J. Blank and Jacob W. Kipp (Westport, CT: Greenwood Press, 1992), 56–59.

87. General-Major V. I. Slipchenko, "Impending Changes from Reform Plans for Employing the Soviet Armed Forces," presentation at the National Defense University, Washington, DC, March 15 and 20, 1991.

88. Mary C. Fitzgerald, "The Soviet Military and the New Air War in the Persian Gulf," *Air Power Journal* (Winter 1991): 22.

89. Maj. M. Pogorelyi, "From a Military Observer's Viewpoint: What the War Showed," *Krasnaya Zvezda* March 8, 1991, as translated by the Parallel History Project, Washington, DC, http://www.isn.ethz.ch/php, accessed October 27, 2009.

90. Leon Aron, "The Foreign Policy Doctrine of Postcommunist Russia and Its Domestic Context," in *The New Russian Foreign Policy*, ed. Michael Mandelbaum (New York: Council on Foreign Relations, 1998), 25–30.

91. Sherman W. Garnett, "The Integrationist Temptation," *The Washington Quarterly* no. 18 (Spring 1995): 35–38.

92. John Warry, *Warfare in the Classical World* (London: Salamander Books, 1998), 7–27.

Chapter 5. Indian Air Power

1. John Slessor, *The Central Blue: Recollections and Reflections by Marshal of the Royal Air Force Sir John Slessor* (London: Cassel and Company Ltd., 1956), 119–139.

2. See debates in the legislative assembly between 1918 and 1940, National Archives, New Delhi.

3. S. C. Gupta, *History of the Indian Air Force, 1933–45* (Delhi: Combined Inter-Services Historical Section, India & Pakistan, 1961), 2.

4. Ibid., 2.

5. The abbreviation AC stands for "army cooperation." Two more squadrons were raised during the war with the same designation. But over the years this was dropped.

6. Ibid.

7. Richard P. Hallion, *Strike from the Sky: History of Battlefield Air Attack, 1911–1945* (Washington, DC: Smithsonian Institution Press, 1999), 65.

8. For the official history of the Indian Air Force, see Gupta, *History of the Indian Air Force, 1933–45*, 1–4.

9. Deteriorating relations with Japan necessitated changes to this plan and the following revised figures were approved on March 12, 1941. Planned Force Level: 21 Squadrons (11 RAF; 10 Indian Air Force) with 325 aircraft; 6 coastal defense flights with 57 aircraft; and 2 conversion and training units with 12 aircraft. Actual Force Level: 3 Squadrons (2 RAF; 1 Indian Air Force); and 5 Indian Air Force Volunteer Reserve Flights.

10. Majumdar was later assigned to the RAF where he flew sixty-eight operational sorties in fighter aircraft totalling 102 hours in the European theater and was awarded a bar to the DFC.

11. See Patrick Turnbull, *Battle of the Box* (London: Ian Allan Ltd., 1979); Geoffrey Evans, "Imphal and Kohima," in *Decisive Battles of the Twentieth Century: Land-Sea-Air*, eds. Noble Frankland and Christopher Dowling (London: Sidgwick & Jackson, 1976); Gupta, *History of the Indian Air Force, 1933–45*.

12. Jasjit Singh, *The ICON: Marshal of the Indian Air Force Arjan Singh* (New Delhi: KW Publishers, 2009), 27–64.

13. In 1750, India's share of global manufacturing output was nearly 25 percent and this had dropped to less than 1 percent by the beginning of the twentieth century. See for example Paul Kennedy, *The Rise and Fall of the Great Powers* (New York: Random House, 1987), 149.

14. Subimal Dutt, *With Nehru in the Foreign Office* (Calcutta: Minerva Associates, 1977), 22.

15. Ancient Indian literature such as Sukranitisara, Arthasastra, and Manusmrti all emphasize this aspect.

16. Sukraniti and other texts.

17. Manusmrti, quoted in Nagendra Singh, *The Theory of Force and Organisation of Defence in Indian Constitutional History* (New Delhi: Asia Publishing House, 1969), 33.

18. Ibid., 267.

19. For example, India adopted a doctrine of credible minimum deterrent against nuclear weapons. See the 1999 draft nuclear doctrine and the press statement, Govt. of India, January 4, 2003, in Manpreet Sethi, *Nuclear Strategy: India's March Towards Credible Deterrence* (New Delhi: KW Publishers, 2009), 341–346.

20. Chairman of the UN Commission for India and Pakistan Josef Korbel recalled that when the commission first visited Karachi in July 1948, the Pakistan Government dropped a "bombshell," telling him that its army had been fighting inside Kashmir for quite some time. See Josef Korbel, *Danger in Kashmir* (Princeton, NJ: Princeton University Press, 1954), 121, cited in C. Dasgupta, *War and Diplomacy in Kashmir, 1947–48* (New Delhi: Sage Publications, 2002), 161–167.

21. See Maj. Gen. Akbar Khan (who planned and directed the covert war), *Raiders in Kashmir* (Delhi: Army Publishers, n.d.); see also Dasgupta, *War and Diplomacy in Kashmir, 1947–48*.

22. For a detailed account of the Indian Air Force's role in the first Kashmir war in 1947–48, see Air Marshal Bharat Kumar, *An Incredible War: Indian Air Force in Kashmir War, 1947–1948* (New Delhi: Knowledge World, 2007).

23. Maj. Gen. D. K. Palit, *War in the High Himalayas: The Indian Army in Crisis, 1962* (New Delhi: Lancer International, 1991).

24. "Record of Conversation between Chinese Premier Zhou Enlai and Mongolian Leader J. Zedenbal, December 26, 1962," Cold War International History Project, http://www.wilsoncenter.org/index.cfm?topic_id=1409&fuseaction, accessed March 10, 2009.

25. Document no. 3, "Memorandum of Conversation of NS Khrushchev with Mao Zedong in Beijing, October 2, 1959," in Cold War International History Project Bulletin issue 12/13 (Fall/Winter 2001), Wilson Center, Washington, DC.

26. Ibid., 262–269.

27. A. K. Dave, "The Real Story of China's War on India, 1962," occasional paper no. 1 (New Delhi: Centre for Armed Forces Research, USI, 2006).

28. Official 1965 War Records of the Indian Air Force, Air HQ, New Delhi, 1966.

29. Singh, *The ICON*. See also Lt. Gen. Gul Hassan Khan, *Memoirs of Lt. Gen. Gul Hassan* (Karachi: Oxford University Press, 1993).

30. The relative performance of air-to-surface operations, adapted from data in Lt. Gen. Harbakhsh Singh, *War Despatches: Indo-Pak Conflict* (New Delhi: Lancer International, 1991); and *Indian Air Force War Records, 1965 War*, Air HQ, New Delhi.

	PAKISTAN'S LOSSES TO THE INDIAN AIR FORCE	INDIAN LOSSES TO THE PAF
Tanks		
Destroyed	123	3
Damaged	20	4
Guns		
Destroyed	56	5
Damaged	3	25

	PAKISTAN'S LOSSES TO THE INDIAN AIR FORCE	INDIAN LOSSES TO THE PAF
Vehicles		
Destroyed	281	172
Damaged	27	50
Rolling stock, etc.		
Trains	8	0
Wagons	64	0
BPI	2	0

31. *The Story of the Pakistan Air Force: A Story of Courage and Honour* (Islamabad: Shaheen Foundation, 1988).

32. B. C. Chakravorty, *History of the Indo-Pak War, 1965* (New Delhi: History Division, Ministry of Defence, 1992).

33. John Fricker, *The Battle for Pakistan: The Air War of 1965* (London: Ian Allen Ltd., 1979), 131–132.

34. Altaf Gauhar, *Ayub Khan: Pakistan's First Military Ruler* (Lahore: Sang-e-Meel Publications, 1993), 343 (emphasis in original).

35. See Lt. Gen. Kamal Matinuddin, *The Tragedy of Errors: East Pakistan Crisis, 1968–1971* (Lahore: Wajidalis, 1994); Maj. Gen. Fazal Muqueem Khan, *Pakistan's Crisis of Leadership* (New Delhi: Alpha & Alpha, 1973); and Dennis Kux, *The United States and Pakistan, 1947–2000* (Karachi: Oxford University Press, 2001).

36. Air Chief Marshal P. C. Lal, *My Years with the Indian Air Force* (New Delhi: Lancer International, 1986), 167.

37. Ibid., 174.

38. The Indian Air Force doctrine was finally written in 1995, a good six decades after it was formed, but it was kept classified, as has its second edition, written in 2008.

39. Lal, *My Years with the Indian Air Force*, 182.

40. Khan, *Memoirs of Lt. Gen. Gul Hassan Khan*, 307.

41. *The Story of the Pakistan Air Force*, 447.

42. Jasjit Singh, *Defence from the Skies: Indian Air Force through 75 Years* (New Delhi: Knowledge World, 2007).

43. Maj. Gen. Shaukat Riza, *The Pakistan Army, 1966–71* (Dehra Dun: Nataraj Publishers, 1977), 166.

44. For a more detailed account of the history of the Indian Navy and its air arm, see Rear Adm. Satyindra Singh, *Blueprint to Blue Water: The Indian Navy, 1951–65* (New Delhi: Lancer International, 1992), and its follow-on book by Vice Adm. G. M. Hiranandani, *Transition to Triumph: History of the Indian Navy, 1965–1975* (New Delhi: Director Personnel Services, NHQ, 2000).

45. For an official assessment, see the Kargil Review Committee, *From Surprise to Reckoning: The Kargil Review Committee Report* (New Delhi: Sage Publications, 2000), and for an independent account, see Jasjit Singh, ed., *Kargil 1999: Pakistan's Fourth War for Kashmir* (New Delhi: Knowledge World, 1999), for details.

46. Gen. V. P. Malik, *Kargil: From Surprise to Victory* (New Delhi: HarperCollins, 2006), 243–244 (emphasis added).

47. This was consistent with other imperial policies. For example, the British Indian government continued to pay £137,000 per year to the Royal Navy for the "defence of India," thus obviating the need to create a regular Indian Navy (although the Indian Navy was created by law in 1934) while subsidizing the Royal Navy's budget.

48. See Singh, *Blueprint to Blue Water*, 188–215.

49. It is relevant to note that during all the wars of the Indian subcontinent, no cities were targeted. And, in January 1994, India offered Pakistan a mutual agreement formalizing this in the shape of an agreement not to attack each other's population centers and economic targets.

50. *Sixth Report of the Standing Committee on Defence* (1995–96), 1–2. These were also outlined by the prime minister in his reply to the demands for grants of the Ministry of Defence in the Lok Sabha on May 16, 1995, cited as the "guidelines" to India's defence forces "strictly followed and observed" by the government (and the defence forces).

51. India's defense commitments included a 16,700 km land border (more than 30 percent of which was disputed with China and Pakistan, who occupied large tracts), a 7,600 km coastline, and the 600-odd islands in the Andaman and Nicobar group, almost 1,200 km from the mainland.

52. Unfortunately, no U.S. combat aircraft were available to India, and the Cold War calculus also led the West to believe and articulate that India had five thousand Soviet military advisers in India. I was a flight commander in the early 1970s and commanded a MiG-21 squadron, the director of flight safety of the Indian Air Force, and later a director of operations at air headquarters in the early 1980s. However, I never saw a Soviet military officer until 1986, when I went to the USSR for the first time to attend an academic seminar with the Institute for Oriental Studies and was invited to lecture at the Soviet War College, the only Indian military officer, if I may say so, to have done so.

53. During the 1971 war with Pakistan, two F-104 Starfighters (which India and many in the Indian Air Force had wanted in the 1960s) were shot down in air combat by MiG-21PFs in low-level dogfights. The Indian Air Force has not lost any MiG-21s in air combat during the past three wars and losses have been due to ground fire (when undertaking close air support) or air-to-ground strikes/bombing by the PAF on aircraft outside hardened shelters.

54. Its unit cost when India stopped manufacturing it in the mid-1980s was Rs. 1.1 crores, which, at that time, was equivalent to $1 million; except that cool air, let alone air conditioning for the cockpit, could not be provided.

55. See *India's Draft Nuclear Doctrine*, August 1999, prepared by the NSAB, and the central government's press release on January 3.

56. Gen. S. Padmanabhan, *A General Speaks* (New Delhi: Manas Publications, 2005), 61.

57. "Mapping the Global Future," Report of the National Intelligence Council's 2020 Project, NIC 2004-13, December 2004, 3.

58. For a text of the statement, see Appendix 3 in Sethi, *Nuclear Strategy*, 349–353.
59. Annual Report on the Military Power of the People's Republic of China, Report to Congress by the Secretary of Defense (Washington, DC: GPO, 2002), 11.
60. Zhogguo Tongxun She News Agency, Hong Kong, November 16, 1999, cited in SWB FE/3695 G/6, dated November 18, 1999.
61. "Air Force Commander Liu Shunyao on Air Force Transformation," FBIS-CHI-1999-1107, November 7, 1999 (emphasis added).
62. "Revolution in Military Affairs with Chinese Characteristics," in *China's National Defence in 2004*, chap. 3, 1, the White Paper published to illustrate China's national defense policies and the progress made in the previous two years, see *China Daily*, December 28, 2004, http://www.chinadaily.com.cn/english/doc/2004-12/28/content_403913.htm (emphasis added), accessed January 2, 2005.
63. China's White Papers on National Defence for various years.
64. Michael D. Swaine and Ashley J. Tellis, *Interpreting China's Grand Strategy: Past, Present, and Future* (Santa Monica, CA: RAND, 2000), 132.
65. See Shalini Chawla, *Pakistan's Military and Its Strategy* (New Delhi: KW Publishers, 2009).
66. Singh, *Kargil 1999*.
67. For details of Pakistan's post-Kargil acquisitions in the air power sector, see Shalini Chawla, "Pakistan Air Force: Seeking Dominance," in *Pakistan's Military and Its Strategy*, 207–228.
68. The table provides a brief overview of the changes that are taking place in the PAF and in Pakistan's naval air capabilities since the Kargil conflict.

	1999	2010	2020
High performance, high-technology, multirole combat aircraft	11%	42%	90%
Airborne early warning aircraft	0	6	14
Aerial refueling tankers	0	4	6?
Attack helicopters	10	40	60+
Maritime patrol and strike force	3	8	12

Source: Military Balance, 1998–99 (London: IISS, 1998), for data for 1999; other figures are from the author's assessment based on current acquisition trends.

69. Daniel Byman and Roger Cliff, *China's Arms Sales: Motivations and Implications* (Santa Monica, CA: RAND, 1999), viii.
70. Cited in *The Gulf Today*, May 19, 1999.
71. Pakistan carried out its first test in Lop Nor in 1983 with the Chinese design of its fourth test. For China's supply of ballistic missiles, see Pakistan prime minister Moeen Qureshi's statement on August 26, 1993, cited in *The Nation*, August 27, 1993. For an earlier confirmation of such supplies, see Wilson and Hua Di, "China's Ballistic Missile Programs," *International Security* 17, no. 2 (Fall 1992): 37.

72. For China's supplies of ballistic missiles to Pakistan, see Pakistan prime minister Moeen Qureshi's statement on August 26, 1993, cited in *The Nation*, August 27, 1993, and Foreign Minister Abdul Sattar's statement to the Senate, August 26, 1993, cited in *The Nation*, August 27, 1993. For an earlier confirmation of Chinese supplies of ballistic missiles to Pakistan, see Chinese ambassador to the United States, Zhu Qizhen's address to the National Press Club, Washington, DC, *Reuters Transcript Report* (June 27, 1991), cited in Wilson and Di, "China's Ballistic Missile Programs," 37.

73. For example, see *Time*'s cover story of March 16, 2009.

74. The war in 1999 was waged by Pakistan under the ruse that the fighters who had secretly infiltrated the Indian side of the line of control were "freedom fighters" and not their army troops. This influenced Indian responses to some extent due to the political decision to restrict the war to the infiltrated region only, and certainly not to cross the line of control in the state of Jammu and Kashmir.

75. "Defence Policy, Planning and Management," Standing Committee of Defence (1995–96), Tenth Lok Sabha, Sixth Report, March 1996, 37.

CHAPTER 6. CHINESE AIR POWER

1. Austin Stephens, "Reds in Korea Challenge U.N. in Air, Vandenberg Declares," *New York Times*, November 22, 1951.

2. Zhang Nongke and Weng Huainan, "The Chinese Air Force Took Historical Strides in Fifty Years," *Zhongxin she* [China News Agency], November 2, 1999.

3. "Glorious Thirty Years: China's Space Continues to Take the Lead to Bring the Dream for the People to Come True," *Renmin ribao* [People's Daily], November 7, 2007.

4. Wang Dinglie, chief ed., *Dangdai Zhongguo kongjun* [Contemporary China's Air Force] (Beijing: Social Science Press, 1989), 2.

5. For Sun's view, see Sun to Homer Lee, November 7, 1910, in *Sun Zhongshan quanji* [Compiled Works of Sun Yat-sen], vol. 1, 290–300; and Sun to Li Jian, May 31, 1911, in History Research Division of Guangdong Provincial Philosophy and Social Science Institute, ed., *Sun Zhongshan nianpu* [The Chronicle of Sun Yat-sen's Life] (Beijing: China Book Bureau, 1980), 118.

6. Wang Suhong and Wang Yubing, *Xuese tiankong: Zhongguo kongjun kongzhan shilu* [Crimson Sky: A True Account of Air Operations by China's Air Force] (Chengdu: Sichuan People's Press, 1996), 24–26; for details, see Xiao Qiang and Li Debiao, *Guofu he kongjun* [The Founding Father and the Air Force] (Taipei: Huata Press, 1983).

7. Ma Yufu, *Zhongguo junshi hangkong* [Chinese Military Aviation] (Beijing: Aviation Industry Press, 1994), 363–364; and Xiao Qiang and Li Debiao, *Guofu he kongjun*, 39.

8. Wang Dinglie, chief ed., *Dangdai Zhongguo kongjun*, 4; and Wang Suhong and Wang Yubing, *Xuese tiankong*, 242–246.

9. Liu Ziming, *Zhongguo jindai junshi sixiang shi* [History of Modern China's Military Thought] (Nanchang: Jiangxi People's Press, 1997), 319–357.

10. By the end of 1937, the number of the Chinese Air Force's aircraft had dramatically decreased from some seven hundred to eighty combat planes. Wang Daoping, *Zhongguo kang Ri zhanzheng shi* [The History of the War of Resistance against Japanese Aggression], vol. 3 (Beijing: PLA Press, 1991), 382.

11. Between October 1937 and June 1941, the Soviet Union furnished China with 885 aircraft. From 1942 to the end of the war, the United States supplied the Nationalist air force with 1,394 aircraft. Wang Zhenghua, *Kangzhan shiqi waiguo duihua junshi yuanzhu* [Foreign Military Assistance to China during the Resistance War] (Taipei: Around the World Book Bureau, 1987), 114, 180–221; and *Kangzhan shengli sishi zhounian lunwenji* [Essays on the Fortieth Anniversary of the Victory of the Resistance War], vol. 1 (Taipei: History Office of the Defense Ministry, 1985), 312–313.

12. For details, see Xiaoming Zhang, *Red Wings over the Yalu: China, the Soviet Union, and the Air War in Korea* (College Station: Texas A & M University Press, 2002), 182–187.

13. Liao Guoliang et al., *Mao Zedong junshi sixiang fazhan shi* [History of Mao Zedong's Military Thought Development] (Beijing: PLA Press, 1991), 476.

14. Mao Zedong, *Mao Zedong xuanji* [Selected Works of Mao Zedong], vol. 2 (Beijing: People's Press, 1991), 511.

15. Shi Zhe, *Zai lishi juren shenbian* [Beside Great Historical Figures: Shi Zhe's Memoirs] (Beijing: Central Archival and Manuscript Press, 1991), 350–351.

16. Mao Zedong, *Mao Zedong junshi wenji* [A Collection of Mao Zedong's Military Papers], vol. 5 (Beijing Military Science Press and Central Archival and Manuscript Press, 1993), 471–477.

17. Wang Dinglie, chief ed., *Dangdai Zhongguo kongjun*, 35.

18. Lü Liping, *Tongtian zhi lu* [The Road to the Sky] (Beijing: PLA Press, 1989), 137, 156.

19. Wang Dinglie, chief ed., *Dangdai Zhongguo kongjun*, 61.

20. Lü Liping, *Tongtian zhi lu*, 65–85.

21. Ibid., 144.

22. For a discussion of Soviet air doctrine in World War II, see John T. Greenwood, "Soviet Frontal Aviation during the Great Patriotic War, 1941–45," in *Russian Aviation and Airpower*, ed. Robin Higham et al. (London: Frank Cass, 1998), 62–90.

23. Lü Liping, *Tongtian zhi lu*, 144.

24. Air Force Headquarters Editorial and Research Office, *Kongjun shi* [History of the Air Force] (Beijing: PLA Press, 1989), 36.

25. Wang Dinglie, chief ed., *Dangdai Zhongguo kongjun*, 87–88.

26. Mark O'Neil, "'The Other Side of the Yalu': Soviet Pilots in the Korean War, Phase One, 1 November 1950–12 April 1951," (PhD dissertation, Florida State University, 1996), 31–39.

27. The Soviet air divisions sent to help with air defense in China were not the same units deployed for air operations in Korea. Wang Dinglie, chief ed., *Dangdai Zhongguo kongjun*, 78–79, 86–88.

28. O'Neil, "The Other Side of the Yalu," 100.

29. Even after the Soviet leadership's intervention, the Soviets continued to persuade the Chinese to purchase from them the remaining MiG-9 engines in stock. Based on the consideration that the MiG-9 still could be used for training and against Taiwan's piston aircraft, as well as out of friendship, the Chinese agreed to this Soviet request. Wang Yazhi, "Some Information before and after Mao Zedong Decided to Send Troops to Korea," *Dang de wenxian* [Party's Historical Documents] no. 5 (1995): 84–89.

30. Wang Dinglie, chief ed., *Dangdai Zhongguo kongjun*, 127.

31. Lü Liping, *Tongtian zhi lu*, 245.

32. Wang Dinglie, chief ed., *Dangdai Zhongguo kongjun*, 127–129; and Liu Zhen, *Liu Zhen huiyilu* [Memoirs of Liu Zhen] (Beijing: PLA Press, 1990), 341–344.

33. Yang Wanqing and Qi Chunyuan, *Liu Yalou jiangjun zhuan* [Biography of General Liu Yalou] (Beijing: CCP Historical Press, 1995), 292; and Zhong Zhao-yuan, *Baizhan Jiangxing Liu Yalou* [One Hundred Battles of Star General Liu Yalou] (Beijing: PLA Literature and Arts Press, 1996), 184–185.

34. Wang Dinglie, chief ed., *Dangdai Zhongguo kongjun*, 128–129.

35. Min Zengfu, chief ed., *Kongjun junshi shixiang gailun* [An Introduction to Air Force Military Thinking] (Beijing: PLA Press, 2006), 73.

36. Zhou Enlai, *Zhou Enlai junshi wenxuan* [Selected Military Papers of Zhou Enlai], vol. 4 (Beijing: People's Press, 1997), 128–129.

37. Wang Dinglie, chief ed., *Dangdai Zhongguo kongjun*, 207.

38. Zhang, *Red Wings over the Yalu*, 102–103.

39. Wang Dinglie, chief ed., *Dangdai Zhongguo kongjun*, 86–88.

40. Ibid., 163–164.

41. Wang Yan, chief ed., *Peng Dehuai zhuan* [Biography of Peng Dehuai] (Beijing: Contemporary China Press, 1993), 428.

42. Liu Pushao et al., "Biography of Zhu Guang," in *Zhongguo renmin zhiyuanjun renwu zhi* [Biographical Notes and Data of Chinese People's Volunteer Army] (Nanjing: Jiangsu People's Press, 1992), 309.

43. Wanfg Yan, chief ed., *Peng Dehuai zhuan*, 502, 506.

44. Liu Yalou, "To Build an Air Force on the Basis of the Army," in *Lantian zhi lu* [The Road to the Blue Sky], ed. The Political Department of the Air Force, vol. 1 (Beijing: Political Department of the Air Force, 1992), 3.

45. Zhang, *Red Wings over the Yalu*, 211.

46. Zhang Aiping, chief ed., *Zhongguo renmin jiefangjun* [Chinese People's Liberation Army], vol. 1 (Beijing: Contemporary China Press, 1994), 160.

47. Mao Zedong, *Jiangguo ylia Mao Zedong wengao* [Mao Zedong's Manuscripts since the Founding of the Nation], vol. 4 (Beijing: Central Archival and Manuscript Press, 1990), 1–2.

48. For details, see Wang Yazhi, Sheng Zhihua and Li Danhui, "Retrospect and Thinking: Some Issues concerning Sino-Soviet Military Relations in the 1950s," Series 2, Part 2, *Guoji zhengzhi zhi yanjiu* [Studies of International Politics] no. 3 (2004): 55–64.

49. Wu Faxian, *Wu Faxian huiyilu* [Memoirs of Wu Faxian] (Hong Kong: Star Books, 2006), 587–588.

50. Hua Renjie et al., *Kongjun xueshu sixiang shi* [The History of the Academic Thinking of the Air Force] (Beijing: PLA Press, 2008), 348.
51. Wang Yazhi, Shen Zhihua, and Li Danhui, "Retrospect and Thinking: Some Issues concerning Sino-Soviet Military Relations in the 1950s, Series 3," *Guoji zhengzhi yanjiu*, no. 1 (2005): 106–125.
52. Ibid., 120.
53. Ibid., 125.
54. Lin Hu, *Baowei zuguo lingkong de zhandou* [Fight to Protect the Motherland's Airspace] (Beijing: PLA Press, 2002), 71; and Wang Dinglie, chief ed., *Dangdai Zhongguo kongjun*, 233.
55. He Tingyi, "The Air Force and the Air Defense Troops Were Merged," *Lantian zhi lu*, vol. 2, 105–112.
56. Wang Dinglie, chief ed., *Dangdai Zhongguo kongjun*, 352.
57. Wu Faxian, *Wu Faxian huiyilu*, 408–409.
58. Xiaoming Zhang, "Air Combat for the People's Republic: The People's Liberation Army Air Force in Action, 1949–1969," in *Chinese Warfighting: The PLA Experience since 1949*, eds. Mark A. Ryan, David M. Finkelstein, and Michael A. McDevitt (Armonk, NY: M.E. Sharpe, 2003), 279–282.
59. Wang Dinglie, chief ed., *Dangdai Zhongguo kongjun*, 341.
60. Zhou Enlai, *Zhou Enlai junshi wenxuan*, vol. 4, 403.
61. Zhang, "Air Combat for the People's Republic," 284.
62. Pei Zhigeng and Xie Bin, "Rushed into Fujian to Combat for Control of the Air," in *Lantian zhi lu*, vol. 2, 139; and Shen Weiping, *8.23 paoji Jinmen* [The August 23 Bombardment of Jinmen] (Beijing: Chinese Art Press, 1998), 657–660.
63. Wang Dinglie, chief ed., *Dangdai Zhongguo kongjun*, 341, 350.
64. According to U.S. data, there were twenty-five air-to-air engagements from August 26 through October 6; the Nationalists destroyed thirty-two aircraft and damaged ten, while losing four of their own aircraft. Kenneth W. Allen, Glenn Krumel, and Jonathan D. Pollack, *China's Air Force Enters the 21st Century* (Santa Monica, CA: RAND, 1995), 68.
65. Ken Allen, "People's Republic of China: People's Liberation Army Air Force," unpublished report (Washington, DC: Defense Intelligence Agency, 1991), section 3, 1.
66. The Chinese did not use the term "air force strategy" until 1979. Guo Jinsuo, "Science of Air Force Strategy," in *Zhongguo kongjun baike quanshu* [Chinese Air Force Encyclopedia], compiled by the Encyclopedia Committee of the Chinese Air Force, vol. 1 (Beijing: Aviation Industry Press, 2005), 59.
67. Cao Lihuai, "Recollection of the Work on the Compilation of Regulations and Teaching Materials," in *Lantian zhi lu*, vol. 2, 158, 165.
68. The best examples are Min Zengfu's *Kongjun junshi shixiang gailun*, and Hua Renjie's *Kongjun xueshu sixiang shi*.
69. Zhao Zhongxin, "Air Force Tactics," in *Zhongguo kongjun baike quanshu*, vol. 1, 108.
70. Min Zengfu, chief ed., *Kongjun junshi shixiang gailun*, 69.

71. Wang Dinglie, chief ed., *Dangdai Zhongguo kongjun*, 129.
72. He Weirong, "Military Thought on the Air Force," in *Zhongguo kongjun baike quanshu*, 5.
73. Cao Lihuai, "Recollection of the Work on the Compilation," 161.
74. Hua Renjie et al., *Kongjun xueshu sixiang shi*, 359.
75. He Weirong, "Military Thought on the Air Force," 3.
76. Min Zengfu, chief ed., *Kongjun junshi shixiang gailun*, 394.
77. He Weirong, "Military Thought on the Air Force," 3.
78. Xiaoming Zhang, "The Vietnam War, 1964–1969: A Chinese Perspective," *The Journal of Military History* 60 (October 1996), 739–746.
79. Lin Hu, *Baowei zuguo lingkong de zhandou*, 96.
80. For example, U.S. pilotless reconnaissance planes had intruded 97 times from August 1964 through 1969. From September 1959 to September 1967, the Nationalist U-2 made 129 overflights, but only 5 suffered setbacks. Wang Dinglie, chief ed., *Dangdai Zhongguo kongjun*, 385–386.
81. Lin Hu, *Baowei zuguo lingkong de zhandou*, 272. For details about the establishment of the air force as the best example for other services to learn, see Wu Faxian, *Wu Faxian huiyilu*, 510–511.
82. In addition to the air force operations field manual, the other major manuals were "PLA Flight Teaching Manual," "Basic Rules for Flying in the People's Republic of China," "Regulations for Air Force Staff Officers," "Air Force Logistics Regulations," "Air Force Navigation Regulations," and "Air Force Training Outlines," etc. Cao Lihuai, "Recollection of the Work on the Compilation," 166–167.
83. Ling Qiang, "Revealing the Secret of China's Jian-12 Fighter," *Xiandai bingqi* [Modern Arsenal] no. 2 (2001): 2–4.
84. Duan Zijun, chief ed., *Dangdai Zhongguo de hangkong gongye* [Contemporary China's Aviation Industry] (Beijing: China Social Science Press, 1998), 163–165.
85. Ling Qiang, "Revealing the Secret of China's Jian-12 Fighter," 3.
86. Liu Yazhou, "Changing Our Air Force Strategy," *Lingdao wencui* [The Best Articles of Leadership] no. 5 (2005), 99. There is no detailed official information about the PLAAF order of battle since its inception. Yao Jun, chief ed., *Zhongguo hangkong shi* [Chinese Aviation History] (Zhengzhou: Daxiang Press, 1997): 655–665, provides an account about when these fifty air divisions were created.
87. Zhang Zhizhi, "The Air Force Troops in the Self-Defensive Counterattacks against Vietnam," in *Lantian zhi lu*, vol. 2, 350–353.
88. Jiang Feng et al., *Yang Yong jiangjun zhuan* [Biography of General Yang Yong] (Beijing: PLA Press, 1991), 497; Zhang Zhizhi, "The Air Force Troops," 355.
89. Daniel Tretiak, "China's Vietnam War and Its Consequences," *The China Quarterly*, no. 80 (December 1979): 751; Edward C. O'Dowd, *Chinese Military Strategy in the Third Indochina War: The Last Maoist War* (New York: Routledge, 2007), 67–68; and King C. Chen, *China's War with Vietnam, 1979*

(Stanford, CA: Hoover Institution Press, 1987), 101–103. The Vietnamese People's Air Force never operated MiG-23s as King Chen wrongly mentioned in his book. In the 1980s, the MiG-23s in Vietnam actually belonged to the Soviet air units stationed at Cam Ranh Bay.

90. Jiang Hao and Lin Pi, "Recollecting His Days in the Air Force: An Interview with 7th Air Division Deputy Chief of Staff Zhou Shouxing," in *Binggong keji* [Armament Science and Technology], no. 10 (October 2006): 30.

91. Zhang Zhizhi, "The Air Force Troops," 355–356.

92. Lin Hu, "The Air Force's Armament Work during the Seventh Five-Year-Plan Period," in *Lantian zhi lu*, vol. 2, 514.

93. Wang Hai, *Wo de zhandou shengya* [My Career in Warfighting] (Beijing: Central Archival and Manuscript Press, 2000), 281.

94. Dai Xu, "Goodbye, Old J-6 Fighters: A Complete Examination of the Service History of the Last Meritorious Fighter in the Chinese Air Force with Combat Victory Record," *Guoji zhanwang* [World Outlook], no. 19 (2005): 21.

95. Zhang Tingfa and Gao Houliang, "Bringing Order out of Chaos: The Air Force Enters into a New Historical Period," in *Lantian zhi lu*, vol. 2, 305.

96. Deng Xiaoping, "Our Strategic Approach Is Active Defense," October 15, 1980, *Deng Xiaoping junshi wenji* [A Collection of Deng Xiaoping's Military Papers], vol. 3 (Beijing: Military Science Press and Central Archival and Manuscript Press, 2004), 177.

97. For a detailed interpretation of these terminologies, see Allen, "People's Republic of China," 4–6.

98. Wu Fushan, "The Air Force Troops Participate in the North Chinese Exercise," in *Lantian zhi lu*, vol. 2, 357–370.

99. Deng Xiaoping, "Modern Warfare Requires the Seizure of Air Superiority," January 18, 1979, *Deng Xiaoping junshi wenji*, vol. 3, 153.

100. Deng Xiaoping, "Downsizing of Troops and Enhancing Troop Quality," *Deng Xiaoping junshi wenji*, vol. 3, 263–265.

101. Zhong Zhaoyun, "Air Force Commander Zhang Tingfa Lives up to the Great Trust of Deng Xiaoping," *Dangshi bolan* [The General Review of the Communist Party of China], no. 11 (2004): 42–45; and Min Zengfu, chief ed., *Kongjun junshi shixiang gailun*, 314.

102. Shao Zhenting, Zhang Zhengping, and Hu Jianping, "Theoretical Thinking on Deng Xiaoping's Views on the Buildup of the Air Force and the Reform of Operational Arts," *Zhongguo junshi kexue* [China Military Science], no. 4 (1996): 46.

103. Lin Hu, "The Air Force's Armament Work during the Seventh Five-Year-Plan Period," 515.

104. John Wilson Lewis and Xue Litai, "China's Search for a Modern Air Force," *International Security* 24, no. 1 (Summer 1999): 72.

105. Shao Zhenting, Zhang Zhengping, and Hu Jianping, "Theoretical Thinking," 44–45.

106. Liu Taihang, "Strengthen Studies on Air Force Military Theory for Quality Construction of the People's Air Force," *Zhongguo junshi kexue* no. 4 (1997): 46.

107. Hua Renjie et al., *Kongjun xueshu sixiang shi*, 362–363.
108. Ibid., 368.
109. Ibid., 369. For the original text, see Giulio Douhet, *The Command of the Air*, trans. Dino Ferrari (Washington, DC: Air Force History and Museum Program, 1998), 250.
110. Hua Renjie et al., *Kongjun xueshu sixiang shi*, 368.
111. Liu Huaqing, *Liu Huaqing huiyilu* [Memoirs of Liu Huaqing] (Beijing: PLA Press, 2004), 581–582; and Ye Huinan, "Four Major Changes of Our Country's National Defense Strategy since Its Founding," *Dangdai Zhongguo shi yanjiu* [Studies of Modern Chinese History], no. 3 (1999): 8.
112. Shao Zhenting, Zhang Zhengping, and Hu Jianping, "Theoretical Thinking," 47.
113. Ge Wenyong, *Feixing zaiji* [Notes about Flying] (Beijing: Lantian Press, 2004), 277–278; and Tang Jianguang and Yang Xinggeng, "From 'Jian-7' to 'Thunder Dragon,'" *Zhongguo xinwen zhoukan* [China Newsweek], no. 149 (2003): 37.
114. Song Yichang, "The Startup of China's Modern Aviation Industry and Reflection on It," *Zhanglue yu guangli* [Strategy and Management], no. 4 (1996): 104.
115. Liu Huaqing, *Liu Huaqing huiyilu*, 573, 593.
116. Si Gu, "Sukhoi Fighters in China," *Bingqi zhishi* [Armament Knowledge], no. 8 (2007): 26–27.
117. Liu Huaqing, "Speed Up Development of Our Country's Aviation Industry," June 3, 1997, in *Liu Huaqing junshi wenji* [A Collection of Liu Huaqing's Military Papers], vol. 2 (Beijing: PLA Press, 2008), 529–531.
118. Liu Huaqing, "Conducting a Thorough Study of the Gulf War Issues," in ibid., 127–129.
119. Dong Wenxian, "An Indignant Warning," *Zhongguo guofang bao* [China National Defense Daily], April 2, 1999.
120. There is no access to these publications, but the *Chinese Air Force Encyclopedia* contains a brief description about these two books.
121. Liu Shunyao, "Follow the Direction Given by Our Party's Three-Generation Collective Leaderships to Build a Powerful Modern People's Air Force," *Zhongguo junshi kexue*, no. 3 (1997): 133.
122. He Weirong, "Military Thought on the Air Force," 4.
123. Dong Wenxian, *Xiandai kongjun lun* (xupian) [On the Modern Air Force (continuation)] (Beijing: Lantian Press, 2005), 47.
124. "China's National Defense Paper in 2008," http://www.gov.cn/english/official/200901/20/ content_1210227.htm, accessed March 25, 2009.
125. Si Gu, "Sukhoi Fighters in China," 27–28.
126. "The Development of the Joint Strike Flying Leopard for the Air Force," *Hangkong shijie* [Aviation World], no. 5 (2005): 38–39.
127. The Western studies wrongly note that the PLAAF is "unenthusiastic" about the JH-7. Phillip C. Saunders and Erik Quam, "Future Structure of Chinese Air Force," in *Right Sizing the People's Liberation Army: Exploring the Contours of China's Military*, eds. Roy Kamphausen and Andrew Scobell (Carlisle, PA: Strategic Studies Institutes, Army War College, 2007), 395.

128. The United States and China do not use the same terminology when discussing generations of aircraft: the former's criterion is that the aircraft produced from 1970 to 1999 are the fourth generation, while the latter refers to them as the third generation.
129. "Flying in the Sky of the New Century: Interview with Air Force Commander Xu Qiliang," *Jiefangjun bao* [PLA Daily], November 2, 2009.
130. Min Zengfu, chief ed., *Kongjun junshi sixiang gailun*, 308.
131. Dong Wenxian, *Xiandai kongjun lun (xubian)*, 327–328, 373, 389.
132. Cai Fengzhen et al., *Kongtian yiti zuozhan xue* [Study of Intergrated Air and Space Operations] (Beijing: PLA Press, 2006), 287–301; and Shan Jinsuo and Li Niguang, "Creative Development of the Party's Guiding Theory of Air Force Building," *Zhongguo junshi kexue* 20, no. 5 (2007): 45.
133. Chinese air power advocates internally argue that any efforts to continue emphasizing the air force's defensive role may cause the PLAAF to lose another opportunity to become an independent striking force. See Liu Yazhou, "Changing Our Air Force Strategy," 95–101.
134. China does not make the PLAAF air order of battle publicly available. China-Defense.com Forum provides some order of battle information. Some information can be found online; see http://www.china-defense.com/forum/index.php?showforum=5, accessed November 10, 2009.
135. Martin Sieff, "Airlift the Key to True Superpower Capability Part One," December 12, 2008, http://www.spacewar.com/report/Airlift_The_Key_To_True_Superpower_Capability_Part_999.html, accessed March 21, 2009; Vladimir Isachenkov, "Russia Faces an Aging Defense Industry," July 20, 2008, http://article.latimes.com/2008/jul/20/world/fg-russia20plr, accessed March 21, 2009. Also see Wang Xiaoxia and He Qingsong, "The Large Military Transport Aircraft of One Unspecified Air Division in the Chengdu Air Force Made Its Debut Flight Successfully," *Jiefangjun bao*, June 24, 2009; and "China's 200 Tons of Military Transport Will Come to Public at the End of This Year," November 5, 2009, *Zhongguo xinwen wang* [China News Web], http://www.chinanews.com.cn/gn/news/2009/11-05/1949953.shtml, accessed October 27, 2009.
136. The PLAAF special operations aircraft units entered service in 2005, but so far there is little evidence to suggest that they are fully operational. According to a *Liberation Army Daily* report on March 31, 2008, for the first time, during a regular training day of an unidentified air division, a regiment commander conducted his command and control role in an AWACS aircraft. See "The Change of Command Mode by One Air Force Division Increases Its Combat Capability: Command Post Flies from the Ground into the Sky," *Jiefangjun bao*, March 31, 2008, 2. Another *Liberation Army Daily* report indicates that Chinese AWACS aircraft flew command and control missions for the 2008 Olympic Games. See Xu Qiliang, "The Dream Flies in Reform and Opening Up," *Jiefangjun bao*, November 4, 2008.
137. Deng Changyou, "Making Consistent Effort to Build Troops with Recruiting Talents," *Jiefangjun bao*, January 8, 2009.

138. Min Zengfu, chief ed., *Kongjun junshi sixiang gailun*, 312.
139. Deng Changyou, "Making Consistent Effort to Build Troops with Recruiting Talents."
140. "PRC: Article Features Chinese Air Force Testing, Training Base," December 1, 2008, CPP20081201004007, Open Sources Center.
141. Wang Jun and Zhang Zhirong, "Air Regiment Commander Fang Bing–A Leading Wild Goose Flying in the Sky," *Jiefangjun bao*, August 29, 2008.
142. Wang Shanhe, "The Air Force Vigorously Promotes the Improvement of Aviation Field Station Supply System," *Jiefangjun bao*, December 8, 2008.
143. Jing Lei and Xia Xiaosheng, "Making the Transition to Train the Single Skilled Personnel with Multi-Skills," *Jiefangjun bao*, December 18, 2004.
144. Ding Yi, "Establishing the Complete Maintenance Standard for the New Type Fighters," *Jiefangjun bao*, December 9, 2008.
145. Ren Lijun, Wang Deshun, and Wang Yehong, "Identify the Major Strategic Direction, Strengthen Air Force Finance Development," *Junshi jinji yanjiu* [Military Economic Study], no. 7 (2008): 52–53.
146. "Sixty Years of the Chinese Air Force and the Capability of Its Equipment Grows," *Wenweipo* (Hong Kong), November 22, 2009.

CHAPTER 7. THE ASIA PACIFIC REGION

1. It is often forgotten that by the early twentieth century Russia was substantially industrialized. See Lionel Kochan, *Russia in Revolution* (London: Granada, 1970), 25–34. During World War II the USSR eventually became a military-industrial power. As far as shared religious and social values are concerned, Communism filled that role in place of Christianity between 1917 and 1991.
2. The term "maneuver" is used in place of "airlift" because it defines the role more accurately as a military action. The same reasoning applies to the use of "close air attack" rather than "close air support."
3. For detailed discussion of the "shape-deter-respond" methodology, see Alan Stephens and Nicola Baker, *Making Sense of War: Strategy for the 21st Century* (Cambridge: Cambridge University Press, 2006), 101–108.
4. "Operation Southern Watch," http://www.globalsecurity.org/military/ops/southern_watch.htm, accessed November 10, 2009. Operation "Northern Watch" similarly enforced a no-fly zone north of latitude 36° north.
5. For an excellent background to the concept of non-nuclear deterrence, see George H. Quester, *Deterrence before Hiroshima* (New Brunswick, NJ: Transaction Books, 1986).
6. For the best commentary on the phenomenon of globalization and its effect on nation-states, see Philip Bobbitt, *The Shield of Achilles: War, Peace, and the Course of History* (New York: Random House, 2003).
7. *The Japanese Air Forces in World War II: The Organization of the Japanese Army and Naval Air Force, 1945* (London: Arms and Armour Press, 1979), 1–4.
8. Mark R. Peattie, *Sunburst: The Rise of Japanese Naval Air Power 1909–1941* (Annapolis, MD: Naval Institute Press, 2001), 3.

9. Naoko Sajima and Kyoichi Tachikawa, *Japanese Sea Power* (Canberra: Sea Power Centre–Australia, 2009), 39–46.

10. Richard Pelvin, "Japanese Air Power 1919–1945: A Case-Study in Military Dysfunction," paper no. 31 (Canberra: Air Power Studies Centre, April 1995); and John J. Whitman, "Japan's Fatally Flawed Air Forces in World War II," *Aviation History* (September 2006).

11. Richard J. Overy, *The Air War 1939–1945* (London: Papermac, 1980), 142.

12. Ibid., 142–145.

13. Stockholm International Peace Research Institute, *SIPRI Yearbook 2008* (Oxford: Oxford University Press, 2008), 222, 224. Defense spending is notoriously hard to calculate because of secrecy, dual-use technologies, varying government accounting practices, and so on. Nevertheless, the source cited here can be accepted as reliably indicative.

14. Those missiles are believed to have been a mix of Scuds, Scud-ERs, and Rodongs. Wendell Minnick, "Japan Weighs THAAD System for Missile Defense," *Defense News*, July 13, 2009, 16; and Bradley Perrett, "Layer on Layer," *Aviation Week & Space Technology*, July 13, 2009, 26–27.

15. *China's National Defense in 2008*, http://www.china.org.cn/government/central_government/2009–01/20/content_17155577.htm, accessed November 10, 2009.

16. All numbers of platforms and systems used in this chapter come from *Aviation Week & Space Technology, 2009 Aerospace Source Book*, January 26, 2009; and *Asia-Pacific Defense Reporter, 2009–2010 Directory/Source Book*, October 2009.

17. "Contract negotiations indicate Japanese step towards JSF," *Jane's Defence Weekly*, October 14, 2009.

18. See for example Masaru Tamamoto, "Japan's Crisis of the Mind," *New York Times*, March 1, 2009.

19. "Crushing end to postwar era: LDP swept from office," *Sydney Morning Herald*, August 31, 2009, 8; and "Japan sets a new course," *Canberra Times*, September 8, 2009, 11.

20. "Japan's defense minister to end Afghan mission," *Jane's Defence Weekly*, September 30, 2009, 19.

21. Greg Lockhart, "Vietnamese Strategy for National Liberation," in *Air Power and Wars of National Liberation*, ed. Keith Brent (Canberra: Air Power Studies Centre, 2002), 43.

22. Vo Nguyen Giap, *Dien Bien Phu: Rendezvous with History* (Hanoi: The Gioi Publishers, 2004), 317.

23. Kenneth P. Werrell, *Archie, Flak, AAA, and SAM* (Maxwell Air Force Base, AL: Air University Press, 1988), 96.

24. These comprised one regiment of 105mm howitzers and one of 75mm howitzers, some forty guns in all. See Bernard Fall, *Street without Joy* (New York: Schocken Books, 1972), 322. Giap, *Dien Bien Phu*, 205–206, 221, also mentions 120mm heavy mortars.

25. Lockhart, "Vietnamese Strategy," 47.

26. Tony Mason, *Air Power: A Centennial Appraisal* (London: Brassey's, 1994), 188; and Werrell, *Archie*, 114–115; "Helicopter losses during the Vietnam War," http://www.vhpa.org/heliloss.pdf, accessed November 5, 2009.

27. William W. Momyer, *The Vietnamese Air Force, 1951–1975*, Monograph-4, USAF Southeast Asia Monograph Series (Washington, DC: Office of Air Force History, 1985).

28. Werrell, *Archie*, 101–102.

29. Ibid., 107–116.

30. Henry Kissinger, *The White House Years* (London: Weidenfeld & Nicholson, 1971), 1461.

31. *Pancasila* democracy was guided by five principles of national ideology: a belief in one god, a just and civilized society, national unity, the guidance of inner wisdom, and social justice for all.

32. "Indonesia's army: going out of business," *The Economist*, October 2, 2008; and "Painful progress," *Jane's Defence Weekly*, October 21, 2009.

33. Alan Stephens, *Going Solo: The Royal Australian Air Force, 1946–1971* (Canberra: AGPS, 1995), 267.

34. *Jane's Defence Weekly*, September 23, 2009, 17.

35. Alan Stephens, *Air Power in Southeast Asia* (Brunei Darussalam: Sultan Haji Hassanal Bolkiah Institute of Defense and Strategic Studies, 2008), 13–14.

36. Other members of the FPDA are Australia, the UK, New Zealand, and Malaysia.

37. Pak Shun Ng, "From 'Poisonous Shrimp' to 'Porcupine': An Analysis of Singapore's Defense Posture Change in the Early 1990s," working paper no. 379 (Canberra: Strategic and Defense Studies Centre, Australian National University, 2005).

38. Australian Bureau of Statistics, *Official Year Book of the Commonwealth of Australia* (Canberra: Australian Bureau of Statistics, 1919–1938).

39. Ibid. Air forces larger than the RAAF were those of the USSR, the USA, and the UK.

40. *Defending Australia in the Asia Pacific Century: Force 2030* (Canberra: Australian Government, Department of Defense, 2009).

41. "Jorn" is an acronym for Jindalee over-the-horizon radar network. The system uses two operational radars.

CHAPTER 8. LATIN AMERICA

1. For an in-depth examination of this period, see Robert Miller, *Mexico: A History* (Norman: University of Oklahoma Press, 1985), chaps. 8–10, passim. Also see Peter Calvert, *Mexico* (New York: Praeger, 1973), 32–201, passim.

2. Friedrich Katz, *The Life and Times of Pancho Villa* (Stanford, CA: Stanford University Press, 1998), 552.

3. Herbert Mason, *The Great Pursuit* (New York: Random House, 1970), 69.

4. There is a significant body of work on the Punitive Expedition. See, for example, Halden Brady, *Pershing's Mission into Mexico* (El Paso: Texas Western College Press, 1966); and Clarence Clendenen, *Blood on the Border: The United*

States and the Mexican Irregulars (New York: Macmillan, 1969). A worthwhile Mexican examination, albeit sympathetic to Villa, can be found in Federico Cervantes, *Francisco Villa y la revolución* (Mexico, D.F.: Ediciones Alonso, 1960).

5. Capt. Benjamin Foulois, *Report of the Operations of the First Aero Squadron, Signal Corps, With Punitive Expedition, USA for Period March 15 to August 15, 1916* (Maxwell Air Force Base, AL: Historical Research Agency, n.d.).

6. Mason, *The Great Pursuit*, 108.

7. Ibid.

8. Foulois, *Report of Operations*, 10–11.

9. For a full treatment, see Robert Heinel, Jr., *Soldiers of the Sea: The United States Marine Corps, 1775–1962* (Annapolis, MD: Naval Institute Press, 1962), 147–290.

10. For a general account of U.S. Navy and Marine air operations in the Caribbean, see James S. Corum and Wray Johnson, *Airpower in Small Wars* (Lawrence: University Press of Kansas, 2003), 21–29.

11. Major General Ford O. Rogers, USMC (Ret.) (Washington, DC: History and Museums Division, Headquarters U.S. Marine Corps, December 3, 1970, Oral History Collection, transcript), 25, cited in Edward Johnson, *Marine Corps Aviation: The Early Years, 1912–1940* (Washington, DC: History and Museums Division, Headquarters U.S. Marine Corps, 1977), 49.

12. On Marine air operations in Nicaragua, see Corum and Johnson, *Airpower in Small Wars*, 29–50.

13. Francis Mulcahy, "Marine Corps Aviation in Second Nicaraguan Campaign," Naval Institute *Proceedings* (August 1933): 1122.

14. Johnson, *Marine Corps Aviation*, 57.

15. "News of the Air Services: First Non-Stop Flight U.S. to Nicaragua," *Aero Digest* (February 1928): 204; and "News of the Air Services: 2d Non-Stop Flight U.S. to Nicaragua," *Aero Digest* (March 1928): 384.

16. Corum and Johnson, *Airpower in Small Wars*, 32–37. According to one account, at least one hundred of the guerrillas who had laid siege to Ocatal were killed by the air attack and, according to another, some three hundred guerrillas had been killed.

17. Mulcahy, "Marine Corps Aviation in Second Nicaraguan Campaign," 1127.

18. Ross Rowell, "The Air Service in Minor Warfare," Naval Institute *Proceedings* (October 1929): 876.

19. Mulcahy, "Marine Corps Aviation in Second Nicaraguan Campaign," 1129.

20. Adrian English, *Armed Forces of Latin America: Their Histories, Development, Present Strength, and Military Strength* (London: Jane's, 1984), 348–349. A good overview of the Paraguayan defense effort is found in Lorenzo Livieres Guggieri, *El Financimiento de la Defensa del Chaco 1924-1934* (Asunción: Arte Nuevo Editores, 1983).

21. English, *The Armed Forces of Latin America*, 76–77.

22. On German military influence in Bolivia and the role of Hans Kundt, see Col. Julio Diaz Arguedas, *Historia del Ejercito de Bolivia* (La Paz, 1940), 751–769.

23. For a good general biography of Estigarribia, see Alfredo Seiferheld, *Estigar-*

ribia (Asunción: Laurel, 1986). See also Estigarribia's memoirs: Jose Felix Estigarribia, *The Epic of the Chaco*, ed. Max Ynsfran (New York: Greenwood Press, 1950), x–xiv and 5–9.

24. See Col. Aquiles Vergara Vicuna, *Bernardino Bilbao Rioja* (La Paz, 1948).

25. On Bolivian military aviation up to the Chaco War, see Col. Julio Diaz Arguedas, *Historia del Ejercito de Bolivia 1825–1932* (La Paz, 1940), 83–110.

26. For detailed information on all the aircraft that flew in the Chaco War, see Dan Hagedorn and Luis Sapienza, *Aircraft of the Chaco War* (Atglen, PA: Schiffer Publishing Co., 1997).

27. On early Paraguayan aviation, see Capt. Felix Zarate Monges, *La Aviación Paraguaya Antes y Durante la Guerra del Chaco* (Asunción, 1985). The French mission stayed from 1926 to 1931 and set the Paraguayan Air Corps on a firm footing and graduated two classes of pilots from the flight school. The French sold Paraguay three Hanriot HD-32 trainers, two Morane–Saulnier 35 trainers, six Potez 25 light bombers/reconnaissance planes, and four Wibault 73 fighters.

28. Bruce Farcau, *The Chaco War* (Westport, CT: Praeger Press, 1996), 51.

29. Ibid., 52.

30. Ibid., 54–61.

31. David Zook, "Airpower in the Chaco War," *The Airpower Historian* (January 1961), 25–26.

32. Hagedorn and Sapienza, *Aircraft of the Chaco War*, 43.

33. The Paraguayans had lost two Wibault fighters to accidents early in the campaign. The Bolivians lost two of their six Vickers Type 143 planes to accidents.

34. On the history of Curtis aircraft in Bolivian service, see Dan Hagedorn, "Curtiss Types in Latin America," *Air Enthusiast*, 67.

35. Johann Jakob, "Bolivian Tin," *Air Enthusiast*, no. 72 (November/December 1997).

36. Bryce Wood, *The United States and Latin American Wars, 1932–1942* (New York: Columbia University Press, 1966), 63–64.

37. Hagedorn, "Curtiss Types in Latin America," 67–70.

38. Georg von Rauch, "The Green Hell Air War," *Air Enthusiast Quarterly* 2 (1976): 210.

39. Bruce Farcau, "The Chaco: War for the Hell of It," *Command Magazine* 12 (September/October 1991): 21.

40. David Zook, *The Conduct of the Chaco War* (New Haven, CT: Bookman Associates, 1960), 163–165.

41. Ibid., 167.

42. Ibid., 209.

43. Pablo E. Tufari Recalde, *La Guerra del Chaco* (Asunción, 1987), 176. During the Chaco War, the Bolivians mobilized 210,000 men. Of these, approximately 60,000 were killed, 10,000 deserted (to Argentina), and 23,250 were taken prisoner. Paraguay mobilized 150,000 men, with 31,500 missing and dead, and 2,500 prisoners.

44. Hagedorn and Sapienza, *Aircraft of the Chaco War*, 49–53. Of the nine Curtiss Falcons ordered by Bolivia, two were lost in combat and four in accidents during the war.

45. Ibid., 270–274.

46. Ibid., 35; von Rauch, "The Green Hell Air War," 212. The most successful Paraguayan attack of the war was conducted against the Bolivian airfield and depot at Ballivían on July 8, 1934. Four Potez 25s escorted by two Fiat fighters dropped forty bombs on the Bolivian airfield and destroyed at least four Curtiss fighters on the ground and damaged others. The Potez bombers also attacked and destroyed the main fuel dump at Ballivían that caused a severe fuel shortage for the Bolivians.

47. On the Paraguayan medical system, see Carlos Jose Fernandez, *La Guerra del Chaco*, vol. VII (Asunción: Editorial Historica, 1987), 143–156.

48. For a good general account of the Peruvian-Ecuadoran wars, see Robert L. Scheina, *Latin America's Wars*, vol. 2, *The Age of the Professional Soldier* (Washington, DC: Brassey's, 2003), 114–126.

49. For details of aircraft supplied to Latin America during World War II, see English, *Armed Forces of Latin America*. Brazil received 946 U.S. aircraft, Mexico 224, Chile 231, Colombia 100+, Venezuela 76+, Peru 67, Uruguay 50+, Paraguay 43, Ecuador 50+, Honduras 30, Haiti 12, Bolivia 100, Guatemala 21, etc.

50. Between February 1941 and September 1942, 173 merchant vessels were sunk in the Western Caribbean, with 33 vessels lost in August alone.

51. Ibid., 120.

52. See James Corum, "La Guerra en el Caribe—el Poderio Aéreo en contra de los Submarinos Alemanes en Aguas Latinoamericanas," *Air and Space Power Journal* (Spanish ed.) (4th Quarter 2003), 89.

53. On the Brazilian First Fighter Group, see L. F. Perdigão, *Missão de Guerra* (Rio de Janeiro: Biblioteca do Exército-Editoa, 1958).

54. On the history of the 201st Fighter Squadron, see Stephen I. Schwab, "The Role of the Mexican Expeditionary Air Force in World War II: Late, Limited, but Symbolically Significant," *Journal of Military History* 66 (October 2002): 1115–1140.

55. See Victor Flintham, *Air Wars and Aircraft* (New York: Facts on File, 1990), 358–359.

56. John Ranelagh, *The Agency: The Rise and Decline of the CIA* (New York: Simon & Schuster, 1987).

57. On the air operations during the Bay of Pigs, see Flintham, *Air Wars and Aircraft*, 345–349.

58. Hugh Thomas, *Cuba or the Pursuit of Freedom* (London: Eyre and Spottiswoode, 1971), 1365.

59. On the air operations in Nicaragua's conflict with the Sandinistas, see Flintham, *Air Wars and Aircraft*, 361–365.

60. Ibid., 359–360.

61. For a general view of the Argentine claims to the Falklands, see Mariano César Bartolomé, *El Conflicto Del Atlántico Sur* (Buenos Aires: Circulo Militar,

1996). See also Carlos Augusto Landaburu, *La Guerra de las Malvinas* (Buenos Aires: Circulo Militar, 1988).

62. For an excellent overview of the Argentine command arrangements and the problems they caused the Argentines, see Alejandro Corbacho, "Improvisation on the March: Argentine Command Structure and Its Impact on Land Operations during the Falklands/Malvinas War (1982)," a paper presented at the Society for Military History Conference, Quantico, VA, April 2000.

63. For a general history of Argentine units and aircraft in the war, see the official history: *Direccion de Estdios Historicos*, ed. Ruben Moro, *Historia de la Fuerza Aerea Argentina*, Tomo VI, vols. 1 and 2 (Buenos Aires: Ejercito Nacional, 1998).

64. There are several books that provide details about the Argentine Air Force's table of organization and aircraft. See Roy Braybrook, *Battle for the Falklands (3) Air Forces* (London: Osprey Publishing, 1982); and Salvador Mafe Huertas and Jesus Romero Briasco, *Argentine Air Forces in the Falklands Conflict* (London: Arms and Armour Press, 1987). On the Argentine Air Force training schools, see Jose Antonio Bautista Castano, "La Escuela de Caza y Bombardeo Argentina," *Revista de Aeronautica y Astronautica* (November 2000), 916–921, esp. 916–917.

65. One of the best discussions of the weapons and technology employed in the Falklands War is found in Lon Nordeen, *Air Warfare in the Missile Age* (Washington, DC: Smithsonian Institution Press, 1985), 191–206. See the table on weapons employed on 233.

66. Much of this information comes from the USAF Armaments Museum personnel, Eglin AFB, FL.

67. Flintham, *Air Wars and Aircraft*, 372–373.

68. Enrique Mariano Ceballos and Jose Raul Buroni, *La Medicina en la Guerra de Malvinas* (Buenos Aires: Circulo Militar, 1992), 23. This work is probably the best source for exact figures for all the army, navy, and air force units that were stationed in the Falklands.

69. The best general history of the war from the Argentine side is Comodoro Ruben Moro, *Historia del Conflicto del Atlantico Sur* (Buenos Aires: Escuela Superior de Guerra Aerea, 1985). On British weapons systems, see 157–158. The British had the new Sea Darts (effective at long range and high altitudes), Sea Wolves (for low-altitude threats), and an array of 20mm and 40mm guns for close defense.

70. Capt. Joseph Udemi, "Modified to Meet the Need: British Aircraft in the Falklands," *Airpower Journal* (Spring 1989): 51–64.

71. The whole issue of combat losses and damage is a very complex one. Both sides exaggerated the damage and losses inflicted upon the other, with the Argentines having a greater degree of exaggeration. The following narrative of combat actions and losses has been pieced together by the author from both the official Argentine histories and reports. For the Argentines, see Comodoro Ruben Moro's works and, for the British, the official after-action report, *The Falklands Campaign: The Lessons* (London: The Secretary of State for Defence,

December 1982). Both sides have carefully documented their own losses and provided aircraft losses by circumstance, tail number, unit, and pilot. I have taken the account of each side per their own losses as the definitive one and have discounted claimed damage for the opponent.

72. Ruben Moro, *La Fuerza Aerea en Malvinas*, vol. 1 and 2 (Buenos Aeres, 1998), the official Argentine Air Force history of the campaign, provides several examples of mission orders complete with maps showing the tanker refueling plans for the FAS strikes. While short on operational narrative, the Argentine Air Force official history offers a wealth of detail on the tactics of each air strike.

73. For an Argentine version of the May 1 battle, see Moro, *Historia del Conflicto del Atlantico Sur*, 176–197.

74. On the Super Étendard, see Christopher Chant, *Super Etendard: Super Profile* (Somerset, UK: Winchmore Publishing, 1983).

75. Braybrook, *Battle for the Falklands (3) Air Forces*, 24.

76. The destroyers *Sheffield* and *Coventry*, the frigates *Ardent* and *Antelope*, the support ship *Atlantic Conveyor*, the landing ship *Sir Galahad*, and the landing craft LCU F4 were all sunk by Argentine bombs and Exocets. The destroyers *Glasgow* and *Antrim*, the frigates *Argonaut* and *Plymouth*, and the landing ship *Sir Tristram* were all heavily damaged and another six ships received minor damage.

77. For an account of the Argentine Air Force's performance in the war, see James S. Corum, "Argentine Airpower in the Falklands War: An Operational View," *Air and Space Power Journal* 16, no. 3 (Fall 2002): 59–77.

78. On the 1995 Peru-Ecuador War, see Scheina, *Latin America's Wars*, 114–126.

79. For a general overview, see James S. Corum, "La Guerra Prolongada de Guatemala: El Papel de la Fuerza Aérea Guatemalteco," *Air and Space Power Journal* (Spanish ed.) (3d Quarter 2004), 81–91.

80. For a general overview of the Guatemalan Air Force's history, see Dan Hagedorn, *Central American and Caribbean Air Forces* (Tonbridge, UK: Air Britain Publications, 1993), 47–59.

81. Mario Overall, "The Lockheed T-33 in Guatemalan Air Force Service," http://www.laahs.com/, 2000, 2–5.

82. Ibid., 6–7.

83. Hagedorn, *Central American and Caribbean Air Forces*, 56.

84. Overall, "The Lockheed T-33 in Guatemalan Air Force Service," 7.

85. George Black, *Garrison Guatemala* (New York: Monthly Review Press, 1984), 22.

86. Marlene Dixon and Susanne Jones, eds., *Revolution and Intervention in Central America* (San Francisco: Synthesis Publications, 1983), 79–83, 95–96.

87. Documents of the Guerrilla Army of the Poor are found in ibid., 126–189.

88. Richard Nyrop, ed., *Guatemala: A Country Study* (Washington, DC: U.S. Army, 1984), 210–211.

89. Kev Darling, Mario Overall, Tulio Soto, and Dan Hagedorn, "Cessna A-37B in Latin America Part I: The Guatemalan A-37Bs," http://www.laahs.com/, Oct. 2000, 1–2. In 1971, the FAG had been modernized with the arrival of eight Cessna A-37B fighter-bombers under the U.S. military aid program.

Guatemala was the first nation in Central America to receive this very capable counterinsurgency aircraft. Seven more A-37s were supplied in 1974 and 1975, which enabled the FAG to establish a full-strength fighter-bomber squadron and retire the P-51s and pull from service some of the T-33s.

90. Nyrop, ed., *Guatemala: A Country Study*, 194–195.

91. Darling et al., "Cessna A-37B in Latin America Part I: The Guatemalan A-37Bs," 2.

92. Hagedorn, *Central American and Caribbean Air Forces*, 57. A good overview of the FAG is also found in *World Airpower Journal* 32 (Spring 1998): 150–151.

93. Hagedorn, *Central American and Caribbean Air Forces*, 56–57. The Israelis were providing arms in this period. Between 1975 and the early 1980s, Israel supplied eleven Arava IAI-201 twin-engine transports to Guatemala. Indeed, the Aravas is ideal transport aircraft for combating an insurgency—it is able to bring several tons of people or equipment into small, rough landing fields and also easily modified as a gunship with the addition of rocket pods and side-mounted machine guns.

94. Information provided by a veteran of the FAG Reserve Command.

95. In 1983, the FAG consisted of eight to ten A-37s, five T-33s, twelve Pilatus PC-7s, three Fouga Magisters as combat and training aircraft, a transport force of one DC-6, one DC-4, nine C-47s, ten IAI Aravas, ten Aerotec T-23 trainers and some Cessna 172s, 180s, and 206s, and a Beech King Air as utility and VIP craft. See Christopher Chant, *Air Forces of the World* (Edison, NJ: Chartwell Books, 1983), 20.

96. Mario Payeras, *Days of the Jungle: The Testimony of a Guatemalan Guerrillero* (New York and Havana: Monthly Review Press, 1983), 29, 77–79, 82, 88.

97. Ibid., 76.

98. For the background to the revolution in El Salvador, see Tommie Sue Montgomery, *Revolution in El Salvador: Origins and Evolution* (Boulder, CO: Westview Press, 1982). See also Liisa North, *Bitter Grounds: Roots of Revolt in El Salvador* (Toronto: Between the Lines, 1982).

99. For very useful works that cover both sides of the conflict, see Marvin Gettleman et al., eds., *El Salvador: Central America in the New Cold War* (New York: Grove Press, 1986). See also Max Manwaring and Courtney Prisk, eds., *El Salvador at War: An Oral History* (Washington, DC: National Defense University Press, 1988).

100. Benjamin Schwarz, *American Counterinsurgency Doctrine and El Salvador: The Frustrations of Reform and the Illusions of Nation Building*, RAND Report R-4042 (Santa Monica, CA: RAND, 1992), 2. During the course of the war, the United States poured $4.5 billion of economic aid into the country and over $1 billion in military aid.

101. Charles Lane, "The Pilot Shark of El Salvador," *New Republic*, September 24, 1990, 27.

102. The most detailed article on the Salvadoran Air Force in the war is James Corum, "The Air War in El Salvador," *Airpower Journal* (Summer 1998): 27–44.

103. Ibid. The main combat force of the FAS consisted of eleven Ouragan ground attack fighters acquired from the Israelis, who had acquired them from the French in the 1950s, and four Fouga Magister trainers modified for combat (another 1950s aircraft). The combat squadrons had six Rallye counterinsurgency aircraft. The rest of the air force consisted of a transport squadron with six C-47s and four Arava transports. The helicopter force amounted to one Alouette III, one FH-1100, one Lama, and ten UH-1s.

104. Schwarz, *American Counterinsurgency Doctrine and El Salvador*, 85.

105. John Waghelstein, *El Salvador: Observations and Experiences in Counterinsurgency* (Carlisle Barracks, PA: U.S. Army War College, 1985), 21–22.

106. Schwarz, *American Counterinsurgency Doctrine and El Salvador*, 2–5; and Waghelstein, *El Salvador*, 36.

107. Gettleman et al., *El Salvador*, 230.

108. *History, U.S. Southern Command 1981* (Maxwell AFB, AL: USAF Historical Research Agency), 10–11.

109. Ibid., 50–54.

110. History, Directorate of International Programs, January 1–June 30, 1982, USAFHRA, K145.01, 5–6.

111. Gen. Fred Woerner, former commander in chief, O.S. Southern Command, interview with author, January 26, 1998.

112. Ibid.

113. Gettleman et al., *El Salvador*, 233.

114. Ibid.

115. Ibid., 234.

116. Elliot Abrams, *Congressional Record*, May 14, 1986, 31.

117. Dr. Judy Gentleman, Air War College Faculty, interview with author, January 19, 1998.

118. Col. Duryea, U.S. defense attaché to El Salvador, cited in Manwaring and Prisk, eds., *El Salvador at War*, 316–319.

119. A. J. Bacevich et al., *American Military Policy in Small Wars: The Case of El Salvador* (Washington, DC: Pergamon-Brassey's, 1998), 29.

120. For an account of this campaign from the rebel viewpoint, see Joe Fish and Cristina Sganga, *El Salvador: Testament of Terror* (New York: Olive Branch Press, 1988), 88–89.

121. Miguel Castellanos and Courtney Prisk, eds., *The Commandante Speaks: Memoirs of an El Salvadoran Guerrilla Leader* (Boulder, CO: Westview Press, 1991), 88–89.

122. *The Military Balance, 1987–1988* (London: International Institute for Strategic Studies, 1987). A good description of El Salvador's helicopter force is Julio Montes, "Las Fuerzas Aéreas Salvadoreñas: Antaño, la mayor flota de helicopteros en América Central," *Revista Aérea* (December 2000/January 2001), 52–55. The airplane force was organized into a fighter squadron, with eight Ouragans, a counterinsurgency squadron with ten A-37Bs, and two AC-47 gunships. A reconnaissance squadron of eleven O-2As supported the counterinsurgency squadron. The transport squadron consisted of five C-47s, one

DC-6, three Aravas, and two C-123Ks. The training squadron had one T-41 and six CM-170 Magisters. The helicopter force had expanded into a force of nine Hughes 500MD attack helicopters, fourteen UH-1H gunships, thirty-eight UH-1H utility helicopters, three SA-315 Lamas, and three SA-316 Alouette IIIs, for a total of sixty-seven helicopters.

123. On the recent efforts of the U.S. and South American nations to cooperate to interdict drug flights see Merrick Krause, "Una Associación para la Seguridad Hemisférica," *Air and Space Power Journal* (Spanish ed.), (4th Quarter 2003), 47–59.

124. Gilles van Nederveen, "EMB-145SA and RS: Brazil's New Eyes in the Sky," *Aerospace Power Journal* (Winter 2000): 87–89. The very capable Brazilian surveillance plan has also been used to support intelligence collection on drug growing in South America.

125. "A New Battlefront Forms for the U.S. in Central America," *Los Angeles Times*, July 9, 2000, 1. This article points out that planes and boats can easily move large quantities of drugs through Central America and its coastline due to lack of assets for interdiction.

126. See *Jane's Defence Weekly*, "Colombia to Order 14 Black Hawks," April 12, 2000. The UH-60 is ideal for Colombian conditions as it is fast (160 knots cruising speed), can carry a heavy load (twelve fully loaded troops or a sling-loaded 105 howitzer), and operates well at high altitudes—the most important factor as most of Colombia is mountainous and counterdrug or counterguerrilla operations will often take place at an eight-thousand-foot elevation.

127. Andrew Selsky, "Colombia Bombs Rebel Territory," *Associated Press*, February 21, 2002.

128. Susannah Nesmith, "Colombian Rebels Step up Violence," *Associated Press*, March 6, 2002.

129. "Colombia elects Uribe: rebels face crackdown," *Washington Times*, May 27, 2002.

130. Jose Higuere, "Latin American Air Forces," *Jane's Defence Weekly*, March 13, 2002, 21.

CHAPTER 9. CONTINENTAL EUROPE

1. The chapter is partly based on the PhD thesis "Continental European Responses to the Air Power Challenges of the Post–Cold War Era," submitted at King's College, University of London, March 2008.

2. Walter J. Boyne, *The Influence of Air Power upon History* (Gretna, LA: Pelican Publishing Company, 2003), 37–38, 44–48.

3. NATO, "Strategic Concept," http://www.nato/int, accessed October 24, 2009.

4. Ibid.

5. NATO, "History of SHAPE and Allied Command Operations" (updated July 7, 2009), http://www.nato.int, accessed October 25, 2009.

6. Tony Mason, *Air Power: A Centennial Appraisal* (London: Brassey's, 2002), 103.

7. Ibid., 97–98.

8. James Sterrett, *Soviet Air Force Theory, 1918–1945* (London: Routledge, 2007), 134.

9. Calculation of percentage is based upon figures from Lewis D. Hill, Doris Cook, and Aron Pinker, *Gulf War Air Power Survey*, vol. V: *A Statistical Compendium and Chronology* (Washington, DC: GPO, 1993), 232–233.

10. Benjamin S. Lambeth, *The Transformation of American Air Power* (Ithaca, NY: Cornell University Press, 2000), 122–123.

11. Mason, *Air Power*, 166.

12. Jérôme de Lespinois, "La participation française à la campagne aérienne de la guerre de libération du Koweït (1991): prolégomènes politico-diplomatiques," *Penser les ailes françaises* 7 (October 2005), 70–75.

13. EAG, "History of the EAG," http://www.euroairgroup.org, accessed May 15, 2007.

14. Louis Gautier, *Mitterrand et son armée, 1990–1995* (Paris: Grasset, 1999), 163–165.

15. Claude Carlier, "L'aéronautique et l'espace, 1945–1993," in *Histoire militaire de la France*, Tome 4—*de 1940 à nos jours*, ed. André Martel (Paris: Presses Universitaires de France, 1994), 465.

16. *Livre blanc sur la défense* (Paris: Ministère de la Defénse, 1994), 59–62.

17. *Une Défense Nouvelle, 1997–2015* (Paris: SIRPA, March 1996).

18. Franz-Josef Meiers, *Zu neuen Ufern? Die deutsche Sicherheits- und Verteidigungspolitik in einer Welt des Wandels, 1990–2000* (Paderborn: Ferdinand Schöningh, 2006), 152.

19. François Bourdilleau, "Évolution de l'Armée de l'Air vers le modèle Air 2015," in *L'Armée de l'Air: les Armées françaises à l'aube du XXIe siècle*, ed. Pierre Pascallon (Paris: L'Harmattan, 2003), 248.

20. "NRF: The French Air Force in Control," *Air actualités* 583 (July/August 2005): 43.

21. Wim H. Lutgert and Rolf de Winter, *Check the Horizon: De Koninklijke Luchtmacht en het conflict in voormalig Joegoslavië, 1991–1995* (The Hague: Sectie Luchtmachthistorie van de Staf van de Bevelhebber der Luchtstrijdkrachten, 2001), 500.

22. Rinus Nederlof, *Blazing Skies: De Groepen Geleide Wapens van de Koninklijke Luchtmacht in Duitsland, 1960–1995* (The Hague: Sectie Luchtmachthistorie van de Staf van de Bevelhebber der Luchtstrijdkrachten, 2002), 534.

23. Lutgert and de Winter, *Check the Horizon*, 500–501.

24. P. J. M. Godderij, "The Evolution of Air Power Doctrine in the Netherlands," speech at the symposium "Air Power: Theory and Application" (Netherlands Defence College, The Hague, April 10–15, 1996).

25. B. A. C. Droste, "Foreword," in *Royal Netherlands Air Force Air Power Doctrine* (Gravenhage: Royal Netherlands Air Force, 1996).

26. M. R. H. Wagevoort, "Materiel Projects in the RNLAF," *Military Technology* 8 (August 1997), 45.

27. André Herbrink, *Bundeswehr Missions Out-of-Area* (Oslo: The Norwegian Institute of International Affairs, 1997), 15.

28. Lawrence Freedman and Efraim Karsh, *The Gulf Conflict, 1990–1991: Diplomacy and War in the New World Order* (London: Faber and Faber, 1994), 355.

29. *Luftwaffendienstvorschrift: Führung und Einsatz von Luftstreitkräften*, LDv 100/1 VS-NfD (Bonn: Ministry of Defense, March 1991), paragraphs 501–515.

30. Michael Graydon, "Foreword," in *AP 3000 Royal Air Force Air Power Doctrine*, 2d ed. (London: HMSO, 1993), v.

31. Meiers, *Zu neuen Ufern?* 335.

32. *Verteidigungspolitische Richtlinien für den Geschäftsbereich des Bundesministers für Verteidigung: Erläuternder Begleittext* (Berlin: Ministry of Defense, May 21, 2003), chap. 2.

33. Walter Jertz, "Unser Schwerpunkt ist der Einsatz: Das Luftwaffenführungskommando auf dem Weg in die Zukunft," *Strategie & Technik* (March 2006), 23.

34. *Weissbuch zur Sicherheit der Bundesrepublik Deutschland und zur Lage und Zukunft der Bundeswehr* (Bonn: Ministry of Defence, 1994), paragraphs 571–572.

35. *Verteidigungspolitische Richtlinien für den Geschäftsbereich des Bundesministers der Verteidigung* (Bonn: Ministry of Defense, 1992), paragraph 38; and Rudolf Scharping, *Die Bundeswehr sicher ins 21. Jahrhundert: Eckpfeiler für eine Erneuerung von Grund auf* (Bonn/Berlin: Ministry of Defense, 2000), paragraphs 3, 15, 20, 22.

36. John Allison, "The Future of Air Power: A European Perspective," *Military Technology* 5 (1999), 10.

37. Wilfrid Schmidt, "Bewaffnung für die Kampfflugzeuge der Luftwaffe: Sachstand-Konzepte-Planungen," *Europäische Sicherheit* 10 (October 1998), 38.

38. "Das Kommando Operative Führung Luftstreitkräfte," *CPM Forum: Luftwaffe 2004* (Sankt Augustin: CPM, 2004), 56–57.

39. Lambeth, *American Air Power*, 178.

40. Mason, *Air Power*, 173.

41. Lambeth, *American Air Power*, 178.

42. Robert C. Owen, "Summary," in *Deliberate Force—A Case Study in Effective Air Campaigning: Final Report of the Air University Balkans Air Campaign Study*, ed. Robert C. Owen (Maxwell AFB, AL: Air University Press, 2000), 483–484.

43. John A. Tirpak, "Deliberate Force," *Air Force—Journal of the Air Force Association* 10 (October 1997), http://www.afa.org, accessed October 27, 2004.

44. Owen, "Summary," 493–494.

45. Lambeth, *American Air Power*, 176.

46. Owen, "Summary," 485.

47. Ibid., 491, 502.

48. Ibid., 471, 511.

49. Sargent, "Aircraft Used in Deliberate Force," 204, 220–222.

50. Richard L. Sargent, "Deliberate Force Combat Air Assessments," in *Deliberate Force*, 332.

51. André Soubirou, "The Account of Lt. Gen. (ret.) André Soubirou, Former Commanding General of the RRF Multinational Brigade in Bosnia from July to October 1995," *Doctrine Special Issue* (February 2007): 28.

52. John Stone, "Air-Power, Land-Power and the Challenge of Ethnic Conflict," *Civil Wars* 3 (Autumn 1999): 29–30.

53. Anthony H. Cordesman, *The Lessons and Non-Lessons of the Air and Missile War in Kosovo*, rev. ed. (Washington, DC: CSIS, July 20, 1999), 4–5.

54. Wesley K. Clark, *Waging Modern War* (Oxford: Public Affairs, 2001), 197–198.

55. Meiers, *Zu neuen Ufern?* 272–273.

56. *Grundgesetz für die Bundesrepublik Deutschland* (May 23, 1949), Article 24, paragraph 2.

57. Walter Jertz, *Im Dienste des Friedens: Tornados über dem Balkan* (Bonn: Bernard & Graefe Verlag, 2000), 14.

58. Bundesverfassungsgericht, AWACS 1: BverfG 2 BvE 3/92 u.a. (Karlsruhe: July 12, 1994).

59. Meiers, *Zu neuen Ufern?* 275–281.

60. Walter Jertz, "Einsätze der Luftwaffe über Bosnien," in *Von Kambodscha bis Kosovo: Auslandseinsätze der Bundeswehr*, ed. Peter Goebel (Frankfurt am Main/Bonn: Report Verlag, 2000), 140.

61. Meiers, *Zu neuen Ufern?* 283–284.

62. Roger Evers, "Transportflieger in humanitärem Auftrag," in *Von Kambodscha bis Kosovo*, 91–92.

63. Hans-Werner Jarosch, *Immer im Einsatz: 50 Jahre Luftwaffe* (Hamburg/Berlin/Bonn: Verlag E. S. Mittler & Sohn, 2005), 139.

64. Evers, "Transportflieger," 86, 100.

65. Meiers, *Zu neuen Ufern?* 291–292.

66. Ibid., 295–296.

67. Jarosch, *Immer im Einsatz*, 141.

68. Meiers, *Zu neuen Ufern?* 301.

69. Jarosch, *Immer im Einsatz*, 145; and Benjamin S. Lambeth, *NATO's Air War for Kosovo: A Strategic and Operational Assessment* (Santa Monica, CA: RAND, 2001), 110.

70. Meiers, *Zu neuen Ufern?* 308.

71. *Lessons Learned from Military Operations and Relief Efforts in Kosovo* (Washington, DC: Hearing of the Senate Committee on Armed Services, October 21, 1999).

72. Ivo H. Daalder and Michael E. O'Hanlon, *Winning Ugly: NATO's War to Save Kosovo* (Washington, DC: Brookings Institution Press, 2000), 198.

73. Ibid., 222; *Lessons Learned*, 14.

74. Daalder and O'Hanlon, *Winning Ugly*, 221.

75. Timothy Garden, "European Air Power," in *Air Power 21: Challenges for the New Century*, ed. Peter W. Gray (London: The Stationery Office, 2000), 114.

76. *NATO Handbook* (Brussels: NATO Office of Information and Press, 2001), 17, 44.

77. NATO, "The Partnership for Peace," http://www.nato.int, accessed November 27, 2009.

78. *NATO Handbook*, 72.

79. "The Alliance's Strategic Concept Approved by the Heads of State and Government Participating in the Meeting of the North Atlantic Council in Washington DC on April 23–24, 1999," NATO Press Release NAC-S(99)65 (April 24, 1999), paragraph 31.

80. NATO, "The NATO Response Force: At the Centre of NATO Transformation," http://www.nato.int, accessed February 28, 2007.

81. NATO, "The NATO Response Force: What Does This Mean in Practice?" http://www.nato.int, accessed February 28, 2007.

82. Meiers, *Zu neuen Ufern?* 143.

83. Jean-Yves Haine, *ESDP: An Overview* (Paris: European Union Institute for Security Studies), http://www.iss-eu.org, accessed May 14, 2005.

84. *Declaration on European Defence*, Franco-British summit, Saint Malo, December 4, 1998, paragraph 2.

85. Meiers, *Zu neuen Ufern?* 161.

86. This specific reality was underlined by a French Air Force officer with ample experience in the African theater at a NATO conference in 2008.

87. John E. Peters, Stuart E. Johnson, Nora Bensahel, Timothy Liston, and Traci Williams, *European Contributions to Operation Allied Force: Implications for Transatlantic Cooperation* (Santa Monica, CA: RAND, 2001), 52.

88. Lambeth, *NATO's Air War for Kosovo*, 33, 66.

89. Garden, "European Air Power," 144.

90. Peters et al., *European Contributions*, 20–24.

91. "Defence Capabilities Initiative," NATO Press Release NAC-S(99)69 (April 25, 1999), paragraph 1.

92. *Headline Goal 2010*, approved by General Affairs and External Relations Council on May 17, 2004, endorsed by the European Council of June 17–18, 2004, 1–3.

93. *Loi de programmation militaire 2003–2008: la politique de défense* (Paris: Journaux officiels, 2003), 28; and Jertz, "Unser Schwerpunkt ist der Einsatz," 22.

94. Henri-Pierre Grolleau, "French Precision: Rafale's Lethal Punch," *Air International* 4 (April 2009), 21–22, 25.

95. "Sixth C-17 Swoops into Britain," *Defence News* (July 3, 2008), http://www.mod.uk, accessed September 24, 2008.

96. Craig Hoyle, "Much-Delayed A400M Edges Closer to Flight," *Flight International* 5205 (September 8–14, 2009), 18.

97. Airbus Military, *A Brief History*, http://www.airbusmilitary.com, accessed December 21, 2009.

98. "NRF: The French Air Force in Control," 43.

99. Michael J. Gething and Bill Sweetman, "Air-to-Air Refuelling Provides a Force Multiplier for Expeditionary Warfare," *Jane's International Defence Review* 39 (February 2006), 45.

100. See *Défense et Sécurité nationale: le Livre blanc* (Paris: La Documentation française, June 2008), 228; and "A330-200 Future Strategic Tanker Aircraft (FSTA)—Multi-Role Tanker Transporter (MRTT), Europe," *Air Force Technology*, www.airforce-technology.com, accessed January 2, 2009.

101. EAG, "European Airlift Centre (EAC)," http://www.euroairgroup.org, accessed March 2, 2007.

102. MCCE, "Background of the MCCE," https://www.mcce-mil.com, accessed August 21, 2009.

103. Bundesministerium der Verteidigung, "Verlässlicher Zugriff auf 'fliegende Güterzüge'" (Leipzig, March 23, 2006), http://www.bmvg.de, accessed August 9, 2006.

104. NATO, "Strategic Airlift Capability: A Key Capability for the Alliance" (updated November 27, 2008), http://www.nato.int, accessed January 2, 2009; and NATO, "First C-17 Plane Welcomed at Papa Airbase" (July 27, 2009), http://www.nato.int, accessed September 7, 2009.

105. Jean-Laurent Nijean, "L'Armée de l'Air et l'Europe: gros plan—au coeur de l'Europe de la défense," *Air actualités* 605 (October 2007): 38; and Luftwaffe, "Startschuss für das Europäische Lufttransportkommando" February 25, 2010, http://www.luftwaffe.de.

106. Lutgert and de Winter, *Check the Horizon*, 511.

107. Interview at the Headquarters Royal Netherlands Air Force, The Hague, June 22, 2004.

108. Dirk Starink, commander in chief of the Royal Netherlands Air Force, briefing at the Shephard Air Power Conference 2005 (London, January 25–27, 2005).

109. B. A. C. Droste, "Decisive Air Power Private: The Role of the Royal Netherlands Air Force in the Kosovo Conflict," *NATO's Nations and Partners for Peace* 2 (1999): 129.

110. Lutgert and de Winter, *Check the Horizon*, 508–510.

111. See Michael E. Thompson, *Political and Military Components of Air Force Doctrine in the Federal Republic of Germany and Their Implications for NATO Defense Policy Analysis* (Santa Monica, CA: RAND, 1987), 22–23.

112. Jan Ahlgren, Anders Linnér, and Lars Wigert, *Gripen: The First Fourth Generation Fighter* (Sweden: Swedish Air Force, FMV, and Saab Aerospace, 2002), 52–53.

113. See Thompson, *Air Force Doctrine in the Federal Republic of Germany*, 22–23, 52.

114. Mason, *Air Power*, 249.

115. E-mail from John R. Kent, F-35 Lightning II Communications, August 3, 2009.

116. Paul Dreger, "JSF Partnership Takes Shape: A Review of the JSF Participation by Australia, Canada, Denmark, Israel, Italy, the Netherlands, Norway, Singapore, Turkey and the UK," *Military Technology* 4 (2003), 28–31.

117. "Norway Commits Funding to Gripen Development," *Gripen News* 1 (2007), 2.

118. Luca Peruzzi, "Italy Details Plans for its JSF Final Assembly Facility," *Flight International* 5167 (December 2–8, 2008): 18.

119. Anno Gravemaker, "Wiring Deal Seals Biggest Dutch F-35 Agreement," *Flight International* 5136 (April 29–May 5, 2008): 22.

120. Craig Hoyle, "Saab Challenges Norway's Rejection," *Flight International* 5169 (December 16, 2008–January 5, 2009): 15.

121. Craig Hoyle, "Gripen Revives Spat over Norwegian Fighter Choice," *Flight International* 5192 (June 9–15, 2009): 28.

122. Ingemar Dörfer, *Arms Deal: The Selling of the F-16* (New York: Praeger, 1983), 96.

123. Wagevoort, "Materiel Projects," 45.

124. Craig Hoyle, "Boeing Ups the Pace with Danish Super Hornet Offer," *Flight International* 5173 (January 27–February 2, 2009): 16.

125. *AP 3000 British Air and Space Power Doctrine*, 4th ed. (London: Air Staff, 2009), 39–40.

126. Shaun Gregory, *French Defence Policy into the Twenty-First Century* (London: Macmillan Press, 2000), 139.

127. *Projet de loi relatif à la programmation militaire pour les années 2009 à 2014 et portant sur diverses dispositions concernant la défense*, rapport annexé (Paris: Assemblée Nationale, October 29, 2008), 6–7.

128. *The Prinsjesdag Letter*, English Translation (The Hague: Ministry of Defence, 2003), 31.

129. *The Military Balance, 1990–1991* (London: published by Brassey's for IISS, 1990), 52.

130. *The Military Balance, 2009* (London: published by Routledge for IISS, 2009), 144.

131. Bartosz Glowacki and Grzegorz Sobczak, "Poland Eyes F-16 Maintenance Deal," *Flight International* 5152 (August 19–25, 2008): 19.

132. Bartosz Glowacki and Grzegorz Sobczak, "Poland Receives Its First of Five C-130E Hercules," *Flight International* 5182 (March 31–April 6, 2009): 20.

133. Bartosz Glowacki and Grzegorz Sobczak, "Poland to Add Advanced Jet Trainers, Tanker Transports," *Flight International* 5156 (September 16–22, 2008): 23.

134. Bartosz Glowacki and Grzegorz Sobczak, "Poland to Put F-16s through First NATO Manoeuvres," *Flight International* 5135 (April 22–28, 2008): 22.

135. "Transforming Poland's Military: A Focus on Western Concepts, Training, and Hardware," *The DISAM Journal* (Fall 2005): 17–18.

136. Craig Hoyle, "Baltic Exchange," *Flight International* 5191 (June 2–8, 2009): 33.

137. Ibid., 30.

138. Ibid., 31.

139. Dino Kucic, "Czech Air Force Outlines Gripen Use," *Flight International* 5143 (June 17–23, 2008): 23.

140. Hoyle, "Baltic Exchange," 32.

141. Ibid., 31–33.

142. Patrice Klein, "Bilan et enseignements des opérations Héraclès et Ammonite," in *L'Armée de l'Air: les Armées françaises à l'aube du XXIe siècle*, ed. Pierre Pascallon (Paris: L'Harmattan, 2003), 203–204; and Benjamin S. Lambeth, *American Carrier Air Power at the Dawn of a New Century* (Santa Monica, CA: RAND, 2005), 17, 19.

143. Frédéric Castel, "Sept mois de coopération exemplaire entre la France et les États-Unis," *Air actualités* 556 (November 2002): 30–32.

144. Stephan de Bruijn, "Rafales in Afghanistan," *Air Forces Monthly* (August 2008): 58–60.

145. Castel, "Sept mois de coopération," 30.

146. "Latest News of the Ministry of Defence," September 26, 2002, http://www.

defensie.nl, accessed June 14, 2004; and Eugénie Baldes, "Les Mirages français passent le relais," *Air actualités* 556 (November 2002): 34–36.

147. Memorandum of Understanding between the Minister of Defence of the Kingdom of Belgium, the Minister of Defence of the Kingdom of Denmark, the Minister of Defence of the Kingdom of the Netherlands, the Minister of Defence of the Kingdom of Norway and the Minister of State and National Defence of the Portuguese Republic concerning the Establishment of the European Participating Air Forces' Expeditionary Air Wing (Istanbul, June 28, 2004).

148. Henk Kamp, minister of defense, speech at the Royal Netherlands Association of Military Science (Nieuwspoort Press Centre, The Hague, March 1, 2004).

149. Jarosch, *Immer im Einsatz*, 161; and Luftwaffe, "'Professionelle Gelassenheit' bei den Immelmännern vor ihrem Auslandseinsatz" (Jagel, March 14, 2007), http://www.luftwaffe.de, accessed March 15, 2007.

150. Jacques Lanxade, "L'opération Turquoise," *Défense nationale* 2 (February 1995), 9–11.

151. United Nations, Peacekeeping Best Practices Unit, Military Division, *Operation Artemis: The Lessons of the Interim Emergency Multinational Force* (UN, October 2004), 3–4; and Marc Joulaud, *Avis présenté au nom de la commission de la défense nationale et des forces armées, sur le projet de loi de finances rectificative pour 2003*, no. 1267 (Paris: Assemblée Nationale, 2003): 20, http://www.assemblee-nationale.fr, accessed July 21, 2006.

152. United Nations, *Operation Artemis*, 12.

153. Gerrard Quille, "Battle Groups to Strengthen EU Military Crisis Management?" *European Security Review* 22 (ISIS Europe, April 2003), 1–2, http://www.forum-europe.com, accessed March 5, 2007.

154. *Headline Goal 2010*, 2.

155. Jan Joel Andersson, *Armed and Ready? The EU Battlegroup Concept and the Nordic Battlegroup* (Stockholm: Swedish Institute for European Policy Studies, March 2006), 39.

156. Jan Jorgensen, "Battle Group Gripens," *AirForces Monthly* (August 2007), 46.

157. Jean Laurent Nijean, "L'Armée de l'Air et l'Europe: opération EUFOR RDC," *Air actualités* 605 (October 2007): 34–35.

158. European Union, *EU Military Operation in Eastern Chad and North Eastern Central African Republic* (Brussels: EU Council Secretariat, updated March 2009), http://www.consilium.europa.eu, accessed September 5, 2009.

159. Lawrence Freedman, "Can the EU Develop an Effective Military Doctrine?" in *A European Way of War*, eds. Steven Everts, Lawrence Freedman, Charles Grant, François Heisbourg, Daniel Keohane, and Michael O'Hanlon (London: Centre for European Reform, 2004), 23.

AFTERWORD

1. William Mitchell, *Winged Defense: The Development and Possibilities of Modern Air Power, Economic and Military* (New York: Dover Publications, Inc., 1925), 3.

2. *World Air Transport Statistics* (*WATS*), 53d ed. (Montreal: International Air Transport Association, 2009), 14–29.

3. Ibid.

4. *The United States Strategic Bombing Surveys* (1945–1946; repr., Maxwell AFB, AL: Air University Press, 1987), 13.

5. National Museum of the U.S. Air Force, "Flight of the Question Mark," http://www.nationalmuseum.af.mil/factsheet.asp?id=743, accessed May 9, 2010.

6. Rebecca Grant, "The Vanishing Arsenal of Airpower," *Air Force Magazine* 93, no. 1 (2010): 46–50.

7. Ibid., 48–50.

8. Australian Government, Department of Defence, "Defending Australia in the Asia Pacific Century: Force 2030," http://www.defence.gov.au/whitepaper/docs/defence_white_paper_2009.pdf, accessed May 9, 2010.

9. Air Force Intelligence Analysis Agency (AFIAA), "Worldwide 5th Generation Programs" (Washington, DC: HQ USAF/A2, 2010).

10. Jane's, *Jane's World Air Forces*, http://www.janes.com/, accessed May 9, 2010.

11. *Threats to Air Dominance*, VHS (Washington, DC: AFIAA, 2008).

12. Ibid.

13. Ibid.

14. George Friedman, *The Next 100 Years: A Forecast for the 21st Century* (New York: Doubleday, 2009), and Colin Gray, "The 21st Century Security Environment and the Future of War," *Parameters* 38, no. 4 (2008): 14–26.

15. For this claim, the author cites the many examples available in Robin Higham and Stephen J. Harris, *Why Air Forces Fail: The Anatomy of Defeat* (Lexington: University Press of Kentucky, 2006).

16. For this claim, the author cites the spring 2010 HQ USAF figure of 168 personnel per remotely piloted aircraft combat air patrol, as contained in the author's presentation "Remotely Piloted Aircraft in the United States Air Force: The Way Ahead."

17. Eliot A. Cohen, "The Mystique of U.S. Air Power," *Foreign Affairs* 73, no. 1 (1994).

SELECTED BIBLIOGRAPHY

Ahlgren, Jan, Anders Linnér, and Lars Wigert. *Gripen: The First Fourth Generation Fighter.* Sweden: Swedish Air Force, FMV, and Saab Aerospace, 2002.

Air Force Studies Board. *Future Air Force Needs for Survivability.* Washington, DC: National Academies Press, 2006.

Allen, Kenneth W., Glenn Krumel, and Jonathan D. Pollack. *China's Air Force Enters the 21st Century.* Santa Monica, CA: RAND, 1995.

Andersson, Jan Joel. *Armed and Ready? The EU Battlegroup Concept and the Nordic Battlegroup.* Stockholm: Swedish Institute for European Policy Studies, March 2006.

Annual Report on the Military Power of the People's Republic of China, report to Congress by Secretary of Defense. Washington, DC: U.S. Government Printing Office, 2002.

AP 3000 British Air and Space Power Doctrine. 4th ed. London: Air Staff, 2009.

Appleman, Roy E. *South to the Naktong, North to the Yalu.* A volume in the *United States Army in the Korean War* series. Washington, DC: Office of the Chief of Military History, 1961.

Armitage, M. J., and R. A. Mason. *Air Power in the Nuclear Age.* London: Macmillan, 1984.

Arnold, Henry H. *Global Mission.* New York: Harper & Brothers, 1949.

Art, Robert J. *The TFX Decision: McNamara and the Military.* Boston, MA: Little, Brown, 1968.

Australian Bureau of Statistics. *Official Year Book of the Commonwealth of Australia.* Canberra, Australia: Australian Bureau of Statistics, 1919–1938.

Bacevich, Andrew J., et al. *American Military Policy in Small Wars: The Case of El Salvador.* Washington, DC: Pergamon-Brassey's, 1998.

Bailey, Bruce M. *"We See All": A History of the 55th Strategic Reconnaissance Wing, 1947–1967.* Privately printed, 55th ELINT Association, 1982.

Bald, Ralph D. *Air Force Participation in Joint Army–Air Force Training Exercises,*

1951–1954. No. 129 in the *USAF Historical Studies* series. Maxwell AFB, AL: Air University, USAF Historical Division, Research Studies Institute, 1957.

Barlow, Jeffrey G. *From Hot War to Cold: The U.S. Navy and National Security Affairs, 1945–1955*. Stanford, CA: Stanford University Press, 2009.

Beard, Edmund. *Developing the ICBM: A Study in Bureaucratic Politics*. New York: Columbia University Press, 1976.

Beattie, Doug. *An Ordinary Soldier*. London: Simon & Schuster, 2009.

Beaufre, A. *NATO and Europe*. London: Faber and Faber, 1967.

Biddle, T. D. *Rhetoric and Reality in Air Warfare*. Princeton, NJ: Princeton University Press, 2002.

Bishop, Chris, ed. *The Encyclopaedia of 20th Century Air Warfare*. Leicester, UK: Silverdale Books, 2001.

Black, George. *Garrison Guatemala*. New York: Monthly Review Press, 1984.

Bobbitt, Philip. *The Shield of Achilles: War, Peace, and the Course of History*. New York: Random House, 2003.

Bowden, Mark. *Black Hawk Down: A Story of Modern War*. New York: Atlantic Monthly Press, 1999.

Bowie, Christopher, Fred Frostic, Kevin Lewis, John Lund, David Ochmanek, and Philip Propper. *The New Calculus: Analyzing Airpower's Changing Role in Joint Theater Campaigns*. Santa Monica, CA: RAND, 1993.

Boyd, Alexander. *The Soviet Air Force*. London: Macdonald and Jane's, 1977.

Boyle, Andrew. *Trenchard*. London: Collins, 1962.

Boyne, Walter J. *The Influence of Air Power upon History*. Gretna: Pelican Publishing Company, 2003.

Brady, Halden. *Pershing's Mission into Mexico*. El Paso, TX: Texas Western College Press, 1966.

Braybrook, Roy. *Battle for the Falklands (3) Air Forces*. London: Osprey Publishing, 1982.

Brooks, Peter W. *The Modern Airliner: Its Origins and Development*. London: Putnam, 1961.

Broughton, Jack. *Rupert Red Two: A Fighter Pilot's Life*. St. Paul, MN: Zenith Press, 2007.

Brown, Archie. *The Gorbachev Factor*. New York: Oxford University Press, 1996.

Budiansky, Stephen. *Air Power: From Kitty Hawk to Gulf War II*. New York: Viking, 2003.

Byman, Daniel, and Roger Cliff. *China's Arms Sales: Motivations and Implications*. Santa Monica, CA: RAND, 1999.

Calvert, Peter. *Mexico*. New York: Praeger, 1973.

Cardwell, Thomas A. III. *Airland Combat: An Organization for Joint Warfare*. Maxwell AFB, AL: Air University Press, December 1992.

Castellanos, Miguel, and Courtney Prisk, eds. *The Commandante Speaks: Memoirs of an El Salvadoran Guerrilla Leader*. Boulder, CO: Westview Press, 1991.

Chakravorty, B. C. *History of the Indo-Pak War, 1965*. New Delhi: History Division, Ministry of Defence, 1992.

Chanoff, David, and Doan Van Toai. *Portrait of the Enemy*. New York: Random House, 1986.

Chant, Christopher. *Air Forces of the World.* Edison, NJ: Chartwell Books, 1983.

———. *Super Étendard: Super Profile.* Somerset, UK: Winchmore Publishing, 1983.

Chawla, Shalini. *Pakistan's Military and Its Strategy.* New Delhi: KW Publishers, 2009.

Chen, King. *China's War with Vietnam, 1979.* Stanford, CA: Hoover Institution Press, 1987.

Christienne, Charles, and Pierre Lissarrague. *A History of French Military Aviation.* Washington, DC: Smithsonian Institution Press, 1986.

Churchill, W. S. *The World Crisis: The Aftermath.* London: T. Butterworth, 1929.

Cichowski, Kurt A. *Aerospace Doctrine Matures through a Storm: An Analysis of the New AFM 1-1.* Maxwell AFB, AL: School of Advance Airpower, Air University, 1992.

Clark, Wesley K. *Waging Modern War: Bosnia, Kosovo, and the Future of Combat.* Oxford: PublicAffairs, 2001.

Clarke, Jeffrey J. *Advice and Support: The Final Years, 1965–1973.* Washington, DC: U.S. Army Center of Military History, 1988.

Clendenen, Clarence. *Blood on the Border: The United States and the Mexican Irregulars.* New York: Macmillan, 1969.

Clodfelter, Mark. *The Limits of Air Power: The American Bombing of North Vietnam.* New York: The Free Press, 1989.

Cohen, Eliot, Michael Eisenstadt, and Andrew Bacevich. *Knives, Tanks, and Missiles: Israel's Security Revolution.* Washington, DC: The Washington Institute for Near East Policy, 1998.

Cohen, Stewart. *The IDF and Israeli Society: A Renewed Analysis.* Tel Aviv: BESA Center for Strategic Studies, Bar Ilan University, 2001.

Condit, Kenneth W. *The Joint Chiefs of Staff and National Policy, 1955–56,* vol. 6 of *History of the Joint Chiefs of Staff.* Washington, DC: Historical Office, Joint Staff, 1992.

Cordesman, Anthony H. *The Lessons and Non-Lessons of the Air and Missile War in Kosovo.* Rev. ed. Washington, DC: Center for Strategic and International Studies, July 20, 1999.

Correll, John T. *The Air Force and the Cold War.* Arlington, VA: Air Force Association, September 2005.

———. *In the Wake of the QDR: The Quadrennial Defense Review and Its Consequences.* Arlington, VA: Air Force Association, September 2006.

Corum, James, and Wray Johnson. *Airpower in Small Wars.* Lawrence, KS: University Press of Kansas, 2003.

Coulam, Robert F. *Illusions of Choice: The F-111 and the Problem of Weapons Acquisition Reform.* Princeton, NJ: Princeton University Press, 1977.

Cox, Sebastian, and Peter Gray, eds. *Air Power History: Turning Points from Kitty Hawk to Kosovo.* London: Frank Cass, 2002.

Crouch, Tom D. *The Eagle Aloft.* Washington, DC: Smithsonian Institution Press, 1983.

Crowell, B., and R. F. Wilson. *The Armies of Industry,* vol. 1. New Haven, CT: Yale University Press, 1921.

Daalder, Ivo H., and Michael E. O'Hanlon. *Winning Ugly: NATO's War to Save Kosovo*. Washington, DC: Brookings Institution Press, 2000.

Dale, Catherine. *Operation Iraqi Freedom: Strategies, Approaches, Results, and Issues for Congress*. Washington, DC: Congressional Research Service, March 28, 2008.

Dasgupta, C. *War and Diplomacy in Kashmir, 1947–48*. New Delhi: Sage Publications, 2002.

Davis, Richard G. *The 31 Initiatives: A Study in Air Force–Army Cooperation*. Washington, DC: Office of Air Force History, 1987.

———. *Anatomy of a Reform: The Expeditionary Aerospace Force*. Washington, DC: Air Force History and Museums Program, 2003.

Dawson, Raymond S. *The Decision to Aid Russia*. Chapel Hill: University of North Carolina Press, 1959.

Day, Dwayne A. *Lightning Rod: A History of the Air Force Chief Scientist's Office*. Washington, DC: Office of the Air Force Chief Scientist, 2000.

Dayan, Moshe. *Diary of the Sinai Campaign*. New York: Schocken Books, 1967.

Dean, M. *The Royal Air Force and Two World Wars*. London: Cassell, 1979.

DeBlois, Bruce M., ed. *Beyond the Paths of Heaven: The Emergence of Space Power Thought*. Maxwell AFB, AL: Air University Press, September 1999.

Defence: Outline of Future Policy. London: Her Majesty's Stationery Office, April 1957.

Defending Australia in the Asia Pacific Century: Force 2030. Canberra: Australian Government, Department of Defense, 2009.

Deptula, David A. *Firing for Effect: Change in the Nature of Warfare*. Arlington, VA: Aerospace Education Foundation, August 1995.

Dixon, Marlene, and Susanne Jones, eds. *Revolution and Intervention in Central America*. San Francisco, CA: Synthesis Publications, 1983.

Doglione, John A., et al. *Air Power and the 1972 Spring Invasion*. Washington, DC: GPO in association with Air University, 1976.

Donnelly, Christopher. *Red Banner: The Soviet Military System in Peace and War*. Coulsdon, UK: Jane's Information Group, 1988.

Doolittle, James, with Carroll V. Glines. *I Could Never Be So Lucky Again: An Autobiography by General James H. "Jimmy" Doolittle*. New York: Bantam Books, 1991.

Dörfer, Ingemar. *Arms Deal: The Selling of the F-16*. New York: Praeger, 1983.

Douhet, Giulio. *The Command of the Air*. Translated by Dino Ferrari. Washington, DC: Air Force History and Museums Program, 1998.

Dubik, James M. *Land Warfare in the 21st Century*. Carlisle Barracks, PA: U.S. Army War College Strategic Studies Institute, February 1993.

Dutt, Subimal. *With Nehru in the Foreign Office*. Calcutta: Minerva Associates, 1977.

Eisenhower, John S. D. *Strictly Personal*. Garden City, NY: Doubleday, 1974.

El Shazly, Saad. *The Crossing of the Suez*. San Francisco, CA: American Mideast Research, 1980.

Emme, Eugene M., ed. *The Impact of Air Power: National Security and World Politics*. Princeton, NJ: D. Van Nostrand, 1959.

English, Adrian. *Armed Forces of Latin America: Their Histories, Development, Present Strength, and Military Strength*. London: Jane's, 1984.

Eschmann, Karl J. *Linebacker: The Untold Story of the Air Raids over North Vietnam.* New York: Ivy Books, 1989.

Estigarribia, Jose Felix. *The Epic of the Chaco: Marshal Estigarribia's Memoirs of the Chaco War, 1932–1935.* Edited by Max Ynsfran. New York: Greenwood Press, 1950.

Fall, Bernard. *Street without Joy.* New York: Schocken Books, 1972.

Farcau, Bruce. *The Chaco War.* Westport, CT: Praeger, 1996.

Faulkenberry, Barbara J. *"Global Reach–Global Power": Air Force Strategic Vision, Past and Future.* Maxwell AFB, AL: Air University, June 1995.

Fedorov, E. K. *The Red Army.* London: Cobbett, 1944.

Field, James A. *History of United States Naval Operations in Korea.* Washington, DC: USN, 1962.

Finletter, Thomas K., et al. *Survival in the Air Age: A Report by the President's Air Policy Commission.* Washington, DC: President's Air Policy Commission, January 1, 1948.

Finney, Robert T. *History of the Air Corps Tactical School, 1920–1940.* Washington, DC: Center for Air Force History, 1992.

Fish, Joe, and Cristina Sganga. *El Salvador: Testament of Terror.* New York: Olive Branch Press, 1988.

Flintham, Victor. *Air Wars and Aircraft.* New York: Facts on File, 1990.

Foulois, Benjamin. *Report of the Operations of the First Aero Squadron, Signal Corps, with Punitive Expedition, USA for Period March 15 to August 15, 1916.* Maxwell AFB, AL: Historical Research Agency, n.d.

Freedman, Lawrence, and Efraim Karsh. *The Gulf Conflict, 1990–1991: Diplomacy and War in the New World Order.* London: Faber and Faber, 1994.

Fricker, John. *The Battle for Pakistan: The Air War of 1965.* London: Ian Allen Ltd., 1979.

Fritz, Oliver, Renee Lajoie, and Barry Bleichman. *A 21st Century Air Force: U.S. Global Engagement and the Use of Air and Space Power.* Washington, DC: DFI International, July 1997.

Frostic, Fred. *Air Campaign against the Iraqi Army in the Kuwaiti Theater of Operations,* Report MR-357-AF. Santa Monica, CA: RAND, 1994.

Futrell, Robert Frank. *The United States Air Force in Korea, 1950–1953.* New York: Duell, Sloan, and Pearce, 1961.

———, et al. *Aces and Aerial Victories: The United States Air Force in Southeast Asia, 1965–1973.* Washington, DC: Office of Air Force History, 1976.

———. *Ideas, Concepts, Doctrine: Basic Thinking in the United States Air Force.* Vol. 1: *1907–1960.* Maxwell AFB, AL: Air University Press, 1989.

Galland, Adolf. *The First and the Last.* Translated by Mervin Savill. New York: Henry Holt and Co., 1954.

Gantz, Kenneth F. *The United States Air Force Report on the Ballistic Missile: Its Technology, Logistics, and Strategy.* Garden City, NY: Doubleday, 1958.

Gauhar, Altaf. *Ayub Khan: Pakistan's First Military Ruler.* Lahore: Sang-e-Meel Publications, 1993.

Gettleman, Marvin, et al., eds. *El Salvador: Central America in the New Cold War.* New York: Grove Press, 1986.

Giap, Vo Nguyen. *Dien Bien Phu: Rendezvous with History*. Hanoi: The Gioi Publishers, 2004.

Gibbs-Smith, Charles H. *The Aeroplane: An Historical Survey*. London: Her Majesty's Stationery Office, 1960.

———. *Clément Ader: His Flight-Claims and His Place in History*. London: Her Majesty's Stationery Office, 1968.

Gilbert, Martin. *Winston S. Churchill*. Vol. 4. London: Heinemann, 1975.

Gilster, Herman L. *The Air War in Southeast Asia: Case Studies of Selected Campaigns*. Maxwell AFB, AL: Air University Press, October 1993.

Global Reach–Global Power: The Evolving Air Force Contribution to National Security. Washington, DC: Office of the Secretary of the Air Force, December 1992.

Golovin, N. N. *The Russian Campaign of 1914*. Fort Leavenworth, KS: The Command and General Staff School Press, 1933.

Gorn, Michael H. *Harnessing the Genie: Science and Technology Forecasting for the Air Force, 1944–1986*. Washington, DC: Office of Air Force History, 1988.

Gouré, Daniel, and Christopher M. Szara, eds. *Air and Space Power in the New Millennium*. Washington, DC: Center for Strategic and International Studies, 1997.

Gowing, M. *Independence and Deterrence*, vol. 1. London: Macmillan, 1974.

Grant, Rebecca. *The Kosovo Campaign: Aerospace Power Made It Work*. Arlington, VA: Air Force Association, September 1999.

———. *The Afghan Air War*. Arlington, VA: Air Force Association, September 2002.

———. *Gulf War II: Air and Space Power Led the Way*. Arlington, VA: Air Force Association, September 2003.

Greer, Thomas H. *The Development of Air Doctrine in the Army Air Arm, 1917–1941*, USAF Historical Study 89. 1955. Reprint, Maxwell AFB, AL: USAF Historical Division, Research Studies Institute, Air University, 1985.

Gregory, Shaun. *French Defence Policy into the Twenty-First Century*. London: Macmillan Press, 2000.

Gunn, William A. *A Study of the Effectiveness of Air Support Operations in Korea*. FEC Project no. 2. Tokyo: Operations Research Office, General Headquarters, Far East Command, January 1952.

Gupta, S. C. *History of the Indian Air Force, 1933–45*. Delhi: Combined Inter-Services Historical Section, India & Pakistan, 1961.

Hagedorn, Dan. *Central American and Caribbean Air Forces*. Tonbridge, UK: Air Britain Publications, 1993.

———, and Luis Sapienza. *Aircraft of the Chaco War*. Atglen, PA: Schiffer, 1997.

Halder, Franz. *The Halder Diaries*, vol. 7. Washington, DC: The Infantry Journal Press, 1950.

Hall, David Ian. *Strategy for Victory: The Development of British Tactical Air Power, 1919–1943*. Westport, CT: Praeger Security International, 2008.

Hallion, Richard P. *The Evolution of Commonality in Fighter and Attack Aircraft Development and Usage*. Edwards AFB, CA: Air Force Flight Test Center History Office, November 1985.

———. *The Naval Air War in Korea*. Baltimore, MD: The Nautical and Aviation Publishing Company of America, 1986.

————, ed. *Air Power Confronts an Unstable World.* Washington, DC: Brassey's, 1997.

————. *Strike from the Sky: History of Battlefield Air Attack, 1911–1945.* Washington, DC: Smithsonian Institution Press, 1999.

Hammond, Grant T. *Paths to Extinction: The U.S. Air Force in 2025,* vol. 4 of the *Air Force 2025* study. Maxwell AFB, AL: Air University, August 1996.

Hansell, Haywood S., Jr. *The Air Plan That Defeated Hitler.* Atlanta, GA: Higgins-McArthur/Longino and Porter, 1972.

Harel, Amos, and Avi Issacharof. *34 Days: Israel, Hezbollah, and the War in Lebanon.* New York: Palgrave Macmillan, 2008.

Harik, Judith Palmer. *Hezbollah: The Changing Face of Terrorism.* London: I.B. Tauris, 2005.

Haydon, Frederick Stansbury. *Aeronautics in the Union and Confederate Armies.* Baltimore, MD: Johns Hopkins University Press, 1941.

Heinl, Robert, Jr. *Soldiers of the Sea: The United States Marine Corps, 1775–1962.* Annapolis, MD: Naval Institute Press, 1962.

————. *Victory at High Tide: The Inchon-Seoul Campaign.* Philadelphia, PA: J. B. Lippincott Co., 1968.

Hennessy, Juliette A. *The United States Army Air Arm, April 1861 to April 1927.* Washington, DC: Office of Air Force History, 1985.

Herbrink, André. *Bundeswehr Missions Out-of-Area.* Oslo: The Norwegian Institute of International Affairs, 1997.

Herzog, Chaim. *The War of Atonement: The Inside Story of the Yom Kippur War.* London: Greenhill Books/Lionel Leventhal, 2003.

Heseltine, Michael. *Life in the Jungle: My Autobiography.* London: Hodder and Stoughton, 2000.

Higham, Robin. *Air Power: A Concise History.* New York: St. Martin's Press, 1972.

————, and Jacob W. Kipp, eds. *Soviet Aviation and Air Power: A Historical View.* London: Brassey's Publishers Ltd., 1978.

Hill, Lewis D., Doris Cook, and Aron Pinker. *Gulf War Air Power Survey.* Vol. 5: *A Statistical Compendium and Chronology.* Washington, DC: U.S. Government Printing Office, 1993.

Hiranandani, G. M. *Transition to Triumph: History of the Indian Navy, 1965–1975.* New Delhi: Director Personnel Services, NHQ, 2000.

Horne, A. *To Lose a Battle: France 1940.* London: Macmillan, 1969.

Hosmer, Stephen T. *The Conflict over Kosovo: Why Milosevic Decided to Settle When He Did.* Santa Monica, CA: RAND, 2001.

Hough, R., and D. Richards. *The Battle of Britain: The Jubilee History.* London: Hodder and Staughton, 1989.

HQ U.S. Air Force. *Global Engagement: A Vision for the 21st Century Air Force.* Washington, DC: HQ USAF, November 21, 1996.

Huertas, Salvador Mafe, and Jesus Romero Briasco. *Argentine Air Forces in the Falklands Conflict.* London: Arms and Armour Press, 1987.

Hurley, Alfred F. *Billy Mitchell: Crusader for Air Power.* Rev. ed. Bloomington, IN: Indiana University Press, 1975.

Jamieson, Perry D. *Lucrative Targets: The U.S. Air Force in the Kuwaiti Theater of Operations.* Washington, DC: Air Force History and Museums Program, 2001.

Johnson, Dana J., Christopher J. Bowie, and Robert P. Haffa. *Triad, Dyad, Monad? Shaping the U.S. Nuclear Force for the Future*, Mitchell Paper 5. Arlington, VA: Air Force Association, December 2009.

Johnson, Herbert A. *Wingless Eagle: U.S. Army Aviation through World War I*. Chapel Hill: University of North Carolina Press, 2001.

Johnson, Stephen B. *The United States Air Force and the Culture of Innovation, 1945–1965*. Washington, DC: Air Force History and Museums Program, 2002.

Jonas, Susanne. *The Battle for Guatemala: Rebels, Death Squads, and U.S. Power*. Boulder, CO: Westview Press, 1991.

Jones, H. A. *The War in the Air*. Oxford: Clarendon Press, 1934.

Joubert de la Ferté, Sir Philip. *The Third Service*. London: Thames and Hudson, 1955.

Kainikara, Sanu. *Red Air: Politics in Russian Air Power*. Boca Raton, FL: Universal Publishers, 2007.

———. *Papers on Air Power*. Canberra: Air Power Development Centre, 2007.

Katz, Friedrich. *The Life and Times of Pancho Villa*. Stanford, CA: Stanford University Press, 1998.

Keaney, Thomas A., and Eliot A. Cohen. *Gulf War Air Power Survey Summary Report*. Washington, DC: GPO, 1993.

Kelley, Jay W., et al. *Spacecast 2020*. Vol. 1. Maxwell AFB, AL: Air University Press, June 1994.

Kent, Glenn A., with David Ochmanek, Michael Spirtas, and Bruce R. Pirnie. *Thinking about America's Defense: An Analytical Memoir*. Santa Monica, CA: RAND, 2008.

Khan, Akbar. *Raiders in Kashmir*. Delhi: Army Publishers, n.d.

Khan, Fazal Muqueem. *Pakistan's Crisis of Leadership*. New Delhi: Alpha & Alpha, 1973.

Khan, Gul Hassan. *Memoirs of Lt. Gen. Gul Hassan*. Karachi: Oxford University Press, 1993.

Khrushchev, Nikita. *Khrushchev Remembers: The Last Testament*. New York: Bantam Books, 1976.

Kilmarx, Robert A. *A History of Soviet Air Power*. London: Faber and Faber, 1962.

Kinnard, Douglas. *The War Managers*. Hanover, NH: University Press of New England, 1977.

Kissinger, Henry. *The White House Years*. London: Weidenfeld & Nicholson, 1971.

Knaack, Marcelle Size. *Post–World War II Fighters, 1945–1973*. Vol. 1 of *Encyclopedia of U.S. Air Force Aircraft and Missile Systems*. Washington, DC: Office of Air Force History, 1978.

———. *Post–World War II Bombers, 1945–1973*. Vol. 2 of *Encyclopedia of U.S. Air Force Aircraft and Missile Systems*. Washington, DC: Office of Air Force History, 1988.

Knott, Richard C. *The American Flying Boat*. Annapolis, MD: Naval Institute Press, 1979.

Knox, MacGregor, and Williamson Murray, eds. *The Dynamics of Military Revolution, 1300–2050*. Cambridge: Cambridge University Press, 2001.

Kochan, Lionel. *Russia in Revolution*. London: Granada, 1970.

————, and Richard Abraham. *The Making of Modern Russia.* 2d ed. Hammondsworth, UK: Penguin Books Ltd., 1983.

Koenig, William, and Peter Scofield. *Soviet Military Power.* Greenwich, CT: Bison Books, 1983.

Komons, Nick A. *Science and the Air Force: A History of the Air Force Office of Scientific Research.* Arlington, VA: Historical Division, USAF Office of Aerospace Research, 1966.

————. *Bonfires to Beacons: Federal Civil Aviation Policy under the Air Commerce Act, 1926–1938.* Washington, DC: GPO, 1978.

Korbel, Josef. *Danger in Kashmir.* Princeton, NJ: Princeton University Press, 1954.

Kugler, Richard L., Michael Baranick, and Hans Binnendijk. *Operation Anaconda: Lessons for Joint Operations.* Washington, DC: Center for Technology and National Security Policy, National Defense University, 2009.

Kumar, Bharat. *An Incredible War: IAF in Kashmir War, 1947–1948.* New Delhi: Knowledge World, 2007.

Kux, Dennis. *The United States and Pakistan, 1947–2000.* Karachi: Oxford University Press, 2001.

Lal, P. C. *My Years with the IAF.* New Delhi: Lancer International, 1986.

Lambert, Andrew P. N. *The Psychology of Air Power.* London: Royal United Services Institute for Defence Studies, 1994.

————. *The Psychology of Air Power.* Whitehall Paper Series. London: Royal United Services Institute for Defence Studies, 1995.

Lambeth, Benjamin S. *NATO's Air War for Kosovo: A Strategic and Operational Assessment.* Santa Monica, CA: RAND, 1998.

————. *The Transformation of American Air Power.* Ithaca, NY: Cornell University Press, 2000.

————. *NATO's Air War for Kosovo: A Strategic and Operational Assessment.* Santa Monica, CA: RAND, 2001.

————. *American Carrier Air Power at the Dawn of a New Century.* Santa Monica, CA: RAND, 2005.

————. *Air Power against Terror: America's Conduct of Operation Enduring Freedom.* Santa Monica, CA: RAND, 2005.

Lasby, Clarence G. *Project Paperclip: German Scientists and the Cold War.* New York: Athenaeum, 1971.

Lawson, Robert L. *The History of U.S. Naval Air Power.* New York: Military Press, 1985.

Leary, William M. *Anything, Anywhere, Anytime: Combat Cargo in the Korean War.* Washington, DC: Air Force History and Museums Program, 2000.

Lee, Asher. *The Soviet Air Force.* London: Duckworth, 1952.

Lenin, V. I. *Works.* Vol. 33, *1941–51.* Moscow: State Political Publishing House, 1951.

Lessons Learned from Military Operations and Relief Efforts in Kosovo. Washington, DC: Hearing of the Senate Committee on Armed Services, October 21, 1999.

Lynch, Kristin F., John G. Drew, Robert S. Tripp, and C. Robert Roll, Jr. *Supporting Air and Space Expeditionary Forces: Lessons from Operation Iraqi Freedom,* Report MG-193. Santa Monica, CA: RAND, 2005.

Makovsky, David, and Jeffrey White. *Lessons and Implications of the Israel-Hizballah War: A Preliminary Assessment.* Policy focus no. 60. Washington, DC: The Washington Institute for Near East Policy, October 2006.

Malik, V. P. *Kargil: From Surprise to Victory.* New Delhi: HarperCollins, 2006.

Mann, David K. *The 1972 Invasion of Military Region I: Fall of Quang Tri and Defense of Hue.* Hickam AFB, HI: HQ Pacific Air Forces, Directorate of Operations Analysis, CHECO/Corona Harvest Division, March 15, 1973.

Manwaring, Max, and Courtney Prisk, eds. *El Salvador at War: An Oral History.* Washington, DC: National Defense University Press, 1988.

Manz, Beatriz. *Refugees of a Hidden War.* Albany: State University of New York Press, 1988.

Mark, Eduard. *Aerial Interdiction: Air Power and the Land Battle in Three American Wars.* Washington, DC: Center for Air Force History, 1994.

———. *Defending the West: The United States Air Force and European Security, 1946–1998.* Washington, DC: Air Force History and Museums Program, 1998.

Mason, Herbert. *The Great Pursuit.* New York: Random House, 1970.

Mason, R. A. *Air Power: A Centennial Appraisal.* London: Brassey's, 1994.

Matinuddin, Kamal. *The Tragedy of Errors: East Pakistan Crisis, 1968–1971.* Lahore: Wajidalis, 1994.

Matloff, Maurice. *Strategic Planning for Coalition Warfare, 1943–1944.* Washington, DC: U.S. Army Center of Military History, 1990.

May, Ernest, John Steinbruner, and Thomas Wolfe. *History of Strategic Arms Competition, 1945–1972.* Edited by Alfred Goldberg. Washington, DC: Historical Office, Office of the Secretary of Defense, March 1981, declassified with deletions, December 1990.

McCall, Gene H., and John A. Corder, eds. *New World Vistas: Air and Space Power for the 21st Century.* 14 vols. Washington, DC: Air Force Scientific Advisory Board, 1996.

McCarthy, James R., and George B. Allison. *Linebacker II: A View from the Rock.* Edited by Col. Robert E. Rayfield. Washington, DC: GPO in association with Air University, 1979.

McClendon, R. Earl. *Unification of the Armed Forces: Administrative and Legislative Developments, 1945–1949.* Maxwell AFB, AL: Air University, USAF Historical Division, Research Studies Institute, 1952.

McCullough, David. *Truman.* New York: Simon & Schuster, 1992.

McFarland, Keith D., and David L. Roll. *Louis Johnson and the Arming of America: The Roosevelt and Truman Years.* Bloomington: Indiana University Press, 2005.

McFarland, Stephen L. *America's Pursuit of Precision Bombing, 1910–1945.* Tuscaloosa: The University of Alabama Press, 1995.

McMaster, H. R. *Dereliction of Duty: Lyndon Johnson, Robert McNamara, the Joint Chiefs of Staff, and the Lies that Led to Vietnam.* New York: HarperCollins, 1997.

McNamara, Robert. *The Essence of Security: Reflections in Office.* New York: Harper & Row, 1968.

———, with Brian Van De Mark. *In Retrospect: The Tragedy and Lessons of Vietnam.* New York: Times Books, 1995.

McPeak, Merrill A. *Selected Works, 1990–1994.* Maxwell AFB, AL: Air University Press, August 1995.

Meid, Pat, and James M. Yingling. *U.S. Marine Operations in Korea, 1950–1953.* Washington, DC: USMC, 1972.

Meilinger, Phillip S. *Ten Propositions on Air Power.* Washington, DC: Air Force History and Museums Program, 1995.

Melhorn, Charles M. *Two-Block Fox: The Rise of the Aircraft Carrier, 1911–1929.* Annapolis, MD: Naval Institute Press, 1974.

Mets, David R. *The Quest for a Surgical Strike: The USAF and Laser Guided Bombs.* Eglin AFB, CA: Air Force Systems Command Armament Division, 1987.

———. *The Air Campaign: John Warden and the Classical Airpower Theorists.* Maxwell AFB, AL: Air University Press, 1999.

Miller, Robert. *Mexico: A History.* Norman: University of Oklahoma Press, 1985.

Miller, Ronald, and David Sawers. *The Technical Development of Modern Aviation.* New York: Praeger, 1970.

Millis, Walter, ed., in collaboration with E. S. Duffield. *The Forrestal Diaries.* New York: Viking Press, 1951.

Mitchell, William. *Our Air Force.* New York: E. P. Dutton & Company, 1921.

Modley, Rudolf, and Thomas J. Cawley, eds. *Aviation Facts and Figures, 1953.* Washington, DC: Lincoln Press, Inc., for the Aircraft Industries Association of America, Inc., 1953.

Momyer, William W. *The Vietnamese Air Force, 1951–1975.* Monograph 4, USAF Southeast Asia Monograph Series. Washington, DC: Office of Air Force History, 1985.

Monney, Chase C. *Organization of Military Aeronautics, 1935–1945 (Executive, Congressional, and War Department Action).* No. 46 of the *Army Air Forces Historical Studies* series. Washington, DC: HQ AAF Historical Office, April 1946.

———, and Martha E. Layman. *Organization of Military Aeronautics, 1907–1935 (Congressional and War Department Action).* No. 25 of the *Army Air Forces Historical Studies* series. Washington, DC: HQ AAF Assistant Chief of Air Staff/Intelligence, Historical Division, December 1944.

Montgomery, Tommie Sue. *Revolution in El Salvador: Origins and Evolution.* Boulder, CO: Westview Press, 1982.

Moody, Walton S. *Building a Strategic Air Force.* Washington, DC: Air Force History and Museums Program, 1996.

———. *Steadfast and Courageous: FEAF Bomber Command and the Air War in Korea, 1950–1953.* Washington, DC: Air Force History and Museums Program, 2000.

Morrow, J. H., Jr. *The Great War in the Air.* Washington, DC: Smithsonian Institution Press, 1993.

Myers, Gary P., ed. *Executive Summary of the Missiles in Cuba, 1962: The Role of SAC Intelligence.* Offutt AFB, NE: Strategic Joint Intelligence Center, 1992.

Nardulli, Bruce R., Walter L. Perry, Bruce Pirnie, John Gordon IV, and John G. McGinn. *Disjointed War: Military Operations in Kosovo, 1999,* Report MR-1406. Santa Monica, CA: RAND, 2002.

NATO Handbook. Brussels: NATO Office of Information and Press, 2001.

Neufeld, Jacob. *Ballistic Missiles in the United States Air Force, 1945–1960.* Washington, DC: Office of Air Force History, 1990.

———, George M. Watson, Jr., and David Chenoweth, eds. *Technology and the Air Force: A Retrospective Assessment.* Washington, DC: Air Force History and Museums Program, 1997.

Nichols, John B., and Barrett Tillman. *On Yankee Station: The Naval Air War over Vietnam.* Annapolis, MD: Naval Institute Press, 1987.

Nordeen, Lon. *Air Warfare in the Missile Age.* Washington, DC: Smithsonian Institution Press, 1985.

North, Liisa. *Bitter Grounds: Roots of Revolt in El Salvador.* Toronto: Between the Lines, 1982.

Norton, Augustus Richard. *Hezbollah: A Short History.* Princeton, NJ: Princeton University Press, 2007.

Nyrop, Richard, ed. *Guatemala: A Country Study.* Washington, DC: U.S. Army, 1984.

O'Dowd, Edward C. *Chinese Military Strategy in the Third Indochina War: The Last Maoist War.* New York: Routledge, 2007.

Odom, William E. *The Collapse of the Soviet Military.* New Haven, CT: Yale University Press, 1998.

Olsen, John Andreas, ed. *A History of Air Warfare.* Washington, DC: Potomac Books, 2010.

Overy, Richard J. *The Air War, 1939–1945.* London: Papermac, 1980.

Owen, Robert C., ed. *Deliberate Force: A Case Study in Effective Air Campaigning.* Maxwell AFB, AL: Air University Press, January 2000.

Padmanabhan, S. *A General Speaks.* New Delhi: Manas Publications, 2005.

Palit, D. K. *War in the High Himalayas: The Indian Army in Crisis, 1962.* New Delhi: Lancer International, 1991.

Pape, Robert A. *Bombing to Win: Air Power and Coercion in War.* Ithaca, NY: Cornell University Press, 1996.

Payeras, Mario. *Days of the Jungle: The Testimony of a Guatemalan Guerrillero.* New York and Havana: Monthly Review Press, 1983.

Peattie, Mark R. *Sunburst: The Rise of Japanese Naval Air Power, 1909–1941.* Annapolis, MD: Naval Institute Press, 2001.

Penkovsky, Olog. *The Penkovsky Papers.* Garden City, NY: Doubleday, 1965.

Perera, Victor. *Unfinished Conquest: The Guatemalan Tragedy.* Berkeley: University of California Press, 1993.

Peters, F. Whitten, and Michael E. Ryan. *Vision 2020.* Washington, DC: HQ Air Force, 2000.

Peters, John E., Stuart E. Johnson, Nora Bensahel, Timothy Liston, and Traci Williams. *European Contributions to Operation Allied Force: Implications for Transatlantic Cooperation.* Santa Monica, CA: RAND, 2001.

Pipes, Richard. *Communism: A Brief History.* London: Weidenfield and Nicholson, 2001.

Polmar, Norman, and John D. Gresham. *Defcon-2: Standing on the Brink of Nuclear War during the Cuban Missile Crisis.* Hoboken, NJ: John Wiley & Sons, 2006.

———, et al. *Aircraft Carriers: A History of Carrier Aviation and Its Influence on World Events.* Vol. 2: *1946–2006.* Washington, DC: Potomac Books, 2008.

Probert, Henry. *Bomber Harris: His Life and Times*. London: Greenhill, 2001.

Quester, George H. *Deterrence before Hiroshima*. New Brunswick, NJ: Transaction Books, 1986.

Raleigh, Walter. *The War in the Air*, vol. 1. Oxford: Clarendon Press, 1922.

———, and H. Jones. *War in the Air*, vol. 6. Oxford: Clarendon Press, 1937.

Ranelagh, John. *The Agency: The Rise and Decline of the CIA*. New York: Simon & Schuster, 1987.

Rice, Donald B. *Toward the Future: Global Reach–Global Power: U.S. Air Force White Papers, 1989–1992*. Washington, DC: Headquarters Air Force, 1992.

Riza, Shaukat. *The Pakistan Army, 1966–71*. Dehra Dun: Nataraj Publishers, 1977.

Robinson, Anthony, ed. *Aerial Warfare*. London: Orbis Publishing, 1982.

Romjue, John L. *From Active Defense to AirLand Battle: The Development of Army Doctrine, 1973–1982*. Washington, DC: GPO, 1984.

Rose, Michael. *Fighting for Peace: Bosnia, 1994*. London: Harvill Press, 1998.

Rostow, Walt W. *Pre-Invasion Bombing Strategy: General Eisenhower's Decision of March 25, 1944*. Austin: University of Texas Press, 1981.

Royal Air Force Manual of Operations, AP 1300. 4th ed. London: Air Ministry, March 1957.

Rubin, Barnett R. *The Fragmentation of Afghanistan*. New Haven, CT: Yale University Press, 1995.

Sadat, Anwar. *In Search of Identity: An Autobiography*. New York: Harper & Row, 1978.

Sajima, Naoko, and Kyoichi Tachikawa. *Japanese Sea Power*. Canberra: Sea Power Centre–Australia, 2009.

Salter, James. *Gods of Tin: The Flying Years*. Edited by Jessica Benton and William Benton. Washington, DC: Shoemaker & Hoard, 2004.

Saunders, Hilary St. George. *Per Ardua: The Rise of British Air Power, 1911–1935*. London: Oxford University Press, 1944.

Schaffel, Kenneth. *The Emerging Shield: The Air Force and the Evolution of Continental Air Defense, 1945–1960*. Washington, DC: Office of Air Force History, 1991.

Scheina, Robert L. *Latin America's Wars*. Vol. 2, *The Age of the Professional Soldier*. Washington, DC: Brassey's, 2003.

Schiff, Ze'ev, and Ehud Ya'ari. *Israel's Lebanon War*. New York: Simon & Schuster, 1984.

Schirmer, Jennifer. *The Guatemalan Military Project: A Violence Called Democracy*. Philadelphia: University of Pennsylvania Press, 1998.

Schwarz, Benjamin. *American Counterinsurgency Doctrine and El Salvador: The Frustrations of Reform and the Illusions of Nation Building*, RAND Report R-4042. Santa Monica, CA: RAND, 1992.

Scott, Harriet Fast, and William F. Scott. *Soviet Military Doctrine: Continuity, Formulation and Dissemination*. London: Westview Press, 1988.

Sethi, Manpreet. *Nuclear Strategy: India's March towards Credible Deterrence*. New Delhi: KW Publishers, 2009.

Shalikashvili, John. *Joint Vision 2010—America's Military: Preparing for Tomorrow*. Washington, DC: Office of the Chairman of the Joint Chiefs of Staff, July 1996.

Sheehan, Neil. *A Fiery Peace in a Cold War: Bernard Schriever and the Ultimate Weapon.* New York: Random House, 2009.

Sherrod, Robert. *History of Marine Corps Aviation in World War II.* Washington, DC: Combat Forces Press, 1952.

Shirer, W. L. *The Rise and Fall of the Third Reich: A History of Nazi Germany.* New York: Simon & Schuster, 1960.

Shtaygman, Yitzhak, ed. *Operation "Kadesh": The Israeli Air Force from 1950–1956: Buildup and Operation.* Tel Aviv: Ministry of Defense, 1986.

Singh, Jasjit, ed. *Kargil 1999: Pakistan's Fourth War for Kashmir.* New Delhi: Knowledge World, 1999.

———. *Defence from the Skies: Indian Air Force through 75 Years.* New Delhi: Knowledge World, 2007.

———. *The ICON: Marshal of the IAF Arjan Singh.* New Delhi: KW Publishers, 2009.

Singh, Nagendra. *The Theory of Force and Organisation of Defence in Indian Constitutional History.* New Delhi: Asia Publishing House, 1969.

Singh, Satyindra. *Blueprint to Blue Water: The Indian Navy, 1951–65.* New Delhi: Lancer International, 1992.

Slessor, J. C. *Air Power and Armies.* Oxford: Oxford University Press, 1936.

———. *The Central Blue: Recollections and Reflections by Marshal of the Royal Air Force Sir John Slessor.* London: Cassel and Company Ltd., 1956.

Smith, Richard K. *Seventy-Five Years of Inflight Refueling: Highlights, 1923–1998.* Washington, DC: Air Force History and Museums Program, 1998.

Sokolovskii, V. D., ed. *Soviet Military Strategy.* Translated and edited by Herbert S. Dinerstein, Leon Gouré, and Thomas W. Wolfe. Englewood Cliffs, NJ: RAND in association with Prentice-Hall, Inc., 1963.

Spector, Ronald H. *Eagle against the Sun: The American War with Japan.* New York: Free Press, 1985.

Spires, David N. *Air Power for Patton's Army: The XIX Tactical Air Command in the Second World War.* Washington, DC: Air Force History and Museums Program, 2002.

Stein, David J. *The Development of NATO Tactical Air Doctrine,* Report R-3385-AF. Santa Monica, CA: RAND, December 1987.

Stephens, Alan. *Going Solo: The Royal Australian Air Force, 1946–1971.* Canberra: AGPS, 1995.

———, ed. *The War in the Air, 1914–1994.* 1994. Reprint, Maxwell AFB, AL: USAF Air University Press, 2001.

———. *Air Power in Southeast Asia.* Brunei Darussalam: Sultan Haji Hassanal Bolkiah Institute of Defense and Strategic Studies, 2008.

———, and Nicola Baker. *Making Sense of War: Strategy for the 21st Century.* Cambridge: Cambridge University Press, 2006.

Sterrett, James. *Soviet Air Force Theory, 1918–1945.* London: Routledge, 2007.

Stewart, James T., ed. *Airpower: The Decisive Force in Korea.* Princeton, NJ: D. Van Nostrand, 1957.

Stockholm International Peace Research Institute. *SIPRI Yearbook 2008.* Oxford: Oxford University Press, 2008.

Sturm, Thomas A. *The USAF Scientific Advisory Board: Its First Twenty Years, 1944–1964.* Rev. ed. Washington, DC: Office of Air Force History, 1986.

Swaine, Michael D., and Ashley J. Tellis. *Interpreting China's Grand Strategy: Past, Present, and Future.* Santa Monica, CA: RAND, 2000.

Sykes, F. H. *From Many Angles.* London: Harrap, 1942.

Tal, Israel. *National Security: The Israeli Experience.* Westport, CT: Praeger, 2000.

Tate, James P. *The Army and Its Air Corps: Army Policy toward Aviation, 1919–1941.* Maxwell AFB, AL: Air University Press, 1998.

Templewood, Lord. *Empire of the Air.* London: Collins, 1957.

Thaler, David E., and David A. Shlapak. *Perspectives on Theater Air Campaign Planning.* Santa Monica, CA: RAND, 1995.

Thatcher, Margaret. *The Downing Street Years.* London: HarperCollins, 1993.

Thayer, Thomas C., ed. *The Air War.* Vol. 5 of *A Systems Analysis View of the Vietnam War, 1965–1972.* Washington, DC: Southeast Asia Intelligence Division, OASD(SA)RP, February 18, 1975.

The Falklands Campaign: The Lessons. The Secretary of State for Defence, London, December 1982.

The Global War on Terrorism: The First 100 Days. Washington, DC: Coalition Information Center, December 2001.

The Japanese Air Forces in World War II: The Organization of the Japanese Army & Naval Air Forces, 1945. London: Arms and Armour Press, 1979.

The Kargil Review Committee. *From Surprise to Reckoning: The Kargil Review Committee Report.* New Delhi: Sage Publications, 2000.

The Military Balance, 1981–1982. London: International Institute for Strategic Studies, 1982.

The Military Balance, 1987–1988. London: International Institute for Strategic Studies, 1987.

The Military Balance, 1990–1991. London: Brassey's for IISS, 1990.

The Military Balance, 2009. London: Routledge for IISS, 2009.

The Prinsjesdag Letter [in Dutch]. The Hague: Ministry of Defence, 2003.

The Story of the Pakistan Air Force: A Story of Courage and Honour. Islamabad: Shaheen Foundation, 1988.

Thomas, Hugh. *Cuba or the Pursuit of Freedom.* London: Eyre and Spottiswoode, 1971.

Thompson, Michael E. *Political and Military Components of Air Force Doctrine in the Federal Republic of Germany and Their Implications for NATO Defense Policy Analysis.* Santa Monica, CA: RAND, 1987.

Thompson, Wayne. *To Hanoi and Back: The USAF and North Vietnam, 1966–1973.* Washington, DC: Smithsonian Institution Press, 2000.

———, and Bernard C. Nalty. *Within Limits: The U.S. Air Force and the Korean War.* Washington, DC: Air Force History and Museums Program, 1996.

Tilford, Earl H., Jr. *National Defense into the 21st Century: Defining the Issues—A Special Report from the Strategic Studies Institute.* Carlisle Barracks, PA: U.S. Army War College Strategic Studies Institute, June 6, 1997.

Till, G. *Air Power and the Royal Navy.* London: Jane's, 1979.

Tira, Ron. *The Limitations of Standoff Firepower-Based Operations: On Standoff Warfare, Maneuver, and Decision.* Memorandum no. 89. Tel Aviv: Institute for National Security Studies, March 2007.

Towle, P. A. *Pilots and Rebels.* London: Brassey's, 1989.

Turnbull, Archibald D., and Clifford L. Lord. *History of United States Naval Aviation.* New York: Arno Press, 1972.

Turnbull, Patrick. *Battle of the Box.* London: Ian Allan Ltd., 1979.

UK MOD. *Kosovo: Lessons from the Crisis.* London: Her Majesty's Stationery Office, Cmd. 4724, June 2000.

Underwood, Jeffery S. *The Wings of Democracy: The Influence of Air Power on the Roosevelt Administration, 1933–1941.* College Station: Texas A&M University Press, 1991.

United Nations, Peacekeeping Best Practices Unit, Military Division. *Operation Artemis: The Lessons of the Interim Emergency Multinational Force.* UN, October 2004.

U.S. Marine Corps. *Small Wars Manual.* Washington, DC: GPO, 1940.

Van Staaveren, Jacob. *Air Operations in the Taiwan Crisis of 1958.* Washington, DC: USAF Historical Division Liaison Office, November 1962.

Vick, Alan, David T. Orletsky, John Bordeaux, and David A. Shlapak. *Enhancing Air Power's Contribution against Light Infantry Targets.* Santa Monica, CA: RAND, 1996.

Von Senger und Etterlin, Frido. *Neither Fear Nor Hope.* New York: E. P. Dutton, 1964.

Waghelstein, John. *El Salvador: Observations and Experiences in Counterinsurgency.* Carlisle Barracks, PA: U.S. Army War College, 1985.

Wakelam, Randall T. *The Science of Bombing: Operational Research in RAF Bomber Command.* Toronto: University of Toronto Press, 2009.

Waller, Douglas. *A Question of Loyalty: Gen. Billy Mitchell and the Court-Martial that Gripped the Nation.* New York: HarperCollins, 2004.

Warden, John A. III. *The Air Campaign: Planning for Combat.* Washington, DC: National Defense University Press, 1988.

Warnock, A. Timothy. *Short of War: Major USAF Contingency Operations.* Maxwell AFB, AL: Air Force History and Museums Program in association with Air University Press, 2000.

Warry, John. *Warfare in the Classical World.* London: Salamander Books, 1998.

Watts, Barry D. *The Foundation of U.S. Air Doctrine: The Problem of Friction in War.* Maxwell AFB, AL: Air University Press, December 1984.

Webster, C., and N. Frankland. *The Strategic Air Offensive against Germany, 1939–1945.* Vol. 1. London: Her Majesty's Stationery Office, 1961.

———. *The Strategic Air Offensive against Germany, 1939–1945.* Vol. 4. London: Her Majesty's Stationery Office, 1961.

Weizman, Ezer. *On Eagles' Wings: The Personal Story of the Leading Commander of the Israeli Air Force.* New York: Macmillan, 1976.

Wells, H. G. *War in the Air.* London: Bell and Son, 1908.

Werrell, Kenneth P. *Archie, Flak, AAA, and SAM.* Maxwell AFB, AL: Air University Press, 1988.

————. *Sabres over MiG Alley: The F-86 and the Battle for Air Superiority in Korea.* Annapolis, MD: Naval Institute Press, 2005.

Westenhoff, Charles M. *Military Air Power: The CADRE Digest of Air Power Opinions and Thoughts.* Maxwell AFB, AL: Air University Press, October 1990.

White, John P., et al. *Directions for Defense: Report of the Commission on Roles and Missions of the Armed Forces.* Washington, DC: Department of Defense, May 24, 1995.

Whiting, Stephen N., and Thomas K. Dale. *Air Force Long-Range Planning: Institutionalization Revisited.* Washington, DC: Air Force History Support Office in association with the Office of the Special Assistant to the Chief of Staff for Long-Range Planning, April 12, 1996.

Wildenberg, Thomas. *All the Factors of Victory: Admiral Joseph Mason Reeves and the Origins of Carrier Airpower.* Washington, DC: Potomac Books, 2003.

Williams, Edwin L., Jr. *Legislative History of the AAF and the USAF, 1941–1951.* No. 84 in the *USAF Historical Studies* series. Maxwell AFB, AL: Air University, USAF Historical Division, Research Studies Institute, September 1955.

Winnefeld, James A., and Dana J. Johnson. *Joint Air Operations: Pursuit of Unity in Command and Control, 1942–1991.* Annapolis, MD: Naval Institute Press and RAND, 1993.

Wolk, Herman S. *Reflections on Air Force Independence.* Washington, DC: Air Force History and Museums Program, 2007.

Wood, Bryce. *The United States and Latin American Wars, 1932–1942.* New York: Columbia University Press, 1966.

Woodward, Sandy, with Patrick Robinson. *One Hundred Days: The Memoirs of the Falklands Battle Group Commander.* London: Fontana, 1992.

Worden, Mike. *Rise of the Fighter Generals: The Problem of Air Force Leadership, 1945–1982.* Maxwell AFB, AL: Air University Press, March 1998.

Wynn, H. *RAF Nuclear Deterrent Forces.* London: Her Majesty's Stationery Office, 1997.

Y'Blood, William T. *MiG Alley: The Fight for Air Superiority.* Washington, DC: Air Force History and Museums Program, 2000.

————. *Down in the Weeds: Close Air Support in Korea.* Washington, DC: Air Force History and Museums Program, 2003.

Yonay, Ehud. *No Margin for Error: The Making of the Israeli Air Force.* New York: Pantheon Books, 1993.

Zhang, Xiaoming. *Red Wings over the Yalu: China, the Soviet Union, and the Air War in Korea.* College Station, TX: Texas A&M University Press, 2002.

Zook, David. *The Conduct of the Chaco War.* New Haven, CT: Bookman Associates, 1960.

A-4 Skyhawks, 149, 153, 325, 327, 355–356,
        358–360, 437
A6M Zeros, 73, 306, 308
A-7s, 53, 98
A-10s, 103, 114, 115, 432, 442
A-12s, 116, 134
A-17s, 70
A-20s, 349
A-29s, 371
A-37s, 361, 363, 364, 367, 371, 484, 486
A400Ms, 391, 392, 394, 408
AAF (Army Air Forces). *See* United States
        Army Air Forces
Abrams, Creighton, 104
Accountability, Operation, 164
acquisition process, U.S., 134–135
Ader, Clément, 63
ADF (Australian Defense Force), 330–332
advanced manned strategic aircraft (AMSA), 102
advanced technology bombers (ATB), 102
aerial medical evacuation, 239, 367, 371
aerial refueling
        Balkans wars, 57, 383, 390, 391
        China, 249, 287, 290, 291
        Czech Republic, 400
        Falklands War, 51, 356, 357–358
        Indonesia, 325
        Israel, 165
        Pakistan, 251
        technology, 86–87, 410
        U.S., 86–87, 88, 94–95, 102, 291, 383, 390
Aerodromes, Langley, 64
Aeromacchi 339s, 356, 357
aerospace power
        China, 176, 248, 249, 260, 289, 292, 293
        Europe, 397–398, 406–407
        U.S., 106, 111, 113, 116, 123, 132, 289, 304
aerospace war, Russian concept of, 214
Afghanistan
        Operation Enduring Freedom (*See* En-
                during Freedom, Operation)

Soviet invasion and occupation, 105,
        206–207, 281
AFM 1s, 92, 106–107, 444
AH-64 Apache helicopters, 162, 327
AH-I Cobra helicopters, 161
Aidid, Mohamed Farrah, 116
AIM-9 missiles, 51, 311, 356, 358, 360
Air Board, British, 11
airborne early warning (AEW), 52, 59, 290, 311
airborne early warning and control
        (AEW&C), 311, 325, 327, 330, 332
airborne standoff radar (ASTOR), 61
airborne warning and control systems
        (AWACS), 55, 93, 249, 288, 290,
        291, 376, 384
Air Commerce Act of 1926, 69
air control strategy of Britain, 220
Air Corps Tactical School (ACTS), 71, 119
aircraft carriers
        Britain, 51–52, 357, 359
        France, 400–401
        India, 236–237
        Japan, 73, 307
        U.S., 72–73, 80, 85–86, 87, 88, 92
Air Defense troops, China, 273–274, 278
Air Effects Committee, 78
Air Expeditionary Forces (AEF), 118, 123–
        124, 452
Air Fighting Committee, Britain, 22–23
Air Force 2025, 118
Air Force Academy, U.S., 66, 111, 117
Air Force Doctrine Document 1 (AFDD-1), 119
Air Force Scientific Advisory Board (SAB),
        U.S., 117
Air France flight kidnapping, 158
AirLand Forces Application (ALFA),
        104, 107
air-launched cruise missiles (ALCM), 102
airlifts
        Argentina, 357
        Britain, 44, 45, 59, 61

China, 282, 290
Continental Europe, 391, 392–393, 399,
  401, 402–403
India, 229, 230, 238, 239
Soviet Union, 206
U.S., 101, 109
Air Materiel Command, U.S., 112, 446
air mobility, 410
Britain, 45, 50, 61
Continental Europe, 387, 391–392, 394,
  402, 405
El Salvador, 368
Soviet Union, 202–203, 207, 210
U.S., 108, 116
air offensive strategies and operations.
  See also specific operations
Australia, 302
Britain, 1, 9–10, 19–22, 38, 40–41, 52, 55
China, 173–177, 259–260, 283, 284–
  286, 289, 292–293
Continental Europe, 29–30, 376, 380, 406
Egypt, 150
India, 232, 243
Israel, 137, 143–145, 151, 155–156, 167,
  168–169
Pakistan, 235–236, 237
Soviet Union, 186–187, 190–194, 200–
  203, 205, 208
U.S., 78–79
Air Operations Command, Germany, 382
air policing
Afghanistan and Iraq, 130, 131, 133
Balkans, 399–400
imperial Britain, 2, 15, 18, 20, 28–29, 30, 44
India, 174
air power
future, 413–415
overview, xv–xviii, 409–410
technology trends, 410–412
air refueling. *See* aerial refueling
air superiority. *See also* air supremacy
Bolivia, 343
Britain, 21, 31, 32, 38, 39, 40
China, 265, 277
Germany, 34, 35, 189, 191
India, 175, 235, 238, 256, 257
Israel, 140, 145, 147, 156, 161, 163
Soviet Union, 189, 191, 192–193, 205,
  210, 265
U.N., 259, 270
U.S., 72, 76, 78, 82, 91, 126, 268, 270
air supremacy. *See also* air superiority
Britain, 8, 12–13, 20, 21, 32
India, 326
Japan, 307
Soviet Union, 205
U.S., 76–77, 81, 82, 85, 315
Vietnam, 321
air-to-air refueling. *See* aerial refueling
Air University, U.S., 83, 92, 93, 113, 116,
  118, 121
Akron, 67
ALA. *See* French Air Force
Albania, 388

Allied Force, Operation, 57–58, 112, 124–
  126, 382–387, 389–391, 394, 405
Almonacid, Vicente, 342
Alon, Mordechai, 138
al-Qaeda, 128–129, 303, 307
AMAL, 162
Amir, Israel, 139
An-124-100s, 392
Anaconda, Operation, 130–131, 401
antiaircraft artillery (AAA), 145, 274, 282,
  318–321
anti-submarine campaign, WWII, 348–349
ANZUS, 128
AP 3000 *British Air and Space Doctrine*, 380, 397
Apache AH-64 helicopters, 162, 327
applying air power, 300–304
Arab-Israeli War of 1967, 4, 144–150, 152,
  156, 198, 307
Arab-Israeli War of 1973, 101–102, 442
Arafat, Yasser, 167
Arbenz, Jacobo, 351
Ardent, HMS, 359, 483–484
Argentina, 50–52, 344, 345, 349, 355–360,
  481, 483–484
Argentine Army, 360
Argentine Navy, 356–357, 359
Armas, Castillo, 351
armement air sol modulaire (AASM), 390–391
Army Cooperation Command, Britain, 30, 32
Army War College, 125
Arnold, Henry H., 68, 79, 81, 84, 87
Arrow air defense systems, 163, 165
Artemis, Operation, 403, 404, 408
Artzav-19, Operation, 161
Asia Pacific region, air power overview, 295–
  296, 299–300, 332–333
Aspin, Leslie, 116–117
asymmetric warfare, 60, 62, 171, 220, 415
Atlantic Conveyor, 51, 359, 483–484
atomic power
Britain, 43–44
NATO, 94
Soviet Union, 84
U.S., 43, 80, 83–84, 85–87, 96, 307, 308
"attack aviation," 76–75, 431–432
Attlee, Clement, 43
attrition, strategy of
against Israel, 150, 155, 169
WWII, 33, 39, 49, 105, 191
Australia, 128. *See also* Royal Australian Air Force
air power, 296, 300, 302, 326, 327,
  328–332, 411
Indonesia, 323–324, 325
Vietnam War, 99
Australian Defense Force (ADF), 330–332
Avia S-199, 138
Avion III, 63–64
Awami National Party, 235
AWPD-1, 72, 75, 106–107

B-1s, 102, 131
B-2s, 102, 135, 452
B-10s, 69
B-17s, 70, 72, 86

B-18s, 70, 349
B-23s, 70
B-24s, 70
B-26s, 351, 352–353, 354, 362–363
B-29s, 81, 86, 89–90, 308, 309
B-36s, 85–86, 98
B-47s, 86
B-50s, 86
B-52s, 86, 100–102, 117, 129, 131, 134, 320–321
B-58s, 87, 238
B-70s, 98
Baker, Newton, 66
Baker Board, 72
Baldwin, Stanley, 19–20, 22, 24, 25
Balkans, the, 382–387
    Britain, 2, 54–58
    France, 377
    Netherlands, 379
    U.S., 111, 121–122, 124–126
ballistic missiles, 412
    Australia, 332
    Britain, 47, 48
    China, 248, 249–250, 311
    Germany, 84
    India, 245–246
    Japan, 312
    North Korea, 310
    Pakistan, 251
    Soviet Union, 195, 202, 204
    U.S., 84, 93, 94, 96, 216
balloon aviation, 63, 182–183
Barratt, A.S., 30
Barry, John L., 127
Battle of Bismarck Sea, 329
Battle of Britain, 2, 24, 32–36, 74–75, 76, 223
Battle of Coral Sea, 329
Battle of Imphal-Kohima, 225
Battle of Kursk, 191
Battle of Midway, 80, 307, 329
Bay of Pigs, 297, 351–353
Bekaa Valley air battle, 105, 256
Belgium, 29, 34, 378, 392, 398, 401, 404
Bell 205 helicopters, 149, 153
Bell 206B helicopters, 364
Ben-Eliyahu, Eitan, 164–165
Bengalis, 234
Ben-Gurion, David, 4, 140–141
Ben-Nun, Avihu, 162, 164
Berlijn, Dick L., 401
Berlin
    blockade, 45, 82, 86, 94
    bombing of, 35, 38, 39, 40
    crisis of 1961, 94, 95
Bevin, Ernest, 44, 45
Biggin Hill Experiments, 23–24
Birmingham, 64
Bismarck Sea, Battle of, 329
Blackhawk helicopters, 164, 312, 331, 369, 371, 487
Blair, Tony, 386
Blériot, Louis, 7
Blue Steels, 49
Bodinger, Herzl, 164

Boeing, 69–70, 72, 150, 158, 324. *See also specific aircraft of*
Boeing 737s, 330, 331
Bolivia, 296, 340–347, 371, 481
Bolivian Air Corps, 341–342, 343–344, 345–347
Bolivian Army, 341, 342–343, 345–346
Bolivian Reserve Corps, 346
Bolshevik Revolution, 182, 185–186
Bomber Command, Britain
    Egypt, 46
    NATO, 95
    nuclear transformation, 44, 48–49
    WWII, 2, 21, 23–24, 26–28, 35–41, 42
Boquerón, 342–344
Bosnia. *See also* Balkans, the
    Britain, 54–58
    Continental Europe, 379, 382–385, 387, 405
    U.S., 112, 121–122, 126
Bouchier, Cecil A., 222
Boutros-Ghali, Boutros, 122, 452
Bradley, Omar, 78, 86, 87
Brazil, 348–350, 368, 371–372, 486
Brazilian Air Force, 350
Breguet, Louis, 68
Breguet XIX bombers, 342
Brezhnev, Leonid, 206
Britain
    aerospace cooperation, 394, 395–396, 397
    Afghanistan and Iraq, 52–54, 58–61
    Africa, 402
    air mobility, 391–392
    air power overview, origins, and definition, 1–2, 7–9, 50, 62, 373
    Australia, 328–329
    Balkans, 54–58, 122, 383
    Bolivia, 342
    Cold War, 46–50
    EU Battle Group, 403
    Falklands War, 2, 50–52, 105, 355–360, 483–484
    first Gulf War, 52–54, 376–377
    France, 388–389
    India, 220–223
    Indonesia, 52–54, 58–60
    Japan, 306
    Kenya and Malaya, 44–45
    nuclear transformation, 2, 43–44, 45, 46–50
    pre-WWII planning, policy, and technology, 16–17, 20–24
    pre-WWII rearmament, 24–28
    WW I, 9–15
    WWII, Bomber Command, 21, 23–24, 26–27, 35–41
    WWII, Coastal Command, 41–43
    WWII, North Africa, 31–32
    WWII, RAF and army cooperation, 28–32
Britain, Battle of, 2, 24, 32–36, 74–75, 76, 223
British army, 8–9, 11–12, 18–21, 28–32
British Burma Army, 224
British Expeditionary Force (BEF), 29–30
British Somaliland, 17
Brook, Norman, 47

Brown, George, 104
Buccaneers, 49, 52, 54
Bulgaria, 388
Burma, 223–225, 300, 432
Bush, George H.W., 109, 111
Bush, George W., 129
Butt, D.M., 37
Butt Report, 37

C14R Ospreys, 344, 345
C-17 Globemasters, 31, 58, 124, 391, 392,
    394, 399
C-46s, 352–353
C-47s, 70, 351, 352, 362–364, 485, 486
C-130 Hercules'
    Argentina, 356, 357
    Australia, 331
    Britain, 51, 54, 58, 61
    Indonesia, 323, 325
    Israel, 150, 158
    Poland, 398
    U.S., 319
Cactus Air Force, 80
Cameron, Neil, 50
Campo Via, 345–346
Canberra, 52
Canberras
    Argentina, 355
    Britain, 45–46, 47, 49, 56, 58
    India, 234, 238
    Peru, 361
CAP (Combat Air Patrol), 52, 55, 57, 109,
    111–112, 357–358, 382
capabilities of air power, 302–303, 304
carriers, aircraft. See aircraft carriers
Carter administration, 117, 363
CAS. See close air support
Casablanca Conference, 38
CASF (Composite Air Strike Forces), 94, 95, 123
Cast Lead, Operation, 171
Castro, Fidel, 351–353
casualties, difference with air/space
    power, 132
Central Intelligence Agency (CIA), 126,
    351–353
Century series fighters, 97. See also specific
    fighters
CFSP (Common Foreign and Security Policy),
    388–389
CH-47 Chinooks, 51, 58, 61, 327, 331
CH-53 helicopters, 150, 152, 153
Chaco War, 296, 340–347, 481
Chamberlain, Neville, 25
Chambers, Washington Irving, 64
Chang'e, 260
Charles de Gaulle, 400–401
Chatfield Committee, 222
Chechnya, 216
Cheney, Dick, 109, 116, 122, 445
Chevènement, Jean-Pierre, 377
Chiefs of Staff Committee (COSC),
    India, 255
Chile, 340, 345, 356, 482
China. See also People's Liberation Army;
    People's Liberation Army Air Force

air doctrine, 276–280, 284–287
air power background and overview, 176–
    177, 259–263, 292–293
    Guatemala, 351
    India, 175, 230–231, 233–234, 245,
        246–250
    Japan, 305, 306, 310–311
    Korean War, 197, 265–271, 470
    military capabilities and modernization,
        254, 304
    Pakistan, 247–248, 251
    revolution in, 260, 261–263
    Vietnam, 280–281, 321
Chinese Air Force. See People's Liberation
    Army Air Force
Chinese Communist Party, 176, 260, 262–265
Chinese Embassy in Belgrade bombing, 126
Chinese People's Volunteers (CPV), 268
Chinooks, 51, 58, 61, 327, 331
Chirac, Jacques, 378, 386
Churchill, Winston
    on air power, xv
    Battle of Britain, 29, 32
    bombing of Germany, 36, 40, 42
    India, 220, 225
    as minister of munitions, 13
    RAF independence and imperial policing,
        16, 17–18, 28, 220
    reputation, 41
    Soviet Union, 189
CIA (Central Intelligence Agency), 126,
    351–353
civil aviation
    Britain, 66
    U.S., 68–70, 94, 127
    use of in Guatemalan War, 364
Civil War, U.S., 63
Clark, Wesley, 124, 126, 386
clash of civilizations, 180–181
Clausewitz, Carl von, 322
Clinton, Bill, 117, 386
close air support (CAS)
    Afghanistan, 61
    Cold War, 103, 104
    India, 220, 222, 231, 246
    Israel, 140, 142–143, 156–157, 158, 168
    Korean War, 90–91
    Vietnam War, 98
    WWII, 75, 191–192, 193, 201
Close Support Tactics - Provisional (Slessor), 222
Coastal Command, Britain, 29, 41–43, 47
Cobras, 161
Cohen, Eliot, 80
Cohen, Richard, 130
Cohen, William, 117, 119–120, 448
Cold War. See also specific aspects of
    Britain, 2, 43–45, 46–50
    China, 271, 292
    Continental Europe, 374–376, 381–382,
        388, 391, 394–395, 402, 404
    India, 226, 228, 466
    Latin America, 297
    Soviet Union, 174, 192, 195–199, 208
    U.S., 92–96, 101–105, 373

Colombia, 297, 368–371, 487
Colombian Air Force, 370–371
Colombian National Police, 297, 369–371
colonialism, 180, 295, 299, 314, 333
Combat Air Patrol (CAP), 52, 55, 57, 109, 111–112, 357–358, 382
Commando Especial Reserva Aerea, 364
command of the air. *See also* air supremacy
    Britain, 7, 8, 12, 15, 20, 29, 52, 62
    China, 249, 277–278, 283
    Soviet Union, 205, 208
Commission on Roles and Missions of the Armed Forces, U.S., 115, 121
Commission on the Use of Aeronautics for Military Purposes, Soviet, 183
commonalities of air forces, xvii
Common Foreign and Security Policy (CFSP), 388–389
Communist ideology and effects on doctrine, Soviet, 173, 179–180, 182, 194–195, 202–203, 477
Composite Air Strike Forces (CASF), 94, 95, 123
Confidential Document (CD) 22, 20
Congo, the, 175, 238, 403–404
Coningham, Arthur, 31–32, 77
Conqueror, HMS, 358
constitution, Japanese, 309–310
Continental Europe air power overview, 297–298
Contras, 353–354
cooperation, EU, 377–378, 389–390, 392–393, 401, 406
Coral Sea, Battle of, 329
Coventry, HMS, 359, 483–484
covert operations
    Pakistan, 229, 231, 237, 246, 250, 252
    U.S. in Latin America, 297, 351–354
CR 20bi's, 344–345
Crespo, Ernesto Horacio, 355, 356, 360
crisis management operations, 376, 379, 388, 389, 393, 404
Croatia, 54, 56, 388
Crowell, Benedict, 66
Cuba, 95–96, 195, 349, 351–353, 366
Cuban Missile Crisis, 95–96, 195
Cultural Revolution of China, 280
culture and effect on war strategy
    China, 176
    India, 226–227
    Japan, 295–296, 307–309, 313, 333
Curtiss, Glenn, 64–65, 68
Curtiss aircraft, 334, 337, 338, 344, 432
Cyprus, 45, 46, 47, 52, 56, 132, 239
Czech Air Force, 374, 399–400
Czechoslovakia, 141
Czech Republic, 387–388, 399–400

Daggers, 157–158, 355, 356, 359, 360–361, 371
Dakotas, 138, 142
Dalai Lama, 230
Dalton, Stephen, 62
Daniel Guggenheim Fund for the Promotion of Aeronautics, 69
Darfur, 404

Dayton Peace Accords, 56, 122, 383
DC-2s, 69–70
DC-3s, 69, 229
de Castries, Christian, 315–317
Defence Committee of the Cabinet (DCC), India, 255
Defence White Papers, British, 46, 47
Defending Australia in the Asia-Pacific Century: Force 2030, 330, 331
Defense Capabilities Initiative (DCI), 390
Deliberate Force, Operation
    Britain, 54–58
    Continental Europe, 379, 382–385, 387, 405
    U.S., 112, 121–122, 126
Democratic Party of Japan (DPJ), 313
Democratic People's Republic of Korea, 88–92, 196–197, 268–269, 310
Deng Xiaonping, 176, 280, 282–283
Denmark, 396, 397, 401, 407
Deny Flight, Operation, 112, 379, 382, 384–385, 393, 405
Department of Defense (DoD), U.S., 66, 82, 98, 115–118, 120, 123, 135
Deptula, David, 53, 103, 108, 445
Dereliction of Duty (McMaster), 120
Desert Fox, Operation, 56, 111
Desert Storm, Operation. *See* Gulf War, first
deterrence policies
    Britain, 2, 25, 46–47
    India, 174–175, 226–228, 242–243, 245–246, 253–254, 464
    Israeli, 142, 169–170
    Japan, 310
    NATO and Warsaw Pact in Cold War, 374–375, 388
    as part of strategic construct, 301, 302, 304
    Singapore, 327–328
    Soviet Union, 195–196, 203–204, 211, 216, 217
    U.S., 97, 108
Devayya, A.B., 233
DH-4 Bombers, 336–337, 338, 339
Diaz, Adolfo, 339
Diaz, Porfirio, 335
Dien Bien Phu, 314–317
disaster relief, 175, 238, 239
dive-bombing, 26, 75–77
doctrinal and operational response, EU, 393–394
doctrine, about, 179
doctrine, Britain, 20, 28–30, 32, 47, 50, 380
doctrine, China, 264–265, 270–271, 273, 276–280
doctrine, Egypt, 153
doctrine, Germany, 380–382
doctrine, India, 226–228, 242–244, 245–246
doctrine, Indonesia, 323
doctrine, Israeli
    air superiority, 145, 147, 156
    coping with SAMs, 157, 160–161
    ground force maneuver, 140, 143, 152, 165
    offensive, 155
    overview, 170–172
    politics over military, 142
    use of fire power, 163–165

doctrine, Japan, 306
doctrine, NATO, 94–95, 375
doctrine, Netherlands, 379
doctrine, Soviet-Russian
    Bolshevik period, 185–187
    Cold War, 195–197, 199–205, 207–210
    overview, 217–218
    post Cold War, 211–216
    WWII, 187–188, 193–194
doctrine, U.S.
    Cold War, 95
    Korean War, 87–88, 90–92
    post Cold War, 105–108, 118–121
    pre-WWII, 70–72
    WWII, 75–78
doctrine, Warsaw Pact, 376
DoD (Department of Defense), U.S., 66, 82,
    98, 115–118, 120, 123, 135
Doenitz, Karl, 42
Dominican Republic, 296, 338
Dongfengs, 311
Doolittle, James H., 79, 81
Doolittle Raid, 80
Douglas Aircraft Company, 69–70. *See also
    specific aircraft of*
Douhet, Guilio, 187, 214, 261, 284–285
Dowding, Hugh, 23, 33, 35, 36
Dresden, destruction of, 40
drug wars, 368–371, 486
Drum Board, 72
dual key command and control, 55
dual-track capability of Soviet Union, 199–
    200, 204–205
dual-use industrial aircraft base, 70, 72
Duarte, José Napoleon, 368
Dugan, Michael, 109, 116, 122
Dugman-5, Operation, 154, 157, 160
Dunkirk, battle at, 9, 29
dwifungsi, 323

E-2C Hawkeyes, 311
E-3 AWACS, 55, 57, 58, 93
E-767 AWACS, 311, 313
EACC (European Airlift Coordination Cell),
    392, 401
Eagle Attack, Operation, 34
Eaker, Ira, 39, 68
East India Company, 239
EATC (European Air Transport Command),
    392–393, 408
Ecuador, 347–348, 360–361
Ecuadorian Air Force, 361
Eden, Anthony, 46
Egypt. *See also* Egyptian Air Force
    Britain, 2, 46
    Israel, 141–142, 144, 146–148, 150–154,
        158–160
    Soviet Union, 198
    WWII, 30
Egyptian Air Force, xvi, 46, 144–145, 147–
    148, 150–151, 156
Egyptian Army, 148, 150, 153–154, 156
Eighth Air Force, U.S., 75, 79
Eisenhower, Dwight D., 68, 78, 82, 83, 87,
    317, 351

Eisenhower administration, 94, 97, 352
Elazar, David, 155
electronic countermeasure (ECM) equipment,
    39, 49, 51, 283, 290, 311, 320
elitism in Japan, 308–309
El Salvador, 297, 354, 365–368, 485, 486
El Salvadoran Armed Forces (ESAF), 366–368
El Salvador and Honduran War, 354
Ely, Eugene, 64
embargo on Bolivia and Paraguay, 345
Embraer 145, 368
Enduring Freedom, Operation
    Australia and Japan, 310, 313, 330
    Europe, 58–61, 62, 377, 400–402
    U.S., 96, 105, 128–131, 133
environments, air power, 413
EPAF EAW (European Participating Air
        Forces' Expeditionary Air Wing),
        401–402
ESDP (European Security and Defense
        Policy), 388–389, 402, 407
Estigarribia, Jose Felix, 341, 342–343,
        345–346
Estonia, 345, 388, 399, 403
EU Battle Groups, 298, 403
EUFOR, Operation, 404
Europe air power overview, Continental,
        297–298
European Air Group, 378, 392
European Airlift Coordination Cell (EACC),
        392, 401
European Air Transport Command (EATC),
        392–393, 408
European Participating Air Forces'
        Expeditionary Air Wing (EPAF
        EAW), 401–402
European Security and Defense Policy
        (ESDP), 388–389, 402, 407
European Union. *See also specific countries of*
    air power overview, 374
    first Gulf War, 297
    maintaining air power, 390–394
    operations, 400, 402–404, 407–408
    precision strike capabilities, 390–391
    security policy, 387, 388–389
    UN, 376
Exocet missiles, 51, 359
Expeditionary Aerospace Force (EAF), 124
expeditionary air forces, 3–4
    Australia, 330
    Brazil, 350
    Britain, 2, 29–30
    Europe, 401
    U.S., 118, 123–124, 337
Expert Committee on the Defence of India, 222

F-2s, 311, 312
F-4s, 98, 102, 149
F4Us, 73, 354
F-5s, 325, 327
F6F Hellcats, 73, 308
F-14s, 53, 102
F-15s
    Israel, 158–159, 164

Japan, 311
  Singapore, 327
  U.S., 52, 102, 105, 109, 117
F-16 Falcons, xvi
  Bolivia, 345, 481
  Continental Europe, 379, 384, 391, 394,
    396–399, 401–402, 405
  Indonesia, 325
  Israel, 158, 159, 162, 165–166, 168
  Japan, 312
  Pakistan, 238
  Singapore, 327
  U.S., 102, 105, 117, 123
F-18s, 52
F-22 Raptors, xv, 119, 124, 134–135, 313,
  331, 414
F-27s, 357, 364
F-35 Lightnings. *See* Joint Strike Fighters
  (JSF)
F-50 bombers, 336
F-51s, 354, 362–363
F-80s, 97, 362
F-84s, 93, 95, 97
F-86 Sabres, 93, 95, 97, 198, 231–232, 233, 235
F-94s, 89–90
F-100s, 93, 95, 97
F-101s, 93, 97
F-102s, 93, 97
F-104s, 95, 97, 231, 233, 467
F-106s, 93, 97
F-110s, 98
F-111s, 49, 98, 102, 302, 397
F/A-18s, 53, 102, 330, 401
FAA. *See* Federal Aviation Administration;
  Fuerza Aerea Argentina
FAG. *See* Fuerza Aerea Guatemala
Falcons. *See* F-16s Falcons
Falklands War, 2, 50–52, 105, 355–360,
  483–484
FAP (Peruvian Air Force), 348, 361, 368
FAR (Revolutionary Air Force), Cuba,
  352–353
Farabundo Martí National Liberation Front
  (FMLN), 365–366, 367–368
FARC (Revolutionary Armed Forces of
  Colombia), 369–371
Far East Air Forces (FEAF), 89, 91
FAS. *See* Fuerza Aerea Sur; Fuerza Aerea
  Salvadorena
FB 11s, 352–353
Federal Aviation Administration (FAA), 66, 69
Feng Ru, 260
Field Manual of Air Force Operations, China,
  276–278, 283
Field Manuals, U.S.
  Air Force, 92, 106–107
  Army Air Corps, 77–78, 87, 90, 92, 97, 432
fifth-generation aircraft, xv–xvi, 3, 331, 414
fighter-bombers, about, xvi, 76, 87, 97, 207,
  431–432. *See also specific fighter-
  bombers*
Fighter Command, Britain, 23–24, 26,
  33–36, 47
Finland, 392, 403
Finletter, Thomas K., 83

Finney, Robert, 71
Firebees, 161
First Five-Year Plan, 270
first Gulf War. *See* Gulf War, first
First Intifada, 4, 162–164
First Kashmir War, 229, 255, 464
First Lebanon War, 4, 160–162, 164
1st Fighter Group, Brazil, 350
Flexible Response doctrine of NATO, 95,
  204, 375, 380, 404
FMLN (Farabundo Marti National Liberation
  Front), 365–366, 367–368
Fogleman, Ronald, 118–120, 122, 451
Fokker Elmo, 396
Fokkers, 324, 339, 342, 357, 364
Follow-on Forces Attack (FOFA), 104, 210,
  375, 376, 377, 414
formation bombing, 27, 28, 33, 269, 282, 306
Forrestal, James, 85, 435
Forrestal, USS, 87, 92
Foulois, Benjamin D., 68, 337
France. *See also* French Air Force; French Navy
  aerospace industry, 394–395, 397–398
  Afghanistan, 400, 404–405
  Africa, 402–404
  airlift cooperation, 392
  Argentina, 355
  background, overview, and capabilities,
    373, 374, 390–392
  the Balkans, 122, 383, 386
  Bolivia, 342
  Britain, 388–389
  Ecuador, 361
  El Salvador, 485
  first Gulf War, 376–378
  Israel, 141, 144, 148, 149
  Japan, 306
  Paraguay, 342, 345
  Peru, 361
  Russia, 184
  Singapore, 326, 327
  Vietnam, 314–317
Franco-British European Air Group, 377–378
Franks, Tommy, 131
Fredendall, Lloyd R., 130
French Air Force (ALA)
  Afghanistan, 400–401
  air mobility and cooperation, 391, 394, 406
  Balkans, 385, 386, 390–391
  EU operations, 402–404
  first Gulf War, 377–378, 379, 404–405
  Paraguay, 342
  pre-WWII, 19
French Navy, 401
Fricker, John, 234
Frontal Aviation, Soviet, 201, 209
Frunze, Mikhail, 186
Fuerza Aerea Argentina (FAA), 355–
  357, 362
Fuerza Aerea Guatemala (FAG), 362–365,
  484, 485
Fuerza Aerea Sur (FAS), 355–360, 485
Fuerza Aerea Salvadorena (FAS), 354, 365–
  368, 485

Furies, 74
"Future of the United Kingdom in World
    Affairs, The" (Brook), 47
future wars, 133–136
F-X program, Japan, 312–313

G3M3 bombers, 306
Gailbraith, John, 231
Galland, Adolf, 188
Gandhi, Rajiv, 238
Garibaldi, 400
Gates, Robert, 135
Gauhar, Altaf, 234
Gayoom, Maumoon Abdul, 238
Gaza Strip, 163, 167, 168, 171
Geddes Committee, 19
Gee navigation aid, 37–38, 45
General Belgrano, 358
Georgia, 216–217
German Air Force, 380–382, 384–386, 402,
    405–406. *See also* Luftwaffe
German Armed Forces, 381
Germany. *See also* German Air Force
    aerospace cooperation, 378, 392, 394
    Afghanistan, 402, 405–406
    Balkans, 383, 384–386, 394
    Bolivia, 344
    current capabilities, 390, 391
    first Gulf War, 380–382
    satellites, 398
    WWI, 373
    WWII, 31–32, 33–36, 77–80, 188–191,
        348–350, 373
Gettysburg, battle of, 63, 113
Giap, Vo Nguyen, 315–317, 319
glasnost, 212
Global Engagement, 116, 121, 448, 451
Global Hawks, 131, 330
global power shift, 247–250, 252–253, 304
Global Presence, 113
Global Presence, 121
Global Reach–Global Power, 3–4, 107–109,
    112–115, 117–118, 120–121, 127
Global Vigilance, Reach, & Power, 127, 454
Glosson, Buster, 103
Goering, Hermann, 33–34, 36
Golan Heights, 148, 154
Gorbachev, Mikhail, 206, 211–212
Gordon, John, 118
Gorrell, Edgar S., 337
Gotha bombers, 10
GR1s, 52, 53, 54, 56, 57
GR3s, 51
GR7s, 57
gradual escalation strategy, 386–387
grammars of warfare, xvii–xviii
Grand Duchy of Muscovy, 181
Grapes of Wrath, Operation, 164
Greater East Asia Co-Prosperity Sphere, 305
Great Patriotic War, 376
Greece, 398
Gripens, 395, 396, 399–400, 404
ground force maneuver doctrine of IDF
    First and Second Lebanon Wars,
        160, 168

overview, 140, 170–171
Sinai War, Six Day War, and War of
    Attrition, 143, 149, 152
wars abroad, 165
Yom Kippur War, 154, 157, 170–171
Grozny, 95–96
Guadalajara, 76
Guadalcanal, 80
Guatemala, 297, 349, 351, 362–365, 484
Guatemalan Army, 363
Guernica, bombing of, 26
Guerrilla Army of the Poor (EGP), 363–365
Gulf War, first
    Britain, 52–57
    China, 284, 286
    Continental Europe, 297, 373, 376–382,
        384, 404–405
    Israel, 162–165
    Russia, 174, 214–215, 218
    U.S., 101, 108–111, 113–114, 117, 122,
        131–132, 445, 447
Gulf War, second, 58–60, 115, 128, 130–
    131, 165

H2S radar, 38
Habibe, B.J., 324
Haganah organization, 138
Hagenbeck, Franklin, 130
Haiti, 296, 338, 482
Haldane, Richard, 8, 17
Halder, Franz, 189
Halifaxes, 25, 37
Halutz, Dan, 165–166, 168
Hamas, 167, 168
Hansell, Haywood S., 75
HARM-firing air defense suppressers, 124
Harriers. *See also specific variants*
    Britain, 51, 56, 57, 59, 61, 356, 357–360, 400
    Italy, 400
Harris, Arthur, 37–40, 42
Hawkers, 23, 29, 30, 34, 73, 74, 76, 225
Hawks, 325, 344, 345, 379
Headline Goals of EU, 390, 403
Heinkel He's, 74
helicopters, 362. *See also specific types*
    Argentina, 357
    Australia, 331
    Britain, 44, 51–52, 58–59, 61, 357
    China, 280
    Colombia, 297, 369–371
    Cuba, 352
    Ecuador, 361
    El Salvador, 366–368, 485, 486
    Guatemala, 362–365
    India, 219, 236–237, 238, 239
    Indonesia, 324
    Israel, 148–150, 152, 153, 156, 161, 162,
        164, 166, 168
    Japan, 312
    Nicaragua, 354
    Pakistan, 238, 468
    Peru, 361
    Singapore, 327
    Soviet-Russian, 183, 206–207

Syria, 159
U.S, 96, 99, 103, 114–115, 124, 130–
    131, 318–319
Vietnam War, 318–319, 321
Helios satellites, 397–398
Henderson, David, 10
Henry, Larry, 103
Hercules aircraft. *See* C-130s Hercules'
Hermes, HMS, 51, 357, 359
Herzegovina, 54, 379, 383
Hewitt, Edgar Ludlow, 26
Hezbollah, 162, 164, 165, 168–170, 171
Himalayas, the, 175, 230–231, 237, 239
Hiroshima atomic attack, 43, 80, 252, 307,
    308, 309, 329
Hitler, Adolf, 33–36, 40
Hoare, Samuel, 18, 85
Ho Chi Minh, 314–315
Hod, Mordechai, 146, 152
Holbrook, Richard, 122
Home Front Command, Israel, 163
Honduras, 297, 351, 353, 354, 482
Horned Owl, 135, 456
Horner, Charles, 52, 53, 103, 109, 114
Houston, SS, 353
HS-2Ls, 338
Huerta, Victoriano, 336
humanitarian aid
    Balkans, 54–55, 57, 111–112
    EU, 388
    India, 239, 384, 385
Hungary, 388, 392, 399
Hunter aircraft, 145, 236–237
Huntington, Samuel, 180
Hurd, Douglas, 55
Hurricanes, 23, 29, 30, 34, 73, 76, 225
Hussein, Saddam, 56, 58, 108–109, 130–132,
    163, 301, 353

IA-58 Pucaras, 355–356, 357, 360
IADS (integrated air defense systems), 93,
    102, 207, 284, 302, 320
IAI (Israel Aviation Industries), 157–158,
    162, 484, 485
ICBMs, 96, 195, 200, 203, 216, 311. *See also*
    ballistic missiles
ideology and Soviet-Russian doctrine,
    179–180, 186, 197, 205, 213, 215,
    217–218
IDF. *See* Israel Defense Forces
Illustrious, HMS, 51, 57
Ilyushin aircraft, 142, 189, 238, 265–266,
    288, 290, 324
Imperial Air Force, 184
Imperial Defence College (IDC), 20–21
Imperial Japanese Navy (IJN), 305, 306–307.
    See also Japanese Naval Air Force
imperial policing, Britain, 2, 15, 18, 20,
    28–29, 30, 44
Imphal-Kohima, Battle of, 225
improvised explosive devices (IEDs), 60, 135
India. *See also* Indian Air Force; Indian Army;
    Indian Navy
    background and overview, 174–175, 464

Britain imperial policing, 18, 28, 239–
    240, 466
Chinese threat, 246–250
defense expenditures and structure, 250,
    254–257
doctrine, 226–228, 235, 242–244, 245–
    246, 464, 466
economy, 253–254
First Kashmir War, 229, 255, 464
force modernization and global power
    shift, 250, 252–255, 304
future military transformation, 227–258
independence, 220–221, 229
India-Pakistan War, 234–237, 467
Iraq, 253
Kargil War, 237–238
Pakistani threat, 250–252
post WWII, 225–226, 240–240
Second Kashmir War, 231–234, 255–256
Sino-Indian War, 231–231, 247
Soviet aircraft and weapons, 244–245
Indian Air Force
    background and overview, xvi, 174–175,
        219–220, 239–240
    doctrine, 235, 243–244, 466
    First Kashmir War, 229
    Indianization, 221–223
    India-Pakistan War, 235–237, 467
    Kargil War, 237–238
    non-war operations, 238–239
    nuclear weapons, 245–246
    post WWII, 240–241
    Second Kashmir War, 231–234, 255–256, 465
    Sino-Indian War, 230–231
    Soviet aircraft and weapons, 199, 244–245
    squadron No. 1, 222, 224–225
    structure, operations, and future transfor-
        mation, 255–258
    training, 208
    Volunteer Reserve, 222
    WWII, 174–175, 223–225, 240, 463
Indian Air Force Act, 221
Indian Army, 219, 229–232, 235–237, 239–
    240, 245–246, 255–257
Indian National Congress Party, 220–221
Indian Navy, 219, 222, 236–237, 239, 255,
    257, 466
Indian Ocean tsunami, 239
Indian Sandhurst Committee, 221
Indian Technical and Followers Corps, 221
India-Pakistan War, 234–237
Indonesia, xvi, 296, 300, 302, 322–326, 333
Indonesian Air Force (TNI-AU), 322–326
Indonesian Army (TNI-AD), 302, 322–326
industry, aviation and aerospace, 409–410
    Australia, 329
    Britain, 35–36
    China, 176, 254, 279–281, 283, 285–
        288, 293, 311
    cooperation among nations, 394–397,
        406–407
    India, 225
    Japan, 306, 312
    Singapore, 327, 328

Soviet-Russian, 183–184, 185, 192–193, 196, 207, 215
Sweden, 376, 395
U.S., 66, 68–69, 70, 72, 73, 123, 430
in-flight refueling (IFR). *See* aerial refueling
Inskip, Thomas, 25
integrated air defense systems (IADS), 93, 102, 207, 284, 302, 320
"Integrity First, Service before Self, Excellence in All We Do," 120, 127, 451, 454
intelligence, Israeli, 153, 156–157, 160–161
intelligence, surveillance, and reconnaissance (ISR). *See* ISR
intelligence, surveillance, target acquisition, and reconnaissance (ISTAR), 59, 62
Inter-Allied Aviation Committee, 13–14
Inter-American Air Force Academy (IAAFA), 366
interdiction
    Cold War, 47, 89, 103, 207
    India, 235
    Korean War, 89
    Latin America, 368–369
    U.S. doctrine, 77–78, 92
    WWII, 32, 191, 193
intermediate-range ballistic missiles (IRBM), 48. *See also* ballistic missiles
Intermediate-Range Nuclear Forces Treaty, 248
international military education and training (IMET) program, 366
International Security Assistance Forces (ISAF), 400–402
interoperability, 96, 378, 387, 396, 398, 404
interservice rivalry, 7, 85–88, 308
investment and future wars, 133–134
Invincible, HMS, 56, 357, 359
IPTN, 324
Iran, 105, 158, 162, 169, 234, 235
Iraq
    air control, 2, 56
    Australia, 330
    Britain, 2, 18, 220
    first Gulf War (*See* Gulf War, first)
    France, 397
    India, 239, 253
    Israel, 147, 155, 159, 163
    Kuwait, 117
    Operation Iraqi Freedom, 58–60, 115, 128, 130–131, 133, 310, 330
    Operation Southern Watch, 111, 123–124, 131, 301
Iraqi Air Force, 53, 147–148
Iraqi Freedom, Operation, 58–60, 115, 128, 130–131, 165
IRBM (intermediate-range ballistic missiles), 48. *See also* ballistic missiles
ISAF (International Security Assistance Forces), 400–402
ISR (intelligence, surveillance, and reconnaissance), 414
    Australia, 328, 330–332
    China, 289
    Continental Europe, 408
    Japan, 311, 313
    Singapore, 327–328

Soviet-Russian, 213, 214
U.S., 135
Israel. *See also* Israel Defense Forces; Israeli Air Force
    air power overview, 4–5
    foreign sales of aircraft, 288, 326, 361, 484, 485
Israel Aviation Industries (IAI), 157–158, 162, 484, 485
Israel Defense Forces (IDF)
    background and overview, 4–5, 137–138, 170–171
    First Intifada, 162–164
    First Lebanon War, 160
    Second Intifada, 166–167
    Second Lebanon War, 166, 168, 170, 171
    Sinai War and after, 141–146, 159
    Six Day War, 146–150, 159, 307
    War of Attrition, 150, 152, 198
    War of Independence, 138–140, 159
    Yom Kippur War and after, 154–155, 157–159, 171
Israeli Air Force
    Arab-Israeli War of 1973, 101–102
    background and overview, 4–5, 137–138, 170–172, 310
    First Gulf War, 163–164
    First Lebanon War, 160–161, 164
    Operations Accountability and Grapes of Wrath, 164
    Second Intifada, 165–167, 171
    Second Lebanon War, 168–169, 170–172
    Sinai War and after, 141–146
    War of Attrition, 150–152
    War of Independence, 138–141
    Yom Kippur War and after, 153–161, 171
ISTAR (intelligence, surveillance, target acquisition, and reconnaissance), 59, 62
Italian Navy, 400
Italy
    aerospace cooperation, 378, 394, 396, 398
    Afghanistan, 400
    Balkans, 56, 57, 383, 390
    first Gulf War, 376
    Latin America, 344, 347
    WWI, 13–14, 373
    WWII, 17, 30–31, 74, 76, 79, 81, 350
Ivory Coast, 403
Ivry, David, 158–159, 162

J-7s, J-8s, J-10s and J-11s, 288, 289
Jaguars, 53, 56, 361
Jammu, 175, 229, 231, 237, 250, 251, 468
Japan
    air power background and overview, 296, 299–300, 304–306, 332–333
    culture and effect on war strategy, 308–309, 313
    doctrine, 306
    future, 312–313
    self defense, 309–312
    WWII, 73–75, 80–81, 223–225, 261–262, 307–309, 314, 329, 350
Japan Air Self-Defense Forces (JASDF), 311–313

Japan Self-Defense Forces, 310, 313. *See also*
     Japan Air Self-Defense Force
Japanese Army Air Force (JAAF), 305–306,
     308–309
Japanese Naval Air Force (JNAF), 305–306,
     307–309
JAS 39s, 399–400, 404
JDAM (Joint Direct Attack Munition), 125,
     129, 166, 331, 447
Jeschonnek, Hans, 34
JH-7s, 287–288, 475
Jiang Zemin, 287
Jian-6s, 281
JNAF (Japanese Naval Air Force), 305–306,
     307–309
Johnson, Dana J., 90
Johnson, Louis, 85–86
Johnson, Lyndon, 99
Johnson administration, 98–99, 100, 102, 126
Joint Chiefs of Staff (JCS), 81, 85, 86, 99, 120
Joint Direct Attack Munition (JDAM), 125,
     129, 166, 331, 447
Joint Helicopter Command, 59
Joint Intelligence Committee (JIC), India, 255
joint operations of U.S. military
     Afghanistan, 129
     Cold War and after, 104–105, 114–116, 119
     Korean War and after, 89–94, 96
     Vietnam War, 100–101
     WWII and after, 80–81, 85–88
Joint Planning Committee (JPC), India, 255
Joint Strike Fighter program, 134, 331,
     395–396
Joint Strike Fighters (JSF)
     about, xv–xvi, 414
     Australia, 330–331, 332
     Europe, 395–396
     Japan, 313
     Singapore, 327
     U.S., 134–135
joint tactical air controller (JTAC), 61
Joint Vision 2010, 115–116, 121, 126
Joint Vision 2020, 126–127, 448, 454
Jones, R.V., 37
Jordan, 124, 145–148, 155, 160, 239, 253
Jordanian Air Force, 145, 147–148
Jordanian Army, 148
Jorn radar system, 330, 332, 479
JSF (Joint Strike Fighters). *see* Joint Strike Fighters
Ju-52s, 344, 347
Ju-87s, 34
Jumper, John, 126, 127, 135

KA-6s, 53
Kalaikunda, 233
kamikazes, 308, 309
Kargil, 237–238, 250–252
Kargil War, 237–238
Kármán, Theodore von, 108
Kashmir
     First and Second Kashmir Wars, 229,
          231–234
     Kargil Wars, 237, 238
     Pakistani occupation and terrorism, 250, 251
Kasserine, 77, 78, 130

Kawasaki, 306, 312
KC-30As, 330, 331
KC-130s, 356
KC-767s, 313, 399
KDC-10s, 401
Keegan, John, 125–126
Kennedy, John F., 95, 352
Kennedy administration, 38
Kenya, 45
Keys, Ronald, 119
Key West Agreement, 85, 96
Kfirs, 157–158, 361, 370
Khan, Ayub, 233–234
Khan, Nur, 233
Khan, Tikka, 236
Khan, Yahya, 235
Khe Sahn, 319
Khrushchev, Nikita, 204
Ki-43s, 73
Kissinger, Henry, 321
Korean War, 87–92, 196–197, 259, 265–271,
     300, 362
Kosovo, 56–58, 124–126, 286, 383–384,
     386. *See also* Balkans
Kroon, Lukas, 384
Kundt, Hans, 341, 345–346
Kursk, Battle of, 191
Kuwait, 52, 56, 108–109, 117, 132, 239,
     253. *See also* Gulf War, first

La-9s and La-11s, 265
Lal, P.C., 235
Lambeth, Benjamin, 126
Lampert, Florian, 68
Langley, Samuel P., 64
Langley Aerodrome, 64
LANTIRN-equipped bombers, 123–
     124, 166
Laos, 99, 300
Lapidot, Amos, 162
Laskov, Haim, 139
Latin America. *See also specific countries*
     air power overview, 296–297, 335, 371–372
     WWII, 348–350
Latvia, 388, 399
Lavi multipurpose aircraft, 158
Law, Bonar, 18–19
LCU F4, 360, 484
leadership as factor in air power effectiveness, xix
League of Nations, 345
Lebanese air forces, 147–148
Lebanon, 4, 147–148, 158–162, 164–165,
     168–172, 239
Lee Kuan Yew, 326
Leigh-Mallory, Trafford, 33
LeMay, Curtis, 84–85, 99
Lend Lease program, 348
Lenin, Vladimir, 182, 202–203
Leninism, 173, 179, 211, 213
Liberal Democratic Party (LDP) of
     Japan, 313
Libya, 31, 105, 373
Linebacker II, Operation, 99, 100–101, 102,
     320–321

Linhard, Robert E., 112, 118
Link, Charles D., 112–113
Litani, Operation, 164
Lithuania, 388, 399
Liu Yalou, 264
Lloyd George, David, 17
Lockheed 14 Super-Electras, 70
Lockheed Martin, xv, 324, 396. *See also specific aircraft of*
logistics as factor in air power effectiveness, xix
Loh, Michael, 103, 451
Lombardo, Juan, 355
London Air Defence Area (LADA), 22
Long-Range Aviation, Soviet Union, 188, 201, 202
Lord Curzon, 220
Luftwaffe. *See also* German Air Force
    Britain, 22–26, 31, 32–36
    North Africa, 32, 77
    Soviet Union, 187, 188, 189–192, 193
    U.S., 79, 194, 220
Luttwak, Edward N., xvii
Lysander reconnaissance aircraft, 224

MacArthur, Douglas, 89, 90, 309
MacDonald, Arthur, 23–24
MacRobertson Air Race, 70
Mad Mullah, 8, 17
Magisters, 146, 485, 486
maintenance as factor in air power effectiveness, xix
Majumdar, K.K., 224, 258, 464
Malaya, 44–45, 306, 323, 330
Malaysia, 300, 323, 326, 479
Maldives, 238
Malik, V.P., 237
Malta, 46, 49, 330
Manchuria, 73, 197, 305
Manu, 227
Mao Zedong
    air power strategy, 262–263, 268, 269–270, 277
    Korean War, 268, 269–270
    political radicalism and influence on military, 279–280
    Sino-Indian War, 230
    Soviet Union, 272
Marshall, George C., 70–71, 82
Martin, 69. *See also* Lockheed Martin; *specific aircraft of*
Marxism, 173, 179, 182, 211, 213
massive retaliation doctrine of NATO, 94–95, 375
M.B. 150s, 73
MC 14/2s, 95
MC-48s, 94, 95
McGiffert, John R., 101
McMahon Act, 43
McMaster, H.R., 99, 120
McNamara, Robert S., 96, 98, 99, 100
McPeak, Merrill
    about, 103, 112–113, 117, 122, 446, 451
    on air power, 63, 116
    first Gulf War, 109, 111

Global Reach–Global Power, 113, 114–115, 118
MD 500 Defender helicopters, 161
Me-110's, 34
Mecozzi, Amedeo, 76, 431
medical evacuation, aerial, 239, 367, 371
Meiji Restoration, 305
Meir, Golda, 101, 155
memorial, U.S. Air Force, 123
Menendez, Mario, 357, 360
Menon, Krishna, 255
Merlin helicopters, 61
Mesopotamia, 16, 17–18
Messerschmitt 109s, 73, 138
Metcalf, Victor, 64
Meteors, 141, 142
Mexican Air Force, 349–350
Mexican Revolution, 335–337
Mexico, 335–337, 348–350
Meyer, Edward C., 132
Mi-8s, 361, 371
Mi-25s, 361
Midway, Battle of, 80, 307, 329
MiG-9s, 265–266, 470
MiG-15s, 89–90, 196–198, 265–266, 269, 325, 352
MiG-17s, 144, 276, 319–320, 352
MiG-19s, 144, 281, 319–320
MiG-21s, 144, 146, 151, 159, 210, 236, 244–245, 319–321, 467
MiG-25s, 159
MiG-29s, 245, 288, 384, 398, 399, 405
Military Transport Aviation, Soviet, 201
Milosevic, Slobodan, 56–58, 122, 125–126, 383, 387
Ministry of Defence (MOD), UK, 61–62
Mirage 3s, 144, 146, 355, 356
Mirage 5 Daggers, 157–158, 355, 356, 359, 360–361, 371
Mirage 2000s, 361, 391, 401
Mirage F1s, 361, 401, 404
missile defense systems, 47, 246, 248, 332
missiles. *See also specific missiles*
    Argentina, 356, 359
    Australia, 330–331, 332
    Britain, 47–49, 51, 357–358
    China, 248–250, 289, 310–311
    Continental Europe, 379, 390
    Egypt, 144–145
    future, 412
    India, 244, 245–246
    Israel, 161, 162, 163, 165–166
    Japan, 311–312
    Korea, 310
    Pakistan, 251
    Soviet Union, 195, 199, 200–201, 202, 203
    U.S., 84, 93–95, 96–99, 101–102
    Vietnam, 320
Mitchell, William, 65–68, 82, 261, 409
Mitsubishi, 306, 312
Mitterrand, François, 377
Mladic, Ratko, 122
Moffett, William, 67, 72
Mohammed, Abdullah, 129
Moked, Operation, 147

Mondragon, Manuel, 335
Mongols, the, 181
Montgomery, Bernard, 31
Montt, Rios, 364
Mooney, Chase, 81
morale, bombing effects on
    Britain, 8, 10, 13–15, 19–21, 35–39, 40, 46
    Spain, 26, 74
Moseley, T. Michael, 124, 135, 401
mottos, air force, xvi
Mountbatten, Louis, 225
Movement Coordination Center Europe, 392
Mozhaiski, Alexander F., 183
M.S. 406s, 73
Mubarakmand, Samar, 251
Mukerjee, Subroto, 222
Mukti Bahini, 235
Muliutin, D.A., 183
multirole combat aircraft (MRCA), 175, 245,
    249, 288, 325, 413–415, 468
Mumbai bombings, 252
MUSIS, 398
Musketeer, Operation, 46
Mustangs, 39, 73, 79, 142, 308, 362, 432
mutually assured destruction, 195, 203
Mystères, 141, 142, 144, 146, 150, 153, 233

NA-50s, 347
Nagasaki atomic attack, 43, 80–81, 307,
    308, 329
Nakajima, 306
Nanawa, 345
Nasser, Gamal Abdel, 46, 141, 146, 148, 150
National Advisory Committee for
    Aeronautics, 69
National Aeronautics and Space
    Administration (NASA), 82
National Defense Act of 1947, 82
National Defense in 2008 (China), 310
Nationalists, Chinese, 259, 261–263, 265,
    271, 274–276
National Security Advisory Board (NSAB),
    India, 245
National Security Council, U.S., 94, 97
National Space Development Agency, Japan, 312
NATO (North Atlantic Treaty Organization)
    Afghanistan, 128, 400–402
    Argentina, 359
    Bosnia, 55–56, 121–122, 382–383, 385
    Britain, 45, 49
    capabilities gaps, 390
    Cold War, 374–376, 382
    European Union, 389, 407
    expansion, 216, 387–388, 399
    Germany, 380
    Kosovo, 57–58, 124–126, 383, 385
    massive retaliation doctrine, 94–95
    NATO Response Force (NRF), 378, 388
    Netherlands, 374
    post Cold War, 387–388
    Strategic Airlift Capability Initiative, 392, 399
    Warsaw Pact threat, 103–104, 195, 210, 404
Navarre, Henri, 314–315, 317
Nehru, Jawaharlal, 230

Neshers, 157–158
Netherlands
    Afghanistan, 401–402, 405
    Balkans, 379, 382–385, 393–394
    EU cooperation, 378, 392–393, 396,
        397, 398
    first Gulf War, 378–379
    Indonesia, 322
    procurement policy, 395, 396
    WWII, 29
New World Vistas, 117
New Zealand, 15, 40, 128, 407, 479
Nicaragua, 76, 338–340, 351, 352, 353–
    354, 366
Nimrods, 49, 51, 53, 56, 57, 58, 61
9/11 attacks, 127, 303, 307
Nixon, Richard, 99, 320
Nixon administration, 99, 320, 442
Noble Eagle, Operation, 128
nonalignment policy of India, 174,
    226, 228
Nordic Battle Group, 403–404
Normandy invasion, 41, 77, 81–82
"normative" war, 132–133
North Africa, 2, 30–32
North Atlantic Council (NAC), 54–55, 57
North Atlantic Treaty Organization (NATO).
    See NATO
North China exercise of 1981, 282
Northern Alliance, 129–130
Northern Watch, Operation, 111, 131
North Korea, 88–92, 196–197, 268–269, 310
North Korean Air Force, 268–269
Northrop, 70
North Vietnam. *See* Vietnam
Northwest Frontier Province of India, 18, 45,
    220, 222, 241
Norway, 41, 396–397, 401, 403
nuclear deterrence
    Britain, 46–50
    India, 245, 253
    NATO, 375
    Pakistan, 246
    Soviet-Russian, 195, 196–197, 211, 216
    U.S., 2, 94
Nuclear Suppliers Group, 247
nuclear transformation and capabilities. *See
    also* atomic power
    Britain, 43–44, 45, 46–50
    China, 247–250, 310–311
    France, 378
    India, 241, 245–246, 247
    NATO, 374–375
    Pakistan, 250–252
    Soviet Union, 199–201, 202, 203–204
    U.S., 92–95, 97–99, 102, 108–109

O-2s, 367, 486
Oboe navigation aid, 37
Obregon, Alvaro, 336
Ochey Wing, 13
Odessa, 95
offensive strategies and operations, air. *See* air
    offensive strategies and operations

Olds, Robert, 68
Omar, Mullah Mohammad, 128
operation research section (ORS), Britain, 37–39
operations as factor in air power effectiveness, xix
Organization of American States (OAS), 128, 354
Oslo Accords, 166
Ospreys, 344, 345
Ouragans, 141, 142, 146, 150, 367, 485, 486
OV-10s, 98
Overlord, Operation, 39, 79

P2Vs, 86
P-8A Poseidons, 331–332
P-38s, 73, 430
P-47s, 73, 350, 351
P-51 Mustangs, 39, 73, 79, 142, 308, 362, 432
Pakistan. *See also* Pakistan Air Force; Pakistan Army
   China, 247–248
   First Kashmir War, 229, 464
   India, 175, 245–246, 468
   India-Pakistan War, 234–237
   Kargil War, 237–238
   nuclear weapons, 250–252, 468
   Second Kashmir War, 231–234, 255–256, 465
   terrorism, 257
Pakistan Air Force
   First Kashmir War, 229
   as future threat to India, 250–252, 467–468
   India-Pakistan War, 235–236
   Kargil War, 238
   Second Kashmir War, 231–234, 465
Pakistan Army, 231, 236, 237–238
Palestine Liberation Organization (PLO), 160, 162, 166
Palestinian Authority, 166–167
Palliser, Operation, 402
pancasila democracy, 322, 478
Paraguay, 296, 340–347, 371, 481
Paraguayan Air Corps, 342, 480
paratroopers
   Guatemala, 365
   Israel, 142, 143, 148–149
   Peru, 347, 348, 371
Park, Keith Rodney, 33, 36
Partnership for Peace (PfP), 387
Pastrana, Andres, 370
Path Finder Force (PFF), 38–39
Patrick, Mason, 68
Patriot missiles, 115, 311, 312, 379
Patton, George, 78
Paveways, 58, 61, 311
PAVN (People's Army of Vietnam), 317, 319, 321
PBY Catalinas, 69, 70
PC-7s, 364, 485
peace accords, Israel and Egypt, 158–160
Peacekeeper missiles, 102
Pearl Harbor, attack on, 73, 80, 223, 307, 329
Peled, Binyamin, 152–153, 155, 156–157, 158
Peng Dehuai, 268
Pennsylvania, 64
People's Army of Vietnam (PAVN), 317, 319, 321
People's Liberation Army (PLA). *See also*

People's Liberation Army Air Force
   background and overview, 248–249, 260, 262–263, 293
   Korean War, 267
   PLAAF subordination, 265, 267, 270–271, 278
   structure and Soviet model, 271–274
People's Liberation Army Air Force (PLAAF).
   See also People's Liberation Army
   background and overview, 176–177, 259–260, 292–293
   current capabilities, 310–311
   doctrine, 276–280
   first half of 20th century, 260–266
   Korean War, 266–271
   late 1970s to 1990, 281–284
   modernization, 246, 248–249, 259
   offensive strategy transition of 1990's, 284–287
   post Korean War identity, 271–276
   Vietnam invasion, 280–281
perestroika, 210–212
Perry, William, 117
Pershing, John J., 65, 337
Peru, 297, 347–348, 360–361, 368, 371
Peru-Ecuador War of 1941, 297, 347–348
Peru-Ecuador War of 1995, 297, 360–361
Peruvian Air Force (FAP), 348, 361, 368
Peters, F. Whitten, 123, 124, 126
PGMs. *See* precision-guided munitions
Phantoms, 52, 105, 149, 151, 153, 158
pilots, commonalities of, xvi
PLA (People's Liberation Army). *See* People's Liberation Army
PLAAF (People's Liberation Army Air Force). *See* People's Liberation Army Air Force
Plan Colombia, 370
Plan Patriota, 370
PLO (Palestine Liberation Organization), 160, 162, 166
Poland, 216, 388, 398–399, 407
Polaris missiles, 49
policing, air. *See* air policing
Polikarpovs, 73, 74, 76
Polish Air Force, 398–399, 407
political systems, importance to strategy, xvii–xviii, 300
political work system, China, 274, 279–280
Popeye missiles, 162
Porokhovschikov, Aleksandr, 183
Portal, Charles, 26, 36, 40, 44
Portugal, 401
Potez 25s, 342, 345, 480, 481
Prague Spring, 103
Prakash, Arun, 237
precision-guided munitions (PGMs). *See also* *specific PGMs*
   Balkans, 383
   Britain, 55, 58–59, 61
   Colombia, 370–371
   Continental Europe, 297, 381, 390–391
   first Gulf War, 297, 376, 379
   Israel, 61, 171
   Soviet Union, 207

precision strike capabilities of EU, 390–391
Predators, 125
Prince of Wales, HMS, 223–224, 307, 329
procurement policies, 28, 69, 395–397, 399, 406–407
Provide Promise, Operation, 111–112
PTDI, 324
Pucaras, 355–356, 357, 360

Q-5s, 288
Quadrennial Defense Review (QDR), 117, 119, 121, 123
Quesada, Elwood R., 87
Quick Reaction Alert (QRA), 48, 399–400

RAAF (Royal Australian Air Force), 46, 219, 302, 323–324, 326, 328–331, 440
Rabin, Yitzhak, 147
Radio Detection Finding (RDF), 53
RAF (Royal Air Force). *See* Royal Air Force
RAF Staff College, 26–27
Rafales, 391, 401
RAH-66 Camanches, 134
Rambouillet Agreement, 57, 58
Ramon, Ilan, 165
RAN (Royal Australian Navy), 328–329
RAND study on first Gulf War, 110
Rann of Kutch conflict, 229
Rao, Narasimha, 242
RAPTOR (Reconnaissance Airborne Pod TORnado), 61
Raptor Recovery Group, 124
Reagan, Ronald, 353, 365
Reapers, 61
reconnaissance, 300, 362. *See also* ISR
    Afghanistan, 61
    Falklands War, 358
    first Gulf War, 54, 56
    Germany, 398
    Japan, 306, 312
    Russia, 213–214
reconnaissance-strike complexes, 214
Red Air Fleet, 185–186
Red Air Force. *See* Soviet Air Force
Red Army, 174, 185–187, 188, 189–191, 192
Remez, Aharon, 139
Republic of Singapore Air Force (RSAF), 301–302, 327–328
Republic of South Korea, 88, 196, 278, 317, 442
Repulse, HMS, 224, 307, 329
respond, ability to, 301, 302, 304
Revolt of the Admirals, 85–86
Revolutionary Air Force (FAR), Cuba, 352–353
Revolutionary Armed Forces of Colombia (FARC), 369–371
Revolutionary Military Committee, 185
RF-4Js, 311
RF-101s, 95
RFC (Royal Flying Corps), 1, 8–13, 29, 65, 67, 219, 221
Rice, Donald B., 107, 112, 116, 117, 120
Rifkind, Malcolm, 55
Rio Escondico, SS, 353
Rioja, Bernardino Bilbao, 341

Riza, Shaukat, 236
RN (Royal Navy). *See* Royal Navy
RNAS (Royal Naval Air Service), 9, 11–13, 67
Robertson, William, 10
robot weapons, 84
Roche, James, 135
ROE (rules of engagement). *See* rules of engagement
Rogers, Bernard W., 375
role specialization, EU, 298, 392, 393, 406
Rolling Thunder, Operation, 99–100
Romania, 388
Rommel, Erwin, 31
Roosevelt, Franklin, 72
Roosevelt, Theodore, 64
Rose, Michael, 121, 452
Rotem alert, 144
Roy, Indra Lal, 221
Royal Air Force (RAF)
    Afghanistan, 58–59, 60–61
    air mobility, 391–392
    bombing policy, 20–22, 77, 432
    Cold War, 46–50
    EU cooperation, 406
    Falklands War, 51–52, 357–360
    first Gulf War, 380, 381
    France, 377–378
    gaining independence, 15–19, 67
    India, 174–175, 219–223, 240, 463
    Iraq, 52–54, 58–60
    Japan, 306
    Kenya and Malaya, 44–45
    nuclear transformation, 43–44, 45, 46–50, 95
    overview and origins, xvi, 1–2, 7, 10–15
    pre-WWII expansion, 24–28
    Singapore, 326
    1990's reduction, 56
    structure and technology, 23–24
    WWII, army cooperation, 28–32
    WWII, Battle of Britain, 32–36
    WWII, Bomber Command, 36–41
    WWII, Coastal Command, 41–43
    WWII, North Africa, 31–32
    Yugoslavia, 54–58
Royal Australian Air Force (RAAF), 46, 219, 302, 323–324, 326, 328–331, 440
Royal Australian Navy (RAN), 328–329
Royal Flying Corps (RFC), 1, 8–13, 29, 65, 67, 219, 221
Royal Naval Air Service (RNAS), 9, 11–13, 67
Royal Navy (RN), 67
    Australia, 328–329
    Burma, 223–224
    Egypt, 46
    Falklands War, 355–360
    India, 466
    Iraq, 59
    Kosovo, 57, 390
    military aviation, 8–9
    nuclear transformation, 49
    RAF, 11–12, 18–21, 59, 85
    WWII, 42, 307
Royal Netherlands Air Force, 378–379, 382, 393, 395, 401, 405

RSAF (Republic of Singapore Air Force),
    301–302, 327–328
rules of engagement (ROE)
    Balkans, 55–56, 57–58, 382, 385
    China, 275
    Iraq, 53, 59
    Vietnam War, 99
Rumsfeld, Donald, 389
Russia. *See* Soviet Union-Russia
Russian Air Force, 173–174, 179–180, 212–217
Rwanda, 402–403
Ryan, Micheal E., 121–124, 126–127, 383

S-55s, 148
SA-2s, 145, 150, 320
SA-3s, 244
SA-7 Strela missiles, 99, 101
SA-10s, 126
SA-12s, 126
Saab, 395, 396
Sadat, Anwar el-, 153, 158
Saint Malo Declaration, 389
Salisbury Committee, 19
SAMs (surface-to-air missiles). *See* surface-to-
    air missiles
Sandinista War, 297, 353–354
Sandino, Augusto, 338–340
SAR-Lupe satellites, 398
satellites, xvi. *See also* aerospace power
    Asia Pacific region, 289, 312, 331, 332
    Britain, 358
    Continental Europe, 397–398, 408
    NATO, 104
Saudi Arabia, 52, 56, 109, 163, 376
SBDs, 70, 432
SCCOA (système de commandement et de
    conduite des opérations aériennes), 378
Schemes, RAF expansion, 24–25, 29
School of Advanced Airpower Studies (SAAS),
    118–119, 449
Schriever, Bernard, 94
Schwalier, Terryl, 119–120
Schwarm fighters, 33
Schwarzkopf, Norman, 110, 114, 131
Schweinfurt, 39, 78
Science of Air Force Campaigns, China, 283–
    284, 286–287
Science of Air Force Tactics, China, 286–287
Scientific Advisory Board (SAB), U.S. Air
    Force, 117
Scouts, 162
Sea Fury FB 11s, 352–353
Sea Harriers (SHARs), 51, 57
Sea Hawks, 236–237, 344, 345
Sealift Coordination Center, 392
Sea Lion, Operation, 33–34, 35
seamless web of air power, 415
Searchers, 162
Second Intifada, 165–167, 171
Second Kashmir War, 231–234, 255–256, 465
Second Lebanon War, 4, 166, 168–170,
    171–172, 239
Second Tactical Air Force (2TAF), 32, 45, 47
self-reliance policy of India, 174, 226

Selfridge, Thomas, 64
Semenov, N., 205
Sentinel R1 radar, 61
Shalikashvili, John, 116
Shamir, Shlomo, 139
shaping, as part of strategic construct, 301–
    302, 304
Sharon, Ariel, 149
Sheffield, HMS, 359, 483–484
Shelton, Henry H., 126–127
Sherman, William, 70
Shkedi, Eliezer, 168
Short, Michael, 126, 386
short-range attack missiles (SRAM), 102
Shrike missiles, 51
Sidewinder missiles, 51, 311, 356, 358, 360
Sierra Leone, 56, 402
Sikorsky, 69. *See also specific aircraft of*
Sikorsky, Igor, 183
Sinai
    peace accord, 158–159
    Sinai War, 141–143, 144, 146, 148, 159
    Six Day War, 146, 148–149
    Yom Kippur War, 153–154, 157
Singapore, 240, 296, 300, 301, 325, 326–
    328. *See also* Republic of Singapore
    Air Force
Singapore Air Defense Command (SADC),
    326–327, 328
Singapore Technologies Aerospace Company,
    328
Singh, Arjan, 225
Singh, Nagendra, 227
Single Integrated Operational Plan (SIOP),
    93, 102
Sino-Indian War, 230–231, 244, 247, 250
SIPRNet, 60
Sir Galahad, 51, 360, 483–484
Sir Tristram, 360, 484
Six Day War, 4, 144–150, 152, 156, 198, 307
Skeen, Andrew, 221
Skeen Committee, 221
Skybolts, 49, 98
Skyhawks, 149, 153, 327, 355, 356, 358–360
Slessor, John, 28, 29, 45–46, 222
Slovakia, 388
Slovenia, 388
Smith, Herbert, 306
Smith, Leighton, 121, 126
Smuts, Jan, 10, 67
Smuts reports, 10–12
Somaliland, British, 17
South Africa, xvi, 15, 31, 219
Southern Watch, Operation, 111, 123–124,
    131, 301
South Korea, 88, 196, 278, 318, 442
South Vietnam, 99, 101, 317–321, 442. *See
    also* Vietnam War
Soviet Air Force
    Afghanistan, 206–207
    background and overview, 173–174,
        179–180, 182, 217
    Bolshevik period, 185, 186–187
    China, 265

Cold War, 198–199, 200–203, 205,
207–210
Korean War, 196–197
WWII, 174, 187–194, 266
Soviet Navy, 182, 200, 201
Soviet Union-Russia. *See also* Red Army;
Russian Air Force; Soviet Air Force;
Soviet Navy
Afghanistan, 105, 206–207
atomic power, 84, 93
background and early history, 173–174,
179–183
Balkans, 387
Bolshevik period, 185–187
China, 176, 262–266, 268–269, 271–
274, 386, 387, 469
Cold War background and overview,
195–196
Cold War doctrine, 199–205, 207–212
Cold War equipment and training effec-
tiveness, 198–199, 209–210
collapse of, 105, 204, 206
Cuba, 352
current challenges, 215–216
Egypt, 144–145, 151
Germany, 381
global power shift, 304
India, 175, 231, 253, 466–467
Indonesia, 323, 325
Israel, 142
Korean War, 196–197
military doctrine, 211–217
NATO, 94–95
Nicaragua, 354
Peru, 360–361
threat in Europe, 103–104
WWI, 184
WWII, 187–194
Spaatz, Carl A., xvi, 39, 68, 83
space based power. *See* aerospace power
Spacecast 2020, 113, 118
Spain, 25–26, 63, 74, 378, 383, 398
Spanish-American War, 63
Spanish Civil War, 25–26
special operations forces (SOF), 129, 133,
411, 447
Spitfiers, 23, 34, 73, 76, 138, 225
springboard air support, 191–192
Srebrenica, massacre at, 121, 382, 385
Sri Lanka, 175, 238–239
SSMs (surface-to-surface missiles), 161, 163, 165
Stalin, Joseph, 94, 189, 194, 196, 202
stealth revolution, 111, 117
Stimson, Henry, 82
Stingers, 105, 379
Stirlings, 25, 37
Straits of Tiran, 146
Strategic Air Command (SAC), Britain, 48
Strategic Air Command (SAC), U.S., 85, 89,
95–96, 103, 112
Strategic Airlift Capability Initiative, NATO,
392, 399
Strategic Airlift Interim Solution (SALIS), 392
strategic bombing

Britain, 14–15, 20–22, 26–28, 37–40, 43, 373
China, 265
Germany, 187
India, 466
nuclear, 241
Soviet/Russian, 187, 193, 197, 202
U.S., 72, 75–79, 89, 119, 320–321, 373
strategic construct of shape-deter-respond,
301–302, 304, 325, 330
Strategic Defence Reviews, British, 59, 61–62
strategy, concept of, xvii–xviii
Stratocruisers, 150
Struck, Peter, 381
Stukas, 34, 76
Sturmovik, the, 183
Su-22s, 321, 361
Su-27s, 249, 286–289, 310–311, 321, 325
Su-30s, 249, 287, 321, 325
Suez Canal, 30, 46, 141, 148, 150–154, 156
Suharto, 322–323
Sukarno, 322–323
Sullivan, Gordon, 114–115, 447
Sun Tzu, 111, 302
Sun Yat-sen, 261, 262, 263
Super-Electras, 70
Super Etendards, 51, 355, 359, 400–401
Super Frelon helicopters, 148, 152, 153
Supermarine Spitfires, 23, 34, 73, 76, 138, 225
Super Mystères, 146, 153
surface-to-air missiles (SAM), 412
Balkans wars, 385, 386
Britain, 47, 52
China, 279, 283
Egypt, 145, 150
India, 244
Israel, 150–152, 154–155, 157, 159,
160–161
Japan, 311
North Vietnam, 320–321
Pakistan, 238
Soviet Union, 199
U.S, 99, 101, 102, 105, 114–115
surface-to-surface missiles (SSMs), 161, 163, 165
Sweden, 376, 392, 395, 397, 399–400,
403–404
Swedish Air Force, 404
Swinton, Viscount, 23, 25
Sykes, Frederick, 10, 15
Symington, Stuart, 85, 112
Syria. *See also* Syrian Air Force; Syrian Army
Ben-Eliyahu years, 165
First Lebanon War, 160–162
Hezbollah, 169
Six Day War and before, 145–148
U.S., 105
War of Attrition, 150
Yom Kippur War and after, 152–156, 159
Syrian Air Force, 145, 147–148, 150, 156
Syrian Army, 148, 154, 156, 160

T-2/F-1s, 312
T-33s, 352, 362–363
T-34B Mentors, 357
TA-2s, 339

Tactical Air Command (TAC), U.S., 87–88,
        95, 104, 107, 109, 112
Tactical Fighter Experimental (TFX), 98, 102
TAC-TRADOC-ALFA initiatives, 104, 107
Taepodong missiles, 310
Taiwan, 250, 262, 263, 265, 270, 278, 285
Taiwan Strait air battles, 274–276
Taliban, 60–61, 128–130, 391
Task Force Hawk, 124
technology, vulnerability of, xvii–xviii
Tedder, Arthur, 31, 39, 44
Telic. *See* Iraqi Freedom, Operation
Ten-Year Rule, 15–16, 22, 28
terrorism
        India, 237, 251–252, 257
        Israel, 141, 146, 162, 166–169
        9/11 attacks, 127, 303, 307
        Russia, 216
        Taliban, 60
Thailand, 175, 224, 300, 330
Thatcher, Margaret, 50, 51
theatre missile defence system (TMD), China, 248
Thermal Imaging Airborne Laser Designators
        (TIALD), 54
three-dimensional warfare, 80, 81, 214, 218
3EA-6Bs, 53
Tibet, 230, 249, 263
Timor-Leste, 302, 324–325
Tizard, Henry, 23, 26
TNI-AD (Indonesian Army), 302, 322–326
TNI-AU (Indonesian Air Force), 322–326
Tolkovsky, Dan, 139–140, 143
Torch, Operation, 32
Tornados. See also specific variants
        about, 391
        Britain, 52–53, 55, 56, 57, 61
        Cold War, 394, 397
        Germany, 381, 385–386, 394, 402
Torpy, Glenn, 59–60
Toward New Horizons (von Kármán), 108
TR 440-15, 71
training, aviation
        Australia, 329
        Britain, 8–9, 16–17, 27
        China, 262, 263–264, 266, 267, 290–291
        El Salvador, 366–367
        as factor in air power effectiveness, xix, 303
        India, 221
        Indonesia, 323–324, 325
        Japan, 306, 309
        Mexico, 350
        Poland, 399
        Soviet Union, 198–199, 208–209
        U.S., 71–72, 85, 98–99, 290–291, 309
Training and Doctrine Command
        (TRADOC), U. S. Army, 104, 107
Trenchard, Hugh
        allied air force, 13–14
        as commander of RFC, 9–10
        foundation of air power, 303
        RAF independence, 2, 15–19
        RAF Staff College, 26–27
        strategy of, 13–15, 20–22, 28, 36–37,
                62, 220
        U.S., 65

Trenchard Memorandum, 29
Trident Conference, 78
Trotsky, Leon, 185, 186
Truman, Harry, 43, 82, 83, 85, 308
Tsushima, battle of, 305
Tu-2s, 265–266
Tu-16s, 144, 325
Tupolev SB, 74
Turkey, 56, 301, 379, 380, 383
Turquiose, Operation, 402–403
Twelfth Army Group, 78, 87, 90
Typhoons, 391, 394, 396

UAVs (unmanned aerial vehicles), 111
        Australia, 330, 331–332
        Britain, 61, 65
        Indonesia, 324
        Israel, 137, 161–162
        Singapore, 328
U-boats, 2, 41–43, 348–350
Uebe, Klaus, 194
UH-1s, 363, 367
UH-19Bs, 362–363
UH-60 Black Hawk helicopters, 164, 312,
        331, 369, 371, 487
Ultra intelligence, 31, 33, 42
unit cohesion as factor in air power effective-
        ness, xix
United Kingdom. *See* Britain
United Nations
        Bosnia, 54–56, 121–122, 382–383, 385, 390
        Congo, 403
        force structure, 301
        India, 175, 232, 238
        Korean War, 89, 197, 259, 268–271, 300
        Kosovo, 124, 126, 390
        Lebanon, 168
        Pakistan, 229
        Security Council Resolutions (*See* United
                Nations Security Council
                Resolutions)
        Timor-Leste, 324–325
United Nations Protection Force
        (UNPROFOR), 55, 121
United Nations Security Council Resolutions
        (UNSCR)
        Balkans, 55, 57
        Iraq, 58, 301
        Israel-Lebanon, 168
United States
        aerospace, 395–396, 397
        Afghanistan, 58, 60–61, 128–130
        air power origins and overview, 2–4,
                63–65
        Argentina, 355
        Asia Pacific region, 304, 317
        Bolivia, 344, 345
        Caribbean interventions, 338–340
        China, 262, 264, 271, 275, 288, 469
        Clinton era, doctrine, 118–121, 126–127
        Clinton era, early, 116–117
        Clinton era, the Balkans, 121–126
        Columbia, 368–370
        current capabilities, 391

Ecuador, 361
El Salvador, 365–367
Falklands War, 51–52, 358
first Gulf War, 52–54, 108–111, 376, 380, 447
future, 133–136
Guatemala, 362–364
India, 244, 247, 253, 466
Indonesia, 325
inter-World War years, 66–73
Iraq (*See* Gulf War, first; Iraqi Freedom, Operation)
Israel, 142, 149, 152, 158
Japan, 306, 313
Korean War, 87–92
Kosovo, 57, 386
Mexico, 336–337
Pakistan, 251
Paraguay, 345
Peru, 361
Poland, 398–399
post Cold War, 105–108, 111–116, 446
post Vietnam War technological transformation, 101–102
post WWII, 82–88, 351–354
Singapore, 326, 327
Soviet Union, 216, 244–245
technological transformation and Cold War, 48–49, 92–96, 104–105
Vietnam War, 96–99, 317–321
WWI, 65–66, 76
WWII, doctrine, 75–79, 432
WWII, foreign air power lessons, 73–75
WWII, joint service operations, 79–81
WWII, Latin American allies, 348–350
United States, USS, 85
United States Air Force (USAF)
Afghanistan, 128–130
Australia, 331
background and origins, 2–4, 64, 66, 68–69, 81–82, 87–88, 430
Britain, 48
Clinton era, doctrine, 118–121, 126–127
Clinton era, early, 116–117
Clinton era, the Balkans, 121–126
Cold War and technological transformation, 102, 104–106
Cuban Missile Crisis, 95–96
education and training, 290–291
first Gulf War, 52–53, 109–111, 113, 377, 447
future, 133–136
Iraqi Freedom, 131–132
Korean War, 88–92, 99, 267–269, 271
post Cold War, 105–108, 111–116, 445, 446
post Korean War transformation, 92–94, 96–99
post WWII, 82–88
space operations, 289
USAFE, 94, 95, 121, 126
Vietnam War, 97–101, 320
United States Air Force in Europe (USAFE), 94, 95, 121, 126
United States Air Force Reserve, 128

United States Air National Guard, 128
United States Army. *See also* United States Army Air Corps; United States Army Air Forces; United States Army Air Service
aviation, 64, 67–69, 77, 80–81
Cold War and after, 104, 114–115
El Salvador, 364
future, 133
Iraqi Freedom, 115, 131–132
Korean War, 88–91
Kosovo, 125
Mexico, 337
post WWII, 85–86, 87–88
Vietnam War, 100–101, 317
United States Army Air Corps, 69, 70, 71–72, 74–75, 430, 432
United States Army Air Forces (USAAF)
air force independence, 81–82, 430
post Korean War transformation, 96
post WWII, 84
WWII, bombing Germany, 37–40, 72, 87, 432
WWII, India, 241
WWII, lessons of Battle of Britain, 74–75
WWII, war against Japan, 80–81, 308, 329
United States Army Air Service, 68–69, 70, 75–78
United States Coast Guard, 369
United States Marine Corps
Afghanistan, 401
Balkans, 391
Caribbean interventions, 338–340
Korean War, 88–91
post Korean War transformation, 88–91, 96
post WWII, 85–86
WWII, 75–76, 77, 80
United States Navy (USN)
Afghanistan, 400
aviation origins, 64
Cuban Missile Crisis, 95
first Gulf War, 53, 445
future, 134
inter-world war years, 67–69, 71, 72–73, 74–75
Korean War, 88–91, 96
post Cold War, 114
post Korean War transformation, 92, 96–98
post WWII, 84–88
Vietnam War, 96–101, 320
WWII, 80–81, 305, 307, 308, 349
United States Strategic Bombing Survey (USSBS), 78, 79–80
unmanned aerial vehicles. *See* UAVs
Uribe, Alvaro, 370
Uruguay, 345, 482
USAAF. *See* United States Army Air Forces
USAF. *See* United States Air Force
USAFE (United States Air Force in Europe), 94, 95, 121, 126
USN. *See* United States Navy
USSBS (United States Strategic Bombing Survey), 78, 79–80
USSR. *See* Soviet Union-Russia

V-1 cruise missiles, 84
V-2 ballistic missiles, 84
Valiants, 44, 46
Vandenberg, Hoyt, 85, 112, 259
Vautours, 146, 150
Vertical Take-off and Landing (VTOL) aircraft, 357
Vespa fighters, 342, 343
V-force, 44, 45, 47–50
Victors, 44, 51
victory, concept of, xvii–xviii
Viet Cong, 100, 101, 317–318, 320
Viet Minh, 314–317
Vietnam
    air power overview, 296, 299–300, 314
    national liberation, 314–317
    post Vietnam War to present, 321
    Vietnam War, 96–102, 199, 259, 271, 278, 314, 317–321
Vietnamese Air Force, 280–281
Vietnamese People's Air Force (VPAF), 319–320, 321
Vietnam War, 96–102, 199, 259, 271, 278, 314, 317–321
Viggens, 395, 397
Vigilant Warrior, Operation, 117, 118
Vikrant, INS, 237
Villa, Pancho, 336–337, 339
von Braun, Wernher, 94
Vought 02U-1s, 339
Vulcan bombers, 44, 51–52, 358
vulnerability analyses, xvii
Vulti Vengeances, 225

Walker, Walton, 89, 90
Wapiti aircraft, 222
Warden, John, 103, 106–107
warfare grammar, xvii–xviii
War of Attrition, 150–152, 198
War of Independence (Israel), 4–5, 138–141, 149, 159
War on Terror, 111, 400–402, 407
Warsaw Pact, 45, 53, 103–104, 195, 374–376, 387, 398
Watson-Watt, Henry Robert, 23
Waziristan campaign, 222
Weizman, Ezer, 143, 146, 154
Welch, Larry, 118, 122, 445
Wells, H.G., 7
West Bank, 148, 163
White, John P., 115
White Commission, 115, 116
Widnall, Sheila, 117
Wild Weasel, 100
Wilson, Woodrow, 336
Winnefeld, James A., 90
Woodward, John, 357

World Conference on Disarmament, 22
World War I
    Britain, 9–15, 29
    Germany, 373
    India, 221
    Italy, 13–14, 373
    Russia, 173, 184
    U.S., 65–66, 76
World War II, 410
    Australia, 328–329
    Battle of Britain, 32–36
    Britain overview, 2
    British Bomber Command, 21–23, 36–40, 42
    British Coastal Command, 41–43
    China, 261–262
    Germany, 31–32, 33–36, 77–80, 188–191, 348–350, 373
    India, 174–175, 223–225, 241, 463
    Italy, 17, 30–31, 74, 76, 79, 81, 350
    Japan, 73–75, 80–81, 223–225, 261–262, 307–309, 314, 329, 350
    Latin America, 348–350
    North Africa, 30–32
    pre-war planning and rearmament, 24–28
    RAF and India, 174–175
    RAF-British army cooperation, 28–32
    Soviet Union, 173–174, 187–194
    U.S. doctrine, 75–79, 432
    U.S. foreign lessons, 73–75
    U.S. military joint operations, 79–81
    Vietnam, 314
Wratten, William, 52–53
Wright Brothers, 3, 64–65
Wynne, Michael, 127, 135

X-20s, 98
XB-15s, 70
XB-70s, 87
XC-2s, 312
XP-1s, 312
Xu Qiliang, 288

Y-7s, 290
Y-8s, 288, 290
Y-20s, 290
Yadin, Yigael, 141
Yak fighter, 192
Ye Jiangying, 281
Yom Kippur War, 4, 101–102, 152–157, 159, 161, 170–171, 442
Yugoslavia, 54–58. See also Bosnia; Kosovo

Zakaria, Fareed, 129
Zeppelin airships, 10
Zhou Enlai, 230, 279

Christian F. Anrig is deputy director of doctrine research and education, Swiss Air Force. From early 2007 until September 2009, he was a lecturer in air power studies in the Defence Studies Department of King's College London, while being based at the Royal Air Force College. He was one of two leading academics who created the new distance-learning master's degree program on Air Power in the Modern World, especially designed for Royal Air Force officers, and served on the editorial board of the *Royal Air Force Air Power Review*. In November 2009 he became a member of the Royal Air Force Centre for Air Power Studies academic advisory panel. Dr. Anrig began his professional career in the field of defense studies as a researcher at the Center for Security Studies at the Swiss Federal Institute of Technology (ETH Zurich) in January 2004. He has published various articles, primarily in German, covering topics from European Union Battle Groups and military transformation to modern air power and its ramifications for small nations. He served in the mountain artillery of the Swiss Army and is currently a reserve captain assigned to the air staff (Swiss Air Force). Dr. Anrig holds a PhD from King's College London. He will publish his first book titled *The Quest for Relevant Air Power: Continental European Responses to the Air Power Challenges of the Post–Cold War Era*, with Air University Press in 2011.

Itai Brun is a brigadier general in the Israeli Air Force and currently heads the Israel Defense Forces' DADO Center for Interdisciplinary Military Studies.

Prior to assuming that position in September 2006, he served as a senior analyst in the Israeli intelligence community, with special responsibility for political-strategic assessments and methodology. From 2001 to 2004, he headed the Analysis Department of Israeli Air Force Intelligence. He has published various articles on intelligence and air power, gives lectures at conferences on military- and security-related issues, and contributes articles and essays to various publications on both the changing character of war and the role of intelligence in warfare. General Brun is a graduate of the IDF Command and Staff College; he also holds a bachelor's degree in law studies from the University of Haifa and a master's degree in political science from the University of Tel Aviv.

James S. Corum is the dean of the Baltic Defence College. From 1991 to 2004, he was a professor at the U.S. Air Force School of Advanced Air and Space Studies at Maxwell Air Force Base, Alabama. In 2005 he was a visiting fellow at All Souls College, Oxford, where he held a Leverhulme Fellowship, and then an associate professor at the U.S. Army Command and General Staff College, Fort Leavenworth, Kansas. Dr. Corum is the author of several books on military history, including *The Roots of Blitzkrieg: Hans von Seeckt and German Military Reform* (1992); *The Luftwaffe: Creating the Operational Air War, 1918–1940* (1997); *The Luftwaffe's Way of War: German Air Doctrine, 1911–1945*, with Richard Muller (1998); *Airpower in Small Wars: Fighting Insurgents and Terrorists*, with Wray Johnson (2003); and *Fighting the War on Terror: A Counterinsurgency Strategy* (2007). He has also authored more than fifty book chapters and journal articles on a variety of subjects related to air power and military history, and was one of the primary authors of *Field Manual 3-24*, the U.S. Army and U.S. Marine Corps doctrine on counterinsurgency. Dr. Corum served in Iraq in 2004 as a lieutenant colonel in the U.S. Army Reserve. He holds a master's degree from Brown University, a Master of Letters from Oxford University, and a PhD from Queen's University, Canada.

David A. Deptula, Lieutenant General (Ret.), United States Air Force (USAF), retired in 2010 from the position of Deputy Chief of Staff for Intelligence, Surveillance, and Reconnaissance. He is a graduate of the University of Virginia, the USAF Fighter Weapons School, Armed Forces Staff College, the Air War College, and the National War College. He holds two master's degrees (in systems engineering and national security strategy) and has flown over three thousand hours (four hundred in combat in the F-15), to include multiple fighter command assignments. He has taken part in operations, planning, and joint warfighting at squadron, wing, major command, service headquarters, and combatant command levels, and has served on two congressional commissions charged with outlining America's future defense posture. He has significant experience in combat and leadership in several major joint contingency operations. He was the principal attack planner for the Desert Storm air campaign and has been a Joint Task Force Commander twice—in 1998–99 for Operation Northern Watch, where he flew eighty-two combat missions over Iraq, and later for Operation Deep Freeze in Antarctica. In 2001 he served as Commander of the Combined Air Operations Center for Operation Enduring Freedom, orchestrating air operations over Afghanistan during the period of decisive combat. In 2005 he commanded all joint air operations for the South Asia tsunami relief effort, and in 2006 he was the standing Joint Force Air Component Commander for Pacific Command, and Vice Commander, Pacific Air Forces. General Deptula has published multiple articles and papers on warfare, air power, and security issues, and lectures internationally. He received the H. H. Arnold Award for 2010 from the Air Force Association for the most significant contribution by a military member for national defense.

Richard P. Hallion, a Smithsonian research associate, retired in 2006 as senior adviser for air and space issues, Directorate for Security, Counterintelligence, and Special Programs Oversight, Office of the Secretary of the Air Force, remaining as senior adviser for aerospace technology (hypersonics and global strike), Office of the Air Force Chief Scientist. Dr. Hallion began his government career at the National Air and Space Museum (NASM) in 1974, serving as curator of science and technology. In 1982 he moved to the U.S. Air Force, where, among other positions, he was a senior issues and policy analyst for Secretary Donald Rice and, for eleven years, the Air Force Historian. He has been a Daniel and Florence Guggenheim Fellow, held the H. K. Johnson Chair at the U.S. Army Military History Institute and the Charles A. Lindbergh Chair at the NASM, was Alfred Verville Fellow at the NASM, and is a Fellow of the Earthshine Institute, the American Institute of Aeronautics and Astronautics, the Royal Aeronautical Society and the Royal Historical Society. He has mission observer flying experience in a variety of aircraft, from biplanes to the F-15E, and teaches, lectures, and consults widely. Dr. Hallion is the author of *Rise of the*

Fighter Aircraft, 1914–1918 (1984); *The Naval Air War in Korea* (1986); *Strike from the Sky: The History of Battlefield Air Attack, 1911–1945* (1989); *Storm over Iraq: Air Power and the Gulf War* (1992); and *Taking Flight: Inventing the Aerial Age, from Antiquity Through the First World War* (2003); and he edited *Air Power Confronts an Unstable World* (1997), among other works. He holds a PhD in history from the University of Maryland, and is a graduate of Federal Executive Institute and the National Security Studies Program, John F. Kennedy School of Government, Harvard University.

Sanu Kainikara is the air power strategist at the Air Power Development Centre of the Royal Australian Air Force. He is also a visiting fellow at the University of New South Wales. Dr. Kainikara is a former fighter pilot of the Indian Air Force who retired as a wing commander after twenty-one years of commissioned service. During his service career, he flew nearly four thousand hours in a number of modern fighter aircraft and held various command and staff appointments. After retirement from active service, he worked for four years as the senior analyst, specializing in air power strategy for a U.S. Training Team in the Middle East. Prior to the current appointment he was the Deputy Director, Wargaming and Doctrine, in the Strategy Group of the Department of Defence. He has also taught aerospace engineering at the Royal Melbourne Institute of Technology University, Melbourne. He is a graduate of the Indian National Defence Academy, the Defence Services Staff College, and the College of Air Warfare. He holds a master's of science degree in defense and strategic studies from the University of Madras, and was awarded a PhD in international politics by the University of Adelaide. Dr. Kainikara is the author of seven books: *Papers on Air Power* (2007), *Pathways to Victory* (2007), *Red Air: Politics in Russian Air Power* (2007), *Australian Security in the Asian Century* (2008), *A Fresh Look at Air Power Doctrine* (2008), *Seven Perennial Challenges to Air Forces* (2009), and *The Art of Air Power: Sun Tzu Revisited* (2010).

R. A. "Tony" Mason, Air Vice-Marshal (Ret.), Royal Air Force (RAF), holds an Honorary Chair at the School of Social Sciences at the University of Birmingham. His academic field of specialization is the interaction of diplomacy and armed forces, with particular reference to air power. His last RAF appointment was as Air Secretary from 1986 to 1989. He is a former director of the Centre for Studies in Security and Diplomacy at the University of Birmingham, was specialist air adviser to the House of Commons Defence Committee from 2001 to 2006, and is a frequent media commentator on defense issues. In 2007 he was appointed an Honorary Fellow of the Royal Aeronautical Society. Professor Mason has contributed to policy studies for the RAF, the U.S. Air Force, and the Australian, New Zealand, German, Swedish, Netherlands, Swiss, Norwegian, Omani, Indian, Thai, South Korean, and Chinese air forces. He has published several books, articles, and papers on air power and related defense subjects, including *Air Power in the Nuclear Age, 1945–1985* (1985) and *Air Power: A Centennial Appraisal* (1994). He is a graduate of the RAF Staff College and the U.S. Air War College, and holds degrees from the University of St. Andrews, the University of London, and the University of Birmingham.

John Andreas Olsen is the deputy commander and chief of the North Atlantic Treaty Organization (NATO) Advisory Team at NATO Headquarters, Sarajevo, and visiting professor of operational art and tactics at the Swedish National Defence College. Previously, he was the dean of the Norwegian Defence University College and head of its division for strategic studies. He is an active-duty colonel in the Norwegian Air Force and a graduate of the German Command and Staff College (2005). Recent assignments include tours as Norwegian liaison officer to the German Operational Command in Potsdam, military assistant to the attaché in Berlin, and tutor and researcher at the Norwegian Air Force Academy. He has a doctorate in history and international relations from De Montfort University, a master's degree in contemporary British literature and politics from the University of Warwick, and a master's degree in English from the University of Trondheim. Professor Olsen is the author of *Strategic Air Power in Desert Storm* (2003) and *John Warden and the Renaissance of American Air Power* (2007). He is also the editor of several books, including *A Second Aerospace Century* (2001), *Asymmetric Warfare* (2002), *On New Wars* (2006), *A History of Air Warfare* (2009), *The Evolution Operational Art: From Napoleon to the Present* with Martin van Creveld (2011), and *Strategy: From Alexander the Great to the Present* with Colin Gray (forthcoming).

Jasjit Singh joined the Indian Air Force in 1954, graduating from the Air Force Academy in 1956. He has held several operational and staff appointments in the air force, having been the commander of a MiG-21 squadron, director of flight safety, and director of operations at Air Headquarters. He retired from the

Indian Air Force in 1988 with the rank of air commodore. He was the director of India's premier think tank on strategic and security issues, the Institute for Defence Studies and Analyses in New Delhi from 1987 to 2001. He currently heads an independent think tank, the Centre for Air Power Studies, in New Delhi. Singh is the author and contributing editor of nearly three dozen books, including *Air Power in Modern Warfare* (1985), *Non-Provocative Defence* (1989), *Nuclear India* (1998), *India's Defence Spending* (2000), *Air Power and Joint Operations* (2003), *Nuclear Deterrence & Diplomacy* (2004), *Defence from the Skies* (2007), and *The ICON: Biography of the Marshal of IAF* (2009). He has lectured regularly at defense and war colleges in India and abroad with special focus on strategic and security issues. He is a fellow of the World Academy of Art and Science and of the Aeronautical Society of India, and has served on various committees and boards, including the National Security Advisory Board from 1990 to 1991 and again from 1998 to 2001.

Alan Stephens is a visiting fellow at the University of New South Wales, Australian Defence Force Academy, where he teaches military history and strategy, with special reference to air power. His previous appointments include adviser on foreign affairs and defense in the Australian federal parliament and visiting fellow at the Strategic and Defence Studies Centre at the Australian National University. He has also served in the Royal Australian Air Force as a pilot, including a tour in Vietnam and the command of an operational squadron. Dr. Stephens has lectured and published extensively. His books include *Going Solo: The Royal Australian Air Force 1946–1971* (1995); *The Australian Centenary History of Defence, Vol. II, The Royal Australian Air Force* (2002); and *Making Sense of War: Strategy for the 21st Century* (2006).

Xiaoming Zhang is associate professor in the Department of Leadership and Strategy at the Air War College, Maxwell Air Force Base, Alabama. He teaches strategy and subjects on China and East Asia. His special areas of expertise include Chinese military history, the People's Liberation Army Air Force, the Korean War, the Vietnam War, and Sino-U.S. and Sino-Soviet relations. Dr. Zhang earned his PhD in history from the University of Iowa in 1994, and taught at Texas Tech University and Texas A&M International University prior to joining the Air War College. He has written several articles on Chinese military involvement in the Korean and Vietnam Wars, and on Sino-Soviet relations during these conflicts. His writings have appeared in *China Quarterly*, *Journal of Cold War Studies*, *The Journal of Conflict Studies*, *China Security*, and *The Journal of Military History*, which has twice awarded him the Moncado Prize for excellence in the writing of military history. He is the author of *Red Wings over the Yalu: China, the Soviet Union, and the Air War in Korea* (2002), and he is currently writing a book on China's 1979 war with Vietnam, as well as articles about the Chinese Air Force.